Quantum Optics
for Experimentalists

Quantum Optics
for Experimentalists

Zhe-Yu Jeff Ou
Indiana University–Purdue University Indianapolis, USA

World Scientific
NEW JERSEY · LONDON · SINGAPORE · BEIJING · SHANGHAI · HONG KONG · TAIPEI · CHENNAI · TOKYO

Published by

World Scientific Publishing Co. Pte. Ltd.

5 Toh Tuck Link, Singapore 596224

USA office: 27 Warren Street, Suite 401-402, Hackensack, NJ 07601

UK office: 57 Shelton Street, Covent Garden, London WC2H 9HE

Library of Congress Cataloging-in-Publication Data
Names: Ou, Zhe-Yu Jeff, author.
Title: Quantum optics for experimentalists / Zheyu Jeff Ou
    (Indiana University - Purdue University Indianapolis, USA).
Description: Singapore ; Hackensack, NJ : World Scientific, [2017] |
    Includes bibliographical references and index.
Identifiers: LCCN 2017018225| ISBN 9789813220195 (hardcover ; alk. paper) |
    ISBN 9813220198 (hardcover ; alk. paper) | ISBN 9789813220201 (pbk. ; alk. paper) |
    ISBN 9813220201 (pbk. ; alk. paper)
Subjects: LCSH: Quantum optics.
Classification: LCC QC446.2 .O93 2017 | DDC 535/.15--dc23
LC record available at https://lccn.loc.gov/2017018225

British Library Cataloguing-in-Publication Data
A catalogue record for this book is available from the British Library.

Printed in Singapore

For my family, Jenny, Anthony, and Nathan

Preface

Optics is an old field of physics. The basis for classical optics is the electromagnetic theory of Maxwell. While quantum optics was developed from classical coherence theory and quantum electrodynamics theory in the 50's of last century, research activities in this field have exploded only since the beginning of the new millennium due to its applications in quantum information and fundamental tests of quantum theory. The technological advances in this field made it possible to implement many ideas that can otherwise only be achieved in theoretical models.

The field of quantum optics has been served quite well by a number of outstanding textbooks, starting with the classic book by Loudon [Loudon (2000)], followed by the comprehensive books by Mandel and Wolf [Mandel and Wolf (1997)], Walls and Milburn [Walls and Milburn (2008)], Scully and Zubairy [Scully and Zubairy (1995)], and most recent one by Agarwal [Agarwal (2013)]. Most of the currently popular textbooks were written by theoreticians, who emphasized on the theoretical aspects of quantum optical fields but left out most of the experimental part. The book by Bachor [Bachor and Ralph (2004)] concentrates mostly on the experiments in quantum optics. Even though the second edition involves Ralph, a theoretician, to include many theoretical aspects of quantum optics, the discussions on experiments and theory are basically separated. So, there is not a textbook that deals with theoretical aspects of quantum optical experiments. When I visited labs around the world, I was frequently asked by students in the lab about how to describe a photon in a real experimental environment, for example, how does a photon pass through a Fabry-Perot filter? Those students who are very skillful and hard working in the lab don't have much time to think deep on this but they cannot find the answer in most of the textbooks either. The reason behind this is quite simple: theory is built on simple and easy-to-describe environment but experiments are usually much more complicated for the theory to cover all. To make things even harder, a complete experimental description involves the concept of modes of the fields, which is not discussed at all in the currently popular textbooks. The mode concept is based on the classical wave theory of electromagnetic fields for which most of the textbooks on quantum optics hardly discuss.

First recognized and promoted by W. Lamb (famous for the Lamb shift of Hydrogen atom) in a paper [Lamb (1995)] with a quite revealing title of "Anti-photon", the concept of modes is the best way of approach in understanding quantum optical phenomena. It is closely related to the experimental reality and leads to the correct physical pictures in the complicated situation of experiments. Yet, most of the textbooks only deal with single-mode or few-mode approaches and are not enough to give a complete description of the real world scenarios, in which field excitations of multiple modes usually occur. As I will show in this book, the approach by the concept of modes also helps in understanding some fundamental aspects in quantum mechanics such as the unification of wave and particle pictures and the relationship between indistinguishability and the visibility of interference. With the concept of modes, the aforementioned questions that were asked by students working in labs are straightforward to answer (see Section 6.3.3).

In a broader sense, after the foundation of quantum optics was laid out around the later part of last century, we have come to an era of engineering of the quantum technologies to make useful quantum devices for practical applications. In this process, a transition from classical to quantum physics is a necessary step because most people working in optics, especially those working in the lab, are quite familiar with the concepts in classical wave theory. As one will see from this book, the presentation of quantum optics based on the mode concept of the classical wave theory makes this transition relatively easy: there is a direct correspondence between classical wave theory and the quantum theory of light.

The current book stems from the lecture notes for a number of lecture series on Quantum Optics that I gave in Tianjin University and East China Normal University. The original notes are in Chinese and here I translate them into English and organize them in the form of a book. There are two parts in this book. In the first part, I try to construct the foundation of Quantum Optics from the perspective of an experimentalist and relate the theory with experiments. I start with a historical introduction about the early development of quantum optics from classical coherent theory and semi-classical theory of light by discussing the Hanbury Brown and Twiss experiment and its significance. This naturally leads to anti-bunching effect of photons and the need for a quantum theory of light. Before the discussion of quantization of light fields, I first introduce the mode theory of classical electromagnetic fields and present a number of examples of optical modes in a variety of situations, from simple to complicated, with which we constantly deal in laboratories of optics. This approach is different from traditional textbooks on quantum optics. From here on, it is straightforward to discuss a variety of quantum states in terms of the multi-mode language, which is the correct way to describe experiments. Then I expand to discuss in depth the theory of quantum optics, including Glauber's quantum coherence and photo-detection theory, the classical-to-quantum correspondence, and the generation and transformation of quantum states. With a solid theoretical framework closely related to experiment, I can discuss in the second part the two most commonly used experimental techniques in quantum optics, that is, photon counting and homodyne detection. These two techniques correspond to the measurement of discrete and continuous variables in quantum information. After introduction of each of the two techniques, I follow with some applications of the experimental technique. Some of more advanced topics in theory are placed at the end of the chapters and marked as further reading sections. Beginning students or those who do not want to go deep in theory can skip these sections without loss of the knowledge for understanding the experimental parts.

The academic significance of the current book lies in the combination between the theory and experiments in quantum optics: the theory has the support of the experiments and in return, the experimental description has a solid theoretical foundations. For theoreticians, this book provides a correct physical picture for photons in real world. For experimentalists, on the other hand, this book will guide them in their daily work in the lab and lead to further discoveries.

I would like to thank Professors Weiping Zhang of East China Normal University and Xiaoying Li of Tianjin University for their hospitality, collaboration, and support during my stay in their universities and valuable discussions and comments on this book.

Indianapolis, May, 2017   *Zhe-Yu Jeff Ou*

Contents

Experimental Techniques in Quantum Optics and Their Applications   191

PART 1
Theoretical Foundations of Quantum Optics

Chapter 1

Historical Development of Quantum Optics and A Brief Introduction

1.1 Historical Background

Light is the most commonly seen matter in the universe, it is also the simplest system in the universe. That is why understanding the behavior of light has played a pivotal role in the development of physics. Many ground breaking concepts in physics originated from optics. For example, Fermat's least action principle was first developed in optics: light propagates between two points in a medium in the shortest time. Then it was extended to other physical systems. The first conceptual revolution in quantum mechanics stemmed from the study of blackbody radiation. Quantum information is the current hot topic in research, yet many of the protocols of quantum information were first realized in optical systems. The reason for all this is straightforward: the simplicity of optical systems makes them easier than others to implement otherwise complicated models in physics. So, understanding of light will help us study other physical systems. Quantum optics is the most complete theory of light. It can explain all the optical phenomena observed so far.

The development of classical optics has a history of a few hundred years. Its highest form is Maxwell's electromagnetic theory of light. Although quantum optics was only developed in recent decades, its origin can be traced back to the start of quantum mechanics. Planck's quantum theory of blackbody radiation [Planck (1900)] is about energy quantization in the atomic emission and absorption of light. It was assumed that the energy of atomic emission or absorption of light can only be integer multiples of some small quantity ϵ. Then Einstein in 1905 introduced the concept of light quanta for the explanation of the photo-electric effect [Einstein (1905)]. This concept is about quantization of the optical fields and is

3

totally independent of atoms. Therefore, it was generally believed that the blackbody radiation theory of Planck in 1900 was the start of the quantum theory whereas the concept of photon was born in 1905 for the start of quantum optics.

It is worth noting that after the introduction of photon, Einstein in 1909 studied the energy fluctuations in blackbody radiation and proposed for the first time the duality of wave and particle for light [Einstein (1909)]. This result is earlier than the duality theory of matter waves by de Broglie. Starting from Planck's blackbody radiation energy spectrum formula, Einstein applied the general argument of thermodynamics to arrive at the following formula for the energy fluctuations:

$$\overline{(\Delta E)^2} = h\nu\overline{E} + \overline{E}^2/Z, \qquad (1.1)$$

where $\overline{A}$ is the average value of quantity A. $h\nu$ is the energy of a photon and Z is some function related to frequency ν and thermal energy kT. In discussing the physical meaning of Eq. (1.1), Einstein first assumed that the field of blackbody radiation consists of independent particles as photons and obtained from the Poisson statistics of random particles the fluctuation of particle number N as: $\overline{(\Delta N)^2} = \overline{N}$. Then using the famous formula for energy quanta from his paper on photoelectric effect: $\epsilon = h\nu$ and total energy formula $E = N\epsilon$, he obtained $\overline{(\Delta E)^2}_p = h\nu\overline{E}$, the first term in Eq. (1.1). For the second term, Einstein assumed that the field of blackbody radiation consists of independent plane waves and obtained $\overline{(\Delta E)^2}_w = \overline{E}^2/Z$. From these, Einstein rewrote Eq. (1.1) as $\overline{(\Delta E)^2} = \overline{(\Delta E)^2}_p + \overline{(\Delta E)^2}_w$. With this, Einstein concluded that the energy fluctuations of blackbody radiation exhibit the behaviors of both particles and waves, i.e., the particle and wave duality. In the concluding remark of the 1909 paper, Einstein gave the following assertion:

"... the next stage of the development of theoretical physics will bring us a theory of light which can be regarded as a kind of fusion of the wave theory and the emission theory ... a profound change in our views of the nature and constitution of light is indispensable."

Einstein was calling for a brand new theory of light in which both particle and wave are unified. Such a theory was constructed by Dirac in 1927 after the completion of the theoretical framework of quantum mechanics [Dirac (1927)]. Later, after Schwinger, Tomonaga, and Feynman solved the difficulties of infinity with renormalization method, this theory became quantum electrodynamics (QED), which is a part of the Standard Model of physics, the most complete and thoroughly tested theory so far. The most famous prediction of QED is the Lamb shift between 2S and 2P energy levels of hydrogen atom, which was confirmed experimentally [Lamb and Retherford (1947)].

However, early studies with QED concentrated mostly on the behaviors of individual photons and electrons. For the case of multiple photons, it was simply assumed that photons are independent of each other. But an experiment performed by Hanbury Brown and Twiss in 1956 showed that it is not the case [Hanbury Brown and Twiss (1956a)]. The collective behavior of multiple photons is quite different from that of multiple independent photons but related to the correlation of optical fields at different locations and times. This is the problem to be solved by the optical coherence theory which was developed by Mandel and Wolf in 1950s [Mandel and Wolf (1997)]. Optical coherence theory stems from the classical wave theory of light. In the early days, it mostly dealt with the phenomena of optical interference and studied the *phase* correlation between the optical fields at different locations and times. The experiment by Hanbury Brown and Twiss, on the other hand, demonstrated for the first time the *intensity* correlation of the optical fields.

The classical coherence theory of light soon evolved into the semi-classical theory of light for the interaction between light and media, in which only the media (atoms or molecules) are described quantum mechanically but the light is still treated as waves with the classical Maxwell electromagnetic wave theory. It simply introduced randomness and fluctuations for the optical fields and became statistical optics. It can explain many optical phenomena including the intensity correlation observed by Hanbury Brown and Twiss. By 1970s, this theory, with the quantization of media only, was successful in explaining many QED effects including photo-electric effect and Lamb shift. A question was thus naturally raised: is it still necessary to quantize the optical fields?

Soon after the establishment of the classical coherence theory of light, Glauber developed the quantum theory of optical coherence in 1963 [Glauber (1963a,b)]. In this theory, Glauber first worked out the theory of photo-detection based on QED and then defined multi-order correlation functions similar to the classical coherence theory. This is a completely quantum mechanical theory of light and is the foundation for quantum optics. Moreover, Glauber's photo-detection theory is also the foundation for the experimental part of the current book. As illustrated in Fig. 1.1, the box on the left contains the optical fields that are described by the theory of quantum optics. The box on the right includes what we observe in the lab

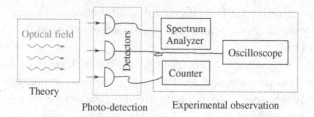

Fig. 1.1 The relationship between the theory of quantum optics and the experiments. Glauber's photo-detection theory connects the theory with the experiments.

with all kinds of modern instruments. Glauber's photo-detection theory on photo-detection processes bridges between the theory in the left box and the experiment in the right box. This will be the main task of the current book.

1.2 Hanbury Brown-Twiss(HBT)Experiment

Hanbury Brown-Twiss (HBT) experiment [Hanbury Brown and Twiss (1956a)] is one of the earliest experiments in quantum optics. Although it demonstrated a classical phenomenon of light, the technique used in the experiment is one of the two major experimental techniques in quantum optics and has been widely applied in experimental measurements. Because of this, we can claim that the HBT experiment laid the foundation of experimental quantum optics. Furthermore, this experiment demonstrated for the first time the fluctuations of intensities or the amplitudes of optical fields. This correlation is different from optical coherence which is about the phase correlations of optical fields.

The experimental setup of the HBT experiment is shown in Fig. 1.2. The light source of the experiment is a mercury vapor lamp. After exiting the source, light first passes through an optical filter to select a single spectral line and filter out stray light. It then is split into two by a beam splitter and sent separately to two photo-detectors. A transverse spatial translation was introduced on one of the detectors to move it in and out of the region of spatial coherence. (A temporal delay may also be used to one of the detectors so that it can be pushed in and out of the region of temporal coherence). Finally, the output photo-electric currents from the two detectors are multiplied for correlation measurement. The result of the observation is that there exists some correlation between the fluctuations of optical intensities at two points within spatial coherence region but no

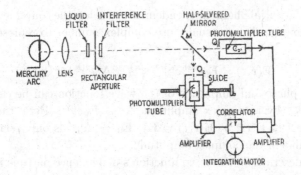

Fig. 1.2 Sketch of the Hanbury Brown-Twiss experiment. Reproduced from [Hanbury Brown and Twiss (1956a)].

correlation for two points outside the region.

After the discovery of the HBT effect, Hanbury Brown and Twiss developed further this newly found spatial correlation technique for the measurement of the size of stars with intensity stellar interferometry [Hanbury Brown and Twiss (1956b)]. However, the real significance of the HBT experiment is that they abandoned the traditional method of optical measurement with a single detector but instead employed two detectors for the direct observation of the correlation between the fluctuations of intensities. This is a breakthrough in experimental methodology and played a pivotal role in the subsequent development of quantum optics.

Another significance of the HBT experiment is the first discovery of intensity or amplitude fluctuations of optical fields. Before this, the phase fluctuations of optical fields have been observed in the interference phenomena of optical waves. So now, both the amplitude and phase of an optical field are not fixed but fluctuate with time. In the following, we will explore further these phenomena to see how to model these fluctuations.

1.3 Fluctuations of Light in Phases and Amplitudes

In the wave description of light, an ideal optical wave can be expressed as a plane wave:

$$\mathbf{E}(\mathbf{r}, t) = \mathbf{E}_0 \cos(\mathbf{k} \cdot \mathbf{r} - \omega t + \varphi_0), \tag{1.2}$$

or as a spherical wave:

$$\mathbf{E}(\mathbf{r}, t) = \frac{\mathbf{E}_0}{r} \cos(kr - \omega t + \varphi_0). \tag{1.3}$$

Here φ_0 is the initial phase independent of time. For the convenience of mathematical expression, we usually use complex number to represent cosine-function:

$$\mathbf{E}(\mathbf{r}, t) = \mathbf{E}_0 e^{i\varphi_0} e^{i(\mathbf{k}\cdot\mathbf{r}-\omega t)} = \mathbf{A}_0 e^{i(\mathbf{k}\cdot\mathbf{r}-\omega t)}. \qquad (1.4)$$

Hence, the phase and amplitude of the wave function can be expressed in a unified way by a complex amplitude $\mathbf{A}_0 \equiv \mathbf{E}_0 e^{i\varphi_0}$. Here, the intensity of the wave $I \equiv \mathbf{E}^*(\mathbf{r}, t) \cdot \mathbf{E}(\mathbf{r}, t) = \mathbf{E}_0 \cdot \mathbf{E}_0 = |\mathbf{A}_0|^2$ is only related to the absolute value of the complex amplitude.

In this description, the wave function's dependence on time is an uninterrupted continuous cosine-function. It is used to describe a monochromatic wave of angular frequency ω. The wave train is infinitely long, as shown in Fig. 1.3.

Fig. 1.3   A continuous and infinitely long monochromatic wave train.

On the other hand, we know that light is emitted when an atom jumps from an excited state to a lower energy level and the excited state of the atom has a finite lifetime ΔT. So, the light wave emitted by the atom is a wave packet of a length traveled by light in the atomic lifetime, as shown in Fig. 1.4(a). Therefore, in order to describe a continuous wave train with a finite wave packet emitted by the atom, we connect every finite wave packet: the wave train is still infinitely long for continuous wave description but there exist interruptions and discontinuity, as shown in Fig. 1.4(b). This is our first model of light that is close to an actual optical field, where the amplitude of the wave is constant but its phase φ_0 is not fixed: $\varphi_0 = \varphi(t)$. It jumps after every period of time ΔT to another random value.

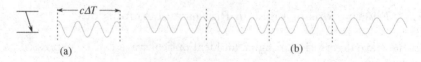

(a)   (b)

Fig. 1.4   (a) The finite wave train emitted by atoms. Its size is a length of $c\Delta T$ traveled by light in the lifetime ΔT of the atom's excited state. (b) The infinitely long wave train formed by connecting finite wave packets emitted by atoms.

The change of the initial phase φ_0 with time will affect the observation of optical interference fringes. First, the phase difference $\varphi_{10} - \varphi_{20}$ of different optical fields is random. This will average out the interference fringes. Secondly, even if the two interfering fields are from the splitting of one field, as in a Michelson interferometer, interference fringes still disappear when the delay between the two fields is larger than the length $c\Delta T$ of the finite wave packet emitted from atoms. This is because when $T > \Delta T$, $\varphi(t + T) - \varphi(t)$ is random. This leads to the concept of optical coherence (more in Section 1.5), where a second-order correlation function:

$$\gamma(\tau) \equiv \frac{\langle E^*(t + \tau)E(t)\rangle}{\langle E^*(t)E(t)\rangle} \tag{1.5}$$

can be used to describe the correlation of the phases. Its absolute value gives the visibility of the interference fringes. The range where $\gamma(\tau)$ is non-zero corresponds to the coherence time $T_c \sim \Delta T$ of the field. Here the average $\langle\rangle$ is taken over the random variable φ_0.

In the model above, the amplitude of the field is unchanged and so is the intensity of the field. But this model cannot explain the intensity fluctuations observed in HBT experiment. For this, we need to introduce amplitude fluctuations. In reality, this is because the time of each atomic emission is random, as shown in Fig. 1.5. So, in our second model of the optical field, the whole complex amplitude $\mathbf{A}_0 \equiv \mathbf{E}_0 e^{i\varphi_0}$ (both magnitude and phase) is a complex random variable. To better understand this, we next briefly describe random variables and random processes.

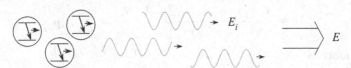

Fig. 1.5   The random emission of atoms.

1.4   Random Variables and Processes

In classical Newtonian mechanics, if the number of particles is extremely large, as in an ensemble of molecules in a regular vapor cell, we are unable to solve the equations of motion for all the particles. In this case, we can only use the method of statistics to take average over these particles so as to obtain macroscopically measurable quantities. This is classical statistical mechanics. In classical optics, when the number becomes large for the light emitting atoms or atoms in the medium of light propagation, we are unable to solve the Maxwell equations to obtain the solutions for the optical waves emitted or scattered by each atom. In this case, we must resort to statistical methods to treat this type of problems. This is classical statistical optics, which is the basis of classical coherence theory. The mathematical foundation of statistical mechanics and statistical optics is the theory of probability. In this section, we will systematically but briefly present some major results and applications of the probability theory and statistics.

1.4.1 *Discrete Random Variables*

Consider a physical quantity A. When we measure it, there are m different outcomes: $A = \{A_1, A_2, ..., A_m\}$. For example, tossing a coin will give only two outcomes: $A = \{\text{head, tail}\}$; but throwing a dice leads to six different values: $A = \{1, 2, ..., 6\}$. We make the measurement N times, in which we have N_1 times the value A_1, N_2 times the value A_2, ..., and N_m times the value A_m. Obviously, $N_1 + N_2 + ... + N_m = N$.

From statistics of mathematics, we know that when $N \to \infty$, N_1/N will approach a limit $p_1 = \lim_{N \to \infty} N_1/N$. Similarly, $p_2 = \lim_{N \to \infty} N_2/N$, ..., $p_m = \lim_{N \to \infty} N_m/N$. Here the limiting values $p_1, p_2, ..., p_m$ are the probabilities of the physical quantity A taking the value of $A_1, A_2, ..., A_m$, respectively. Obviously, $0 \le p_i \le 1 (i = 1, 2, ..., m)$ and they satisfy normalization condition: $p_1 + p_2 + ... + p_m = \sum_i p_i = 1$.

The average of the physical quantity A is defined as

$$\langle A \rangle \equiv \frac{A_1 N_1 + A_2 N_2 + ... + A_m N_m}{N} = \sum_i A_i p_i. \qquad (1.6)$$

Its variance is

$$\text{var}(A) \equiv \langle (A - \langle A \rangle)^2 \rangle = \langle \Delta^2 A \rangle = \langle A^2 \rangle - \langle A \rangle^2 \qquad (1.7)$$

and the standard deviation is $\sigma_A \equiv \sqrt{\text{var}(A)} = \sqrt{\langle \Delta^2 A \rangle}$. For example, for dice throwing, if the dice is a regular cube, we have $p_1 = p_2 = ...p_6 = 1/6$. Then $\langle A \rangle = 3.5$, $\text{var}(A) = 2.9$, and $\sigma_A = 1.7$.

Higher order moments are defined as

$$\langle A^r \rangle \equiv \frac{A_1^r N_1 + A_2^r N_2 + ... + A_m^r N_m}{N} = \sum_i A_i^r p_i \equiv \nu_r. \qquad (1.8)$$

They can be obtained from the moment generating function $M_A(\xi) \equiv \langle e^{\xi A} \rangle = \sum_r \langle A^r \rangle \xi^r / r!$:

$$\nu_r = \langle A^r \rangle = \frac{d^r}{d\xi^r} M(\xi)|_{\xi=0}. \qquad (1.9)$$

Another important function is the characteristic function: $C_A(\xi) \equiv \langle e^{j\xi A} \rangle$.

1.4.2 *Continuous Random Variables*

For a continuous variable $X = [x_1, x_2]$, we first discretize it: we divide the range $[x_1, x_2]$ into M small regions, i.e., $X = [x_1, x_1 + \Delta] + [x_1 + \Delta, x_1 + 2\Delta] + ... + [x_1 + (M-1)\Delta, x_2]$, where $\Delta \equiv (x_2 - x_1)/M$. Then, similar to the discrete case, we measure N times the quantity X and record the number of times of the outcomes falling in each region: N_1 times in $[x_1, x_1 + \Delta]$, N_2 times in $[x_1 + \Delta, x_1 + 2\Delta]$, etc. So, similar to the discrete case, we can obtain probabilities $p_1, p_2, ..., p_M$ for each of the M small regions as $N \to \infty$. Obviously, when M is large, $p_i \propto \Delta$. Then, we can obtain a density of probability for region i:

$$p(x) = \lim_{M \to \infty} \frac{p_i}{\Delta} \quad \text{or} \quad \Delta P = p_i = p(x)\Delta. \qquad (1.10)$$

When $\Delta \to 0$, $\Delta P \to dP = p(x)dx$ is the probability of finding X in the region $[x, x + dx]$. For a finite range $[a, b]$, the probability of finding X in this region is

$$P_{ab} = \int_a^b p(x)dx. \qquad (1.11)$$

Obviously, we have the normalization relation

$$P_{x_1 x_2} = \int_{x_1}^{x_2} p(x)dx = 1. \qquad (1.12)$$

Similar to the discrete probabilities $p_1, p_2, ...$, the probability density $p(x)$ determines the properties of the continuous random variable X, whose average and variance are given by

$$\langle X \rangle = \int_{x_1}^{x_2} xp(x)dx, \quad \langle \Delta^2 X \rangle = \int_{x_1}^{x_2} (x - \langle X \rangle)^2 p(x)dx. \qquad (1.13)$$

For the moment of any order, we can find it from the moment generating and characteristic functions. They are related to the probability density $p(x)$ by

$$M_X(\xi) = \int e^{\xi x} p(x)dx, \quad C_X(\xi) = \int e^{j\xi x} p(x)dx. \qquad (1.14)$$

The normalization relation in Eq. (1.12) gives $M_X(0) = 1 = C_X(0)$. Notice that $p(x)$ and $C_X(\xi)$ are a pair of Fourier transformation. So, we can obtain the probability density $p(x)$ from the characteristic function $C_X(\xi)$:

$$p(x) = \frac{1}{2\pi} \int e^{-j\xi x} C_X(\xi)d\xi. \qquad (1.15)$$

Probabilities $\{p_i\}$ or probability density $p(x)$ are also known as probability distributions. Some well-known probability distributions are

(i) Gaussian normal distribution:

$$p(x) = \frac{1}{\sqrt{2\pi\sigma^2}}e^{-(x-\mu)^2/2\sigma^2}, \quad (\langle X \rangle = \mu, \ \text{var}(X) = \sigma^2), \quad (1.16)$$

(ii) Poisson distribution (discrete variable) $N = \{0, 1, 2, ..., n, ...\}$:

$$p(k) = \frac{\lambda^k}{k!}e^{-\lambda}, \quad (\langle n \rangle = \lambda, \ \text{var}(n) = \lambda). \quad (1.17)$$

(iii) Binomial distribution (discrete variable) $n \leq N, p \leq 1$:

$$p(n) = \frac{N!p^n(1-p)^{N-n}}{n!(N-n)!}, \quad (\langle n \rangle = pN, \ \text{var}(n) = Np(1-p)). \quad (1.18)$$

1.4.3   *Joint Probability of Multiple Random Variables*

Now we deal with multiple random variables. Consider first the case of tossing two coins. There are four possibilities: $A = \{$ (H1H2), (H1T2), (T1H2), (T1T2)$\}$. For the case of throwing two dices, there are totally $6 \times 6 = 36$ possibilities: $A = \{11, 12, 13, ..., 21, 22, 23, ..., 66\}$. We may ask what are the probabilities for each possibility.

1.4.3.1   *Joint Probability*

For the convenience of discussion, we consider two discrete random variables: $X = \{x_1, x_2\}, Y = \{y_1, y_2, y_3\}$. We denote the probability of having simultaneously $X = x_i, Y = y_j$ as p_{ij}, where i, j are the indices for the first and second random variables, respectively. Here, the simultaneous probability p_{ij} is called joint probability. In the current example, it has $2 \times 3 = 6$ values. Table 1.1 lists all 6 possibilities and their joint probabilities.

Table 1.1   List of joint probability for X, Y.

(X, Y)	y_1	y_2	y_3	p_X
x_1	p_{11}	p_{12}	p_{13}	$p_X(x_1)$
x_2	p_{21}	p_{22}	p_{23}	$p_X(x_2)$
p_Y	$p_Y(y_1)$	$p_Y(y_2)$	$p_Y(y_3)$	

Of course, we can still ask about the probability $p_{x_1} = p_X(x_1)$ of $X = x_1$ or the probability $p_{y_3} = p_Y(y_3)$ of $Y = y_3$ as before. From Table 1.1, we find $p_X(x_1) = p_{11} + p_{12} + p_{13}$ and $p_Y(y_3) = p_{13} + p_{23}$. Probability normalization condition requires $\sum p_{ij} = 1 = p_X(x_1) + p_X(x_2) = p_Y(y_1) + p_Y(y_2) + p_Y(y_3)$.

For continuous variables, consider a 2-dimensional vector $\mathbf{R} = (X, Y)$. If variables X, Y are both random variables, vector $\mathbf{R} = (X, Y)$ is a 2-dim random vector. We can likewise write the joint probability of finding $X = [x, x + dx], Y = [y, y + dy]$ as $dP(x, y) = p(x, y)dxdy$ with $p(x, y)$ as the joint probability density. The normalization condition is

$$\int p(x, y)dxdy = 1. \qquad (1.19)$$

From $p(x, y)$, we may obtain the probability densities $p_X(x), p_Y(y)$ for X and Y, respectively:

$$p_X(x) = \int p(x, y)dy, \qquad p_Y(y) = \int p(x, y)dx. \qquad (1.20)$$

1.4.3.2 *Conditional Probability*

With the joint probability, we may ask further: what is the probability of finding $X = x_2$ given $Y = y_3$? This is the conditional probability with the condition $Y = y_3$. We write it as $p_X(x_2|Y = y_3)$ or abbreviate it as $p_X(x_2|y_3)$. Since we ask about the probability of random variable X and it takes only two values x_1, x_2 in the example above, the probabilities should be normalized: $p_X(x_1|y_3) + p_X(x_2|y_3) = 1$. In other words, with $Y = y_3$, we have either $X = x_1$ or $X = x_2$ and the total probability for the two possibilities must be 1. Similarly, we have $p_X(x_1|y_j) + p_X(x_2|y_j) = 1 (j = 1, 2, 3)$. To summarize, when talking about conditional probability, we are only concerned about the subset that satisfies the condition, e.g., $y = y_3$. In the example above, the subset of concern is $(x_1, y_3), (x_2, y_3)$ and the conditional probability must be normalized in this subset. Notice the difference between conditional probabilities $p_X(x_1|y_3), p_X(x_2|y_3)$ and the joint probabilities $p(x_1, y_3) = p_{13}, p(x_2, y_3) = p_{23}$: the former is normalized whereas the latter is not. They are certainly related. Obviously, we have $p_X(x_1|y_3) \propto p(x_1, y_3) = p_{13}$, $p_X(x_2|y_3) \propto p(x_2, y_3) = p_{23}$, that is, $p_X(x_i|y_3) = Cp_{i3}$. Normalization condition gives $C = 1/(p_{13} + p_{23})$. Therefore, we arrive at

$$p_X(x_1|y_3) = \frac{p_{13}}{p_{13} + p_{23}}, \qquad p_X(x_2|y_3) = \frac{p_{23}}{p_{13} + p_{23}}. \qquad (1.21)$$

From what we had earlier: $p_Y(y_3) = p_{13} + p_{23}$, we obtain Bayes' theorem $p(x_1, y_3) = p_X(x_1|y_3)p_Y(y_3)$ or the more general form

$$p(x_i, y_j) = p_X(x_i|y_j)p_Y(y_j) = p_Y(y_j|x_i)p_X(x_i). \qquad (1.22)$$

If X, Y are independent of each other, the outcomes of measuring X has nothing to do with Y. Hence $p_X(x_i|y_j) = p_X(x_i)$. Similarly, $p_X(y_j|x_i) =$

$p_Y(y_j)$. Then, from Bayes' theorem, we have $p(x_i, y_j) = p_X(x_i)p_Y(y_j)$ for independent random variables X, Y.

The discussion above also applies to continuous variables. We only need to drop the subscripts i, j in the corresponding equations.

1.4.3.3 *Coefficient of Correlation*

In general, $p(x_i, y_j) \neq p(x_i)p(y_j)$, that is, $p(x_i|y_j)$ depends on the outcomes of Y. Conditional probability $p_X(x_i|y_j)$ then describes the relationship between X and Y. For example, if $p_X(x_1|y_3) = 1$, that means whenever $Y = y_3$, we must have $X = x_1$. The reverse is not always true though: when $X = x_1$, we may not have $Y = y_3$. But if $p(x_1|y_3) = 1 = p(y_3|x_1)$, x_1 and y_3 are completely correlated. On the other hand, if $p(x_i, y_j) = p(x_i)p(y_j)$, X and Y are completely uncorrelated.

For the case of partial correlation, we use the covariance quantity to describe, which is defined as

$$\langle \Delta X \Delta Y \rangle = \langle XY \rangle - \langle X \rangle \langle Y \rangle. \tag{1.23}$$

By Cauchy-Schwarz inequality, we have $|\langle \Delta X \Delta Y \rangle|^2 \leq \langle \Delta^2 X \rangle \langle \Delta^2 Y \rangle$. After normalization to the right side of the inequality, we obtain the dimensionless correlation coefficient:

$$\rho_{XY} \equiv \frac{|\langle \Delta X \Delta Y \rangle|}{\sqrt{\langle \Delta^2 X \rangle \langle \Delta^2 Y \rangle}}. \tag{1.24}$$

Obviously, if X and Y are completely uncorrlated, i.e., $p(x_i, y_j) = p(x_i)p(y_j)$, we have $\rho_{XY} = 0$. On the other hand, if $\rho_{XY} = 1$, it can be proved that X and Y has a linear dependence: $x_i = Cy_i + D$, that is $p(x_i) = p(y_i) \equiv p_i$ or $p(x_i, y_j) = p_i\delta_{ij}$. Then from Bayes' theorem, we have $p_X(x_i|y_j) = \delta_{ij} = p_Y(y_j, x_i)$. This means X and Y are completely correlated. For the general case, we have $0 \leq \rho_{XY} \leq 1$ by Cauchy-Schwarz inequality. So, correlation coefficient ρ_{XY} quantitatively describes the degree of correlation between X and Y.

1.4.3.4 *Gaussian Distributions*

For Gaussian distribution of two variables, we start with the independent case where the joint probability density is a product of two Gaussian distributions:

$$p(x, y) = \frac{1}{2\pi\sigma_X\sigma_Y} e^{-(x-\mu_X)^2/2\sigma_X^2 - (y-\mu_Y)^2/2\sigma_Y^2}. \tag{1.25}$$

Treating x, y as the coordinates of a point, we rotate $X - Y$ coordinates and obtain the general form of Gaussian distribution for two variables:

$$p(x, y) = \frac{1}{2\pi\sqrt{\mathcal{N}}} e^{A(x-\mu_X)^2 + B(x-\mu_X)(y-\mu_Y) + C(y-\mu_Y)^2}, \qquad (1.26)$$

where $B^2 - 4AC < 0, A, C < 0$ and $\mathcal{N} = 1/|B^2 - 4AC|$ is a normalization factor. The correlation coefficient between X and Y is $\rho_{XY} = B/\sqrt{4AC}$.

The general form of Gaussian distribution for k variables $X_1, ..., X_k$ is

$$p(x_1, ..., x_k) = \frac{1}{\sqrt{(2\pi)^k |\mathbf{\Sigma}|}} \exp\left[-\frac{1}{2}(\mathbf{x} - \mu)^{\mathrm{T}} \mathbf{\Sigma}^{-1}(\mathbf{x} - \mu) \right], \qquad (1.27)$$

where $\mathbf{x}^{\mathrm{T}} = (x_1, ..., x_k)$ is a k-dimensional real vector variable, μ is a k-dim real constant vector, $|\mathbf{\Sigma}|$ is the determinant of matrix $\mathbf{\Sigma}$, and $\mathbf{\Sigma}^{-1}$ is a symmetric positive definite $k \times k$ matrix. The correlation coefficient between X_i and X_j is $\rho_{ij} = \mathbf{\Sigma}_{ij}^{-1}/\sqrt{\mathbf{\Sigma}_{ii}^{-1}\mathbf{\Sigma}_{jj}^{-1}}$.

For a multiple of variables with Gaussian distribution, we have the Isserlis Theorem about the higher order moments:

$$\langle x_1 x_2 ... x_{2n} \rangle = \sum \prod \langle x_i x_j \rangle \qquad (1.28)$$

$$\langle x_1 x_2 ... x_{2n+1} \rangle = 0 \qquad (1.29)$$

where $\sum$ is the sum over all possible ways of pairing $x_1 x_2 ... x_{2n}$ and $\prod$ is the product over all pairs from $x_1 x_2 ... x_{2n}$. $\langle x_i x_j \rangle = \mathbf{\Sigma}_{ij}^{-1}$ can be obtained directly from the symmetric matrix $\mathbf{\Sigma}^{-1}$.

1.4.3.5   Central Limit Theorem of Infinite Number of Random Variables

Consider the sum of M random variables $X_1, ..., X_M$: $\bar{X} \equiv (X_1 + X_2 + ... + X_M)$, which is also a random variable. If all M random variables are independent and identical with the same probability distribution $p(x)$, the Central Limit Theorem of probability theory states that when M is very large, the probability distribution of $\bar{X}$ is a Gaussian with its average equal to $M\langle X_i \rangle = M \int xp(x)dx$ and its variance as $M\langle \Delta^2 X_i \rangle$. This conclusion is independent of the original distribution $p(x)$. This is the reason why we encounter Gaussian statistics so often and Gaussian distribution is also called normal distribution.

1.4.4   Random Processes

When the random variable changes with time, we need to investigate the dynamic behavior of the random variable. This is the random process.

Consider a time-dependent function of the random variable X: $Y_X(t) = f(X,t)$. At a specific time $t = t'$, $Y_X(t') = f(X,t')$ is another random variable related to X: $Y_X(t') = \{y = f(x,t'), x = [x_1, x_2]\}$. But for a specific value $x = x'$ of X, $y_X(t) = f(x',t)$ is a function of time. x' is a possible outcome when we measure X, or a sample. Then $y_X(t) = f(x',t)$ is one possible observed process of the random process $Y_X(t) = f(X,t)$, or a realization. A random process is realized with an ensemble of many sample processes. For example, $Y_X(t) = X\cos\omega t$ describes a harmonic process with a random amplitude. Figure 1.6(a) shows two observed samples.

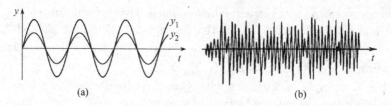

(a) (b)

Fig. 1.6 (a) Two observed samples of random process $Y_X(t) = X\cos\omega t$. (b) One sample process for the random process $Y_X(t) = \sum_i X_i \cos\omega_i t$.

The two sample processes in Fig. 1.6(a) do not look random at all. That is because it only depends on one random variable. If it is a function of multiple random variables: $Y_X(t) = \sum_i X_i \cos\omega_i t$, one sample of the observed realization will look more random, as shown in Fig. 1.6(b).

Similar to random variables, we can take an average of a random process:

$$\langle Y_X(t)\rangle = \int f(x,t)p_X(x)dx. \tag{1.30}$$

We can also obtain the correlation at two different times:

$$\langle Y_X(t_1)Y_X(t_2)\rangle = \int f(x,t_1)f(x,t_2)p_X(x)dx. \tag{1.31}$$

Because it belongs to the same process, the definition above is also called auto-correlation function. When a continuous random process does not have starting and ending time, it is not sensitive to the time shift. This process is known as a stationary process, which satisfies

$$\langle Y_X(t_1+\tau)Y_X(t_2+\tau)...Y_X(t_k+\tau)\rangle = \langle Y_X(t_1)Y_X(t_2)...Y_X(t_k)\rangle. \tag{1.32}$$

1.5 Classical Coherence Theory of Light

In the classical coherence theory of light, the field function $E(\mathbf{r},t)$ is a random process that satisfies Maxwell equations. It is a function of some random variables and time: $E(\mathbf{r},t) = f(\{\alpha\},t)$. In later chapters (Chapters 4 and 5), we will specify these random variables $\{\alpha\}$ explicitly. The quantities we measure in the experiment are the average over these random variables. For example, optical intensity $\langle I(\mathbf{r},t)\rangle_{\{\alpha\}} = \langle E^*(\mathbf{r},t)E(\mathbf{r},t)\rangle_{\{\alpha\}}$.

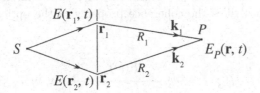

Fig. 1.7   Sketch of Young's double slit experiment.

As an example, let us consider interference experiment of light. The first one is Young's double slit experiment (Fig. 1.7). At observation point P, the optical field is written as

$$E_P(\mathbf{r},t) = \frac{E(\mathbf{r}_1, t - R_1/c)}{R_1}e^{i[\mathbf{k}_1\cdot(\mathbf{r}-\mathbf{r}_1)-\omega t]} + \frac{E(\mathbf{r}_2, t - R_2/c)}{R_2}e^{i[\mathbf{k}_2\cdot(\mathbf{r}-\mathbf{r}_2)-\omega t]}.$$
$$(1.33)$$

Here the two slits act as two point sources with amplitudes determined by the field $E(\mathbf{r},t)$ right before the slits at positions $\mathbf{r}_1, \mathbf{r}_2$. Time delays originate from different propagation times. If the distances from P to the two slits are equal: $R_1 = R_2 = R$, the intensity of light at P is

$$\langle I_P(\mathbf{r},t')\rangle = \langle|E(\mathbf{r}_1,t)|^2\rangle + \langle|E(\mathbf{r}_2,t)|^2\rangle$$
$$+\left[\langle E^*(\mathbf{r}_1,t)E(\mathbf{r}_2,t)\rangle e^{i(\mathbf{k}_1-\mathbf{k}_2)\cdot\mathbf{r}+i\varphi_0} + c.c.\right]$$
$$\propto 1 + V|\gamma_{12}|\cos[(\mathbf{k}_1 - \mathbf{k}_2)\cdot\mathbf{r} + \Delta\varphi], \qquad (1.34)$$

where $t' = t + R/c$. The cosine-function in the expression above gives the interference effect. $\varphi_0 \equiv \mathbf{k}_2\cdot\mathbf{r}_2 - \mathbf{k}_1\cdot\mathbf{r}_1$ and $\Delta\varphi$ is some phase difference related to φ_0. V is related to the optical intensities at $\mathbf{r}_1, \mathbf{r}_2$:

$$V \equiv \frac{2\sqrt{\langle I(\mathbf{r}_1,t)\rangle\langle I(\mathbf{r}_2,t)\rangle}}{\langle I(\mathbf{r}_1,t)\rangle + \langle I(\mathbf{r}_2,t)\rangle}, \qquad (1.35)$$

where $\langle I(\mathbf{r}_i,t)\rangle = \langle|E(\mathbf{r}_i,t)|^2\rangle$ $(i = 1,2)$. γ_{12} is the spatial coherence function defined as

$$\gamma_{12} \equiv \frac{\langle E^*(\mathbf{r}_1,t)E(\mathbf{r}_2,t)\rangle}{\sqrt{\langle I(\mathbf{r}_1,t)\rangle\langle I(\mathbf{r}_2,t)\rangle}}. \qquad (1.36)$$

Here, because phases change with time, $\langle E \rangle$ usually is zero. So, the quantity γ_{12} above is the same as the correlation coefficient between random variables $E^*(\mathbf{r}_1, t)$ and $E(\mathbf{r}_2, t)$ defined in Eq. (1.24). From Eqs. (1.34) and (1.35), we find that when the two intensities are equal, the absolute value of the spatial coherence function γ_{12} is the visibility of the interference fringes. Therefore, it describes the coherence property of the optical field at two points $\mathbf{r}_1, \mathbf{r}_2$.

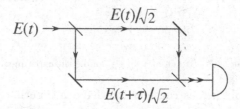

Fig. 1.8   Sketch of Mach-Zehnder interferometer.

Next, we take a look at Mach-Zehnder interferometer (Fig. 1.8). We can show similarly that its interference visibility is given via field correlation function as

$$\gamma(\tau) \equiv \frac{\langle E^*(\mathbf{r}, t+\tau) E(\mathbf{r}, t) \rangle}{\sqrt{\langle I(\mathbf{r}, t+\tau)\rangle \langle I(\mathbf{r}, t)\rangle}}. \tag{1.37}$$

This is the temporal coherence function which describes the coherence property of the optical field at two different times. Due to cosine- or sine-function, field function $E(\mathbf{r}, t)$ is more sensitive to the fluctuation of its phase than the fluctuation of its absolute value or intensity $|E(\mathbf{r}, t)|^2$. So, what coherence function γ_{12} describes is typically the correlation of the phases of the optical fields.

In HBT experiment, the correlation measured by two photo-detectors gives the correlation function between the intensities of the optical field at two locations or two times:

$$g^{(2)}(\tau) \equiv \frac{\langle I(\mathbf{r}, t+\tau) I(\mathbf{r}, t) \rangle}{\langle I(\mathbf{r}, t+\tau)\rangle \langle I(\mathbf{r}, t)\rangle}. \tag{1.38}$$

Here, when $\tau = 0$, we obtain the auto-correlation of intensity: $g^{(2)}(0) = \langle I^2(0)\rangle / \langle I(0)\rangle^2$. Obviously, $g^{(2)}(\tau)$ depends on the fluctuations of intensity (amplitude) of the optical field but is independent of the phase fluctuations of the field, which is closely related to $\gamma_{12}(\tau)$ as we have just seen.

1.6 Classical Interpretation of HBT Effect

The light bunching effect first discovered by Hanbury Brown and Twiss is about the auto-correlation function $g^{(2)}(\tau)$. Figure 1.9 shows the result measured by Hanbury Brown and Twiss in 1958. Normalized correlation function $\Gamma^2(\nu_0, d)$, which is $g^{(2)}(\tau) - 1$, is plotted against separation d (equivalent to $c\tau$). The bunching effect appears at $d = 0$ or $\tau = 0$ with $g^{(2)}(0) = 2$, which also indicates that there exists intensity fluctuation for the detected optical field, for otherwise we must have $\langle I^2(\mathbf{r}, t) \rangle = \langle I(\mathbf{r}, t) \rangle^2$ or $g^{(2)}(0) = 1$.

Next, we will explain the HBT effect using what we learned earlier in statistical optics. First, when $\tau \to \infty$, $I(\tau)$ and $I(0)$ are completely irrelevant to each other or uncorrelated and we have $\langle I(\tau)I(0) \rangle = \langle I(\tau) \rangle \langle I(0) \rangle$ or $g^{(2)}(\infty) = 1$, which is consistent with Fig. 1.9 at $\tau \to \infty$. For the case of $\tau = 0$, it is related to the probability distribution $p(I)$ of the intensity of the optical field. To find $p(I)$, we need to establish a model for the optical field. For the atomic vapor cell, we assume that there are N identical atoms emitting light at $t = 0$. So, the observed optical field can be written as

$$E(0) = \sum_i A_i e^{j\varphi_i} = \sum_i (X_i + jY_i), \qquad (1.39)$$

where $X_i = A_i \cos \varphi_i, Y_i = A_i \sin \varphi_i$. Because of the randomness of the time of atomic emission, the amplitude and the phase of each atom at $t = 0$ are all random, as shown in Fig. 1.5. This means X_i, Y_i are random variables for each atom.

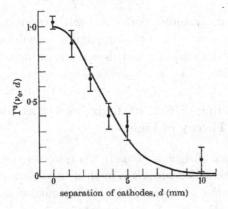

Fig. 1.9 The bunching effect observed in Hanbury Brown-Twiss experiment. Reproduced from [Hanbury Brown and Twiss (1958)].

For the atomic vapor, $N \sim 10^{12 \sim 14}$ and we can apply the Central Limit Theorem to $E(0) \equiv X + jY$ so that

$$X = \sum_i X_i, \quad Y = \sum_i Y_i \qquad (1.40)$$

both are Gaussian random variables. Because of the randomness of φ_i, we have $\langle X \rangle = 0 = \langle Y \rangle, \langle X^2 \rangle = \langle Y^2 \rangle = \sigma^2$ where σ^2 is determined by the intensity of the optical field: $\langle I(0) \rangle = \langle |E(0)|^2 \rangle = \langle X^2 \rangle + \langle Y^2 \rangle = 2\sigma^2$. For the higher order moment $\langle I^2(0) \rangle$, we can use Isserlis theorem in Eq. (1.28) of Section 1.4.3 for multiple Gaussian variables to deduce

$$\langle I^2(0) \rangle = \langle (X^2 + Y^2)^2 \rangle = \langle X^4 \rangle + \langle Y^4 \rangle + 2\langle X^2 \rangle \langle Y^2 \rangle$$
$$= 3\sigma^4 + 3\sigma^4 + 2\sigma^2\sigma^2 = 8\sigma^4 = 2\langle I(0) \rangle^2. \qquad (1.41)$$

Hence, we obtain the HBT bunching effect:

$$g^{(2)}(0) \equiv \frac{\langle I^2(0) \rangle}{\langle I(0) \rangle^2} = 2. \qquad (1.42)$$

For the dependence on the separation d or delay τ of $g^{(2)}(\tau)$, we will present a more detailed discussion based on multi-mode description of the optical field in Section 4.2.4.

The derivation process above shows that HBT bunching effect is simply a result of Gaussian distribution. We can further show that random variable $I = X^2 + Y^2$ has an exponential distribution: $p(I) = e^{-I/I_0}/I_0$ with $I_0 = \langle I(0) \rangle$ (see Problem 1.4). Therefore, we obtain higher order bunching effect:

$$g^{(n)}(0) \equiv \frac{\langle I^n(0) \rangle}{\langle I(0) \rangle^n} = \frac{n!I_0^n}{I_0^n} = n!. \qquad (1.43)$$

In later chapters on quantum optics, we will give an explanation of the bunching effect in terms of the photon picture. It will be shown that the photon bunching effect in Eq. (1.43) is a result of multi-photon quantum interference (Section 8.3.1).

1.7   Anti-Bunching Effect of Light and the Need for A Quantum Theory of Light

The bunching effect of light is actually the result of the Cauchy-Schwarz inequality in mathematics when it is applied to statistical optics. For two arbitrary real random variables X, Y, this inequality states: $|\langle XY \rangle|^2 \leq \langle X^2 \rangle \langle Y^2 \rangle$. Taking $X = I(0), Y = 1$, we have

$$\langle I(0) \rangle^2 \leq \langle I^2(0) \rangle, \qquad (1.44)$$

which gives $g^{(2)}(0) = \langle I^2(0) \rangle / \langle I(0) \rangle^2 \geq 1$, i.e., the photon bunching effect. Therefore, the wave theory of light, after the introduction of random variable description to become statistical optics, must satisfy the bunching conditions: $g^{(2)}(0) > 1$.

However, if we assume the particle nature for light, as Einstein did in explaining photo-electric effect, the energy of each photon is already minimum and cannot be divided further. So when a stream of light in the form of single photons hits a beam splitter, each photon can only go to one of the two sides of the beam splitter, as shown in Fig. 1.10. If we perform a HBT experiment on this stream of light, each photon cån only be detected by one of the two detectors while the other detector has none, which leads to zero outcome in photo-current correlation, i.e., $\langle I^2(0) \rangle = 0$ or $g^{(2)}(0) = 0 < 1$. This is the anti-bunching effect and does not satisfy the bunching condition. So, it cannot be explained by the wave theory of light. The photon anti-bunching effect was first observed by Kimble, Dagenais, and Mandel in 1977 [Kimble et al. (1977)]. To explain this phenomenon, we must resort to quantum optics where the optical field is quantized.

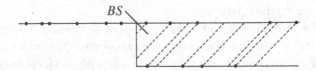

Fig. 1.10   A stream of light with single photons is separated by a beam splitter: each photon can only go to one of the two sides. The dashed lines connect events of the same time ($\tau = 0$) and no two photons appear on the two sides at the same time and this leads to the anti-bunching effect of light.

1.8   The Topics of Quantum Optics

In the following chapters, we will study the topics of quantum optics in the form of the following questions:

(1) How to describe an optical field quantum mechanically? We will introduce the quantum state description of light in Chapters 3 and 4.

(2) How to describe the evolution of an optical field? We will discuss the transformation of quantum state. This is done in Chapter 6.

(3) What is the relationship between the quantum theory and the wave theory of light? We will categorize quantum states into classical states and non-classical states in Chapter 5.

(4) How do optical fields interact with each other? We introduce nonlinear optics in Chapter 6 to generate a variety of quantum states of light.

(5) How to observe the optical field? We will discuss this in the second part of this book. We will introduce Glauber's photo-detection theory and make a connection with the experimental observations. In this part, we will discuss two basic experimental measurement techniques of quantum optics and find their applications.

1.9   Problems

Problem 1.1 In Section 1.3, we presented the first model of light fluctuations. In Fig. 1.4, the phases of the optical field jump to another value after every period of ΔT whereas the amplitude of the optical wave is fixed. Assume the phase change is completely random. Find the temporal coherence function $\gamma(\tau)$ and coherence time T_c.

Problem 1.2 Stellar Interferometry is a technique to find the size of a star by measuring the spatial coherence property of the light after it travels from the star to the earth.

(i) Consider two independent point sources of separation a. Draw a line perpendicular to the line connecting the two points. Now move a distance $D >> a$ along this line and consider two points with one located on the line and the other a distance $d(<< D)$ to the line. Assume all the points are in a plane. Find the normalized second order correlation function $\gamma_{12}(d)$ between these two points.

(ii) Do the same if the two point sources are replaced by a line source of length a. Assume the amplitude $A(x)$ for a point on the line source satisfies $\langle A^*(x)A(x')\rangle = A^2\delta(x - x')$ because of the independence of the points on the line source.

(iii) Do the same if the line source is replaced by a two-dimensional rectangular source of size $a \times b$. Assume the amplitude $A(x,y)$ for a point on the source satisfies $\langle A^*(x,y)A(x',y')\rangle = A^2\delta(x - x')\delta(y - y')$. Show that $\gamma_{12}(c,d) = \mathrm{sinc}(\pi ac/D)\mathrm{sinc}(\pi bd/D)$ where c,d are the coordinates in the observation plane.

(iv) Do the same if the line source is replaced by a uniform disk of radius $R << D$. Show that

$$\gamma_{12}(d) = 2J_1(\beta)/\beta, \tag{1.45}$$

where J_1 is the first order Bessel function with $\beta = kdR/D$.

From this problem, we can see that the size R of a star can be found by measuring $\gamma_{12}(d)$ with interferometric method on earth. This technique was first proposed [Michelson (1890, 1920)] and implemented [Michelson and Pease (1921)] by Michelson.

It is clear from the results that γ_{12} follows the diffraction pattern of the corresponding opening. This is actually a special case of the far-zone form of the van Cittert-Zernike theorem [van Cittert (1934); Zernike (1938)].

Problem 1.3 Intensity correlation measurement can be used to reveal higher-order interference between independent sources. We will investigate this in the following.

Consider two *independent* point sources, which are separated by a distance of d. Let the x-axis bisect the line connecting the two point sources. Move a distance D along the x-axis and place two detectors in such a way that the x-axis is perpendicular to the line connecting the two detectors and the distance to the x-axis is y_1, y_2 for the two detectors, respectively. Find (a) intensities at the two detectors and (b) intensity correlation between the two detectors as a function of D, d, y_1, y_2 and wavelength λ, assuming $D \gg d, y_1, y_2$.

Problem 1.4 Assume X, Y are two independent Gaussian random variables, whose averages are zero and standard deviations are σ.
(i) Calculate $C(r) \equiv \langle \exp[jr(X^2 + Y^2)] \rangle$ where r is a parameter.
(ii) The characteristic function of random variable $I = X^2 + Y^2$ is exactly the function calculated in (i). Find therefrom the probability distribution $p(I)$ for I.

Problem 1.5 Proof of the Central Limit Theorem.

We will provide a non-rigorous proof of the Central Limit Theorem in this exercise. We start by considering N identical but independent random variables of $x_i (i = 1, 2, ..., N)$ with an average of $\langle x_i \rangle = a$ and a variance of $\langle (x_i - a)^2 \rangle = \sigma^2$ for $i = 1, 2, ..., N$. Define a new random variable

$$X_N \equiv \left(\frac{1}{N} \sum_i x_i \right) - a. \tag{1.46}$$

(i) Show that X_N has an average of zero and a variance of $\mathrm{var}(X_N) = \sigma^2/N$.
(ii) Use the independent property of $\{x_i\}$ to show that the characteristic

function of X_N can be written as

$$C_{X_N}(r) \equiv \langle \exp(jrX_N) \rangle = \prod_i C_i(r) \qquad (1.47)$$

with $C_i(r) = \langle \exp[jr(x_i - a)/N] \rangle$.

(iii) Expand $C_i(r)$ to the first non-zero order of $1/N$.

(iv) Show that for large N, $C_{X_N}(r) \approx \exp(-r^2\sigma^2/2N)$, which gives the Gaussian distribution with a variance of $\text{var}(X_N) = \sigma^2/N$ independent of the distribution of x_i.

The result in (iv) shows why N multiple measurements in an experiment will improve the precision from single measurement by a factor of $1/\sqrt{N}$.

Chapter 2

Mode Theory of Optical Fields and Their Quantization

As we found out at the end of last chapter, the classical wave theory of light cannot explain the anti-bunching effect of light. We have to resort to the photon picture in quantum theory of light for explanation. Unlike the classical wave theory, quantum theory of light can cover all known optical phenomena and is therefore a more general theory. But as we will see in this chapter, the quantum theory of light is constructed with Maxwell's classical theory of electromagnetic wave as its foundation.

In a broader perspective, the quantum theory of light is the part about photon in the general quantum field theory. In this general theory, the following is how the wave and particle are unified to give rise to the duality: different particles of matter, such as electron, muon, photon, phonon, etc., have each a description by a field of its own through its own wave equation (or the field equation because the spatial spread of the wave provides a field for the particle). Different particles have different field equations. For example, the wave equation for electrons is the Dirac equation and Schrödinger equation is the wave equation for spinless non-relativistic particles. When we make a second quantization of the wave equation, we then create the particle for this field by energy excitation. The reason we call this process a "second" quantization is because we already call the process of going from classical Newtonian equation of motion for a particle to the wave equation (such as Schrödinger equation and Dirac equation) as a quantization process, in which energy is quantized, e.g., the energy levels of hydrogen atom. So, the first quantization process makes a wave (or field) for a classical particle but the second quantization goes back to particle in return. However, this cycle does not go back to the old classical particle but leads to a new duality picture in which wave and particle are unified. We will discuss this unification further at the end of this chapter after we introduce the quantum field theory of light.

For light field, its quantization process is a bit more special than other fields. The classical theory that we use so often is already a wave theory with Maxwell equations as its wave equations. So, quantization of the optical fields described by Maxwell equations is equivalent to a second quantization process. If we still want to make a comparison with other particles in the quantization process as a whole, we may consider the process from Newton's corpuscular theory of light or geometric theory of light to Huygens' wave theory and further to Maxwell's electromagnetic theory as the first quantization process for light. But it seems unclear where the quantization occurs in this process since there is no concept of quantization in Maxwell's electromagnetic theory. It turns out, as we will see in this chapter, that as we solve Maxwell equations with boundary conditions, the solutions for the electromagnetic waves only allow discrete values for the frequency of light, which are similar to the energy eigenvalues obtained when solving the Schrödinger equation. This is the mode theory of electromagnetic waves, which is the foundation for the quantum theory of light.

In this chapter, we will start by reviewing systematically Maxwell's electromagnetic wave theory with emphasis on the mode theory of light fields. Based on this, we can then quantize light fields with the model of simple harmonic oscillators. In fact, the mode theory of light is very useful even for just understanding the wave aspect of light.

2.1 The Classical Theory of Light

2.1.1 *Maxwell Equations*

Light fields consist of electric and magnetic fields. In Maxwell's electromagnetic theory of light, the electric and magnetic fields are generated by charges. But electromagnetic fields and charges are totally different types of matter. So, we must separate the two in the quantization processes. To do so, let us only consider pure electromagnetic fields without charges, or the free fields. Charges can be considered as another field of matter that are totally independent of the electromagnetic fields. The interaction between electromagnetic fields and matter fields with charges gives rise to electromagnetic interaction and leads to the generation and absorption of electromagnetic waves. The related phenomena are discussed in the theory for the interaction between light and matter and are not the subject of the current chapter (see Chapter 6). So, here we only deal with free electromagnetic fields without charges.

Maxwell equations without charges have the following form:

$$\nabla \times \mathbf{E} + \frac{1}{c}\frac{\partial \mathbf{B}}{\partial t} = 0, \tag{2.1}$$

$$\nabla \cdot \mathbf{B} = 0, \tag{2.2}$$

$$\nabla \times \mathbf{B} - \frac{1}{c}\frac{\partial \mathbf{E}}{\partial t} = 0, \tag{2.3}$$

$$\nabla \cdot \mathbf{E} = 0. \tag{2.4}$$

Here, we used the cgs unit system, in which electric field and magnetic field have the same unit and the Maxwell equations are symmetric with respect to these two quantities.

From Eq. (2.2), we can introduce a vector potential $\mathbf{A}$: $\mathbf{B} = \nabla \times \mathbf{A}$. Substituting into Eq. (2.1), we obtain

$$\nabla \times \left(\mathbf{E} + \frac{1}{c}\frac{\partial \mathbf{A}}{\partial t}\right) = 0. \tag{2.5}$$

The above equation allows us to introduce a scalar potential ϕ: $\mathbf{E} + \frac{1}{c}\frac{\partial \mathbf{A}}{\partial t} = -\nabla\phi$. Hence, the electric and magnetic fields can be expressed in terms of the scalar and the vector potentials $\phi, \mathbf{A}$:

$$\mathbf{E} = -\frac{1}{c}\frac{\partial \mathbf{A}}{\partial t} - \nabla\phi, \quad \mathbf{B} = \nabla \times \mathbf{A}. \tag{2.6}$$

Substituting Eq. (2.6) into Eqs. (2.3) and (2.4), we obtain

$$\frac{1}{c^2}\frac{\partial^2 \mathbf{A}}{\partial t^2} - \nabla^2\mathbf{A} + \nabla\left(\nabla \cdot \mathbf{A} + \frac{1}{c}\frac{\partial \phi}{\partial t}\right) = 0, \tag{2.7}$$

$$\nabla^2\phi + \frac{1}{c}\nabla \cdot \frac{\partial \mathbf{A}}{\partial t} = 0. \tag{2.8}$$

But ϕ and $\mathbf{A}$ are not unique for $\mathbf{E}$ and $\mathbf{B}$. For example, the following transformation

$$\phi' = \phi + \frac{1}{c}\frac{\partial \chi}{\partial t}, \quad \mathbf{A}' = \mathbf{A} - \nabla\chi \tag{2.9}$$

will leave $\mathbf{E}$ and $\mathbf{B}$ unchanged. So, we can freely choose χ to obtain different ϕ and $\mathbf{A}$. They give the same $\mathbf{E}$ and $\mathbf{B}$. Every different set of ϕ and $\mathbf{A}$ is called a "gauge" in field theory and $\mathbf{E}$ and $\mathbf{B}$ are unchanged under gauge transformation. Hence, electromagnetic fields have the property of gauge invariance or gauge symmetry. They are therefore called gauge fields. Two commonly used gauges are

(1) Lorentz Gauge:

$$\nabla \cdot \mathbf{A} + \frac{1}{c}\frac{\partial \phi}{\partial t} = 0, \tag{2.10}$$

(2) Coulomb Gauge:

$$\nabla \cdot \mathbf{A} = 0. \tag{2.11}$$

Obviously, Lorentz gauge satisfies the Lorentz invariance of relativity and is suitable for the electromagnetic fields of high energy, such as γ-rays. Coulomb gauge is more suitable for the electromagnetic fields of low energy, such as the commonly used visible band of light, for which the photon energy is around a few eVs. The speed of electrons at this energy level is non-relativistic. Therefore, we will in general use the Coulomb gauge in quantum optics. Furthermore, we learned from electromagnetic theory that the scalar potential ϕ is generated from charges, which are absent for a free field. So, we can set $\phi = 0$ for free fields. In this case, Lorentz gauge and Coulomb gauge are the same and Eqs. (2.6)–(2.8) change to

$$\nabla^2 \mathbf{A} - \frac{1}{c^2}\frac{\partial^2 \mathbf{A}}{\partial t^2} = 0, \tag{2.12}$$

$$\nabla \cdot \mathbf{A} = 0, \tag{2.13}$$

$$\mathbf{E} = -\frac{1}{c}\frac{\partial \mathbf{A}}{\partial t}, \quad \mathbf{B} = \nabla \times \mathbf{A}. \tag{2.14}$$

Equation (2.12) is in the form of wave equation that must be satisfied for any classical waves. From Eq. (2.12), we conclude that electromagnetic fields can propagate in the form of waves and the speed of the electromagnetic wave is the speed of light c. This is what led Maxwell to postulate the existence of electromagnetic wave, which was later confirmed by Hertz, and light wave as a form of electromagnetic wave. We now call Eq. (2.12) as Maxwell wave equation or simply Maxwell equation for short.

2.1.2   The Eigen-Solutions of Maxwell Equation

With wave equation, let us find the exact expression for the light fields by solving it. We will use the method of separation of variables in solving partial differential equations: Eq. (2.12) has separate spatial and temporal differentiations. So, let us write separately the space and temporal part of $\mathbf{A}(\mathbf{r}, t)$ as $\mathbf{A}(\mathbf{r}, t) = q(t)\mathbf{A}(\mathbf{r})$. Equations (2.12) and (2.13) become

$$\nabla^2 \mathbf{A}(\mathbf{r}) + k^2 \mathbf{A}(\mathbf{r}) = 0, \tag{2.15}$$

$$\nabla \cdot \mathbf{A}(\mathbf{r}) = 0, \tag{2.16}$$

$$\frac{d^2 q(t)}{dt^2} + \omega^2 q(t) = 0. \tag{2.17}$$

Here k and ω are some constants to be determined and they satisfy $k = \omega/c$. Equation (2.15) is also known as Helmholtz equation. It can be considered as the eigen-equation for operator ∇^2. By operator theory, for a certain boundary condition (e.g. $\mathbf{A}(\mathbf{r}) = 0, \mathbf{r} =$ boundary), we can always find a set of values for k and a group of orthonormal eigenfunctions $\{\mathbf{A}_\lambda(\mathbf{r})\}$, such that they satisfy:

$$\nabla^2 \mathbf{A}_\lambda(\mathbf{r}) + k_\lambda^2 \mathbf{A}_\lambda(\mathbf{r}) = 0, \qquad (2.18)$$

$$\nabla \cdot \mathbf{A}_\lambda(\mathbf{r}) = 0, \qquad (2.19)$$

$$\int d^3\mathbf{r} \mathbf{A}_\lambda^*(\mathbf{r}) \cdot \mathbf{A}_\mu(\mathbf{r}) = \delta_{\lambda\mu}. \qquad (2.20)$$

The solution for the temporal part (Eq. (2.17)) is that for a simple harmonic oscillator. It has two independent solutions: $q_\lambda^{(\pm)}(t) = q_\lambda(0)e^{\pm i\omega_\lambda t}$. Here, we use complex number to represent the harmonic motion with its real part corresponding to the form of motion for the harmonic oscillator, i.e., cosine-function. In the two independent solutions, $q_\lambda^{(-)}(t) = q_\lambda(0)e^{-i\omega_\lambda t}$ corresponds to the plane wave solution that will be discussed later while $q_\lambda^{(+)}(t) = q_\lambda(0)e^{i\omega_\lambda t}$ is its complex conjugate, which will be included when converting back to the representation in real numbers. In the following we only use $q_\lambda^{(-)}(t) = q_\lambda(0)e^{-i\omega_\lambda t}$ as the solution to the temporal part.

Combining the spatial and temporal parts, we arrive at the special eigen-solutions for Maxwell equations: $\mathbf{A}_\lambda(\mathbf{r},t) = q_\lambda(0)e^{-i\omega_\lambda t}\mathbf{A}_\lambda(\mathbf{r})$. Since Eqs. (2.12) and (2.13) are linear and homogeneous, any superposition of their solutions should also be a solution. From operator theory, we can prove further that the eigenfunctions of the operator ∇^2 are complete, i.e., any arbitrary solution of Eqs. (2.12) and (2.13) can be written as a linear superposition of the eigenfunctions:

$$\mathbf{A}(\mathbf{r},t) = \sum_\lambda q_\lambda(t)\mathbf{A}_\lambda(\mathbf{r}) = \sum_\lambda q_\lambda(0)e^{-i\omega_\lambda t}\mathbf{A}_\lambda(\mathbf{r}), \qquad (2.21)$$

or, if we use real numbers to represent optical fields, we have

$$\mathbf{A}(\mathbf{r},t) = \sum_\lambda \left[q_\lambda(t)\mathbf{A}_\lambda(\mathbf{r}) + q_\lambda^*(t)\mathbf{A}_\lambda^*(\mathbf{r}) \right]$$

$$= \sum_\lambda q_\lambda(0)e^{-i\omega_\lambda t}\mathbf{A}_\lambda(\mathbf{r}) + c.c., \qquad (2.22)$$

where $q_\lambda(0)$ is some constant. Since $\mathbf{A}_\lambda(\mathbf{r})$ is normalized, the size of $q_\lambda(0)$ will then determine the strength of the optical fields.

2.1.3 *Special Eigen-Solutions: Box Model and Plane Waves*

As we have discussed in the previous section, the solution of the Helmholtz equation is determined by the boundary conditions. Let us start with

the simplest border, i.e., a cubic box. We use the continuity boundary
conditions in which the field function $\mathbf{A}(x, y, z)$ is continuous from one side
to the opposite side of the box, that is,

$$\mathbf{A}(0, y, z) = \mathbf{A}(L, y, z), \quad \mathbf{A}(x, 0, z) = \mathbf{A}(x, L, z),$$
$$\mathbf{A}(x, y, 0) = \mathbf{A}(x, y, L). \tag{2.23}$$

Here, L is the side length of the cubic box. We can easily obtain the plane
wave solutions for the Helmholtz equation under the continuity boundary
condition in Eq. (2.23)

$$\mathbf{A}_{\mathbf{k},s}^{(d)}(\mathbf{r}) = \hat{\epsilon}_s e^{i\mathbf{k}\cdot\mathbf{r}}/L^{3/2}, \tag{2.24}$$

where $\hat{\epsilon}_s$ is the unit vector and the eigen wave vectors are $\mathbf{k} =
2\pi(n_x, n_y, n_z)/L$. n_x, n_y, n_z are integers. So, the eigen wave vectors $\mathbf{k}$
cover the whole $\mathbf{k}$-space but they have discrete values and the minimum
separation between two values is $\Delta k = 2\pi/L$. Since the eigen wave vector
$\mathbf{k}$ is discrete, we place a superscript "d" in field function $\mathbf{A}_{\mathbf{k},s}^{(d)}$ to distinguish
from the continuous case to be discussed in the next section.

Unit vector $\hat{\epsilon}_s$ determines the direction of the vector function $\mathbf{A}(\mathbf{r})$. For
a 3-dim vector, it has three independent directions, that is, $\hat{\epsilon}_s$ corresponds
to three independent polarization states with subscript $s = 1, 2, 3$. But the
field function $\mathbf{A}_{\mathbf{k},s}^{(d)}$ must satisfy the Coulomb gauge condition in Eq. (2.19).
Substituting the field function $\mathbf{A}_{\mathbf{k},s}^{(d)}$ in Eq. (2.24) into Eq. (2.19), we find
that $\hat{\epsilon}_s$ must satisfy the transverse wave condition: $\mathbf{k} \cdot \hat{\epsilon}_s = 0$, that is, the
polarization of the optical field must be perpendicular to the wave vector
$\mathbf{k}$. So, $\hat{\epsilon}_s$ can only take two independent and orthogonal polarization states
such as linearly polarized $\hat{x}, \hat{y}$ or circularly polarized $\hat{\epsilon}_{\pm} = (\hat{x} \pm i\hat{y})/\sqrt{2}$.
Because of the transverse wave condition, the polarization of an optical
field is related to the wave vector: $\hat{\epsilon} = \hat{\epsilon}_{\mathbf{k},s}(s = 1, 2)$. Let $\hat{\mathbf{k}} = \mathbf{k}/k$ be
the unit vector for wave vector $\mathbf{k}$. Since $\hat{\mathbf{k}}, \hat{\epsilon}_{\mathbf{k},1}, \hat{\epsilon}_{\mathbf{k},2}$ are three unit vectors
perpendicular to each other, they form the base vectors of the 3-dim vector
space.

It can be easily proved that the set of eigen solutions in Eq. (2.24)
satisfies the discrete orthonormal relations:

$$\int_{L^3} d^3\mathbf{r} \mathbf{A}_{\mathbf{k},s}^{(d)*}(\mathbf{r}) \cdot \mathbf{A}_{\mathbf{k}',s'}^{(d)}(\mathbf{r}) = \delta_{\mathbf{k},\mathbf{k}'}\delta_{s,s'}. \tag{2.25}$$

Combining with the temporal part, we obtain the plane wave solution
for Maxwell equation in Eq. (2.12):

$$\mathbf{A}(\mathbf{r}, t) = \hat{\epsilon}_{\mathbf{k},s} e^{i(\mathbf{k}\cdot\mathbf{r}-\omega t)}/L^{3/2}.$$

Notice that the discreteness of the wave vector $\mathbf{k}$ leads to the quantized angular frequency $\omega = ck = (2\pi c/L)\sqrt{n_x^2 + n_y^2 + n_z^2}$. This is analogous to the energy quantization from Schrödinger equation in the first quantization process. So, Maxwell wave equation is equivalent to Schrödinger wave equation for the first quantization of optical fields, as we discussed in the beginning of this chapter.

2.1.4   Special Eigen-Solutions: Continuous k-Space and Plane Wave Solutions for Three-Dimensional Free Space

When L goes to infinity, the finite box space becomes infinite three-dimensional free space and the original discrete $\mathbf{k}$-space is transformed into a continuous $\mathbf{k}$-space. For this transition, let us re-define the spatial part of the field function in Eq. (2.24):

$$\mathbf{A_{k,s}}(\mathbf{r}) \equiv (L/2\pi)^{3/2}\mathbf{A}_{\mathbf{k,s}}^{(d)}(\mathbf{r}) = \hat{\epsilon}_{\mathbf{k},s}e^{i\mathbf{k}\cdot\mathbf{r}}/(2\pi)^{3/2}, \qquad (2.26)$$

and in the meantime, the orthonormal relation is changed to

$$\int_{L^3} d^3\mathbf{r}\,\mathbf{A}_{\mathbf{k},s}^*(\mathbf{r}) \cdot \mathbf{A}_{\mathbf{k'},s'}(\mathbf{r}) = \delta_{s,s'}\delta_{\mathbf{k},\mathbf{k'}}(L/2\pi)^3 = \delta_{s,s'}\delta_{\mathbf{k},\mathbf{k'}}/(\Delta k)^3. \quad (2.27)$$

Here, $\Delta k \equiv 2\pi/L$ is the minimum separation between two adjacent discrete wave vectors $\mathbf{k} = 2\pi(n_x, n_y, n_z)/L$. In the limit of $L \to \infty$ and $\Delta k \to 0$, we obtain the following transition:

$$1 = \sum_{\mathbf{k}} \Delta k^3 \delta_{\mathbf{k},\mathbf{k'}}/(\Delta k)^3 \to \int d^3\mathbf{k}\,\delta^{(3)}(\mathbf{k}-\mathbf{k'}) = 1 \qquad (2.28)$$

Hence, the discrete δ-function is changed to a continuous δ-function:

$$\lim_{L\to\infty} \delta_{\mathbf{k},\mathbf{k'}}/(\Delta k)^3 = \delta^{(3)}(\mathbf{k}-\mathbf{k'}). \qquad (2.29)$$

We will use again later the above transition from discrete to continuous cases.

We then obtain Eq. (2.26) as the plane wave solution for three-dimensional free space with orthonormal relation:

$$\int d^3\mathbf{r}\,\mathbf{A}_{\mathbf{k},s}^*(\mathbf{r}) \cdot \mathbf{A}_{\mathbf{k'},s'}(\mathbf{r}) = \delta_{s,s'}\delta^{(3)}(\mathbf{k}-\mathbf{k'}), \qquad (2.30)$$

where the wave vector $\mathbf{k} = (k_x, k_y, k_z)$ is an arbitrary vector in a three-dimensional continuous $\mathbf{k}$-space and the transverse wave condition is the same: $\mathbf{k} \cdot \hat{\epsilon}_{\mathbf{k},s} = 0$.

2.1.5   The Concept of Modes and the Decomposition in Terms of Modes

Because of the orthonormal relation in Eq. (2.20), the eigen-solution of Maxwell equation has its special meaning. First, the general solution of Maxwell equation can be written as the superposition of the eigen solutions, that is, in the form of Eq. (2.21) or (2.22). When we obtain a complete set of eigen solutions, we will have the arbitrary solution. Secondly, although an arbitrary solution is a superposition of a number of eigen-solutions, the orthonormal relation of the eigen solutions ensures the independence among all the eigen solutions and their addition will not lead to the energy exchange among them, as we will see in the following decomposition of energy in terms of the energy of eigen-solutions.[1] Let us start with the total intensity of the field. In the second part of this book on experiment, we will discuss about intensity of an optical field whose expression in terms of field amplitude is $I(\mathbf{r}, t) \propto \mathbf{E}^*(\mathbf{r}, t) \cdot \mathbf{E}(\mathbf{r}, t)$, where $\mathbf{E}$ can be obtained from Eq. (2.14) with $\mathbf{A}$ given in Eq. (2.21). Then the total intensity is

$$I_{tot} = \int d^3\mathbf{r} I(\mathbf{r}, t) \propto \int d^3\mathbf{r} \mathbf{E}^*(\mathbf{r}, t) \cdot \mathbf{E}(\mathbf{r}, t). \qquad (2.31)$$

Using Eqs. (2.14) and (2.21) together with the orthonormal relation in Eq. (2.20), we obtain

$$
\begin{aligned}
I_{tot} &\propto \frac{1}{c^2} \int d^3\mathbf{r} \left[\sum_\lambda \dot{q}_\lambda^*(t) \mathbf{A}_\lambda^*(\mathbf{r}) \right] \cdot \left[\sum_{\lambda'} \dot{q}_{\lambda'}(t) \mathbf{A}_{\lambda'}(\mathbf{r}) \right] \\
&= \frac{1}{c^2} \sum_{\lambda, \lambda'} \dot{q}_\lambda^*(t) \dot{q}_{\lambda'}(t) \int d^3\mathbf{r} \mathbf{A}_\lambda^*(\mathbf{r}) \cdot \mathbf{A}_{\lambda'}(\mathbf{r}) \\
&= \frac{1}{c^2} \sum_{\lambda, \lambda'} \dot{q}_\lambda^*(t) \dot{q}_{\lambda'}(t) \delta_{\lambda, \lambda'} = \frac{1}{c^2} \sum_\lambda |\dot{q}_\lambda(t)|^2 \\
&= \sum_\lambda I_\lambda. \qquad (2.32)
\end{aligned}
$$

Here, $I_\lambda \equiv |\dot{q}_\lambda(t)|^2 / c^2$.

Similarly, we will prove in the following that the total energy U_{tot} of the optical field can also be decomposed into the sum of the energy of individual eigen solutions:

$$U_{tot} = \sum_\lambda U_\lambda. \qquad (2.33)$$

---

[1]One may argue that interference effect due to superposition is energy re-distribution. But that is a result of interaction between optical fields and detectors and no energy re-distribution occurs for superposition of free fields.

To prove the above, we start from the electromagnetic theory and write the expression for the total energy of the electromagnetic field as

$$U_{tot} = \frac{1}{8\pi} \int d^3r \left[\mathbf{E}(\mathbf{r},t) \cdot \mathbf{E}(\mathbf{r},t) + \mathbf{B}(\mathbf{r},t) \cdot \mathbf{B}(\mathbf{r},t) \right], \qquad (2.34)$$

where $\mathbf{E}, \mathbf{B}$ are electric and magnetic fields and must have real values. So, they must be derived from Eq. (2.14) via the real expression of $\mathbf{A}$ in Eq. (2.22). Let us first calculate the first term in Eq. (2.34), that is, the energy of electric field. From Eq. (2.14) and orthonormal relation in Eq. (2.20), we obtain

$$\begin{aligned}
U_{tot}^{(E)} &= \frac{1}{8\pi} \int d^3r \mathbf{E}(\mathbf{r},t) \cdot \mathbf{E}(\mathbf{r},t) \\
&= \frac{1}{8\pi c^2} \int d^3r \sum_{\lambda,\lambda'} \left[\dot{q}_\lambda(t)\mathbf{A}_\lambda(\mathbf{r}) + c.c. \right] \cdot \left[\dot{q}_{\lambda'}(t)\mathbf{A}_{\lambda'}(\mathbf{r}) + c.c. \right] \\
&= \frac{1}{8\pi c^2} \sum_{\lambda,\lambda'} \left[\dot{q}_\lambda(t)\dot{q}_{\lambda'}^*(t) + \dot{q}_\lambda^*(t)\dot{q}_{\lambda'}(t) \right] \delta_{\lambda,\lambda'} \\
&\quad + \frac{1}{8\pi c^2} \sum_{\lambda,\lambda'} \left[\dot{q}_\lambda(t)\dot{q}_{\lambda'}(t) \int d^3r \mathbf{A}_\lambda(\mathbf{r}) \cdot \mathbf{A}_{\lambda'}(\mathbf{r}) + c.c. \right] \\
&= \frac{1}{8\pi c^2} \sum_{\lambda} \left[\dot{q}_\lambda(t)\dot{q}_\lambda^*(t) + \dot{q}_\lambda^*(t)\dot{q}_\lambda(t) \right] \\
&\quad + \frac{1}{8\pi c^2} \sum_{\lambda,\lambda'} \left[\dot{q}_\lambda(t)\dot{q}_{\lambda'}(t) I_{\lambda,\lambda'} + c.c. \right]. \qquad (2.35)
\end{aligned}$$

Here, $I_{\lambda,\lambda'} \equiv \int d^3r \mathbf{A}_\lambda(\mathbf{r}) \cdot \mathbf{A}_{\lambda'}(\mathbf{r})$. For the convenience of introducing operator in the future, we kept the original order of product for $\dot{q}_\lambda(t), \dot{q}_\lambda^*(t)$ in the expression above. The evaluation of the energy of magnetic field needs the following equation of vector differentiation:

$$\begin{aligned}
(\nabla \times \mathbf{A}) \cdot (\nabla \times \mathbf{A}') &= \sum_m \left[(\nabla A_m) \cdot (\nabla A_m') - (\nabla A_m) \cdot (\partial_m \mathbf{A}') \right] \\
&= \nabla \cdot \left(\sum_m A_m \nabla A_m' \right) - \mathbf{A} \cdot \nabla^2 \mathbf{A}' \\
&\quad - \nabla \cdot \left(\sum_m A_m \partial_m \mathbf{A}' \right) + \mathbf{A} \cdot \nabla(\nabla \cdot \mathbf{A}'). \ (2.36)
\end{aligned}$$

Here $\{\mathbf{A}, \mathbf{A}'\} = \{\mathbf{A}_\lambda, \mathbf{A}_\lambda^*\}$. After using Eqs. (2.18) and (2.19) in the equation above, we obtain

$$\begin{aligned}
(\nabla \times \mathbf{A}) \cdot (\nabla \times \mathbf{A}') \\
= \nabla \cdot \left(\sum_m A_m \nabla A_m' \right) - \nabla \cdot \left(\sum_m A_m \partial_m \mathbf{A}' \right) + k^2 \mathbf{A} \cdot \mathbf{A}'. \ (2.37)
\end{aligned}$$

Using the Gauss divergence theorem:

$$\int d^3\mathbf{r}(\nabla \cdot \mathbf{F}) = \oint dS(\hat{n} \cdot \mathbf{F}) \tag{2.38}$$

and boundary condition $\mathbf{A}(\mathbf{r}) = 0$, we change the spatial integral of Eq. (2.37) to[2]

$$\int d^3\mathbf{r}(\nabla \times \mathbf{A}) \cdot (\nabla \times \mathbf{A'}) = k^2 \int d^3\mathbf{r}\mathbf{A} \cdot \mathbf{A'}. \tag{2.39}$$

Now we can calculate the magnetic part in Eq. (2.34):

$$
\begin{aligned}
U_{tot}^{(B)} &= \frac{1}{8\pi} \int d^3\mathbf{r}\mathbf{B}(\mathbf{r},t) \cdot \mathbf{B}(\mathbf{r},t) \\
&= \frac{1}{8\pi} \int d^3\mathbf{r} \sum_{\lambda,\lambda'} \left[q_\lambda(t)\nabla \times \mathbf{A}_\lambda(\mathbf{r}) + c.c. \right] \cdot \left[q_{\lambda'}(t)\nabla \times \mathbf{A}_{\lambda'}(\mathbf{r}) + c.c. \right] \\
&= \frac{1}{8\pi} \sum_{\lambda,\lambda'} \left[q_\lambda(t)q_{\lambda'}^*(t) \int d^3\mathbf{r}(\nabla \times \mathbf{A}_\lambda(\mathbf{r})) \cdot (\nabla \times \mathbf{A}_{\lambda'}^*(\mathbf{r})) + c.c. \right] \\
&\quad + \frac{1}{8\pi} \sum_{\lambda,\lambda'} \left[q_\lambda(t)q_{\lambda'}(t) \int d^3\mathbf{r}(\nabla \times \mathbf{A}_\lambda(\mathbf{r})) \cdot (\nabla \times \mathbf{A}_{\lambda'}(\mathbf{r})) + c.c. \right] \\
&= \frac{1}{8\pi} \sum_\lambda k_\lambda^2 \left[q_\lambda(t)q_\lambda^*(t) + q_\lambda^*(t)q_\lambda(t) \right] \\
&\quad + \frac{1}{8\pi} \sum_{\lambda,\lambda'} k_\lambda k_{\lambda'} \left[q_\lambda(t)q_{\lambda'}(t)I_{\lambda,\lambda'} + c.c. \right]. \tag{2.40}
\end{aligned}
$$

Here, we used the orthonormal relation in Eq. (2.20). Combining Eqs. (2.35) and (2.40) and using $k_\lambda = \omega_\lambda/c$, we obtain the total energy of the optical field (electromagnetic field) as

$$
\begin{aligned}
U_{tot} &= \frac{1}{8\pi c^2} \sum_\lambda \left\{ \left[\dot{q}_\lambda(t)\dot{q}_\lambda^*(t) + \dot{q}_\lambda^*(t)\dot{q}_\lambda(t) \right] + \omega_\lambda^2 \left[q_\lambda(t)q_\lambda^*(t) + q_\lambda^*(t)q_\lambda(t) \right] \right\} \\
&\quad + \frac{1}{8\pi c^2} \sum_{\lambda,\lambda'} \left\{ \left[\dot{q}_\lambda(t)\dot{q}_{\lambda'}(t) + \omega_\lambda \omega_{\lambda'} q_\lambda(t)q_{\lambda'}(t) \right] I_{\lambda,\lambda'} + c.c. \right\}. \tag{2.41}
\end{aligned}
$$

We next use the solution for the temporal part in Eq. (2.17): $q_\lambda(t) = q_\lambda(0) e^{-i\omega_\lambda t}$ to obtain $\dot{q}_\lambda(t) = -i\omega_\lambda q_\lambda(t)$. Substituting into Eq. (2.41), we have

$$U_{tot} = \frac{1}{4\pi c^2} \sum_\lambda \omega_\lambda^2 \left[q_\lambda(t)q_\lambda^*(t) + q_\lambda^*(t)q_\lambda(t) \right] = \sum_\lambda U_\lambda. \tag{2.42}$$

---

[2]It can be shown directly that the plane wave solutions in Sections 2.1.3 and 2.1.4 also satisfy Eq. (2.39).

Now, we just proved the decomposition of total energy in Eq. (2.33), where

$$U_\lambda = \frac{1}{4\pi c^2}\omega_\lambda^2\left[q_\lambda(t)q_\lambda^*(t) + q_\lambda^*(t)q_\lambda(t)\right] = \frac{1}{2\pi c^2}\omega_\lambda^2|q_\lambda(0)|^2. \quad (2.43)$$

We see from Eqs. (2.22), (2.42), and (2.43) that the optical field can be decomposed into the sum of the eigen solutions and is determined by them independently. This is very similar to the normal mode description in a mechanical oscillator system. The eigen solutions of the optical field is similar to the normal modes of the oscillator system. We likewise define each eigen solution as a mode of the optical field which can be described completely by all the modes. Compared to quantum mechanics, the modes of the optical field is analogous to the eigen energy states for Schrödinger equation in mathematical formalism. So, solving the modes for Maxwell equation is equivalent to the first quantization process involving Schrödinger equation.

We can also find from Eq. (2.43) that the energy in each mode is only determined by the initial condition $q_\lambda(0)$ but is completely unrelated to the mode function $\mathbf{A}_\lambda(\mathbf{r})$. So, quantity $q_\lambda(0)$ is also known as the excitation of mode λ. Since $q_\lambda(0)$ is independent from other different modes, each mode can be excited separately in an independent way. If there is only one mode being excited, we have a single-mode field. If more than one mode is excited, the optical field is of multi-mode. We will see later in the coherence theory of light, the number of excited modes will affect the coherence property of the optical field.

On the other hand, the mode function $\mathbf{A}_\lambda(\mathbf{r})$ is independent of the mode excitation $q_\lambda(0)$. As can be seen from the process that the mode function $\mathbf{A}_\lambda(\mathbf{r})$ is derived, it is only determined by the boundary conditions which are usually related to the geometry of the border. So, it is unrelated to whether the mode is excited or not and it exists even for vacuum where there is no mode excitation at all ($q_\lambda(0) = 0$). However, when the boundary conditions or the geometry of the border change, it may result in the change in the mode function as well as the eigen-values $\mathbf{k}_\lambda$, which will lead to field energy change even for vacuum if it has non-zero energy as in quantum theory of light. This is the famous Casimir effect (see Section 2.4).

In general, mode function $\mathbf{A}_\lambda(\mathbf{r})$ of the optical field can be written as

$$\mathbf{A}_\lambda(\mathbf{r}) = \hat{\epsilon}_{\mathbf{k},s}u_{\mathbf{k},s}(\mathbf{r}). \quad (2.44)$$

It can be determined by the following quantities of independent degrees of freedom:
(1) Polarizations of light wave, such as $\hat{\epsilon}_s = \hat{x}, \hat{y}$. Different orthogonal polarization states correspond to different polarization modes.

(2) Frequencies of light wave $\omega_\lambda = c|\mathbf{k}|$. Different colors of light correspond to different frequency modes.

(3) Directions of light wave $\hat{\mathbf{k}} = \mathbf{k}/|\mathbf{k}|$ or spatial functions $u_{\mathbf{k},s}(\mathbf{r})$. Different directions of propagation or spatial functions give rise to different spatial modes.

In the next section, let us introduce some commonly used modes of optical fields.

2.2 More Modes of Optical Fields

2.2.1 Paraxial Rays and Gaussian Beams – the Modes of Optical Resonators

When $k_x, k_y \ll k_z$, we can make the paraxial approximation. As an example, let us first look at a spherical wave:

$$\mathbf{A}_{\text{sph}}(\mathbf{r}) = \frac{\hat{\epsilon}A_0}{r}e^{ikr}, \qquad (2.45)$$

where $r = \sqrt{x^2 + y^2 + z^2}$. In paraxial approximation, the field does not leave far from the z-axis: $(x^2 + y^2)/z^2 \equiv \theta^2 \ll 1$. So, we can make the following approximation:

$$r = z\sqrt{1 + \frac{x^2 + y^2}{z^2}} = z\sqrt{1 + \theta^2} \approx z\left(1 + \frac{\theta^2}{2}\right) = z + \frac{x^2 + y^2}{2z}. \quad (2.46)$$

Then, we obtain the paraxial approximation of the spherical wave in the Fresnel form:

$$\mathbf{A}_{\text{sph}}(\mathbf{r}) \approx \frac{\hat{\epsilon}A_0}{z}e^{ikz}\exp\left(ik\frac{x^2 + y^2}{2z}\right). \qquad (2.47)$$

For arbitrary waves, we can write the vector wave function $\mathbf{A}(\mathbf{r})$ as

$$\mathbf{A}(x, y, z) = \hat{\epsilon}u(x, y, z)e^{ikz}, \qquad (2.48)$$

where $u(x, y, z)$ satisfies the conditions for paraxial approximation: $|\partial u/\partial z| \ll ku, |\partial^2 u/\partial z^2| \ll k^2 u$. Then Helmholtz equation in Eq. (2.15) can be approximated as

$$\frac{\partial^2 u}{\partial x^2} + \frac{\partial^2 u}{\partial y^2} + 2ik\frac{\partial u}{\partial z} = 0. \qquad (2.49)$$

It can be easily shown that the Fresnel form of the spherical wave in Eq. (2.47) satisfies the paraxial Helmholtz equation in Eq. (2.49).

It is not so easy to find a general solution for Eq. (2.49) but we can obtain a special solution, that is, the Gaussian beam, by a simple transformation.

For this, we make a translation along the z-axis: $z' = z - z_0$. It is easy to see that Eq. (2.49) is unchanged. There is a simple physical picture for this transformation: the center of the spherical wave is moved to $z = z_0$ by the transformation. Now, let us make a translation of an imaginary number: $z'' = z - ib$ and the Fresnel form of the spherical wave is changed to

$$u_G(\mathbf{r}) = \frac{A_0}{q(z)} \exp\left(ik\frac{x^2 + y^2}{2q(z)} \right). \tag{2.50}$$

Here we only write down the part of u-function with $q(z) = z - ib$. Since ib is a constant, a simple derivation shows that this transformation does not change the form of Eq. (2.49), either. So, the transformed u-function in (2.50) is still a solution of Eq. (2.49). A simple substitution can verify this.

In order to find the physical meaning of Eq. (2.50), we change its form to:

$$\frac{1}{q(z)} = \frac{z}{z^2 + b^2} + i\frac{b}{z^2 + b^2} = \frac{1}{R(z)} + i\frac{\lambda}{\pi w^2(z)}, \tag{2.51}$$

where

$$R(z) \equiv z\left[1 + \left(\frac{b}{z}\right)^2\right], \quad w(z) \equiv w_0\left[1 + \left(\frac{z}{b}\right)^2\right]^{1/2}, \tag{2.52}$$

with λ as the wavelength and $w_0^2 \equiv \lambda b/\pi$. Defining $\zeta(z) \equiv \tan^{-1}(z/b)$,[3] we have $q(z) = -i|q(z)|e^{i\zeta(z)}$. Equation (2.50) then becomes

$$u_G(\mathbf{r}) = A_0'\frac{w_0}{w(z)} \exp\left[-\frac{x^2 + y^2}{w^2(z)} \right] \exp\left[ik\frac{x^2 + y^2}{2R(z)} - i\zeta(z) \right]. \tag{2.53}$$

Compared with (2.47), the wave front of $u_G(\mathbf{r})$ is the same as a spherical wave with radius $R(z)$, but the transverse distribution of its amplitude or intensity is a Gaussian distribution:

$$I_G(\mathbf{r}) = |\mathbf{A}(\mathbf{r})|^2 = \frac{2P}{\pi w^2(z)} \exp\left[-\frac{2(x^2 + y^2)}{w^2(z)} \right], \tag{2.54}$$

where P is the total power of the beam $P = \int I_G(\mathbf{r})dxdy$. This beam is called "Gaussian beam" because of the Gaussian distribution for the transverse intensity, whose cross-section is a circle of radius around $w(z)$. Figure 2.1(a) shows $w(z)$ as a function of z, where it reaches minimum value of w_0 at $z = 0$ and the divergent angle is $\theta_0 = \tan^{-1}(w_0/b)$. w_0 is also known as the "waist" of the Gaussian beam. When θ_0 is small, $\theta_0 \sim w_0/b = \lambda/\pi w_0$, which corresponds to the diffraction angle by a circular

---

[3]Phase $\zeta(z)$ changes slowly with z from $-\pi/2$ to $\pi/2$ and is known as Guoy phase.

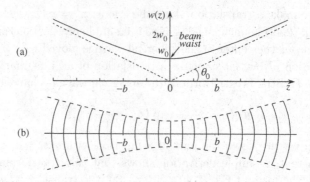

Fig. 2.1   (a) The radius $w(z)$ of the Gaussian beam cross-section vs. the propagation distance z. (b) The wave front of the Gaussian beam vs. the propagation distance z.

opening of radius w_0. So, we can say that Gaussian beams are diffraction-limited. As can be seen, the parameter b uniquely determines the waist and the divergent angle of the Gaussian beam and thus defines the whole Gaussian beam. Figure 2.1(b) shows the change of the wave front of the Gaussian beam as propagation distance z changes. In the interval $(-b, b)$, the wave front is nearly plane wave-like. This interval is also known as the Rayleigh region.

Equation (2.53) is a special and most simple Gaussian beam solution to the paraxial Helmholtz equation in (2.49). The paraxial Helmholtz equation also has higher order Gaussian beam solutions in a general form [Siegman (1986)]

$$u_{lm}(x, y, z) = \frac{F_{lm}(x, y, z)}{q(z)} \exp[ik(x^2 + y^2)/q(z)], \qquad (2.55)$$

where

$$F_{lm}(x, y, z) = u_0 H_l(x/w(z)) H_m(y/w(z)) e^{-i(l+m)\zeta(z)}. \qquad (2.56)$$

Here $H_l(x)$ is the lth order Hermite function. All the Gaussian beams with different l, m form the Gaussian modes of the optical field. Figure 2.2 shows the images of the spatial distribution of a number of Gaussian modes.

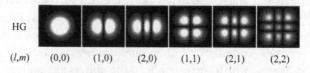

Fig. 2.2   Spatial distributions for a number of different Hermite-Gaussian modes.

The Gaussian modes discussed above can be produced from optical resonator formed by spherical reflection mirrors. The Gaussian modes so produced are determined by the geometry of the optical resonator (see Fig. 2.3). The boundary conditions require the wave fronts of the Gaussian beam coincide with the shape of the two reflectors, that is,

$$R_1 = z_1 + \frac{b^2}{z_1}, \quad R_2 = z_2 + \frac{b^2}{z_2}, \quad L = z_2 - z_1, \tag{2.57}$$

where L is the length of the resonator, z_1, z_2 are respectively the locations of the two reflectors relative to the waist of the Gaussian beam. Then we have

$$z_2 = \frac{L(L + R_1)}{2L - R_2 + R_1}, \quad z_1 = \frac{L(R_2 - L)}{2L - R_2 + R_1}, \tag{2.58}$$

$$b^2 = \frac{L(R_2 - L)(R_1 + L)(L + R_1 - R_2)}{(2L + R_1 - R_2)^2}. \tag{2.59}$$

So, the waist of the Gaussian beam $w_0 = \sqrt{\lambda b/\pi}$ and locations z_1, z_2 are uniquely determined by the radii R_1, R_2 of the two reflectors and their separation L. From Eq. (2.57), the signs of R_1, R_2 depend on z_1, z_2.

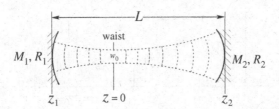

Fig. 2.3 The Gaussian modes produced by the optical resonator consisted of two spherical reflectors.

The eigen values k_{lm} are determined by the resonance condition:

$$k_{lm}(z_2 - z_1) - (l + m + 1)[\zeta(z_2) - \zeta(z_1)] = N\pi, \tag{2.60}$$

Then the resonant frequencies of the optical resonator are

$$\nu_{lm} = ck_{lm}/2\pi = N\Delta\nu + (l + m + 1)\delta\nu, \tag{2.61}$$

where $\Delta\nu = c/2L, \delta\nu = \Delta\nu[\zeta(z_2) - \zeta(z_1)]/\pi = \Delta\nu \cos^{-1}(\sqrt{g_1 g_2})/\pi$ with $g_1 \equiv 1 + L/R_1, g_2 \equiv 1 - L/R_2$. The first term in Eq. (2.61) corresponds to the fundamental modes given by Eq. (2.53) and is known as the longitudinal mode while the second term is for the transverse modes corresponding to higher-order Gaussian modes in Eq. (2.55). Cavity stability requires a real $\sqrt{g_1 g_2}$ or $0 \le g_1 g_2 \le 1$.

To analyze the mode structure of a resonator experimentally, we direct a beam of single-frequency laser onto the resonator. When the laser frequency coincides with one of the resonant frequencies of the resonator, the corresponding Gaussian mode is excited and built up. Some of the excited field can also leak out of the resonator and be detected by a photo-detector. According to Eq. (2.61), Fig. 2.4 shows the leaked field intensity as the laser frequency is scanned: all the resonant modes can be scanned out. The height for each mode depends on the mode match of the incident laser beam to each transverse mode of the cavity. On the other hand, if there is a gain medium in the resonator and for some modes, if the gain is higher than the losses, laser oscillation can start and a laser beam is produced. Since only Gaussian modes can be excited, the output modes of the laser beam are simply the Gaussian modes of the resonator. Which mode will lase is determined by the gains and losses of the modes. In general, several modes can be excited and have laser output. This leads to a multi-mode (multi-frequency) laser. To obtain a single-mode (single-frequency) laser, we need a Fabre-Perot etalon to select mode (frequency).

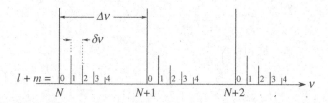

Fig. 2.4 The frequencies of the modes of the resonator are obtained as the laser frequency is scanned.

For more discussions about optical resonator cavities and Gaussian modes, the book by Siegman is very comprehensive and covers many of the topics [Siegman (1986)]. Besides Hermite-Gaussian modes discussed here, Laguerre-Gaussian modes have spatial profiles which are circularly symmetric and are used to describe orbital angular momentum (OAM) of optical fields [Andrews and Babiker (2013)]. OAM of an optical field describes structured and twisted light such as optical vortices which have a high degree of freedom for information carrying.

In addition to the Gaussian modes, other frequently used spatial modes are those from optical fibers, which are widely used for optical communications. Their mode structures are similar to the Gaussian modes so they are commonly used in the lab to clean up spatial modes of optical fields. For details of optical fiber modes, refer to [Buck (1995)].

2.2.2 Temporal Modes and Generalized Definition of Single-Mode Fields

What we discussed in the previous section is the spatial mode. It gives rise to the spatial distribution of the optical field in the transverse direction that is perpendicular to the direction of propagation. In this section, we will discuss the longitudinal distribution of the optical field along the direction of propagation. For one-dimensional field (see Section 2.3.5), the longitudinal direction is equivalent to the temporal degree of freedom. In the mode decomposition of the optical field discussed earlier, the temporal part is $e^{-i\omega t}$. So, a single frequency field corresponds to a frequency mode and spectral analysis is equivalent to decomposition into longitudinal modes. On the other hand, in the application to ultra-short pulses, it is more convenient to work in temporal domain. Consider a pulse of a one-dimensional field:

$$E(z,t) = \frac{1}{\sqrt{2\pi}} \int d\omega \mathcal{E}(\omega) e^{i(kz-\omega t)}. \tag{2.62}$$

If $\mathcal{E}(\omega)$ is a well-behaved function such as a Gaussian function, the spectral bandwidth $\Delta\omega$ of $\mathcal{E}(\omega)$ and the pulse width Δt of $E(z,t)$ satisfy the relation for Fourier transformation:

$$\Delta\omega\Delta t = 2\pi. \tag{2.63}$$

Such a pulse is known as the transform-limited pulse. It can be generated from a mode-locked laser where each frequency mode has the same phase. So, such a pulse has really good coherence and its waveform is a well-defined function of time. From function theory in mathematics, we learn that for a well-defined temporal function $f(t)$, we can always construct a set of functions $g_1(t), g_2(t), \ldots$ so that the whole function set of $\{f(t), g_1(t), g_2(t), \ldots\}$ forms a complete and orthonormal base. This set has the same properties as the mode function set $\{u_\lambda(\mathbf{r})\}$ discussed earlier. The function set $\{f(t), g_1(t), g_2(t), \ldots\}$ then forms a set of temporal modes and function $f(t)$ is one of the modes. Hence, a transform-limited pulse is a single temporal mode.

Notice that the single mode defined by Eq. (2.62) is a superposition of many frequency modes. When $\mathcal{E}(\omega)$ is a well-defined function, $E(z,t)$

in Eq. (2.62) is a solution of Maxwell equation. From this property, we may define a generalized single-mode field as any one of the solutions of Maxwell equation with a well-defined function of $(\mathbf{r}, t)$. It can be a linear superposition of many modes:

$$\mathbf{A}(\mathbf{r}, t) = \sum_\lambda c_\lambda \mathbf{A}_\lambda(\mathbf{r}) e^{-i\omega_\lambda t}, \qquad (2.64)$$

where $\{c_\lambda\}$ is a set of definite constants. The generalized field mode so defined will be applied to the quantization of optical fields in the next section and as we will see later (Section 8.4), this becomes important when we discuss the photon indistinguishability: if a group of photons can be described by the same well-defined solution of Maxwell equation, they are labeled as in the same mode and are completely indistinguishable. Photons having this property will exhibit the largest effect of quantum interference.

We can also understand the concept of generalized mode from the perspective of linear algebra. A complete set of modes is like a complete base vector in a vector space. One mode corresponds to a base vector. A generalized mode can be considered of as a vector that is a linear superposition of the base vectors. This vector can form another complete base set with other vectors orthogonal to it so that it is one of the base vectors in the newly formed complete bases. In this analogy, the generalized mode is also a base mode and belongs to a single mode in a new set of complete base modes. Take an example of polarizations: $\hat{x}, \hat{y}$ form a base set for the polarization vectors of light field in which $\hat{x}$ describes a single polarization mode. But polarization vectors in 45 degree and 135 degree can also form a complete set of polarization base vectors. So, even though polarization vector in 45 degree is a superposition of $\hat{x}, \hat{y}$, it actually is a single polarization mode. Other example such as circularly polarized light $\hat{\epsilon}_\pm = (\hat{x} \pm i\hat{y})/\sqrt{2}$ is also a single polarization mode. In fact, any elliptically polarized light $\hat{\epsilon} = \hat{x} \cos\theta + \hat{y} e^{i\varphi} \sin\theta$ is also in a single polarization mode.

The concept of single mode will become crucial in understanding and describing multi-photon indistinguishability. We will come back to this later in Chapter 8.

2.3   Quantization of Optical Fields

2.3.1   *Description of Modes by Simple Harmonic Oscillators*

When the modes of an optical field are fixed, the time evolution of the field is determined by the mode excitation $q_\lambda(t)$. Let us now go back to the energy expression in Eq. (2.43). It is the starting point for the quantization of the optical field. The coefficient $1/4\pi c^2$ in the expression is a fixed constant. In order to be consistent with the formulae after quantization, we combine it with $q_\lambda(t)$. So, Eq. (2.43) changes to

$$U_\lambda = \omega_\lambda^2 \left[q_\lambda(t) q_\lambda^*(t) + q_\lambda^*(t) q_\lambda(t) \right]. \tag{2.65}$$

The field function in Eq. (2.22) becomes

$$\mathbf{A}(\mathbf{r},t) = \sqrt{4\pi c^2} \sum_\lambda \left[q_\lambda(t) \mathbf{A}_\lambda(\mathbf{r}) + q_\lambda^*(t) \mathbf{A}_\lambda^*(\mathbf{r}) \right]$$

$$= \sqrt{4\pi c^2} \sum_\lambda \left[q_\lambda(0) \mathbf{A}_\lambda(\mathbf{r}) e^{-i\omega_\lambda t} + q_\lambda^*(0) \mathbf{A}_\lambda^*(\mathbf{r}) e^{i\omega_\lambda t} \right]. \tag{2.66}$$

Because $q_\lambda(t)$ is a complex number, the expression in Eq. (2.65) for single mode energy is not in a form that looks familiar to us. Let us use the real and imaginary parts of $q_\lambda(t)$ as new variables:

$$Q_\lambda \equiv q_\lambda(t) + q_\lambda^*(t)$$
$$P_\lambda \equiv \omega_\lambda \left[q_\lambda(t) - q_\lambda^*(t) \right] / i, \tag{2.67}$$

where we add a factor of ω_λ for the imaginary part P_λ so that it has the meaning of a generalized momentum for the generalized coordinate Q_λ. Equation (2.67) can be rewritten as

$$q_\lambda(t) = (Q_\lambda + iP_\lambda/\omega_\lambda)/2. \tag{2.68}$$

Substituting Eq. (2.68) into Eq. (2.65), we have

$$U_\lambda = \frac{1}{2} \left(P_\lambda^2 + \omega_\lambda^2 Q_\lambda^2 \right). \tag{2.69}$$

Since $\dot{q}_\lambda(t) = -i\omega_\lambda q_\lambda(t)$, we obtain from Eq. (2.67)

$$P_\lambda = \dot{Q}_\lambda. \tag{2.70}$$

The energy expression in Eq. (2.69) is exactly the same as that for a simple harmonic oscillator of mass $m = 1$. The momentum P of the oscillator is given by Eq. (2.70). Indeed, if we treat Eq. (2.69) as the Hamiltonian of the oscillator, i.e.,

$$H_\lambda = \frac{1}{2} \left(P_\lambda^2 + \omega_\lambda^2 Q_\lambda^2 \right), \tag{2.71}$$

with Q_λ as the generalized coordinate and P_λ as the generalized momentum, Hamiltonian mechanics gives rise to an equation of motion same as Eq. (2.17), i.e.,

$$\dot{P}_\lambda = -\frac{\partial H_\lambda}{\partial Q_\lambda}$$

$$\dot{Q}_\lambda = \frac{\partial H_\lambda}{\partial P_\lambda}. \tag{2.72}$$

The last line of the above equations is the same as Eq. (2.70).

It is well-known that the time evolution of a system is determined by the Hamiltonian of the system. So, each mode of the optical field can be described by its corresponding simple harmonic oscillator with mass $m = 1$ and its oscillating frequency as that of the mode. In fact, many forms of wave motion in nature can be described by harmonic oscillation of some particles such as water waves, the standing wave on a violin string. However, different from these waves that are formed by real particles, the harmonic oscillators for the description of the modes of optical fields are not real objects but virtual oscillators. So are their positions and momenta.

The classical description of the virtual harmonic oscillator is given by Eq. (2.72). Next, we will present its quantum description. Since the optical field is decomposed into harmonic oscillators of each modes, the quantum description of the harmonic oscillators will lead to the quantum description of the whole field.

2.3.2 *Quantization of Simple Harmonic Oscillators*

The quantization of harmonic oscillators is a typical example that can be found in almost every textbook of quantum mechanics. Here, we will only present the results. In the quantum description of a harmonic oscillator, the generalized coordinate Q and momentum P become operator $\hat{Q}, \hat{P}$ with a commutation relation

$$[\hat{Q}, \hat{P}] = i\hbar.$$

Since different modes of the optical field are independent of each other, the operators corresponding to different modes commute with each other. Hence, the commutation relations for all the operators of the modes are

$$[\hat{Q}_\lambda, \hat{P}_{\lambda'}] = i\hbar\delta_{\lambda,\lambda'}, \quad [\hat{Q}_\lambda, \hat{Q}_{\lambda'}] = 0, \quad [\hat{P}_\lambda, \hat{P}_{\lambda'}] = 0. \qquad (2.73)$$

Introduce the annihilation and creation operators:

$$\hat{a}_\lambda \equiv \left(\hat{Q}_\lambda + i\frac{\hat{P}_\lambda}{\omega_\lambda}\right)\sqrt{\frac{\omega_\lambda}{2\hbar}}, \quad \hat{a}_\lambda^\dagger \equiv \left(\hat{Q}_\lambda - i\frac{\hat{P}_\lambda}{\omega_\lambda}\right)\sqrt{\frac{\omega_\lambda}{2\hbar}}, \qquad (2.74)$$

These a-quantities are proportional to the q-quantities in Eq. (2.68): $a = q\sqrt{2\omega_\lambda/\hbar}$. We will later use this relation to establish the correspondence between quantum and classical descriptions. The Hamiltonian in Eq. (2.71) changes to

$$\hat{H}_\lambda = \frac{1}{2}\hbar\omega_\lambda(\hat{a}_\lambda^\dagger\hat{a}_\lambda + \hat{a}_\lambda\hat{a}_\lambda^\dagger), \qquad (2.75)$$

and the Hamiltonian of the optical field with all modes included takes the form of

$$\hat{H} = \sum_\lambda \hat{H}_\lambda = \sum_\lambda \frac{1}{2}\hbar\omega_\lambda(\hat{a}_\lambda^\dagger \hat{a}_\lambda + \hat{a}_\lambda \hat{a}_\lambda^\dagger), \tag{2.76}$$

where $\hat{H}_\lambda$'s are independent of each other for different λ and the commutation relation of operators $\hat{a}_\lambda, \hat{a}_\lambda^\dagger$ can be obtained from Eq. (2.73):

$$[\hat{a}_\lambda, \hat{a}_{\lambda'}^\dagger] = \delta_{\lambda,\lambda'}. \tag{2.77}$$

The δ-function in the expression above is the result of the independence between different modes. With the commutation relation above, the Hamiltonian in Eq. (2.75) is now in a form familiar to us:

$$\hat{H}_\lambda = \hbar\omega_\lambda \hat{a}_\lambda^\dagger \hat{a}_\lambda + \hbar\omega_\lambda/2. \tag{2.78}$$

Since operator $\hat{a}_\lambda^\dagger \hat{a}_\lambda$ is a positive definite, meaning that $\langle\psi|\hat{a}_\lambda^\dagger \hat{a}_\lambda|\psi\rangle \geq 0$ for any state $|\psi\rangle$, the energy $E_\lambda \equiv \langle\psi|\hat{H}_\lambda|\psi\rangle$ has a minimum value of $\hbar\omega_\lambda/2$. Later in Section 3.1, we will see that $\hbar\omega_\lambda/2$ is the energy of vacuum for mode λ.

In the Heisenberg picture, the equation of motion for $\hat{a}_\lambda$ is

$$\frac{d\hat{a}_\lambda}{dt} = \frac{1}{i\hbar}[\hat{a}_\lambda, \hat{H}_\lambda], \tag{2.79}$$

or

$$\frac{d\hat{a}_\lambda}{dt} = -i\omega_\lambda \hat{a}_\lambda. \tag{2.80}$$

It has a solution of $\hat{a}_\lambda(t) = \hat{a}_\lambda(0)e^{-i\omega_\lambda t}$.

2.3.3 *Field Operators for an Optical Field*

Annihilation and creation operators are the basic operators for the description of the optical field. Any other physical quantity can be expressed in terms of them. From Eq. (2.66), we have the operator form for the vector potential:

$$\hat{\mathbf{A}}(\mathbf{r}, t) = \sqrt{4\pi c^2} \sum_\lambda \sqrt{\frac{\hbar}{2\omega_\lambda}} \left[\hat{a}_\lambda e^{-i\omega_\lambda t} \mathbf{A}_\lambda(\mathbf{r}) + \hat{a}_\lambda^\dagger e^{i\omega_\lambda t} \mathbf{A}_\lambda^*(\mathbf{r})\right]. \tag{2.81}$$

Here, $\hat{a}_\lambda = \hat{a}_\lambda(0)$ is the operator in Schrödinger picture. From Eq. (2.14), we obtain the operator forms for the electric and magnetic fields:

$$\hat{\mathbf{E}}(\mathbf{r}, t) = \sqrt{4\pi} \sum_\lambda i\sqrt{\frac{\hbar\omega_\lambda}{2}} \hat{a}_\lambda e^{-i\omega_\lambda t} \mathbf{A}_\lambda(\mathbf{r}) + h.c.$$

$$\hat{\mathbf{B}}(\mathbf{r}, t) = \sqrt{4\pi c^2} \sum_\lambda \sqrt{\frac{\hbar}{2\omega_\lambda}} \hat{a}_\lambda e^{-i\omega_\lambda t} \nabla \times \mathbf{A}_\lambda(\mathbf{r}) + h.c. \tag{2.82}$$

Here, *h.c.* stands for Hermitian conjugate.

2.3.3.1   Field Operators for Discrete **k**-Space

In the box model, $\mathbf{A}_\lambda(\mathbf{r})$ is given by Eq. (2.24). We then obtain the operator form of the vector potential for the discrete **k**-space:

$$\hat{\mathbf{A}}(\mathbf{r},t) = \sqrt{4\pi c^2} \sum_{\mathbf{k},s} \sqrt{\frac{\hbar}{2\omega}} \, \hat{a}_{\mathbf{k},s} \hat{\epsilon}_{\mathbf{k},s} \frac{e^{i(\mathbf{k}\cdot\mathbf{r}-\omega t)}}{L^{3/2}} + h.c. \qquad (2.83)$$

Similarly, the operator forms for the electric and magnetic fields are

$$\hat{\mathbf{E}}(\mathbf{r},t) = i\sqrt{4\pi} \sum_{\mathbf{k},s} \sqrt{\frac{\hbar\omega}{2}} \, \hat{a}_{\mathbf{k},s} \hat{\epsilon}_{\mathbf{k},s} \frac{e^{i(\mathbf{k}\cdot\mathbf{r}-\omega t)}}{L^{3/2}} + h.c.$$

$$\hat{\mathbf{B}}(\mathbf{r},t) = i\sqrt{4\pi} \sum_{\mathbf{k},s} \sqrt{\frac{\hbar\omega}{2}} \, \hat{a}_{\mathbf{k},s} (\hat{\mathbf{k}} \times \hat{\epsilon}_{\mathbf{k},s}) \frac{e^{i(\mathbf{k}\cdot\mathbf{r}-\omega t)}}{L^{3/2}} + h.c. \qquad (2.84)$$

2.3.3.2   Field Operators for Continuous **k**-Space
          and 3-Dimensional Free Space

When L goes to infinity, the discrete **k**-space changes to continuous **k**-space. However, the mode function $\mathbf{A}_\lambda(\mathbf{r})$ given by Eq. (2.24) will go to zero and the sum in Eqs. (2.81)–(2.84) will change to integral. So, this process cannot be trivially done. We will handle this process in the following.

In the discrete **k**-space, the smallest volume is $\Delta^3 k = (2\pi/L)^3$. When L goes to infinity, $\Delta^3 k = (2\pi/L)^3 \to d^3\mathbf{k}$, i.e., 3-dimensional differential volume element for integration. So, the transition from a summation to an integral is:

$$\left(\frac{2\pi}{L}\right)^3 \sum_{\mathbf{k}} = \sum_{\mathbf{k}} \Delta^3 k \to \int d^3\mathbf{k}. \qquad (2.85)$$

For this transition, we rewrite the vector potential in Eq. (2.83) as

$$\hat{\mathbf{A}}(\mathbf{r},t) = \sqrt{4\pi c^2} \left(\frac{2\pi}{L}\right)^3 \sum_{\mathbf{k},s} \sqrt{\frac{\hbar}{2\omega}} \, \hat{a}_{\mathbf{k},s} \left(\frac{L}{2\pi}\right)^{\frac{3}{2}} \hat{\epsilon}_{\mathbf{k},s} \frac{e^{i(\mathbf{k}\cdot\mathbf{r}-\omega t)}}{(2\pi)^{3/2}} + h.c. \qquad (2.86)$$

But there is still a factor of $(L/2\pi)^{3/2}$ in the sum that will diverge as $L \to \infty$. To solve this difficulty, we introduce the annihilation operator for the continuous **k**-space:

$$\hat{a}_s(\mathbf{k}) \equiv \hat{a}_{\mathbf{k},s}(L/2\pi)^{3/2} = \hat{a}_{\mathbf{k},s}/(\Delta k)^{3/2}. \qquad (2.87)$$

From the commutation relation in Eq. (2.77), we obtain

$$[\hat{a}_s(\mathbf{k}), \hat{a}_{s'}^\dagger(\mathbf{k}')] = \delta_{s,s'}\delta_{\mathbf{k},\mathbf{k}'}/\Delta^3 k \to \delta_{s,s'}\delta^{(3)}(\mathbf{k} - \mathbf{k}')$$

or

$$[\hat{a}_s(\mathbf{k}), \hat{a}_{s'}^\dagger(\mathbf{k}')] = \delta_{s,s'}\delta^{(3)}(\mathbf{k} - \mathbf{k}'). \tag{2.88}$$

Hence, the annihilation operator $\hat{a}_s(\mathbf{k})$ defined in Eq. (2.87) for the continuous $\mathbf{k}$-space satisfies the commutation relation for continuous variables in Eq. (2.88). Equation (2.86) then changes to the vector potential for the continuous $\mathbf{k}$-space:

$$\hat{\mathbf{A}}(\mathbf{r}, t) = \sqrt{4\pi c^2} \sum_{s=1,2} \int d^3\mathbf{k} \sqrt{\frac{\hbar}{2\omega}} \, \hat{a}_s(\mathbf{k})\hat{\epsilon}_{\mathbf{k},s} \frac{e^{i(\mathbf{k}\cdot\mathbf{r}-\omega t)}}{(2\pi)^{3/2}} + h.c. \tag{2.89}$$

Similarly, the electric and magnetic field operators in Eq. (2.84) change to:

$$\hat{\mathbf{E}}(\mathbf{r}, t) = i\sqrt{4\pi} \sum_{s=1,2} \int d^3\mathbf{k} \sqrt{\frac{\hbar\omega}{2}} \, \hat{a}_s(\mathbf{k})\hat{\epsilon}_{\mathbf{k},s} \frac{e^{i(\mathbf{k}\cdot\mathbf{r}-\omega t)}}{(2\pi)^{3/2}} + h.c.$$

$$\hat{\mathbf{B}}(\mathbf{r}, t) = i\sqrt{4\pi} \sum_{s=1,2} \int d^3\mathbf{k} \sqrt{\frac{\hbar\omega}{2}} \, \hat{a}_s(\mathbf{k})(\hat{\mathbf{k}} \times \hat{\epsilon}_{\mathbf{k},s}) \frac{e^{i(\mathbf{k}\cdot\mathbf{r}-\omega t)}}{(2\pi)^{3/2}} + h.c. \tag{2.90}$$

Note that the mode function is now given by Eq. (2.26) with an orthonormal relation given by Eq. (2.30) for the continuous $\mathbf{k}$-space.

2.3.4 The Quasi-Monochromatic Field Approximation

When the spectral width $\Delta\omega_F$ of the excited modes of the optical field is much smaller than the central frequency ω_0 of the optical field, i.e., $\Delta\omega_F \ll \omega_0$, this optical field is called quasi-monochromatic field. On the other hand, there is a finite detection bandwidth $\Delta\omega_D$ for any photo-detector. This means that detectors will not respond to the spectral modes of the optical field outside the spectral bandwidth of the detectors. So, we can ignore the frequency modes outside the detection spectral bandwidth in our calculation. In general, the spectral bandwidth of the detectors is usually quite narrow so that $\Delta\omega_D \ll \omega_0$. Therefore, the optical fields seen by the detectors are simply quasi-monochromatic fields.

From Eqs. (2.81) and (2.82), we find that the operators of the optical field can all be written in the form of

$$\hat{\mathbf{F}}(\mathbf{r}, t) = \sum_{\mathbf{k},s} l(\omega) \, \hat{a}_{\mathbf{k},s}\hat{\epsilon}_{\mathbf{k},s}u_{\mathbf{k},s}(\mathbf{r})e^{-i\omega t} + h.c. \tag{2.91}$$

where $l(\omega) = \sqrt{2\pi\hbar c^2/\omega}$ for $\mathbf{A}$ and $l(\omega) = \sqrt{2\pi\hbar\omega}$ for the electric and magnetic fields. A quasi-monochromatic field has $\Delta\omega_F \ll \omega_0$ for which the change in $l(\omega)$ is much slower than $e^{-i\omega t}$. In this case, we can make a quasi-monochromatic approximation: $l(\omega) \approx l(\omega_0)$. Then Eq. (2.91) can be approximated as

$$\hat{\mathbf{F}}(\mathbf{r},t) \approx l(\omega_0) \sum_{\mathbf{k},s} \hat{a}_{\mathbf{k},s}\hat{\epsilon}_{\mathbf{k},s}u_{\mathbf{k},s}(\mathbf{r})e^{-i\omega t} + h.c. \qquad (2.92)$$

So, different field operators only differ by a constant. We can then define a new field operator:

$$\hat{\mathbf{V}}(\mathbf{r},t) \equiv \sum_{\mathbf{k},s} \hat{a}_{\mathbf{k},s}\hat{\epsilon}_{\mathbf{k},s}u_{\mathbf{k},s}(\mathbf{r})e^{-i\omega t} + h.c. \equiv \hat{\mathbf{V}}^{(+)}(\mathbf{r},t) + \hat{\mathbf{V}}^{(-)}(\mathbf{r},t), \quad (2.93)$$

where we write $\hat{\mathbf{V}}$ in terms of the sum of the positive and negative frequency parts with the positive negative and frequency parts as

$$\hat{\mathbf{V}}^{(+)}(\mathbf{r},t) = \sum_{\mathbf{k},s} \hat{a}_{\mathbf{k},s}\hat{\epsilon}_{\mathbf{k},s}u_{\mathbf{k},s}(\mathbf{r})e^{-i\omega t} = \left[\hat{\mathbf{V}}^{(-)}(\mathbf{r},t)\right]^{\dagger}. \qquad (2.94)$$

From this, we may define the operator

$$\hat{n}(\mathbf{r},t) \equiv \hat{\mathbf{V}}^{(-)}(\mathbf{r},t) \cdot \hat{\mathbf{V}}^{(+)}(\mathbf{r},t). \qquad (2.95)$$

Its physical meaning can be obtained from the following calculation:

$$\int d^3r \hat{n}(\mathbf{r},t) = \sum_{\mathbf{k},s} \sum_{\mathbf{k}',s'} \hat{a}^{\dagger}_{\mathbf{k},s}\hat{a}_{\mathbf{k}',s'} e^{i(\omega-\omega')t}\left(\hat{\epsilon}^{*}_{\mathbf{k},s} \cdot \hat{\epsilon}_{\mathbf{k}',s'}\right) \int d^3r u^{*}_{\mathbf{k},s}u_{\mathbf{k}',s'}$$

$$= \sum_{\mathbf{k},s} \hat{a}^{\dagger}_{\mathbf{k},s}\hat{a}_{\mathbf{k},s} \equiv \sum_{\mathbf{k},s} \hat{n}_{\mathbf{k},s} = \hat{N}_{TOT}. \qquad (2.96)$$

From Section 3.1.2, we find that $\hat{a}^{\dagger}_{\mathbf{k},s}\hat{a}_{\mathbf{k},s} \equiv \hat{n}_{\mathbf{k},s}$ is the photon number operator of mode $\{\mathbf{k},s\}$. So, $\hat{N}_{TOT}$ is the total photon number operator and $\hat{n}(\mathbf{r},t) = \hat{\mathbf{V}}^{(-)} \cdot \hat{\mathbf{V}}^{(+)}$ is the photon number density operator of the optical field.

2.3.5   *One-Dimensional Approximation of Optical Fields*

In the future chapters, we will encounter numerous situations where the optical field only propagates in one fixed direction, that is, only modes with one fixed direction are excited while modes in other directions are otherwise in vacuum. In this case, the three-dimensional free field operators can be further simplified to one-dimensional field operators. To do this, let us go back to the box model of Section 2.1.3. The difference is that the cubic box is replaced by a rectangular box with length L and a cross-section S and the field function $\mathbf{A}(\mathbf{r})$ only depends on one coordinate z along the length of the box: $\mathbf{A}(\mathbf{r}) = \mathbf{A}(z)$. Here, the cross-section S is usually chosen as the beam size of the one-dimensional field. Our selection of the coordinate system is such that z-direction is the direction of propagation of the beam $\hat{z} = \hat{\mathbf{k}} = \mathbf{k}/k$. The boundary condition is that $\mathbf{A}$ is continuous only on the two opposite sides in z-direction:

$$\mathbf{A}(0) = \mathbf{A}(L). \tag{2.97}$$

The orthonormal field function satisfying this condition is then

$$\mathbf{A}(z) = \hat{\epsilon}_s e^{ikz}/\sqrt{SL}, \tag{2.98}$$

with $k = 2\pi m/L$ (m is integer). So, the one-dimensional form of the vector potential is

$$\hat{\mathbf{A}}(z,t) = \sqrt{4\pi c^2} \sum_{k,s} \sqrt{\frac{\hbar}{2\omega}}\, \hat{a}_{k,s}\hat{\epsilon}_s e^{i(kz-\omega t)}/\sqrt{SL} + h.c. \tag{2.99}$$

In the photo-detection theory discussed later, we will learn that the output photo-electric signal from the detector is usually proportional to the square of the field function, i.e., $|\mathbf{A}|^2$ and the photo-electric signal is the contributions from all points on the cross-section of the detector, i.e., an integration over the cross-section of the detector. But the contribution only comes from the part with the illumination of the optical field. So, the area of integration is the beam size S and the photo-electric signal is proportional to $|\mathbf{A}|^2 S$. From Eq. (2.99), we find that the cross-section area S is canceled. Hence, we can drop out S from the expression for the vector potential in Eq. (2.99). Setting $L \to \infty$, we make a transition to a one-dimensional form with a continuous k-value:

$$\hat{\mathbf{A}}(z,t) = \sqrt{4\pi c^2} \sum_{s=1,2} \int dk \sqrt{\frac{\hbar}{2\omega}}\, \hat{a}_s(k)\hat{\epsilon}_s e^{i(kz-\omega t)}/(2\pi)^{1/2} + h.c.$$

$$= \sqrt{4\pi} \sum_{s=1,2} \int d\omega \sqrt{\frac{c\hbar}{2\omega}}\, \hat{a}_s(\omega)\hat{\epsilon}_s e^{-i\omega t'}/(2\pi)^{1/2} + h.c. \tag{2.100}$$

where one-dimensional sum is changed to one-dimensional integral by $(2\pi/L)\times \sum_k = \sum_k \Delta k \to \int dk$ and $\hat{a}_s(k) \equiv \hat{a}_{k,s}(L/2\pi)^{1/2} = a_{k,s}/\Delta k^{1/2}$, which satisfies the commutation relation for continuous variables:

$$[\hat{a}_s(k), \hat{a}_{s'}^\dagger(k')] = (L/2\pi)[\hat{a}_{k,s}, \hat{a}_{k,s'}^\dagger] = \delta_{s,s'}\delta_{k,k'}/\Delta k$$

$$\to \delta_{s,s'}\delta(k-k'). \tag{2.101}$$

In the integral of the second line in Eq. (2.100), we made a change of variable: $k \to \omega/c$ and $\hat{a}_s(\omega) \equiv \hat{a}_s(k)/\sqrt{c}$, which satisfies

$$[\hat{a}_s(\omega), \hat{a}_{s'}^\dagger(\omega')] = (1/c)[\hat{a}_{k,s}, \hat{a}_{k,s'}^\dagger] = \delta_{s,s'}(1/c)\delta(k-k')$$

$$= \delta_{s,s'}\delta(\omega-\omega'). \tag{2.102}$$

Note that in the second line of Eq. (2.100), the time variable is changed to $t' = t - z/c$. So, in one-dimensional case, we only need the time variable. Spatial translation is equivalent to a delay or advance in time: $\Delta t = -\Delta z/c$.

Similar to Eq. (2.100), we have the one-dimensional expressions for the electric and magnetic fields:

$$\hat{\mathbf{E}}(t) = i\sqrt{4\pi} \sum_{s=1,2} \int d\omega \sqrt{\frac{\hbar\omega}{2}} \, \hat{a}_s(\omega)\hat{\epsilon}_s \frac{e^{-i\omega t}}{\sqrt{2\pi c}} + h.c. \tag{2.103}$$

$$\hat{\mathbf{B}}(t) = i\sqrt{4\pi} \sum_{s=1,2} \int d\omega \sqrt{\frac{\hbar\omega}{2}} \, \hat{a}_s(\omega)(\hat{\mathbf{k}} \times \hat{\epsilon}_s) \frac{e^{-i\omega t}}{\sqrt{2\pi c}} + h.c. \tag{2.104}$$

For the quasi-monochromatic field, we have the one-dimensional expression for the field operator

$$\hat{\mathbf{V}}^{(+)}(t) = \frac{1}{\sqrt{2\pi}} \sum_{s=1,2} \int d\omega \, \hat{a}_s(\omega)\hat{\epsilon}_s \, e^{-i\omega t}. \tag{2.105}$$

Since there is only time variable, the physical meaning of the quantity $\hat{\mathbf{V}}^{(-)}(t) \cdot \hat{\mathbf{V}}^{(+)}(t) \equiv \hat{R}(t)$ is no longer the photon number density but the photon number rate in time. This can be seen from the time integral of $\hat{R}(t)$:

$$\int dt\hat{R}(t) = \sum_{s,s'} \int d\omega d\omega' \hat{a}_s^\dagger(\omega)\hat{a}_{s'}(\omega')(\hat{\epsilon}_s^* \cdot \hat{\epsilon}_{s'}) \frac{1}{2\pi} \int dt e^{i(\omega - \omega')t}$$

$$= \sum_s \int d\omega \hat{a}_s^\dagger(\omega)\hat{a}_s(\omega) = \hat{N}_{TOT}. \tag{2.106}$$

We will use the one-dimensional expression in Eq. (2.105) for the field operator quite often in the later chapters.

2.4 Further Reading: Casimir Effect, A Quantum Effect of Vacuum due to Mode Change

From Eqs. (2.76) and (2.78) in Section 2.3.2, we find the minimum energy of the system is $E_{vac} = \sum_\lambda \hbar\omega_\lambda/2$, which is also the energy for vacuum from Section 3.1 and hence is labeled with a subscript "vac". From Sections 2.1.3 and 2.2.1, we find ω_λ depends on the mode structure. So, the energy of vacuum depends on the mode structure. If we change the mode structure, the energy of vacuum will change. Energy conservation means that some work needs to be done for the change. This leads to a force. This effect of vacuum energy is known as the Casimir effect [Casimir (1948)]. In this section, we will demonstrate how this effect arises from vacuum energy change and derive a formula for the Casimir force.

Consider the box model for mode structure. But different from the periodic box in Section 2.1.3, the box here is an $L \times L \times L$ cube with perfectly conducting walls, in which the electric field is zero. To change the mode structure, we insert another perfectly conducting plate of size $L \times L$ and negligible thickness at $x = a(\ll L)$ and parallel to the yz plane. We compare the total energy of the system with and without the plate.

For the case without the plate, the simple boundary condition of $\mathbf{E} = 0$ at the walls $(x, y, z = 0, L)$ gives rise to a mode function of the electric field as

$$\mathbf{E_k(r)} = \hat{\epsilon}_\mathbf{k}\left(\frac{2}{L}\right)^3 \sin k_x x \sin k_y y \sin k_z z \qquad (2.107)$$

with $\mathbf{k} = (k_x, k_y, k_z)$ and $k_x = n_x\pi/L, k_y = n_y\pi/L, k_z = n_z\pi/L$ (n_x, n_y, n_z = positive integers). Polarization vector $\hat{\epsilon}_\mathbf{k}$ satisfies the transverse wave condition $\hat{\epsilon}_\mathbf{k} \cdot \mathbf{k} = 0$, which leads to two independent polarization modes. Furthermore, the mode function of the form

$$\mathbf{E(r)} = \hat{\epsilon}_\mathbf{k}\left(\frac{4}{L^3}\right) \sin k_x x \sin k_y y \qquad (2.108)$$

is allowed so long as $\hat{\epsilon}_\mathbf{k} = \hat{z}$ because $E_\perp$ may not but $E_\parallel$ must be zero at the conducting walls due to the general boundary conditions of electromagnetic fields and likewise,

$$\hat{x}\left(\frac{4}{L^3}\right) \sin k_y y \sin k_z z, \qquad \hat{y}\left(\frac{4}{L^3}\right) \sin k_x x \sin k_z z \qquad (2.109)$$

are also allowed.

With the modes given above and $\omega = ck$, we find the vacuum energy of the system without the plate as

$$E_{vac}^{NP} = \sum_\lambda \hbar\omega_\lambda/2 = \sum_\lambda c\hbar k_\lambda/2$$

$$= \frac{c\hbar}{2}\left\{2 \sum_{n_x,n_y,n_z=1}^{\infty} \sqrt{\left(\frac{n_x\pi}{L}\right)^2 + \left(\frac{n_y\pi}{L}\right)^2 + \left(\frac{n_y\pi}{L}\right)^2} \right.$$

$$\left. + 3 \sum_{n_y,n_z=1}^{\infty} \sqrt{\left(\frac{n_y\pi}{L}\right)^2 + \left(\frac{n_z\pi}{L}\right)^2}\right\}. \qquad (2.110)$$

Here NP denotes the case without plate. The factors of 2 and 3 in front of the summations are for the two independent polarizations and 3 special solutions, respectively. The three special modes have the same contribution. With the transition from discrete to continuous k-space discussed in Section 2.3.3, we have for $L \to \infty$

$$E_{vac}^{NP} = \frac{c\hbar}{2} \left\{ 2 \left(\frac{L}{\pi} \right)^3 \int_0^\infty dk_x dk_y dk_z \sqrt{k_x^2 + k_y^2 + k_z^2} \right.$$
$$\left. +3 \left(\frac{L}{\pi} \right)^2 \int_0^\infty dk_y dk_z \sqrt{k_y^2 + k_z^2} \right\}. \quad (2.111)$$

Note that $\Delta k = \pi/L$ here.

The expression in Eq. (2.111) diverges so we need to compare it with the case with plate, where there are two regions with different modes: the left side has $k_x = n_x \pi/a, k_y = n_y \pi/L, k_z = n_z \pi/L$ while the right side has $k_x = n_x \pi/(L-a), k_y = n_y \pi/L, k_z = n_z \pi/L$, except the mode $\hat{x} \left(\frac{4}{L^3} \right) \sin k_y y \sin k_z z$, which is the same for both regions. So, we have for the case with plate

$$E_{vac}^{P} = \frac{c\hbar}{2} \left\{ 2 \sum_{n_x,n_y,n_z=1}^{\infty} \sqrt{\left(\frac{n_x \pi}{a} \right)^2 + \left(\frac{n_y \pi}{L} \right)^2 + \left(\frac{n_z \pi}{L} \right)^2} \right.$$
$$+2 \sum_{n_x,n_y,n_z=1}^{\infty} \sqrt{\left(\frac{n_x \pi}{L-a} \right)^2 + \left(\frac{n_y \pi}{L} \right)^2 + \left(\frac{n_z \pi}{L} \right)^2}$$
$$+2 \sum_{n_x,n_y=1}^{\infty} \sqrt{\left(\frac{n_x \pi}{a} \right)^2 + \left(\frac{n_y \pi}{L} \right)^2} + 2 \sum_{n_x,n_y=1}^{\infty} \sqrt{\left(\frac{n_x \pi}{L-a} \right)^2 + \left(\frac{n_y \pi}{L} \right)^2}$$
$$\left. +2 \sum_{n_y,n_z=1}^{\infty} \sqrt{\left(\frac{n_y \pi}{L} \right)^2 + \left(\frac{n_z \pi}{L} \right)^2} \right\}$$
$$= \frac{c\hbar}{2} \left\{ 2 \sum_{n_x,n_y=1,n_z=0}^{\infty} \sqrt{\left(\frac{n_x \pi}{a} \right)^2 + \left(\frac{n_y \pi}{L} \right)^2 + \left(\frac{n_y \pi}{L} \right)^2} \right.$$
$$+2 \sum_{n_x,n_y=1,n_z=1}^{\infty} \sqrt{\left(\frac{n_x \pi}{L-a} \right)^2 + \left(\frac{n_y \pi}{L} \right)^2 + \left(\frac{n_z \pi}{L} \right)^2}$$
$$+2 \sum_{n_x,n_y=1}^{\infty} \sqrt{\left(\frac{n_x \pi}{L-a} \right)^2 + \left(\frac{n_y \pi}{L} \right)^2}$$
$$\left. +2 \sum_{n_y,n_z=1}^{\infty} \sqrt{\left(\frac{n_y \pi}{L} \right)^2 + \left(\frac{n_z \pi}{L} \right)^2} \right\}. \quad (2.112)$$

Here in the second equation, we absorb the second sum from the first equation into the first sum as the term with $n_z = 0$. With a finite and $L \to \infty$, the remaining two terms due to the special solutions are the same and we make a transition to the continuous k-space:

$$E_{vac}^P = \frac{c\hbar}{2}\left\{2\left(\frac{L}{\pi}\right)^2 \sum_{n_x=1}^{\infty} \int_0^{\infty} dk_y dk_z \sqrt{\left(\frac{n_x\pi}{a}\right)^2 + k_y^2 + k_z^2}\right.$$

$$+2\left(\frac{L}{\pi}\right)^2\left(\frac{L-a}{\pi}\right) \int_0^{\infty} dk_x dk_y dk_z \sqrt{k_x^2 + k_y^2 + k_z^2}$$

$$\left.+4\left(\frac{L}{\pi}\right)^2 \int_0^{\infty} dk_y dk_z \sqrt{k_y^2 + k_z^2}\right\}. \quad (2.113)$$

Note that the integration differential is $dk_x = \Delta k_x \equiv \pi/(L-a)$ for the right side of the plate and we only keep the terms up to L^2, which will be the largest non-zero contributing terms to the energy difference (see below).

The energy difference between the cases with and without the plate is then

$$\Delta E = c\hbar\left(\frac{L}{\pi}\right)^2\left\{\int_0^{\infty} dk_y dk_z \left[\frac{1}{2}\sqrt{k_y^2 + k_z^2} + \sum_{n_x=1}^{\infty} \sqrt{\left(\frac{n_x\pi}{a}\right)^2 + k_y^2 + k_z^2}\right]\right.$$

$$\left.-\frac{a}{\pi} \int_0^{\infty} dk_x dk_y dk_z \sqrt{k_x^2 + k_y^2 + k_z^2}\right\}. \quad (2.114)$$

Now making a change of variables: $k_x = n\pi/a, k_y = u\pi/a, k_z = v\pi/a$, we arrive at

$$\Delta E = c\hbar\left(\frac{L}{\pi}\right)^2\left(\frac{\pi}{a}\right)^3 \mathcal{A} \quad (2.115)$$

with

$$\mathcal{A} \equiv \int_0^{\infty} dudv\left(\sum_{n=(0),1}^{\infty} \sqrt{n^2 + u^2 + v^2} - \int_0^{\infty} dn\sqrt{n^2 + u^2 + v^2}\right). \quad (2.116)$$

Here (0) denotes the extra factor of $1/2$ for the term of $n = 0$ in the sum. In practice, there is a high frequency cut-off because for high frequency, say, γ-ray, the conducting metal plate becomes transparent and will not support those modes. With this, the integral in Eq. (2.116) has an upper cut-off limit and it can be shown that $\mathcal{A} = -\pi/720$ [Casimir (1948)]. So, the energy difference per unit area by introducing the plate is

$$\Delta E/L^2 = -\frac{c\hbar\pi^2}{720a^3}, \quad (2.117)$$

which increases with the separation a between the plate and the wall. For this increase, an external force has to do the work. This leads to an attractive force per unit area between the plate and the wall:

$$F = \frac{\partial}{\partial a}(\Delta E/L^2) = \frac{c\hbar\pi^2}{240a^4}. \quad (2.118)$$

As has been demonstrated, this attractive force stems from the energy of vacuum due to the change of the mode structure so it is independent of the material of the plates.

The effect of mode structure change of vacuum can also modify the rate of spontaneous emission of atoms. This is because atoms radiate light into the vacuum modes of the surrounding optical fields and when the available modes change due to the modification of the surrounding geometry, the rate of radiation will change. The mode structure can be easily modified in a cavity environment and this gives rise to the cavity QED effect for atoms [Haroche and Kleppner (1989)].

2.5   Some Remarks about the Unification of Particles and Waves in Quantum Theory of Light

In the quantum theory of light, the particle and wave pictures are unified in the expressions of the field operators in Eqs. (2.81) and (2.82), which have a general form of

$$\hat{V}(\mathbf{r},t) = \sum_\lambda \mathbf{u}_\lambda(\mathbf{r},t)\hat{a}_\lambda + \mathbf{u}_\lambda^*(\mathbf{r},t)\hat{a}_\lambda^\dagger. \tag{2.119}$$

Here, $\mathbf{u}_\lambda(\mathbf{r},t)$ is the mode function satisfying Maxwell's wave equations and thus has all the properties of waves. On the other hand, the creation and annihilation operators $\hat{a}_\lambda^\dagger, \hat{a}_\lambda$ concern the particles of photons as we will see in the next chapter. The classical wave phenomena such as interference can be explained through the mode functions which are related to spatial and temporal behaviors of the optical field, whereas the quantum behaviors associated with particles is attributed to the quantum mechanical operators of $\hat{a}_\lambda^\dagger, \hat{a}_\lambda$ via the expectation (average) values of the operators. Notice that during the second quantization process, the classical quantities q_λ, q_λ^* in Eq. (2.22) or (2.66), which determine the mode excitation strength, are replaced by operators $\hat{a}_\lambda^\dagger, \hat{a}_\lambda$. So, operators $\hat{a}_\lambda^\dagger, \hat{a}_\lambda$ will likewise determine in quantum theory the excitation and thus the strength of the optical field via their expectation (average) values but in a way that the energy is quantized as particles of photons.

Although it seems that the quantum behaviors of the optical field are all borne in operators $\hat{a}_\lambda^\dagger, \hat{a}_\lambda$, the mode function $\mathbf{u}_\lambda(\mathbf{r},t)$ nevertheless plays an important role in quantum interference via the concept of photon indistinguishability thanks to the complementarity principle of quantum mechanics (see Sections 8.2.3 and 8.4). When photons are created in a common mode described by $\mathbf{u}_\lambda(\mathbf{r}, t)$, they are completely indistinguishable from each other and will give rise to the maximum effect in quantum interference. On the other hand, if two photons are respectively in two orthogonal modes of $\mathbf{u}_\lambda(\mathbf{r}, t), \mathbf{u}_{\lambda'}(\mathbf{r}, t)$ with $\int dt d^3\mathbf{r} \, \mathbf{u}_\lambda(\mathbf{r}, t) \cdot \mathbf{u}_{\lambda'}^*(\mathbf{r}, t) = 0$, they become completely distinguishable and produce no quantum interference effect, as we will demonstrate in Section 8.4. They will have partial indistinguishability and some quantum interference effect if $\int dt d^3\mathbf{r} \, \mathbf{u}_\lambda(\mathbf{r}, t) \cdot \mathbf{u}_{\lambda'}^*(\mathbf{r}, t) \neq 0$ but $\mathbf{u}_\lambda(\mathbf{r}, t) \neq \mathbf{u}_{\lambda'}(\mathbf{r}, t)$, i.e., their modes are partially overlapped. Since interference effect is characterized by coherence function, the mode functions will also be important in determining the coherence function (see Section 8.4.4). All these discussions apply equally well to the generalized mode where the generalized mode function is a superposition of a set of orthogonal modes: $\mathbf{u}(\mathbf{r}, t) = \sum_\lambda c_\lambda \mathbf{u}_\lambda(\mathbf{r}, t)$.

From the discussion above, we find that the mode function is basically the identity of the photon and we cannot talk about the photon without its mode function, or the wave aspect of the optical field. This was first pointed out by Lamb in a paper with a provocative title of "Anti-Photon" [Lamb (1995)], where he argued that "Photons cannot be localized in any meaningful manner, and they do not behave at all like particles, whether described by a wave function or not." So in this sense, Lamb considered the word "photon" as a bad description of the quantum radiation field which does not resemble a particle at all but is actually a wave in the classical limit, i.e., the mode function of waves.

To make a more close connection with our daily experience, in some sense, the mode function resembles a house that people live in and photons are the people in it. Houses, whose shapes are highly dependent on the environment, are fixed in locations with physical addresses while people move in and out of the houses. Modes of the optical field exist even in vacuum just like empty houses without people living in. The difference is that there are still activities for modes in vacuum due to (real) quantum fluctuations and the activities in empty houses can only be (unreal) ghostly.

Since we cannot discuss photons without its wave description and waves are intrinsically of nonlocal nature – mode function $\mathbf{u}(\mathbf{r}, t)$ is spread in both space and time, it is not surprising that the particle of photon – if we insist using this term to describe the radiation field as Newtonian particles – will be nonlocal and the violations of the Bell's inequalities [Bell (1964)], which are satisfied by local realistic theories, are inevitable.

2.6 Problems

Problem 2.1 Any arbitrary physical quantity can be expressed in terms of the creation and annihilation operators $\hat{a}_s(\mathbf{k}), \hat{a}_s^\dagger(\mathbf{k})$, for example, the energy expression in Eq. (2.78). Starting from the 3-dimensional free space electromagnetic field operators in Eq. (2.90), derive the total momentum of the optical field:

$$\mathbf{P} \equiv \frac{1}{8\pi} \int d^3\mathbf{r}(\mathbf{E} \times \mathbf{B} - \mathbf{B} \times \mathbf{E}). \tag{2.120}$$

Problem 2.2 Prove that the three-dimensional free space electromagnetic field operators in Eq. (2.90) have an equal-time commutation relation:

$$[\hat{E}_j(\mathbf{r}, t), \hat{B}_k(\mathbf{r}', t)] = -4i\pi\hbar\epsilon_{jkl}\frac{\partial}{\partial r_l}\delta^3(\mathbf{r} - \mathbf{r}'), \tag{2.121}$$

where

$$\epsilon_{ijk} = \begin{cases} 1, & \text{if } i, j, k \text{ are even permutation of } 1, 2, 3; \\ -1, & \text{if } i, j, k \text{ are odd permutation of } 1, 2, 3; \\ 0, & \text{if any two indices are equal.} \end{cases} \tag{2.122}$$

Problem 2.3 Define the spin operator for the optical field as

$$\hat{\Omega}_j^S \equiv -\frac{1}{4\pi c} \int d^3\mathbf{r}\epsilon_{jkl}\frac{\partial \hat{A}_k}{\partial t}\hat{A}_l \tag{2.123}$$

(i) Use the operator form of the vector potential $\hat{\mathbf{A}}$ to prove that the spin of the optical field takes the form of

$$\hat{\Omega}_j^S = i\hbar \sum_{s,s'} \int d^3\mathbf{k} \ \hat{a}_s^\dagger(\mathbf{k})\hat{a}_{s'}(\mathbf{k})(\hat{\epsilon}_{\mathbf{k},s'} \times \hat{\epsilon}_{\mathbf{k},s})_j. \quad (s, s' = 1, 2) \tag{2.124}$$

(ii) Prove that, if we define $\hat{\epsilon}_+\hat{a}_+(\mathbf{k}) + \hat{\epsilon}_-\hat{a}_-(\mathbf{k}) \equiv \hat{\epsilon}_1\hat{a}_1(\mathbf{k}) + \hat{\epsilon}_2\hat{a}_2(\mathbf{k})$ where $\hat{\epsilon}_\pm = (\hat{\epsilon}_1 \pm i\hat{\epsilon}_2)/\sqrt{2}$, we have

$$\hat{a}_\pm(\mathbf{k}) = [\hat{a}_1(\mathbf{k}) \mp i\hat{a}_2(\mathbf{k})]/\sqrt{2},$$
$$[\hat{a}_\pm(\mathbf{k}), \hat{a}_\mp^\dagger(\mathbf{k}')] = 0, \quad [\hat{a}_\pm(\mathbf{k}), \hat{a}_\pm^\dagger(\mathbf{k}')] = \delta(\mathbf{k} - \mathbf{k}').$$

This means that $\hat{a}_\pm(\mathbf{k})$ is the annihilation operator for the circularly polarized light $\hat{\epsilon}_\pm$.

(iii) Prove

$$\hat{\Omega}_j^S = \int d^3\vec{k} \sum_{s=+,-} s\hbar\hat{k}_j\hat{a}_s^\dagger(\mathbf{k})\hat{a}_s(\mathbf{k}), \qquad (2.125)$$

where $\hat{k} \equiv \hat{\epsilon}_1 \times \hat{\epsilon}_2$ is the direction of the propagation of the optical wave. So, the spin of photon only takes two values of $\pm\hbar$ with the sign determined by the state of the circularly polarized photon (left or right). The photon spin states have $S = 1, m = \pm1$. But the state with $m = 0$ does not exist because of the transverse property of light waves.

Chapter 3

Quantum States of Single-Mode Fields

We start with the simplest case, i.e., a single-mode field to discuss about the description of the quantum states of an optical field. In this case, only one mode of the optical field is excited while all other modes are in vacuum. This is of course an ideal case. It is hard to produce a purely single-mode field in experiments. In general, we mostly use a multi-mode description for the optical field in experiments. But under some special circumstances, we can treat the optical field in the experiment approximately as a single-mode field, when, for example, the spectral width of the field is much narrower than that of the detectors. In this case, we will find that the single-mode description gives rise to the same result as the multi-mode description. Here, we may regard the field approximately as a single-mode field with only one frequency component. A transform-limited pulse, even though its bandwidth is much wider than that of the detectors, can be regarded as a single temporal mode field. The single-mode approach is of course much simpler than the multi-mode treatment.

3.1 Energy Eigen-States and Number States

3.1.1 *Energy Eigen-States of a Simple Harmonic Oscillator and the Concept of Photon*

In previous chapter, we learned that a single-mode field can be described as a simple harmonic oscillator. Quantization of the field is basically the quantization of the harmonic oscillator. This subject can be found in any standard textbook on quantum mechanics. Here, we will only present the results. The Hamiltonian of a simple harmonic oscillator is given in Eq. (2.78) and together with the commutation relation in Eq. (2.77) for one λ, we can derive the energy eigen-states $|E_n\rangle$ of the Hamiltonian as:

$$\hat{H}|E_n\rangle = \hbar\omega(\hat{a}^\dagger\hat{a} + 1/2)|E_n\rangle = \hbar\omega(n + 1/2)|E_n\rangle, \qquad (3.1)$$

where we drop the mode-labeling subscript "λ" for the single-mode field and $n = 0, 1, 2, \dots$. The energy of the harmonic oscillator takes discrete values with equal spacing between adjacent energy levels. When the oscillator absorbs energy of $\hbar\omega$, it will jump to the next higher energy level becoming more excited whereas when it lowers its energy by jumping from a higher energy level to the next lower level, it will release energy of $\hbar\omega$. So, the minimum amount of energy that can be absorbed or released by the oscillator is $\hbar\omega$. "Energy quanta" is the early name that Einstein gave to the minimum amount of energy $\hbar\omega$ [Einstein (1905)]. It was later commonly known as the energy of a "photon". Whenever the oscillator absorbs or releases $\hbar\omega$, we will say in terms of the photon language that the field acquires or loses a photon.

From Eq. (3.1), we find that $|E_0\rangle$ or the state with $n = 0$ is the state of the harmonic oscillator with the least energy, or the ground state. It corresponds to the case with no photon, or the vacuum state of the field. But from Eq. (3.1), we see that the vacuum state of the field has a non-zero energy of $\hbar\omega/2$. This is one of the major differences between the quantum and classical theory of light. The energy of the vacuum state of the field will show up in the form of Casimir effect [Casimir (1948)] (see Section 2.4).

3.1.2 *Photon Creation and Annihilation Operators and Photon Number States*

When the harmonic oscillator is in the first excited state $|E_1\rangle$, the energy of the field is one photon's energy more than the ground state or the vacuum state. Then, the field has an occupation of one photon. The first excited state $|E_1\rangle$ is the single-photon state of the field. When the field is in the n-th excited state $|E_n\rangle$, the field has the energy of n photons more than the vacuum state. This is the n-photon state of the field. Since number n uniquely determines the state, we use $|n\rangle$ instead of $|E_n\rangle$ to represent the n-photon state. This is also known as the number state of the field. So from the description above, we see that a photon is simply an energy excitation of $\hbar\omega$. In terms of the new labels, Eq. (3.1) can be rewritten as

$$\hat{H}|n\rangle = \hbar\omega(\hat{a}^\dagger\hat{a} + 1/2)|n\rangle = \hbar\omega(n + 1/2)|n\rangle, \qquad (3.2)$$

where $|0\rangle$ is the ground state of the field and satisfies $\hat{a}|0\rangle = 0$. It corresponds to the vacuum state of no photon excitation. From Eq. (3.2), we see that $|n\rangle$ is the eigen-state of operator $\hat{a}^\dagger\hat{a}$ with an eigen-value of n:

$$\hat{a}^\dagger\hat{a}|n\rangle = n|n\rangle. \tag{3.3}$$

Therefore, $\hat{n} \equiv \hat{a}^\dagger\hat{a}$ is called the photon number operator.

From the commutation relation (single-mode) in Eq. (2.77) of the creation and annihilation operators $\hat{a}^\dagger, \hat{a}$, we can deduce (see Problem 3.1)

$$|n\rangle = \frac{\hat{a}^{\dagger n}}{\sqrt{n!}}|0\rangle. \tag{3.4}$$

From the above and the commutation relation of $\hat{a}^\dagger, \hat{a}$, we can easily obtain

$$\hat{a}^\dagger|n\rangle = \sqrt{n+1}\,|n+1\rangle,$$
$$\hat{a}|n\rangle = \sqrt{n}\,|n-1\rangle. \tag{3.5}$$

The expressions above show that the action of operators $\hat{a}^\dagger, \hat{a}$ on the number state will increase or decrease the photon number by one. So, they are also known as the photon number raising or lowering operators.

3.1.3 q-Space Representation of Photon Number States: Wavefunction of a Single-Photon State

We learned from Section 2.3.2 that for the virtual harmonic oscillator in the description of a single-mode field, its generalized coordinate (position) operator is $\hat{Q}$ and the corresponding generalized momentum operator is $\hat{P} = -i\hbar d/dQ$. From quantum mechanics textbooks, we know that the eigen-states $|q\rangle$ of position operator $\hat{Q}$ form a set of complete bases for the state space of the oscillator. The wave function $\psi(q)$ is the projection of an arbitrary state $|\psi\rangle$ on the $|q\rangle$-base: $\psi(q) = \langle q|\psi\rangle$, i.e., the representation of state $|\psi\rangle$ in the q-space. Then, what is the wave function $\psi_n(q)$ for the number state $|n\rangle$? We will derive it in the following.

First of all, for the vacuum state of $n = 0$, we have $\hat{a}|0\rangle = 0$. From Eq. (2.74), we obtain

$$\hat{a} = \sqrt{\frac{\omega}{2\hbar}}\Big(\hat{Q} + i\frac{\hat{P}}{\omega}\Big), \quad \hat{a}^\dagger = \sqrt{\frac{\omega}{2\hbar}}\Big(\hat{Q} - i\frac{\hat{P}}{\omega}\Big). \tag{3.6}$$

Then $\hat{a}|0\rangle = 0$ becomes

$$\sqrt{\frac{\omega}{2\hbar}}\langle q|\hat{Q} + i\frac{\hat{P}}{\omega}|0\rangle = 0 \tag{3.7}$$

in the Q-space. Acting operators $\hat{Q}, \hat{P}$ on $\langle q|$, we have

$$\Big(q + \frac{\hbar}{\omega}\frac{d}{dq}\Big)\langle q|0\rangle = 0. \tag{3.8}$$

Introducing the dimensionless quantity $x = q/q_0$ with $q_0 \equiv \sqrt{\hbar/\omega}$, Eq. (3.8) is changed to

$$\left(x + \frac{d}{dx} \right) \psi_0(x) = 0. \tag{3.9}$$

The solution of this differential equation is

$$\psi_0(x) = Ce^{-x^2/2}. \tag{3.10}$$

C is a constant determined by the normalization condition. So, the normalized wave function for the vacuum state is

$$\psi_0(q) = \frac{1}{\sqrt{q_0\sqrt{\pi}}} e^{-q^2/2q_0^2}, \tag{3.11}$$

$q_0 \equiv \sqrt{\hbar/\omega}$ is the amplitude of a classical oscillator whose mass is 1 and total energy is $\hbar\omega/2$ or the energy of the vacuum state.

For the n-photon state $|n\rangle$, using Eqs. (3.4) and (3.6), we have its wave function as

$$\begin{aligned}
\psi_n(q) &= \langle q|n\rangle \\
&= \frac{1}{\sqrt{2^n n! q_0 \sqrt{\pi}}} \left(\frac{q}{q_0} - q_0 \frac{d}{dq} \right)^n e^{-q^2/2q_0^2} \\
&= \frac{1}{\sqrt{2^n n! q_0 \sqrt{\pi}}} H_n(q/q_0) e^{-q^2/2q_0^2},
\end{aligned} \tag{3.12}$$

where $H_n(x) = (-1)^n e^{x^2} \frac{d^n}{dx^n}(e^{-x^2})$ is the nth order Hermite polynomial. Particularly for $n = 1$, we have the single-photon wave function as

$$\psi_1(q) = \frac{\sqrt{2}q}{\sqrt{q_0^3\sqrt{\pi}}} e^{-q^2/2q_0^2}. \tag{3.13}$$

In Section 10.4, we will discuss quantum tomography technique for the measurement of the quantum state of the optical field. When the field is in a single-photon state, Eq. (3.13) is the wave function of the field to be measured.

3.1.4 Photon Number States as the Base States of the State Space

Because the Hamiltonian of the harmonic oscillator is a Hermitian operator, its energy eigen-states $\{|E_n\rangle\}$ or the photon number states $\{|n\rangle\}$ are orthonormal:

$$\langle n|m\rangle = \delta_{nm}. \tag{3.14}$$

Furthermore, since the energy eigen-states $\{|E_n\rangle\}$ are non-degenerate, that is, its eigen-value uniquely determines the eigen-state, this set of eigen-states forms a complete base set for the quantum state space of the harmonic oscillator. Hence, the photon number states $\{|n\rangle\}$ satisfy the completeness relation:

$$\sum_n |n\rangle\langle n| = \hat{I}, \tag{3.15}$$

where $\hat{I}$ is the unit operator.

For any operator $\hat{\rho}$, we have after using the completeness relation in Eq. (3.15)

$$\begin{aligned} \hat{\rho} &= \hat{I}\hat{\rho}\hat{I} \\ &= \sum_n |n\rangle\langle n|\hat{\rho}\sum_m |m\rangle\langle m| \\ &= \sum_{m,n} \rho_{nm}|n\rangle\langle m|, \end{aligned} \tag{3.16}$$

where $\rho_{nm} = \langle n|\hat{\rho}|m\rangle$. So, in the number state representation $\{|n\rangle\}$, operator $\hat{\rho}$ can be uniquely determined by matrix $\{\rho_{nm}\}$. For example, the position and momentum operators have their matrices as

$$\{Q_{nm}\} = \sqrt{\frac{\hbar}{2\omega}}\begin{pmatrix} 0 & 1 & 0 & \cdots \\ 1 & 0 & \sqrt{2} & \cdots \\ 0 & \sqrt{2} & 0 & \cdots \\ \vdots & \vdots & \vdots & \ddots \end{pmatrix}, \tag{3.17}$$

$$\{P_{nm}\} = \sqrt{\frac{\hbar\omega}{2}}\begin{pmatrix} 0 & -i & 0 & \cdots \\ i & 0 & -i\sqrt{2} & \cdots \\ 0 & i\sqrt{2} & 0 & \cdots \\ \vdots & \vdots & \vdots & \ddots \end{pmatrix}. \tag{3.18}$$

For the creation and annihilation operators, we have

$$\{a_{nm}\} = \begin{pmatrix} 0 & 1 & 0 & \cdots \\ 0 & 0 & \sqrt{2} & \cdots \\ 0 & 0 & 0 & \cdots \\ \vdots & \vdots & \vdots & \ddots \end{pmatrix}, \quad \{a^\dagger_{nm}\} = \begin{pmatrix} 0 & 0 & 0 & \cdots \\ 1 & 0 & 0 & \cdots \\ 0 & \sqrt{2} & 0 & \cdots \\ \vdots & \vdots & \vdots & \ddots \end{pmatrix}. \tag{3.19}$$

The relation in Eq. (3.15) allows an arbitrary quantum state of the harmonic oscillator or a single-mode field to be expressed in terms of the superposition of the photon number states $\{|n\rangle\}$:

$$|\psi\rangle = \hat{I}|\psi\rangle = \sum_n |n\rangle\langle n|\psi\rangle = \sum_n c_n|n\rangle, \tag{3.20}$$

where

$$c_n = \langle n|\psi\rangle. \tag{3.21}$$

Next, we will discuss a special superposition state of number states – the coherent state.

3.2 Coherent States $|\alpha\rangle$

The expression for the coherent state was first given by Schrödinger in 1926 [Schrodinger (1926)]. He proved that the coherent state gives the minimum value in the Heisenberg uncertainty inequality and used it to describe the classical trajectory of a quantum mechanical harmonic oscillator. But later, it was Glauber who discovered that the coherent state is the best quantum state to describe the coherence properties of an optical field and developed therefrom the quantum theory of optical coherence, which was the foundation of quantum optics [Glauber (1963a,b, 1964)].

3.2.1 *Definition of the Coherent State and Its Number State Representation*

The definition of the coherent state given by Glauber is

$$\hat{a}|\alpha\rangle = \alpha|\alpha\rangle, \tag{3.22}$$

That is, the coherent state is the eigen-state of the annihilation operator $\hat{a}$. But since the annihilation operator $\hat{a}$ is not a Hermitian operator, its eigenvalue α may not be a real number. The coherent state is thus not a projection state of any physical measurement. But this is not what this state is meant for.

From the definition of the coherent state in Eq. (3.22), we can deduce its specific expression in terms of the superposition of the photon number states. There are many ways for this. Among them, the most direct way is to use the expression of photon number states in Eq. (3.4), the definition of the coefficients of a superposition state in Eq. (3.20), and the definition of the coherent state in Eq. (3.22). The derivation is as follows:

$$|\alpha\rangle = \sum_n c_n|n\rangle, \tag{3.23}$$

with

$$\begin{aligned}
c_n &= \langle n|\alpha\rangle \\
&= \langle 0|\frac{\hat{a}^n}{\sqrt{n!}}|\alpha\rangle = \frac{1}{\sqrt{n!}}\langle 0|\hat{a}^n|\alpha\rangle = \frac{1}{\sqrt{n!}}\langle 0|\alpha^n|\alpha\rangle \\
&= \frac{\alpha^n}{\sqrt{n!}}\langle 0|\alpha\rangle \equiv c_0\frac{\alpha^n}{\sqrt{n!}}.
\end{aligned} \tag{3.24}$$

Except a constant phase term, $c_0 \equiv \langle 0|\alpha\rangle$ can be determined by the normalization condition as $c_0 = e^{-|\alpha|^2/2}$. Then we obtain the coherent state in the number state representation:

$$|\alpha\rangle = e^{-|\alpha|^2/2} \sum_n \frac{\alpha^n}{\sqrt{n!}}|n\rangle. \tag{3.25}$$

With Eq. (3.4) for the number states, the expression above is changed to

$$|\alpha\rangle = e^{-|\alpha|^2/2} \sum_n \frac{\alpha^n}{n!}\hat{a}^{\dagger n}|0\rangle = e^{-|\alpha|^2/2+\alpha\hat{a}^\dagger}|0\rangle$$

$$\equiv \hat{D}(\alpha)|0\rangle, \tag{3.26}$$

where $\hat{D}(\alpha)$ is the displacement operator. Its final form $\hat{D}(\alpha) = e^{\alpha\hat{a}^\dagger - \alpha^*\hat{a}}$ can be obtained from Eq. (3.120) by $\hat{a}$ and $\hat{a}^\dagger$ algebra in Section 3.6.

3.2.2 Photon Statistics Distribution and Photon Number Fluctuations of a Coherent State

The coherent state is a superposition state of many number states. The probability for each photon number state, that is, the photon number statistical distribution is

$$P_n = |c_n|^2 = \frac{(|\alpha|^2)^n}{n!}e^{-|\alpha|^2}. \tag{3.27}$$

This is the Poisson distribution with an average of $\bar{n} = |\alpha|^2$. From the above, we can find the variance of the photon number distribution as

$$\overline{(\Delta n)^2} = \sum_n (n - \bar{n})^2 P_n = \bar{n}. \tag{3.28}$$

On the other hand, we can also use the method in quantum mechanics to calculate the expectation value of an operator. For photon number operator $\hat{n} \equiv \hat{a}^\dagger\hat{a}$, we obtain

$$\langle\hat{n}\rangle = \langle\alpha|\hat{a}^\dagger\hat{a}|\alpha\rangle = |\alpha|^2 = \bar{n} \tag{3.29}$$

and

$$\langle\hat{n}^2\rangle = \langle\alpha|(\hat{a}^\dagger\hat{a})^2|\alpha\rangle$$
$$= \langle\alpha|\hat{a}^\dagger\hat{a}\hat{a}^\dagger\hat{a}|\alpha\rangle = \langle\alpha|\hat{a}^\dagger(\hat{a}^\dagger\hat{a}+1)\hat{a}|\alpha\rangle$$
$$= |\alpha|^4 + |\alpha|^2 = \bar{n}^2 + \bar{n}$$
$$= \langle\hat{n}\rangle^2 + \langle\hat{n}\rangle. \tag{3.30}$$

The expression above is exactly Eq. (3.28).

3.2.3   Classical Trajectory and Quantum Uncertainty of Simple Harmonic Oscillator

We mentioned in the introduction of the coherent state that Schrödinger derived the expression of the coherent state in order to find a quantum state that gives a description that is closest to the classical trajectory of a harmonic oscillator. That is, the position average $\langle \hat{x}(t) \rangle$ as a function of time is consistent with the classical trajectory, and its quantum uncertainty is the minimum. In Problem 3.2, we will prove that the number state is not such a state. But the coherent state is. To prove it, we start with the Hamiltonian in Eq. (2.78) and obtain the time evolution of the coherent state:

$$
\begin{aligned}
|\alpha(t)\rangle &= e^{-i\hat{H}t/\hbar}|\alpha\rangle \\
&= e^{-|\alpha|^2/2} \sum_n \frac{\alpha^n}{\sqrt{n!}} e^{-i\hat{H}t/\hbar}|n\rangle \\
&= e^{-|\alpha|^2/2} \sum_n \frac{\alpha^n}{\sqrt{n!}} e^{-i(n+1/2)\omega t}|n\rangle \\
&= e^{-i\omega t/2} e^{-|\alpha|^2/2} \sum_n \frac{(\alpha e^{-i\omega t})^n}{\sqrt{n!}} |n\rangle \\
&= e^{-i\omega t/2} |\alpha e^{-i\omega t}\rangle.
\end{aligned}
\tag{3.31}
$$

So, apart from a general phase, the harmonic oscillator initially in a coherent state will still be in a coherent state later, but the value α is multiplied by a phase factor $e^{-i\omega t}$: $\alpha(t) = \alpha e^{-i\omega t}$. Now we find the average value for the position $\langle \hat{x}(t) \rangle$. For the virtual harmonic oscillator corresponding to a single-mode field, $\hat{x} = \hat{Q} = \sqrt{\hbar/2\omega}(\hat{a} + \hat{a}^\dagger)$. We then have

$$
\begin{aligned}
\langle \hat{Q} \rangle(t) &= \sqrt{\hbar/2\omega}\langle \alpha(t)|(\hat{a} + \hat{a}^\dagger)|\alpha(t)\rangle \\
&= \sqrt{\hbar/2\omega}\left[\langle \alpha(t)|\hat{a}|\alpha(t)\rangle + \langle \alpha(t)|\hat{a}^\dagger|\alpha(t)\rangle\right] \\
&= \sqrt{\hbar/2\omega}[\alpha(t) + \alpha^*(t)] = \sqrt{2\hbar/\omega}\,|\alpha|\cos(\varphi_\alpha - \omega t).
\end{aligned}
\tag{3.32}
$$

We find from the above that the average is similar to the expected classical oscillatory trajectory of the position of a harmonic oscillator. Similarly, we obtain the oscillation of the velocity of the harmonic oscillator:

$$
\begin{aligned}
\langle \dot{\hat{Q}} \rangle(t) = \langle \hat{P} \rangle(t) &= \sqrt{\hbar\omega/2}\langle \alpha(t)|(\hat{a} - \hat{a}^\dagger)/i|\alpha(t)\rangle \\
&= \sqrt{\hbar\omega/2}[\alpha(t) - \alpha^*(t)]/i \\
&= \sqrt{2\hbar\omega}\,|\alpha|\sin(\varphi_\alpha - \omega t).
\end{aligned}
\tag{3.33}
$$

Therefore, the quantum harmonic oscillator in a coherent state has its average trajectory same as the classical trajectory: in the $x - p$ phase space, the trajectory of the virtual harmonic oscillator, after some proper normalization to each quantity in Eqs. (3.32) and (3.33), is a circle. From the result above, we obtain the equivalent energy of the classical oscillator as

$$U = \frac{1}{2}\left[\langle\hat{P}\rangle^2(t) + \omega^2\langle\hat{Q}\rangle^2(t)\right]$$
$$= \hbar\omega|\alpha|^2 = \bar{n}\hbar\omega. \tag{3.34}$$

which, apart from a constant of vacuum energy, is the same as the quantum counterpart.

As for the proof of minimum uncertainty for the coherent state, let us first consider the dimensionless quantities of quadrature-phase amplitudes.

3.2.4 *Quadrature-Phase Amplitudes and Quantum Noise*

For the virtual harmonic oscillator in the description of a single-mode field, the position $\hat{Q}$ and momentum $\hat{P}$ do not have a straightforward physical meaning. But let us check out the following dimensionless quantities:

$$\hat{X} = \hat{a} + \hat{a}^\dagger, \quad \hat{Y} = (\hat{a} - \hat{a}^\dagger)/i. \tag{3.35}$$

These are the quadrature-phase amplitudes of the optical field. The phrase "quadrature-phase" comes from the decomposition of the Hermitian field operator $\hat{E}(t) \equiv \hat{a}e^{-i\omega t} + \hat{a}^\dagger e^{i\omega t}$ in Heisenberg picture in orthogonal phases:

$$\hat{E}(t) \equiv \hat{a}e^{-i\omega t} + \hat{a}^\dagger e^{i\omega t} = \hat{X}\cos\omega t + \hat{Y}\sin\omega t. \tag{3.36}$$

$\hat{X}, \hat{Y}$ represent the amplitudes of the phase-orthogonal terms of $\cos\omega t, \sin\omega t$, respectively. For a quantized harmonic oscillator, they correspond to the position and momentum operators $\hat{Q}, \hat{P}$ after some normalization: $\hat{X} = \hat{Q}\sqrt{2\omega/\hbar}, \hat{Y} = \hat{P}\sqrt{2/\omega\hbar}$.

For a coherent state, the average of these two quantities are

$$\langle\hat{X}\rangle_\alpha = \langle\alpha|\hat{a} + \hat{a}^\dagger|\alpha\rangle = \alpha + \alpha^*, \quad \langle\hat{Y}\rangle_\alpha = (\alpha - \alpha^*)/i. \tag{3.37}$$

In order to find the fluctuations $\langle\Delta^2\hat{X}\rangle, \langle\Delta^2\hat{Y}\rangle$ of $\hat{X}, \hat{Y}$, we may first calculate $\langle\hat{X}^2\rangle, \langle\hat{Y}^2\rangle$ as follows:

$$\langle\hat{X}^2\rangle_\alpha = \langle\alpha|(\hat{a} + \hat{a}^\dagger)^2|\alpha\rangle$$
$$= \langle\alpha|\hat{a}^2 + \hat{a}^{\dagger 2} + \hat{a}\hat{a}^\dagger + \hat{a}^\dagger\hat{a}|\alpha\rangle$$
$$= \langle\alpha|\hat{a}^2 + \hat{a}^{\dagger 2} + 2\hat{a}^\dagger\hat{a} + 1|\alpha\rangle$$
$$= (\alpha + \alpha^*)^2 + 1. \tag{3.38}$$

Similarly, we have

$$\langle \hat{Y}^2 \rangle_\alpha = |\alpha - \alpha^*|^2 + 1. \tag{3.39}$$

Hence,

$$\langle \Delta^2 \hat{X} \rangle_\alpha = 1 = \langle \Delta^2 \hat{Y} \rangle_\alpha. \tag{3.40}$$

The quadrature-phase amplitude for arbitrary phase angle is defined as

$$\hat{X}(\varphi) \equiv \hat{a} e^{-i\varphi} + \hat{a}^\dagger e^{i\varphi}. \tag{3.41}$$

It is easy to see that $\hat{X} = \hat{X}(0)$, $\hat{Y} = \hat{X}(\pi/2)$. Actually, $\hat{X}(\varphi)$ is related to $\hat{X}, \hat{Y}$ by a rotation of angle φ: $\hat{X}' = \hat{X}\cos\varphi + \hat{Y}\sin\varphi = \hat{X}(\varphi)$ (see Fig. 3.1). The average of $\hat{X}(\varphi)$ is

$$\langle \hat{X}(\varphi) \rangle_\alpha = \langle \hat{X} \rangle_\alpha \cos\varphi + \langle \hat{Y} \rangle_\alpha \sin\varphi = 2|\alpha| \cos(\varphi - \varphi_\alpha). \tag{3.42}$$

Similar to Eq. (3.40), it is easy to prove

$$\langle \Delta^2 \hat{X}(\varphi) \rangle_\alpha = 1. \tag{3.43}$$

Notice that the result above is independent of the rotation angle φ as well as α. In the phase space of X-Y (similar to the classical x-p phase space, i.e., the Wigner phase space, see Section 3.7.2), the coherent state is described as a vector with a circular pattern of radius of 1 at its tip, as shown in Fig. 3.1, where the unit size of the circle represents that the variance of $\hat{X}(\varphi)$ is 1 as given in Eq. (3.43), and the vector from the origin of the coordinates to the center of the circle represents the complex number $\langle \hat{X} \rangle_\alpha + i\langle \hat{Y} \rangle_\alpha = 2\alpha = 2|\alpha|e^{i\varphi_\alpha}$. Note that this coordinate system rotates clockwise with an angular speed of ω so that the time-varying quantity $2\alpha(t) = 2|\alpha|e^{i(\varphi_\alpha - \omega t)}$ is changed to time-independent quantity $2\alpha = 2|\alpha|e^{i\varphi_\alpha}$. From Eq. (3.35), we have $[\hat{X}, \hat{Y}] = 2i$. This leads to the Heisenberg uncertainty relation:

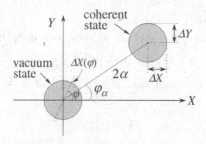

Fig. 3.1   The representation of a coherent state $|\alpha\rangle$ and the vacuum state $|0\rangle$ in X-Y phase space. The circle of radius 1 represents the size of the quantum noise. The coherent state is obtained by displacing the vacuum state with a complex value 2α.

$\Delta X \Delta Y \geq 1$. Therefore, since $\Delta X \Delta Y = 1$ for a coherent state, it is the state that Schrödinger required to give rise to the minimum value in Heisenberg uncertainty inequality.

When $\alpha = 0$, the coherent state becomes the vacuum state, which is represented by a circle with its center at the origin and its radius as 1, as shown in Fig. 3.1. The variance of X, Y in Eq. (3.40) represents the fluctuations of vacuum field, i.e., the vacuum noise. In classical physics, vacuum has nothing in it. So, vacuum noise is caused by the quantization of the electromagnetic fields and is a kind of quantum noise. The squeezed states that we will discuss later in Section 3.4 are a kind of states in which this type of vacuum quantum noise is squeezed. From Eq. (3.26), we find that the coherent state can also be considered as being displaced by 2α from the vacuum state (see Fig. 3.1).

3.2.5 *Non-Orthogonality and Over-Completeness of Coherent States*

For two different coherent states $|\alpha\rangle, |\beta\rangle$, we obtain their inner product from Eq. (3.25) as

$$\langle\alpha|\beta\rangle = e^{\alpha^*\beta-(|\alpha|^2+|\beta|^2)/2}, \tag{3.44}$$

or

$$|\langle\alpha|\beta\rangle|^2 = e^{-|\alpha-\beta|^2} \neq 0. \tag{3.45}$$

So, different coherent states are not orthogonal. This reflects the non-Hermitian property of the annihilation operator $\hat{a}$. But when $|\alpha-\beta|^2 >> 1$, we have $\langle\alpha|\beta\rangle \approx 0$.

On the other hand, using Eq. (3.25) and the completeness relation in Eq. (3.15) for the bases of number states, we can directly prove

$$\int d^2\alpha|\alpha\rangle\langle\alpha| = \pi\hat{I}. \tag{3.46}$$

The expression above indicates that the whole set of the coherent states is overcomplete. This can be viewed as a direct result of the non-orthogonality of the coherent states demonstrated in Eq. (3.44). Nevertheless, we can rewrite Eq. (3.46) as

$$\frac{1}{\pi}\int d^2\alpha|\alpha\rangle\langle\alpha| = \hat{I}, \tag{3.47}$$

and use it for the expansion of an arbitrary quantum state of a single-mode field in terms of the coherent states:

$$|\psi\rangle = \hat{I}|\psi\rangle = \frac{1}{\pi}\int d^2\alpha|\alpha\rangle\langle\alpha|\psi\rangle = \int d^2\alpha\psi(\alpha)|\alpha\rangle, \tag{3.48}$$

where $\psi(\alpha) \equiv \langle\alpha|\psi\rangle/\pi$. So, the whole set of the coherent states forms an overcomplete set of bases with which any quantum state can be expressed. But the non-orthogonality of the coherent states in Eq. (3.44) makes function $\psi(\alpha)$ in the expansion in Eq. (3.48) non-unique. This can also be considered as the result of the overcompleteness of the coherent state bases.

The consequence of non-uniqueness of the coherent state expansion is not all negative. We will show later in Section 3.7.1 that the overcompleteness of the coherent state bases makes it possible to express the density operator of any quantum state in Glauber-Sudarshan P-representation. This is the basis for Glauber's quantum coherence theory as well as the foundation for quantum optics. Later in Chapter 5 when we introduce Glauber's quantum coherence theory, we will discuss more properties of coherent states.

3.3   Further Reading: Schrödinger Cat States

The Schrödinger cat state is a quantum superposition of two completely exclusive classical macroscopic states (such as the dead and alive states of a cat). Even though the state concerns two classical states, its existence is allowed in quantum mechanics. Since it has the superposition property of the microscopic world, it leads to the famous Schrödinger cat paradox in the explanation of quantum mechanics [Schrodinger (1935)]. So, whether or not we can create it in the lab is a test for the suitability of quantum mechanics for the macroscopic world. It is just because we usually do not find Schrödinger cat states in the macroscopic world that our macroscopic world can be safely described by the classical theory of physics most of the time. This also shows on the other hand that it is extremely hard to realize a Schrödinger cat state. Nevertheless, scientists recently produced a Schrödinger cat state in the lab [Brune et al. (1996); Monroe et al. (1996)] and therefore proved the suitability of quantum mechanics in macroscopic world and the feasibility of quantum computers. More importantly in their experiment is that the scientists observed the process of de-coherence of the Schrödinger cat state, that is, the transition process from quantum superposition to classical mixture states [Myatt et al. (2000); Deléglise et al. (2008)]. This explained why there are very few quantum phenomena in our macroscopic world.

We mentioned earlier that a coherent state is a quantum state that is closest to a classical description of the electromagnetic waves. So, two coherent states with large separation and thus very small overlap can be considered as two classical macroscopic states. The superposition state of these two states is the Schrödinger cat state of an optical field:

$$|\psi\rangle_{cat} = \mathcal{N}\big(|\alpha\rangle + e^{i\phi}|\beta\rangle\big). \qquad (3.49)$$

Here, $|\alpha\rangle, |\beta\rangle$ are nearly orthogonal: if $|\alpha - \beta|^2 \gg 1$, then $|\langle\alpha|\beta\rangle|^2 = e^{-|\alpha-\beta|^2} \approx 0$ and the normalization constant $\mathcal{N}^{-2} = 2[1 + \mathrm{Re}(\langle\alpha|\beta\rangle e^{i\varphi})] \approx 2$. The commonly discussed Schrödinger cat state has $\beta = \alpha e^{i\theta}$, i.e., state $|\beta\rangle$ is rotated from $|\alpha\rangle$ by an angle θ. So, $|\alpha - \beta|^2 = 4|\alpha|^2 \sin^2 \theta/2$. When $\theta \gg 1/|\alpha|$, we have $|\alpha - \beta|^2 \gg 1$.

Normally, the consequence of quantum superposition is interference phenomena. But different from the interference phenomena due to the superposition between two optical fields, the quantum superposition state in Eq. (3.49) does not give rise to the traditional interference fringes in which the intensity (photon number) changes with the phase difference. A direct calculation gives the average photon number for the Schrödinger cat state in Eq. (3.49) as

$$\langle\hat{n}\rangle \approx |\alpha|^2 + |\beta|^2. \qquad (3.50)$$

It is independent of the phase difference between the two states. This is because the two states $|\alpha\rangle, |\beta\rangle$ are nearly orthogonal.

The consequence of the quantum superposition in Eq. (3.49) is exhibited in the interference effect in the probability distribution of the quadrature-phase amplitude X. To see this, consider the following Schrödinger cat state:

$$|\psi\rangle_{cat} = N_r(|-ir\rangle + |ir\rangle), \qquad (3.51)$$

where r is real and $N_r^{-2} = 2(1 + e^{-2r^2})$ so that $|\psi\rangle_{cat}$ is normalized. Since $\hat{X} = \hat{a} + \hat{a}^\dagger = \sqrt{2}\hat{Q}/q_0$ with $q_0 \equiv \sqrt{\hbar/\omega}$, the distribution of X is that of Q. For this, we calculate the wave function $\psi_{cat}(q)$ for state $|\psi\rangle_{cat}$:

$$\psi_{cat}(q) = \langle q|\psi\rangle_{cat} = \big(\langle q| - ir\rangle + \langle q|ir\rangle\big)N_r. \qquad (3.52)$$

From Problem 3.3, we find the wave function for the coherent state $|\alpha\rangle$ as

$$\psi_\alpha(x) = \frac{1}{(\pi)^{1/4}} e^{-\mathrm{Im}^2(\alpha) - (x - \alpha\sqrt{2})^2/2} \qquad (3.53)$$

with $x = q/q_0$. Substituting into Eq. (3.52), we then obtain the wave function for the Schrödinger cat state:

$$\psi_{cat}(x) = \frac{2N_r}{(\pi)^{1/4}} e^{-x^2/2} \cos \sqrt{2}rx. \qquad (3.54)$$

So, the probability distribution for the position Q of the virtual harmonic oscillator in the Schrödinger cat state is

$$P_{cat}(x) = \frac{2N_r^2}{\sqrt{\pi}} e^{-x^2} (1 + \cos 2\sqrt{2}rx). \tag{3.55}$$

It can be seen that this probability distribution shows an interference fringe as the position x changes.

The visibility of the interference fringe in Eq. (3.55) is 100%. Classically, the two states can only be in a statistically mixed state which we will discuss later in Section 3.5. It gives no interference, i.e., the visibility is zero (see Section 3.5.1). The quantum superposition in the Schrödinger cat state is extremely sensitive to losses. In Chapter 6, we will treat the effect of losses quantum mechanically and prove that when the loss is γ, the distribution in Eq. (3.55) is changed to [Walls and Milburn (1985)] (see Problem 6.9)

$$P_{cat}(x) \approx \frac{1}{\sqrt{\pi}} e^{-x^2} \left(1 + \mathcal{V} \cos 2\sqrt{2}rx\sqrt{1-\gamma} \right), \tag{3.56}$$

where $\mathcal{V} = e^{-2\gamma r^2}$ is the visibility of the fringe. From the above, we find that for a macroscopic Schrödinger cat state with $r^2 \gg 1$, when $\gamma \gg 1/r^2 \sim 0$, $\mathcal{V} \approx 0$, that is, a very small loss can make the Schrödinger cat state de-cohere to the classical mixture state. The larger the macroscopic quantity is, the smaller the required loss is for de-coherence. This explains why it is hard to find a quantum superposition Schrödinger cat state in a real macroscopic world.

3.4   Squeezed Vacuum States and Squeezed Coherent States

As we discussed in Sections 3.1.1 and 3.2.4, vacuum fluctuations of electromagnetic fields is a kind of quantum noise, i.e., it is the inevitable consequence of quantization of the optical fields. Although it is impossible to avoid the quantum noise, can we rearrange and redistribute it so that some of it will be reduced while other part will increase? The squeezed states that will be discussed in this section are such kind of quantum states, which have some of their noise below vacuum noise level. Let us see how to squeeze the quantum noise in the following.

3.4.1   *Squeezing of Quantum Noise*

From Problem 3.2, we find that the photon number state has the uncertainties $\langle \Delta^2 \hat{X} \rangle_n$, $\langle \Delta^2 \hat{Y} \rangle_n$ larger than the minimum value of the Heisenberg inequality whereas from Section 3.2.4, we proved that the coherent states (including the vacuum state) has the minimum value of the Heisenberg inequality, i.e.,

$$\langle \Delta^2 \hat{X} \rangle_\alpha \langle \Delta^2 \hat{Y} \rangle_\alpha = 1. \tag{3.57}$$

Therefore, it seems we already reached the minimum value of the Heisenberg inequality with the coherent states and cannot go even further. This argument is correct if both $\langle \Delta^2 \hat{X} \rangle$ and $\langle \Delta^2 \hat{Y} \rangle$ are reduced together.

However, the inequality in Eq. (3.57) is about the product of the two quantities, that is, quantum mechanics only sets the limit on the product of the two uncertainties but not on individual quantities. So, we can rearrange the two quantities in such a way that only one of them is reduced at the expense of the increase in the other while keeping the product unchanged, i.e., Eq. (3.57) is unchanged.

Suppose some state $|r\rangle$ satisfies the requirement above, that is,

$$\langle \Delta^2 \hat{X} \rangle_r = \langle \Delta^2 \hat{X} \rangle_\alpha e^{2r} = e^{2r}, \quad \langle \Delta^2 \hat{Y} \rangle_r = \langle \Delta^2 \hat{Y} \rangle_\alpha e^{-2r} = e^{-2r}, \tag{3.58}$$

where r is a positive number to determine the degree of squeezing. In the following, we will derive such a quantum state $|r\rangle$.

3.4.2 Squeezing Operators

Suppose we can obtain $|r\rangle$ by applying a unitary operator $\hat{S}$ on the coherent state:

$$|r\rangle = \hat{S}|\alpha\rangle. \tag{3.59}$$

Substituting into Eq. (3.58), we have

$$\langle \hat{S}^\dagger \Delta^2 \hat{X} \hat{S} \rangle_\alpha = \langle \Delta^2 \hat{X} \rangle_\alpha e^{2r}, \quad \langle \hat{S}^\dagger \Delta^2 \hat{Y} \hat{S} \rangle_\alpha = \langle \Delta^2 \hat{Y} \rangle_\alpha e^{-2r}. \tag{3.60}$$

The sufficient condition for the above is that the unitary operator $\hat{S}$ must satisfy the following operator relations

$$\hat{S}^\dagger \hat{X} \hat{S} = \hat{X} e^r, \quad \hat{S}^\dagger \hat{Y} \hat{S} = \hat{Y} e^{-r}. \tag{3.61}$$

Notice that the operator transformation above keeps commutation relation $[\hat{X}, \hat{Y}] = 2i$ unchanged. From Eq. (3.61), we can find the transformation for $\hat{a}, \hat{a}^\dagger$:

$$\hat{S}^\dagger \hat{a} \hat{S} = \hat{a} \cosh r + \hat{a}^\dagger \sinh r, \quad \hat{S}^\dagger \hat{a}^\dagger \hat{S} = \hat{a}^\dagger \cosh r + \hat{a} \sinh r. \tag{3.62}$$

Similarly, the commutation relation $[\hat{a}, \hat{a}^\dagger] = 1$ is unchanged for this transformation.

From the operator algebra for $\hat{a}$ and $\hat{a}^\dagger$ (see Section 3.6), we can deduce from Eq. (3.62) the unitary operator $\hat{S}$ as [Stoler (1970)]

$$\hat{S}(r) = e^{r(\hat{a}^{\dagger 2} - \hat{a}^2)/2}. \tag{3.63}$$

This is the squeezing operator. Its action on the vacuum state gives the squeezed vacuum state:

$$|r\rangle = \hat{S}(r)|0\rangle, \tag{3.64}$$

and when it acts on the coherent states, we obtain the squeezed coherent states:

$$|r, \alpha\rangle = \hat{S}(r)|\alpha\rangle. \tag{3.65}$$

For the quadrature-phase amplitude $\hat{X}(\varphi) \equiv \hat{a}e^{-i\varphi} + \hat{a}^\dagger e^{i\varphi}$ with an arbitrary angle φ, we can use Eq. (3.62) to prove

$$\langle \Delta^2 \hat{X}(\varphi) \rangle_r = e^{2r} \cos^2 \varphi + e^{-2r} \sin^2 \varphi. \tag{3.66}$$

In the phase space of X-Y as shown in Fig. 3.2, the quantum fluctuation of $\hat{X}(\varphi)$ is represented by an ellipse, which is the area in which the Wigner quasi-probability density is appreciably non-zero (see Section 3.7.2 for more). This is the noise ellipse. For vacuum state and coherent states, the ellipse becomes a circle of unit radius.

Using Eq. (3.62), we have

$$\langle r, \alpha | \hat{X} | r, \alpha \rangle = (\alpha + \alpha^*)e^r, \quad \langle r, \alpha | \hat{Y} | r, \alpha \rangle = (\alpha - \alpha^*)e^{-r}/i. \tag{3.67}$$

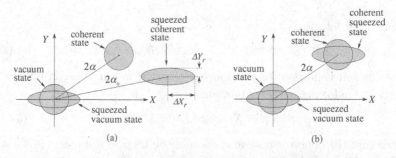

(a)   (b)

Fig. 3.2   (a) The representation of squeezed coherent states and squeezed vacuum state in the X-Y phase space. The squeezed ellipse represents the size of the quantum noise in the polar coordinates. Quantum noise is squeezed in Y-direction and is smaller than the vacuum noise but is increased in X-direction. The coherent components are also squeezed and stretched in Y- and X-directions, respectively. (b) The representation of the coherent squeezed states in X-Y phase space: coherent squeezed states is shifted by 2α from the squeezed vacuum state.

Using the above and the noise ellipse of the squeezed vacuum state, we obtain in Fig. 3.2(a) the representation of the squeezed coherent state in the X-Y phase space. The squeezed vacuum state is represented by the noise ellipse with its center at the origin. It can be seen that the noise along Y-direction (i.e., $\langle \Delta^2 \hat{Y} \rangle_r$) is smaller than that of the vacuum state (the circle with its center at origin and radius of 1). The squeezed coherent state is shifted away from center by the amount given in Eq. (3.67). Note that the original coherent components are squeezed in Y-direction and stretched in X-direction just like the noise part.

From Eq. (3.63), we obtain $\hat{S}^\dagger(-r) = \hat{S}$. Hence,

$$\hat{S}\hat{a}\hat{S}^\dagger = \hat{a}\cosh r - \hat{a}^\dagger \sinh r. \tag{3.68}$$

If we define operator $\hat{b} \equiv \hat{a}\cosh r - \hat{a}^\dagger \sinh r$, then from Eqs. (3.68) and (3.22), we obtain

$$\hat{b}|r, \alpha\rangle = \hat{S}\hat{a}\hat{S}^\dagger \hat{S}|\alpha\rangle = \hat{S}\hat{a}|\alpha\rangle = \alpha|r, \alpha\rangle, \tag{3.69}$$

that is, the squeezed coherent states are the eigen-states of operator $\hat{b}$ with an eigen-value of α. Similar to Eq. (3.22), Eq. (3.69) can also be considered as the definition of squeezed coherent states [Yuen (1976)].

3.4.3 Coherent Squeezed States and Squeezed Coherent States

From Fig. 3.2(a), we find that when the squeezing operator acts on the coherent state, not only its noise in Y or X is squeezed or amplified but also are its averages. This is because the squeezed coherent state can be written as

$$|r, \alpha\rangle = \hat{S}(r)\hat{D}(\alpha)|0\rangle, \tag{3.70}$$

where $\hat{D}(\alpha)$ is the displacement operator defined in Eq. (3.26). So, the squeezed coherent state is subject to displacement first and then squeezing so that the displacement amount is also squeezed. If we switch the order of the squeezing and displacement operators, we obtain the coherent squeezed state:

$$|\alpha, r\rangle = \hat{D}(\alpha)\hat{S}(r)|0\rangle. \tag{3.71}$$

This is different from the squeezed coherent state $|r, \alpha\rangle$ defined in Eq. (3.70). From Section 3.6 on the algebra of $\hat{a}, \hat{a}^\dagger$, we can prove

$$\hat{D}^\dagger(\alpha)\hat{a}\hat{D}(\alpha) = \hat{a} + \alpha. \tag{3.72}$$

With the above, we have

$$\langle \alpha, r | \hat{a} | \alpha, r \rangle = \langle 0 | \hat{S}^\dagger(r) \hat{D}^\dagger(\alpha) \hat{a} \hat{D}(\alpha) \hat{S}(r) | 0 \rangle$$
$$= \langle 0 | \hat{S}^\dagger(r) \hat{a} \hat{S}(r) | 0 \rangle + \alpha$$
$$= \alpha. \tag{3.73}$$

Similarly,

$$\langle \alpha, r | \hat{a}^\dagger | \alpha, r \rangle = \alpha^*. \tag{3.74}$$

Hence,

$$\langle \alpha, r | \hat{X} | \alpha, r \rangle = \alpha + \alpha^*, \quad \langle \alpha, r | \hat{Y} | \alpha, r \rangle = (\alpha - \alpha^*)/i. \tag{3.75}$$

So, different from the squeezed coherent state in Eq. (3.67), the averages of X, Y for the coherent squeezed state are the same as those of the coherent state. Using Eq. (3.72), we can prove

$$\langle \alpha, r | \Delta^2 \hat{X} | \hat{\alpha}, r \rangle = e^{2r}, \quad \langle \alpha, r | \Delta^2 \hat{Y} | \hat{\alpha}, r \rangle = e^{-2r}, \tag{3.76}$$

that is, the fluctuations of the coherent squeezed state are the same as those of the squeezed coherent state. Therefore, as shown in Fig. 3.2(b), a coherent squeezed state is displaced by 2α from a squeezed vacuum state.

Although the squeezed coherent state is obtained by squeezing the coherent state, we can see from Fig. 3.2(a) that it can also be obtained by displacing the squeezed vacuum state. The displacement amount is not 2α but depends on both α, r. Using Eq. (3.62), we can prove

$$\hat{S}^\dagger(\pm r) \hat{D}(\alpha) \hat{S}(\pm r) = \hat{D}(\alpha_\mp), \tag{3.77}$$

where $\alpha_\pm = \alpha \cosh r \pm \alpha^* \sinh r$. Using $\hat{S}(-r) = \hat{S}^\dagger(r) = \hat{S}^{-1}(r)$, we have

$$\hat{D}(\alpha) \hat{S}(r) = \hat{S}(r) \hat{D}(\alpha_-), \tag{3.78}$$
$$\hat{S}(r) \hat{D}(\alpha) = \hat{D}(\alpha_+) \hat{S}(r). \tag{3.79}$$

So, the relation between a squeezed coherent state and a coherent squeezed state is

$$|\alpha, r\rangle = |r, \alpha_-\rangle, \quad |r, \alpha\rangle = |\alpha_+, r\rangle. \tag{3.80}$$

For an arbitrary squeezing parameter $\xi = r e^{i\theta}$, we can generalize the squeezing operator in Eq. (3.63) as

$$\hat{S}(\xi) = e^{(\xi \hat{a}^{\dagger 2} - \xi^* \hat{a}^2)/2}. \tag{3.81}$$

Making a transformation of $\hat{a}' = \hat{a} e^{-i\theta/2}$, Eq. (3.81) is then changed back to Eq. (3.63). So, we can easily obtain

$$\hat{S}^\dagger(\xi) \hat{a} \hat{S}(\xi) = \hat{a} \cosh r + e^{i\theta} \hat{a}^\dagger \sinh r. \tag{3.82}$$

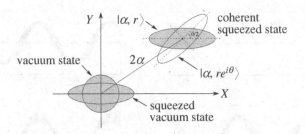

Fig. 3.3   The representation in the X-Y phase space of the coherent squeezed state with an arbitrary squeezing parameter $\xi = re^{i\theta}$: the noise ellipse is rotated counterclockwise by $\theta/2$.

Using this, we can prove that for a coherent squeezed state $|\alpha, \xi\rangle \equiv \hat{D}(\alpha)\hat{S}(\xi)|0\rangle$,

$$\langle \Delta^2 \hat{X}(\varphi) \rangle_\xi = e^{2r} \cos^2(\varphi - \theta/2) + e^{-2r} \sin^2(\varphi - \theta/2). \qquad (3.83)$$

Comparing with $|\alpha, r\rangle$ ($\theta = 0$), $|\alpha, \xi\rangle$ simply rotates counterclockwise the noise ellipse by an angle of $\theta/2$ in the X-Y phase space (see Fig. 3.3).

Next, let us see how the electric field strength changes with time for an electromagnetic field in a coherent state or a coherent squeezed state. This allows us to have a direct picture of the quantum field in a similar way to the picture for a classical electromagnetic wave (see Fig. 1.3 in Section 1.3). From Eq. (2.84), we find the electric field operator as

$$\hat{E}(\mathbf{r}, t) = Ci\hat{a}e^{-i\omega t'} + h.c. = C\hat{X}(\omega t' - \pi/2). \qquad (3.84)$$

where $t' \equiv t - \mathbf{k} \cdot \mathbf{r}/c$ and $C = \sqrt{4\pi\hbar\omega/2L^3}$ is a constant.

For a coherent state $|\alpha\rangle$ ($\alpha = |\alpha|e^{i\varphi_\alpha}$), we have the average value of the electric field as

$$\langle \hat{E}(\mathbf{r}, t) \rangle_\alpha = Ci\alpha e^{-i\omega t'} + c.c. = 2C|\alpha| \sin(\omega t' - \varphi_\alpha). \qquad (3.85)$$

This corresponds to the change in space and time of the electric field strength for a classical field, as shown in Fig. 3.4(a). In the meantime, the fluctuation of the electric field is

$$\sqrt{\langle \Delta^2 \hat{E}(\mathbf{r}, t) \rangle_\alpha} = C\sqrt{\langle \Delta^2 \hat{X}(\omega t' - \pi/2) \rangle_\alpha} = C. \qquad (3.86)$$

So, the uncertainty, i.e., fluctuation, of the electric field at every moment is the same, i.e., C. Figure 3.4(b) shows the change of the electric field with time for an electromagnetic field in the coherent state. The fluctuation of the electric field is represented by a band with a width of C in Fig. 3.4(b). For comparison, Fig. 3.4(a) shows the change of the electric field with time

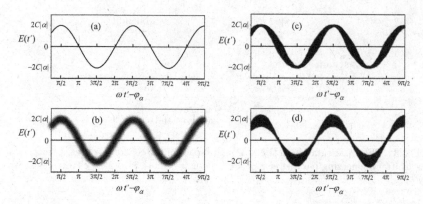

Fig. 3.4   The wave of the electric field as it changes with time: (a) a classical field; (b) a field in a coherent state; (c) a field in an amplitude-squeezed state; (d) a field in a phase-squeezed state.

for a classical wave, i.e., Eq. (3.85), which has a definite value at every moment.

For the coherent squeezed state $|\alpha, r\rangle (\alpha = |\alpha|e^{i\varphi_\alpha})$, the average value of the electric field strength is the same as Eq. (3.85) but the fluctuation of the electric field is

$$\sqrt{\langle \Delta^2 \hat{E}(\mathbf{r}, t)\rangle_r} = C\sqrt{\langle \Delta^2 \hat{X}(\omega t' - \pi/2)\rangle_r}$$

$$= C\sqrt{e^{2r}\cos^2(\omega t' - \pi/2) + e^{-2r}\sin^2(\omega t' - \pi/2)}. \quad (3.87)$$

It is a function of time and changes as the wave propagates. Set $\varphi_\alpha = \pi/2$ or $\alpha = i|\alpha|$. The peak of the electric field occurs at $\omega t' - \varphi_\alpha = \omega t' - \pi/2 = N\pi \pm \pi/2(N = \text{integer})$. At this moment, the fluctuation of the electric field given in Eq. (3.87) is minimum, i.e., Ce^{-r}. Figure 3.4(c) shows the evolution of the average value and the fluctuation of the electric field as time changes. We find that the uncertainty of the electric field at the peak of the electric field is smaller than the corresponding part in Fig. 3.4(b) for the coherent state. So, the amplitude fluctuation of the electric field is squeezed for a field in the coherent squeezed state $|i|\alpha|, r\rangle$. On the other hand, when $\omega t' - \varphi_\alpha = \omega t' - \pi/2 = N\pi$, the electric field is zero, i.e., at the trough of the wave. At this moment, the fluctuation of the electric field is the maximum, i.e., Ce^r by Eq. (3.87). We know that for a sine function, it is most sensitive to phase change at zero crossing so its zero position is used to determine the phase. But the phase uncertainty at zero position in Fig. 3.4(c) is larger than that in Fig. 3.4(b). Therefore, the phase fluctuation of the field in the coherent squeezed state $\big|i|\alpha|,r\big\rangle$ is larger than a coherent state.

We find from Eq. (3.83) that the squeezing position for the fluctuation of the electric field can be adjusted by the phase θ of ξ. So, for the coherent squeezed state $\big|i|\alpha|,-r\big\rangle$ with $\theta=\pi$, we can use a similar analysis as before and find that the phase fluctuation of the electric field is squeezed while the amplitude fluctuation becomes large, as shown in Fig. 3.4(d). That is why $\big|i|\alpha|,-r\big\rangle$ is called the phase-squeezed state whereas $\big|i|\alpha|,r\big\rangle$ is the amplitude-squeezed state. For an arbitrary phase φ_α of the coherent component, we find that the coherent squeezed state $|\alpha,-re^{2i\varphi_\alpha}\rangle$ corresponds to the amplitude-squeezed state and $|\alpha,re^{2i\varphi_\alpha}\rangle$ is for the phase-squeezed state. From the diagram of the amplitude-squeezed state $|\alpha,-re^{2i\varphi_\alpha}\rangle$ in the X-Y phase space [Fig. 3.5(a)], we can see that as the phase φ_α of α increases from zero, the tilt angle of the noise ellipse also changes. The noise squeezing direction always keeps at the amplitude direction (radial direction). Figure 3.5(b) shows the diagram for the phase-squeezed state $|\alpha,re^{2i\varphi_\alpha}\rangle$.

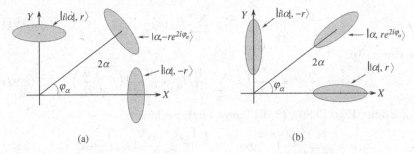

(a)   (b)

Fig. 3.5 The diagrams in the X-Y phase space for (a) the amplitude-squeezed state $|\alpha,-re^{2i\varphi_\alpha}\rangle$ and (b) the phase-squeezed state $|\alpha,re^{2i\varphi_\alpha}\rangle$.

3.4.4 *Further Reading: Photon Statistics of a Squeezed State and the Oscillation Effects*

We have just seen that the squeezed coherent state can squeeze the quantum fluctuation in the quadrature-phase amplitudes. This is what we get if we make a homodyne detection (see Chapter 9 for more). On the other hand, if we measure photon number as in photon counting technique (see Chapter 7), the behavior of the squeezed coherent state is completely different.

In order to find the photon number behavior, we need to write the squeezed coherent state in the number state base:

$$| -r, \alpha \rangle = \hat{S}(-r)\hat{D}(\alpha)|0\rangle = \sum_n s_n |n\rangle, \qquad (3.88)$$

where $s_n = \langle n|\hat{S}(-r)\hat{D}(\alpha)|0\rangle = \langle n|\hat{S}^\dagger(r)|\alpha\rangle$. Here, for the ease of calculation and interpretation, we consider the squeezed state with squeezing in X-direction. This coefficient can be calculated from the following quantity

$$e^{|\beta|^2/2}\langle\alpha|\hat{S}(r)|\beta\rangle = e^{|\beta|^2/2}\langle\alpha|\hat{S}(r)\left(e^{-|\beta|^2/2}\sum_n \frac{\beta^n}{\sqrt{n!}}|n\rangle\right)$$

$$= \sum_n \frac{\beta^n}{\sqrt{n!}}s_n^* \qquad (3.89)$$

by expanding the left-hand side in the series of β. $\langle\alpha|\hat{S}(r)|\beta\rangle$ is obtained from Eq. (3.139) in Section 3.6 with the operator algebra method as

$$\langle\alpha|\hat{S}(r)|\beta\rangle = \frac{1}{\sqrt{\mu}} \exp\left[\frac{\nu}{2\mu}(\alpha^{*2} - \beta^2) + \frac{\alpha^*\beta}{\mu} - \frac{1}{2}(|\alpha|^2 + |\beta|^2)\right]. \quad (3.90)$$

Using the generating function of Hermite function:

$$\exp(2xt - t^2) = \sum_n H_n(x)\frac{t^n}{n!}, \qquad (3.91)$$

we can write

$$\exp\left[-\frac{\nu}{2\mu}\beta^2 + \frac{\alpha^*\beta}{\mu}\right] = \sum_n \frac{1}{n!}H_n\left(\frac{\alpha^*}{\sqrt{2\mu\nu}}\right)\left(\beta\sqrt{\frac{\nu}{2\mu}}\right)^n \qquad (3.92)$$

Combining Eqs. (3.89), (3.90), and (3.92), we have

$$s_n = \frac{1}{\sqrt{\mu n!}}\left(\frac{\nu}{2\mu}\right)^{n/2}\exp\left(-\frac{1}{2}|\alpha|^2 + \frac{\nu\alpha^2}{2\mu}\right)H_n\left(\frac{\alpha}{\sqrt{2\mu\nu}}\right), \qquad (3.93)$$

Stoler was the first to derive this coefficient in unnormalized form [Stoler (1970)]. The absolute value square of this gives the probability of n photons in the squeezed coherent state:

$$P_n = |s_n|^2 = \frac{e^{-|\alpha|^2}}{\mu n!}\left(\frac{\nu}{2\mu}\right)^n \exp\left[\frac{\nu}{2\mu}(\alpha^2 + \alpha^{*2})\right]\left|H_n\left(\frac{\alpha}{\sqrt{2\mu\nu}}\right)\right|^2. \quad (3.94)$$

The photon number distribution of the squeezed coherent state shows some interesting oscillatory behavior as a function of both n and α. To see this and for the ease of numerical calculation, we take α as real and positive. Then Eq. (3.94) becomes

$$P_n = \frac{e^{\beta^2/(\mu+\nu)^2}}{\mu n!}\left(\frac{\nu}{2\mu}\right)^n e^{-\beta^2}\left|H_n(\beta)\right|^2 \qquad (3.95)$$

with $\beta \equiv \alpha/\sqrt{2\mu\nu}$. For moderate $r \gtrsim 1$, the shape of $P_n(\beta)$ depends mostly only on β and n. In Fig. 3.6(a), we plot P_n as a function of β with $r = 2$ for $n = 5$(i), 10(ii), 20(iii). It shows an oscillatory behavior with the number of oscillation given by $[(n+1)/2]$. This behavior is attributed to a multi-photon interference effect that will be the subject of Chapter 8 (See Eq. (8.122) of Problem 8.5).

Further oscillatory behavior occurs in P_n as a function of n, as shown in Fig. 3.6(b), where we plot P_n as a function of n for $r = 5$ and $\beta = 20$. This oscillatory behavior is a result of quantum interference in the X-Y phase space [Schleich and Wheeler (1987)]. Interestingly, the interference effect in Fig. 3.6(a) can also be explained as interference in the X-Y phase space [Schleich (2001)].

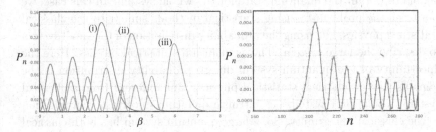

Fig. 3.6 The oscillatory behavior of photon number probability P_n of the squeezed coherent state: (a) $P_n(\beta)$ as a function of $\beta = \alpha/\sqrt{2\mu\nu}$ for $r = 2$ or $\mu = 3.76, \nu = 3.63$ with $n = 5$(i), 10(ii), 20(iii) and (b) $P_n(\beta)$ as a function of n for $\beta = 20$ and $r = 5$.

3.5 Mixed Quantum States

In classical physics, if we know the initial states of a system, we can find the evolution path for the system through its Hamiltonian. Similarly in quantum physics, the initial states of a system uniquely determine the evolution of the system via its Hamiltonian. The same determinism and causality principles apply to both classical and quantum systems. The difference between quantum and classical physics is in the description of the states of the physical system. On the other hand, if the system is very large and contains many degrees of freedom, it is impossible for us to know every detail. Particularly when a large system contains many identical small systems such as an ensemble of gas system consisting of many identical molecules, we are unable to know the exact states of each molecule. But since the number of the small system (the number of molecules for gas system) is usually quite large ($\sim 10^{23}$), we can use statistical methods to describe them. The identical nature of the small systems allows us to describe each of them as a sample in statistics. The whole collection of the samples is known as an ensemble in statistics and it is just the large system (the gas system). Although the states of each sample are undetermined or unknown to us, we can use probability distribution to describe the ensemble. The state of the large system can be described as the statistical average of the ensemble of the small systems. This is classical statistical physics. The statistical fluctuations of the states of the small systems may lead to the uncertainty of the state of the large system. This is the origin of the classical uncertainty. For a large quantum system, we will encounter similar situation, that is, due to the largeness of the system size, we are unable to obtain the quantum state for the whole system. In this case, we can treat the problem by using statistical methods similar to the classical statistical physics, i.e., using the probability distribution of the small system to describe the large system. This is quantum statistical physics. Here the uncertainty of the quantum system due to probability distribution is the same as that in classical statistical physics. But different from a classical system, a quantum system has intrinsic quantum uncertainty in addition to the classical uncertainty. So, a large quantum system has both classical uncertainty due to its many degrees of freedom and the intrinsic quantum uncertainty.

In Chapter 2, we find an optical field has many modes and thus many degrees of freedom. If many modes are excited but we are unable to know exact excitation for each excited mode, this can lead to classical uncertainty, as described above. This is the basis for statistical optics to be discussed in Chapter 4. Even for a single mode field, the uncertainty in the emission process leads to uncertainty in phase and amplitude of an optical field and statistical description is needed, as we found in Section 1.3. When the optical field is treated quantum mechanically, we need to use a quantum statistical method to describe it. This is the mixed state description of optical fields.

3.5.1   *Density Operator for the Description of a Mixed Quantum State*

Before we discuss mixed states in quantum statistical description, let us introduce the density operator representation of a quantum state. When a system is in a definite quantum state $|\psi\rangle$, we say the system is in a pure quantum state. We can associate it with an operator: $\hat{\rho}_\psi \equiv |\psi\rangle\langle\psi|$, named as the density operator. This operator is also known as the projection operator because its action on any state projects it onto this $|\psi\rangle$ state as a result: $\hat{\rho}_\psi|\Phi\rangle = (\langle\psi|\Psi\rangle)|\psi\rangle$ with $|\langle\psi|\Psi\rangle|^2$ as the projection probability. Since it is a projection operator, it must satisfy $\hat{\rho}_\psi^2 = \hat{\rho}_\psi$. This is the characteristic of the density operator of a pure state. For the system in this state, the measurement of a physical observable with operator $\hat{O}$ will result in an expectation value as

$$\langle\hat{O}\rangle_\psi = \langle\psi|\hat{O}|\psi\rangle = \text{Tr}(\langle\psi|\hat{O}|\psi\rangle) = \text{Tr}(|\psi\rangle\langle\psi|\hat{O}) = \text{Tr}(\hat{\rho}_\psi\hat{O}). \quad (3.96)$$

So, any quantum state $|\psi\rangle$ can be equivalently represented by a density operator $\hat{\rho}_\psi = |\psi\rangle\langle\psi|$.

Now let us discuss mixed states. Consider two quantum states $|\psi_1\rangle, |\psi_2\rangle$ for a system. Suppose that we do not know due to some reason which state the system is in but we know that the probability is p_1 for the system in $|\psi_1\rangle$ and p_2 in $|\psi_2\rangle$ with $p_1 + p_2 = 1$. Taking the polarization states of light as an example, let us consider two independent beams of light with polarizations of $\hat{x}$ and $\hat{y}$, respectively. Now combine the two with a 50:50 beam splitter, as shown in Fig. 3.7. What is the polarization of the combined beam? It cannot be $\hat{x}$ or $\hat{y}$ for sure. Can it be 45°-polarized? But then why can't it be 135°-polarized? We know that both 45° and 135° polarization states are superposition of $\hat{x}$ and $\hat{y}$ polarizations and require a fixed phase between $\hat{x}$ and $\hat{y}$ polarizations: $\hat{\epsilon}_{45} = (\hat{x} + \hat{y})\sqrt{2}$, $\hat{\epsilon}_{135} = (\hat{x} - \hat{y})\sqrt{2}$. It turns out that all other polarization states such as circularly polarized states require fixed phase relation between $\hat{x}$ and $\hat{y}$ polarizations. But there is no fixed phase relation between two independent fields. Therefore, the combined beam after the beam splitter (BS) does not have a definite polarization state. Since we use a 50:50 BS, we can claim that it has 50% of probability of being $\hat{x}$-polarized and 50% of probability of being $\hat{y}$-polarized. This is a statistical mixture of $\hat{x}$ and $\hat{y}$ polarizations and is a mixed state of polarization.

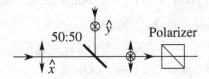

Fig. 3.7 Mixed polarization states: independent $\hat{x}$-polarized and $\hat{y}$-polarized fields are combined with a 50:50 beam splitter.

Let us next look at the result of a polarizer on this light field of mixed polarization. Assume the polarizer has a passing direction of θ relative to x and the intensity of the combined field after the BS is I_0. After the polarizer, the contribution from the $\hat{x}$-polarization is $I_0 \cos^2 \theta$ while that from the $\hat{y}$-polarization is $I_0 \sin^2 \theta$. Since both have 50% probability due to the BS, the total contribution is

$$I^{out} = I_0 \cos^2 \theta \times \frac{1}{2} + I_0 \sin^2 \theta \times \frac{1}{2} = I_0/2. \tag{3.97}$$

Hence, the intensity after the polarizer is independent of the angle θ of the polarizer. This result is the same as that for a natural light. On the other hand, for 45°-polarized light, the output intensity is $I_{45}^{out} = I_0 \cos^2(\theta - 45)$, which depends on the angle of the polarizer.

Now we come back to the case of the mixed state of two quantum states: $|\psi_1\rangle, |\psi_2\rangle$. Similar to the case of polarization measurement, the result of the measurement on the physical observable $\hat{O}$ is

$$\begin{aligned}
\langle \hat{O} \rangle &= \langle \hat{O} \rangle_{\psi_1} \times p_1 + \langle \hat{O} \rangle_{\psi_2} \times p_2 \\
&= p_1 \langle \psi_1 | \hat{O} | \psi_1 \rangle + p_2 \langle \psi_1 | \hat{O} | \psi_2 \rangle \\
&= p_1 \mathrm{Tr}(\hat{\rho}_{\psi_1} \hat{O}) + p_2 \mathrm{Tr}(\hat{\rho}_{\psi_2} \hat{O}) \\
&= \mathrm{Tr}(\hat{\rho}_m \hat{O}), \tag{3.98}
\end{aligned}$$

where $\hat{\rho}_m \equiv p_1 |\psi_1\rangle\langle\psi_1| + p_2 |\psi_2\rangle\langle\psi_2|$ is the density operator for the mixed state of $|\psi_1\rangle, |\psi_2\rangle$.

On the other hand, for a pure state: $|\psi\rangle = \sqrt{p_1}|\psi_1\rangle + \sqrt{p_2}|\psi_2\rangle$, it also has p_1 probability in state $|\psi_1\rangle$ and p_2 probability in state $|\psi_1\rangle$. What is the difference between this pure superposition state and the mixed state $\hat{\rho}_m \equiv p_1 |\psi_1\rangle\langle\psi_1| + p_2 |\psi_2\rangle\langle\psi_2|$? For comparison, we write the density operator for the pure state:

$$\begin{aligned}
\hat{\rho}_\psi &= \big(\sqrt{p_1}\langle\psi_1| + \sqrt{p_2}\langle\psi_2|\big)\big(\sqrt{p_1}|\psi_1\rangle + \sqrt{p_2}|\psi_2\rangle\big) \\
&= \hat{\rho}_m + \sqrt{p_1 p_2}|\psi_1\rangle\langle\psi_2| + \sqrt{p_1 p_2}|\psi_2\rangle\langle\psi_1|. \tag{3.99}
\end{aligned}$$

It has two more cross terms than the mixed state $\hat{\rho}_m$. If we make a measurement of physical observable $\hat{O}$, the results for the two states are

$$\begin{aligned}
\langle \hat{O} \rangle_{\rho_\psi} &= p_1 \langle\psi_1|\hat{O}|\psi_1\rangle + p_2 \langle\psi_1|\hat{O}|\psi_2\rangle \\
&\quad + \sqrt{p_1 p_2}\langle\psi_2|\hat{O}|\psi_1\rangle + \sqrt{p_1 p_2}\langle\psi_1|\hat{O}|\psi_2\rangle \\
\langle \hat{O} \rangle_{\rho_m} &= p_1 \langle\psi_1|\hat{O}|\psi_1\rangle + p_2 \langle\psi_1|\hat{O}|\psi_2\rangle. \tag{3.100}
\end{aligned}$$

Again, the pure state $|\psi\rangle$ has two more cross terms known as the quantum coherence terms, which keep a fixed phase relation (coherence) between

$|\psi_1\rangle, |\psi_2\rangle$ and give rise to quantum interference phenomena. The mixed state $\hat{\rho}_m$ does not produce such coherence terms. So, the pure state $|\psi\rangle$ is called a quantum superposition state of $|\psi_1\rangle, |\psi_2\rangle$ whereas the state $\hat{\rho}_m$ is a classical statistical mixed state of $|\psi_1\rangle, |\psi_2\rangle$.

A typical example is the Schrödinger cat state discussed in Section 3.3. In this case, $|\psi_1\rangle, |\psi_2\rangle$ are two mutually exclusive macroscopic classical states, i.e., $|\text{dead}\rangle$ or $|\text{alive}\rangle$. As we just mentioned, the cat state $|\psi_{cat}\rangle = (|\text{dead}\rangle + |\text{alive}\rangle)/\sqrt{2}$ will produce some interference phenomena whereas the mixed state $\hat{\rho}_m = (|\text{dead}\rangle\langle\text{dead}| + |\text{alive}\rangle\langle\text{alive}|)/2$ will not. Mixed state $\hat{\rho}_m$ is what we see in the classical world since we do not see interference effect in our daily life.

The transition from a quantum Schrödinger cat state $|\psi_{cat}\rangle$ to the classical mixed state $\hat{\rho}_m$ is a de-coherence process caused by the interaction of the system with the environment. We will see in Section 6.3 of Chapter 6 an example of photon dissipation in a single cavity mode through its coupling to an outside vacuum environment with a continuous spectrum of modes. This dissipation mechanism is what causes the loss of the coherent cross terms in the pure state density operator in Eq. (3.99). The more degrees of freedom the system is coupled to, the faster the de-coherence is. So, the largeness of a macroscopic system makes it coupled to more degrees of freedom and thus more likely to lose the quantum coherence than a small microscopic system. Eventually, a quantum system quickly de-coheres and becomes a classical system.

In the more general case, a quantum system may be in any one of n different quantum states. Suppose the n different states are labeled as $|\psi_1\rangle, |\psi_2\rangle, ..., |\psi_n\rangle$. Note that these states may or may not be orthogonal to each other. If we are not certain which state the system is in, we can describe them with probability: the probability is p_1 in state $|\psi_1\rangle$, p_2 in state $|\psi_2\rangle$, ..., p_n in state $|\psi_n\rangle$ with $p_1 + p_2 + ... + p_n = 1$. If we make a measurement of physical observable $\hat{O}$ in this system, the expectation value of the measurement is

$$\langle\hat{O}\rangle = p_1\langle\psi_1|\hat{O}|\psi_1\rangle + p_2\langle\psi_2|\hat{O}|\psi_2\rangle + ... + p_n\langle\psi_n|\hat{O}|\psi_n\rangle$$

$$= \sum_{j=1}^{n} p_j\langle\psi_j|\hat{O}|\psi_j\rangle. \tag{3.101}$$

With the introduction of density operator:

$$\hat{\rho} \equiv p_1|\psi_1\rangle\langle\psi_1| + p_2|\psi_2\rangle\langle\psi_2| + ... + p_n|\psi_n\rangle\langle\psi_n|$$

$$= \sum_{j=1}^{n} p_j|\psi_j\rangle\langle\psi_j|, \tag{3.102}$$

Eq. (3.101) becomes

$$\langle \hat{O} \rangle = \text{Tr}(\hat{\rho}\hat{O}). \qquad (3.103)$$

The mixed polarization state discussed earlier can be obtained from the general polarization state $|\epsilon_\phi\rangle = (|x\rangle + e^{i\phi}|y\rangle)/\sqrt{2}$. If the two incident fields of x and y polarizations are independent, the phase ϕ is random, i.e., it has a uniform probability density $p(\phi) = 1/2\pi$ in $[0, 2\pi]$ range. So, Eq. (3.102) gives the density operator for the mixed polarization state as

$$\hat{\rho}_p = \int_0^{2\pi} d\phi \, p(\phi)|\epsilon_\phi\rangle\langle\epsilon_\phi| = (|x\rangle\langle x| + |y\rangle\langle y|)/2, \qquad (3.104)$$

which properly describes the respective 50% probability for x and y polarizations for the combined light field.

Next, we will discuss some commonly encountered mixed state of light.

3.5.2 Density Operator for Lasers with Random Phases

The coherent state is a quantum state derived from theory. Besides quantum noise, it has a well-defined amplitude and phase, which is the closest quantum counterpart for a classical monochromatic infinite electromagnetic wave train. The output of a single-mode laser is a lab-produced light source that is closest to a coherent state. But the phase of a single-mode laser is not completely fixed. Due to the inevitable spontaneous emission with a random phase, the phase of a single-mode laser can diffuse with time. The characteristic diffusion time determines the upper bound on the coherence time of the laser and leads to the Schawlow-Townes line width of a single-mode laser [Schawlow and Townes (1958)], which is of quantum nature.[1] The other technical imperfection can result in broader line width and shorter coherence time. In a time interval much smaller than the coherence time of the laser, its phase diffusion is negligibly small and the output of the laser can be approximated as a coherent state. But for a time interval much longer than the coherence time, the diffused phase ranges from 0 to 2π, i.e., the phase of the laser output can be completely random with a uniform probability distribution in the range of $[0, 2\pi]$: $p(\varphi) = 1/2\pi (0 \leq \varphi \leq 2\pi)$. Therefore, similar to Eq. (3.104), we have its density operator as

$$\hat{\rho}_{\text{laser}} = \int_0^{2\pi} \frac{d\varphi}{2\pi} ||\alpha|e^{i\varphi}\rangle\langle|\alpha|e^{i\varphi}|. \qquad (3.105)$$

---

[1]In Section 8.3.1, we will present a simple derivation of the Schawlow-Townes line width based on photon indistinguishability in multi-photon interference.

Substituting the number state representation of the coherent state as in Eq. (3.25) into the expression above and carrying out the phase integral, we obtain

$$\hat{\rho}_{\text{laser}} = \sum_m P_m |m\rangle\langle m|, \tag{3.106}$$

where $P_m = (|\alpha|^2)^m e^{-|\alpha|^2}/m!$ is the Poisson distribution, which is the same as the photon number distribution of the coherent state.

3.5.3 *Density Operator for a Thermal State*

Thermal light sources are the most common light sources in nature. The quantum state describing this kind of sources is called a thermal state. Any object with a finite temperature is a radiation source of electromagnetic waves and the emitted electromagnetic waves have the same photon statistical property as the thermal state.

For a simple harmonic oscillator corresponding to a single-mode optical field, when it reaches thermal equilibrium with a heat reservoir of temperature T, it is in the thermal state. The density operator for the description of the thermal state can be derived from statistical quantum mechanics as

$$\hat{\rho}_{th} = e^{-\beta \hat{H}}/Z, \tag{3.107}$$

where $\beta \equiv 1/kT$ with k as the Boltzmann constant, $\hat{H} = \hbar\omega(\hat{a}^\dagger\hat{a} + 1/2)$, and $Z = \text{Tr}(e^{-\beta\hat{H}})$. In the number state representation, we have

$$
\begin{aligned}
\hat{\rho}_{th} &= e^{-\beta\hat{H}}\left(\sum_n |n\rangle\langle n|\right)/Z = \sum_n \left(e^{-\beta\hat{H}}|n\rangle\langle n|\right)/Z \\
&= \frac{1}{Z}\sum_n e^{-\beta\hbar\omega(n+1/2)}|n\rangle\langle n| \\
&= (1-\zeta)\sum_n \zeta^n |n\rangle\langle n| \tag{3.108}
\end{aligned}
$$

with $\zeta = e^{-\hbar/kT}$. The average photon number of the thermal state is

$$\bar{n} = \langle\hat{n}\rangle = \text{Tr}(\hat{a}^\dagger\hat{a}\hat{\rho}_{th}) = 1/(e^{\hbar\omega/kT} - 1). \tag{3.109}$$

From Eq. (3.108), we obtain the photon statistical distribution of the thermal state:

$$P_n = (1-\zeta)\zeta^n = (1 - e^{-\hbar/kT})e^{-n\hbar/kT} = \frac{\bar{n}^n}{(1+\bar{n})^{n+1}}. \tag{3.110}$$

This is exactly the Bose-Einstein statistics of the blackbody radiation. From this, we can find

$$\langle \hat{n}^2 \rangle = \sum_n n^2 P_n = \bar{n} + 2\bar{n}^2, \tag{3.111}$$

which can be rewritten as

$$\langle \hat{a}^{\dagger 2} \hat{a}^2 \rangle = \langle \hat{n}^2 \rangle - \langle \hat{n} \rangle = 2\bar{n}^2. \tag{3.112}$$

Here $\hat{a}^{\dagger 2} \hat{a}^2 = \ :\hat{n}^2: \ $ is the normal ordering operation on $\hat{n}^2$. In Chapter 5 when we discuss about photodetection, we will find the intensity of the optical field is proportional to the average photon number: $\langle I \rangle = \eta \langle \hat{n} \rangle$ (η is the quantum efficiency of the detector) and the auto-correlation of the intensity is proportional to $\langle :\hat{n}^2: \rangle$, i.e., $\langle I^2 \rangle = \eta^2 \langle :\hat{n}^2: \rangle$. From Eq. (3.112), we obtain

$$g^{(2)} \equiv \langle I^2 \rangle / \langle I \rangle^2 = \langle :\hat{n}^2: \rangle / \langle \hat{n} \rangle^2 = 2, \tag{3.113}$$

where $g^{(2)}$ is the normalized second-order intensity correlation function. It is the intensity correlation function measured in Hanbury Brown-Twiss experiment. So, the value of 2 for $g^{(2)}$ in Eq. (3.113) gives a direct quantum explanation of the photon bunching effect observed in Hanbury Brown-Twiss experiment. In Chapter 1, we provided the classical wave explanation of the photon bunching effect. The explanation here is directly from the photon statistics of the thermal state and is the most direct explanation of the photon bunching effect.

From Eq. (3.111), we can find the photon number fluctuation of the thermal state:

$$\langle \Delta^2 \hat{n} \rangle = \langle \hat{n}^2 \rangle - \langle \hat{n} \rangle^2$$
$$= \bar{n} + \bar{n}^2. \tag{3.114}$$

The expression above was first derived by Einstein in 1909 [Einstein (1909)]. The formula he derived is about the energy fluctuation of blackbody radiation:

$$\langle \Delta^2 E \rangle = \hbar \omega \langle E \rangle + \langle E \rangle^2 / Z. \tag{3.115}$$

If we use Planck equation $E = n\hbar\omega$ for the energy of blackbody radiation, the two terms in Eq. (3.115) correspond exactly to the two terms in Eq. (3.114) (the extra factor Z is due to multi-mode consideration). In discussing the physical meaning of the two terms in Eq. (3.115), Einstein discovered that the first term is due to the particle nature of the radiation whereas the second term originates from its wave nature. He derived the first term by using Poisson statistics of random particle. Indeed, Poisson distribution of the photon number of the coherent state leads to this term, as shown in Eq. (3.28). For the second term, Einstein was able to derive it by using the plane wave model of the radiation. Indeed, we can see from Eq. (3.112) that this term gives rise to the photon bunching effect in HBT experiment and as we showed in Section 1.6 of Chapter 1, wave nature alone can explain the photon bunching effect. Furthermore, we will show in Section 8.3.1 of Chapter 8 that photon bunching effect can also be explained by multi-photon constructive interference principle of quantum waves, which indicates the wave origin of the second term.

From Eq. (3.114), we can see that, when the average photon number is much smaller than 1, the first term dominates. In this case, the optical field exhibits particle nature (quantum). But when the average photon number is much larger than 1, the second term dominates, leading to a classical wave field. This is consistent with our intuition about the relationship between quantum and classical worlds.

The thermal sources from blackbody radiation have a wide spectrum. Besides blackbody radiation, other methods to obtain a thermal source include the light emitted from ionization processes in hot vapor of atoms such as hydrogen lamps and mercury lamps and the spontaneous emission from optical amplifiers such as erbium-doped fiber amplifiers and parametric amplifiers. In early experiments, atomic ionization is very common such as the source used in the HBT experiment. After the invention of lasers, a commonly used thermal source is a kind of quasi-thermal light source produced by passing a laser beam through a fast rotating grounded glass plate, as shown in Fig. 3.8. As the grounded glass plate rotates, the laser beam hits different positions of the grounded glass. But the surface of the grounded glass is very rough and scatters incident light to all directions randomly. So, the amplitude transmissivity coefficient t at different place is a random complex number, i.e., its amplitude and phase are both random variables:

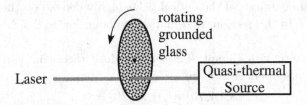

Fig. 3.8 The schematic for producing a quasi-thermal light source by rotating a grounded glass plate with a coherent state illumination.

$t = |t|e^{i\varphi_t}$. When the rotating speed is fast, the observed transmissivity is an average of a large number of random variables from different locations. We learned from the central limit theorem in Section 1.4.3 of Chapter 1 that the probability distribution of average transmissivity t is approximated by a Gaussian distribution: $P(t) = e^{-|t|^2/\sigma_t^2}/\pi\sigma_t^2$. If the incident laser beam is described by a coherent state, $|\alpha_0\rangle$, the instantaneous state of the transmitted light is also a coherent state of $|t\alpha_0\rangle$. But for a time period long enough, we take t as random variable with a Gaussian distribution and the transmitted state is a mixed state of different coherent states with a density operator:

$$\hat{\rho}_{gr} = \int d^2t P(t)|t\alpha_0\rangle\langle t\alpha| \tag{3.116}$$

where $d^2t = d\text{Re}(t)d\text{Im}(t)$. Substituting the number state expansion of the coherent state in Eq. (3.25) and carrying out the integration of t, we have

$$\hat{\rho}_{Qth} = \sum_n \frac{(\sigma_t^2|\alpha_0|^2)^n}{(1 + \sigma_t^2|\alpha_0|^2)^{n+1}}|n\rangle\langle n|. \tag{3.117}$$

The state above is in the same form as the density operator in Eq. (3.108) for a thermal state with an average photon number $\langle n \rangle = \sigma_t^2|\alpha_0|^2$. The reason we call this type of source a quasi-thermal source is because in the integration in Eq. (3.116), the range of $|t|$ is from 0 to ∞ but in reality, $|t|$ only takes values from 0 to 1. Only when $\sigma_t \ll 1$, this approximation is true. Moreover, the transmissivity is periodic for a uniformly rotated grounded glass. The average transmissivity σ_t depends on the roughness of the surface and the speed of rotation. The coherence time of the thermal source is determined also by the speed of rotation.

3.6   Further Reading: The Operator Algebra of $\hat{a}$ and $\hat{a}^\dagger$

All the physical observables of optical fields can be written in terms of the operators $\hat{a}$ and $\hat{a}^\dagger$. Since $\hat{a}$ and $\hat{a}^\dagger$ do not commute, the commutators of the physical quantities of the optical fields highly depend on the behavior of $\hat{a}$ and $\hat{a}^\dagger$. In this section, we will discuss about how to handle functions of $\hat{a}$ and $\hat{a}^\dagger$.

Let us start with Campbell-Baker-Hausdorff theorem. For two non-commuting operators $\hat{A}, \hat{B}$ with

$$[\hat{A}, [\hat{A}, \hat{B}]] = [\hat{B}, [\hat{A}, \hat{B}]] = 0, \tag{3.118}$$

then

$$e^{\hat{A}+\hat{B}} = e^{\hat{A}}e^{\hat{B}}e^{-[\hat{A},\hat{B}]/2} = e^{\hat{B}}e^{\hat{A}}e^{[\hat{A},\hat{B}]/2}. \tag{3.119}$$

Applying the above to $\hat{a}$ and $\hat{a}^\dagger$, we have

$$\hat{D}(\alpha) = e^{\alpha\hat{a}^\dagger - \alpha^*\hat{a}} = e^{-\alpha^*\hat{a}} e^{\alpha\hat{a}^\dagger} e^{|\alpha|^2/2} = e^{\alpha\hat{a}^\dagger} e^{-\alpha^*\hat{a}} e^{-|\alpha|^2/2}. \quad (3.120)$$

This is the normal ordering form of the displacement operator $\hat{D}(\alpha)$.

The more general theorem is

$$e^B A e^{-B} = A + [B, A] + \frac{1}{2!}[B, [B, A]]$$

$$+ \ldots + \frac{1}{n!}[B, [B, \cdots [B, A] \cdots]] + \ldots. \quad (3.121)$$

Applying the above to $\hat{D}(\alpha)$ and $\hat{a}$, we have

$$\hat{D}^\dagger(\alpha)\hat{a}\hat{D}(\alpha) = \hat{a} + \alpha. \quad (3.122)$$

Applying to the squeezing operator $\hat{S}(r) = \exp[r(\hat{a}^{\dagger 2} - \hat{a}^2)/2]$ in Eq. (3.63) and $\hat{a}$, we have

$$\hat{S}^\dagger \hat{a} \hat{S} = \hat{a} \cosh r + \hat{a}^\dagger \sinh r,$$

$$\hat{S}^\dagger \hat{a}^\dagger \hat{S} = \hat{a}^\dagger \cosh r + \hat{a} \sinh r, \quad (3.123)$$

which is just Eq. (3.62). So, we proved that the squeezing operator $\hat{S}(r)$ in Eq. (3.63) can give rise to the required relations in Eq. (3.62). But the argument in Section 3.4.1 requires the reverse, i.e., derive the squeezing operator $\hat{S}(r)$ in Eq. (3.63) from the relations in Eq. (3.62). For this, we need a different approach.

Suppose operator $\hat{S}(r)$ satisfies relations in Eq. (3.123). We see that $\hat{S}(0) = \hat{I}$, the identity operator. For infinitesimal $r = \delta r \ll 1$, we can make a linear expansion of $\hat{S}(r)$ in terms of δr:

$$\hat{S}(r) \approx \hat{I} + i\delta r f(\hat{a}, \hat{a}^\dagger) \quad (3.124)$$

where $f(\hat{a}, \hat{a}^\dagger)$ is some function of $\hat{a}, \hat{a}^\dagger$ to be determined. Since $\hat{S}(r)$ is unitary, i.e., $\hat{S}^\dagger(r)\hat{S}(r) = \hat{S}(r)\hat{S}^\dagger(r) = \hat{I}$, substituting Eq. (3.124) into these and keeping only the linear terms in δr, we have $f^\dagger = f$, i.e., f is a Hermitian operator.

Next, write Eq. (3.123) for infinitesimal $r = \delta r \ll 1$:

$$(\hat{I} - i\delta r f)\hat{a}(\hat{I} + i\delta r f) = \hat{a} + \delta r \hat{a}^\dagger, \quad (3.125)$$

or,

$$\hat{a} + i\delta r[\hat{a}, f] = \hat{a} + \delta r \hat{a}^\dagger, \quad (3.126)$$

where we only keep up to linear terms in δr. Comparing the two sides of the above, we have

$$[\hat{a}, f] = -i\hat{a}^\dagger. \quad (3.127)$$

Similarly,

$$[\hat{a}^\dagger, f] = -i\hat{a}. \tag{3.128}$$

In Problem 3.1, we can prove $[\hat{a}, \hat{a}^{\dagger n}] = n\hat{a}^{\dagger n-1}, [\hat{a}^\dagger, \hat{a}^n] = -n\hat{a}^{n-1}$. It is straightforward to extend these two relations to any ordered function of $\hat{a}, \hat{a}^\dagger$: $f(\hat{a}^\dagger, \hat{a}) = \sum c_{kl}\hat{a}^{\dagger k}\hat{a}^l$ as follows

$$[\hat{a}, f(\hat{a}^\dagger, \hat{a})] = \frac{\partial f}{\partial \hat{a}^\dagger}, \quad [\hat{a}^\dagger, f(\hat{a}^\dagger, \hat{a})] = -\frac{\partial f}{\partial \hat{a}}. \tag{3.129}$$

Hence, Eqs. (3.127) and (3.128) become

$$\frac{\partial f}{\partial \hat{a}^\dagger} = -i\hat{a}^\dagger, \quad \frac{\partial f}{\partial \hat{a}} = i\hat{a}. \tag{3.130}$$

Apart from a c-number, we have the solution of f as

$$f(\hat{a}^\dagger, \hat{a}) = i(\hat{a}^2 - \hat{a}^{\dagger 2})/2. \tag{3.131}$$

So, $\hat{S}(\delta r) = \hat{I} - \delta r(\hat{a}^2 - \hat{a}^{\dagger 2})/2$. For a finite r, we can divide r into N small δr's with $\delta r = r/N$:

$$\hat{S}(r) = [\hat{S}(r/N)]^N = \lim_{N\to\infty} \left[\hat{I} - \frac{r}{N}(\hat{a}^2 - \hat{a}^{\dagger 2})/2\right]^N$$
$$\to \exp[-r(\hat{a}^2 - \hat{a}^{\dagger 2})/2], \tag{3.132}$$

which is exactly the squeezing operator in Eq. (3.63).

The relations in Eq. (3.129) are especially useful to calculate the normally ordered form $f^{(n)}(\hat{a}^\dagger, \hat{a})$ of some function $f(\hat{a}^\dagger, \hat{a})$ of $\hat{a}^\dagger, \hat{a}$. Here normal ordering is such that $\hat{a}^\dagger$ is moved to the left of $\hat{a}$. For example, the normally ordered form of $\hat{n}^2 = (\hat{a}^\dagger\hat{a})^2$ is $\hat{n}^2 = \hat{a}^{\dagger 2}\hat{a}^2 + \hat{a}^\dagger\hat{a}$. We will discuss more of normal ordering and its applications in Chapter 5. The usefulness of the normally ordered form is its action on coherent states: $\langle\alpha|f^{(n)}(\hat{a}^\dagger, \hat{a})|\alpha\rangle = f^{(n)}(\alpha^*, \alpha)$. To see how we can make use of Eq. (3.129), we consider the squeezing operator $\hat{S}(r)$ and calculate $\langle\alpha|\hat{S}(r)|\beta\rangle = S^{(n)}(\alpha^*, \beta)\langle\alpha|\beta\rangle$.

Take a partial derivative of $\hat{S}(r)$ with respect to r:

$$\frac{\partial \hat{S}(r)}{\partial r} = \frac{1}{2}(\hat{a}^{\dagger 2} - \hat{a}^2)\hat{S}(r) = \frac{1}{2}\hat{a}^{\dagger 2}\hat{S}(r) - \frac{1}{2}\hat{a}^2\hat{S}(r). \tag{3.133}$$

Using relations in Eq. (3.129) twice, we have

$$\hat{a}^2\hat{S}(r) = \hat{S}(r)\hat{a}^2 + 2\frac{\partial \hat{S}(r)}{\partial \hat{a}^\dagger}\hat{a} + \frac{\partial^2 \hat{S}(r)}{\partial \hat{a}^{\dagger 2}}. \tag{3.134}$$

Now take the normally ordered form $S^{(n)}(\hat{a}^\dagger, \hat{a})$ for $\hat{S}(r)$ in Eqs. (3.133) and (3.134) and apply $\langle\alpha|$ from the left side and $|\beta\rangle$ from the right side:

$$\frac{\partial S^{(n)}(\alpha^*, \beta, r)}{\partial r} = \frac{1}{2}\left(\alpha^{*2} - \beta^2 - 2\beta\frac{\partial}{\partial\alpha^*} - \frac{\partial^2}{\partial\alpha^{*2}}\right)S^{(n)}(\alpha^*, \beta, r). \tag{3.135}$$

Here, we divide both sides by $\langle\alpha|\beta\rangle$. Take a trial solution of the form: $S^{(n)}(\alpha^*,\beta,r) = e^{G(\alpha^*,\beta,r)}$ with $G(\alpha^*,\beta,r) = A(r) + B(r)\alpha^{*2} + C(r)\beta^2 + D(r)\alpha^*\beta$. Here, $A(r), B(r), C(r), D(r)$ are to be determined. Substituting into Eq. (3.135) and carrying out the derivatives, we have the right side as

$$\text{right} = \frac{1}{2}\Big[\alpha^{*2} - \beta^2 - 4B\beta\alpha^* - 2D\beta^2$$
$$-2B - (2B\alpha^* + D\beta)^2\Big]S^{(n)}(\alpha^*,\beta,r). \qquad (3.136)$$

Comparing both sides of Eq. (3.135), we have

$$\frac{dA}{dr} = -B, \quad \frac{dB}{dr} = \frac{1}{2} - 2B^2,$$
$$\frac{dC}{dr} = -\frac{1}{2}(D+1)^2, \quad \frac{dD}{dr} = -2B - 2BD. \qquad (3.137)$$

The initial conditions are $A(0) = B(0) = C(0) = D(0) = 0$. We can solve B immediately as $B(r) = (1/2)\tanh r$, then $A(r) = -(1/2)\ln(\cosh r)$, and $D(r) = 1/\cosh r - 1$, and $C(r) = -B$. The final form of $S^{(n)}(\alpha^*,\beta,r)$ is

$$S^{(n)}(\alpha^*,\beta,r) = \frac{1}{\sqrt{\mu}}\exp\Big[\frac{\nu}{2\mu}(\alpha^{*2} - \beta^2) + \Big(\frac{1}{\mu} - 1\Big)\alpha^*\beta\Big], \qquad (3.138)$$

where $\mu = \cosh r, \nu = \sinh r$. With this, we have

$$\langle\alpha|\hat{S}(r)|\beta\rangle = S^{(n)}(\alpha^*,\beta)\langle\alpha|\beta\rangle$$
$$= \frac{1}{\sqrt{\mu}}\exp\Big[\frac{\nu}{2\mu}(\alpha^{*2} - \beta^2) + \frac{\alpha^*\beta}{\mu} - \frac{1}{2}(|\alpha|^2 + |\beta|^2)\Big]. \qquad (3.139)$$

3.7 Glauber-Sudarshan *P*-Distribution and Wigner Distribution

As we have seen in Section 3.2.5, the coherent states form an over-complete set of bases with which any arbitrary state can be represented. We can do the same for any operator $\hat{O}$ by using the completeness relation in Eq. (3.47):

$$\hat{O} = \int \frac{d^2\beta}{\pi}|\beta\rangle\langle\beta|\hat{O}\int\frac{d^2\alpha}{\pi}|\alpha\rangle\langle\alpha| = \int\frac{d^2\alpha d^2\beta}{\pi^2}\langle\beta|\hat{O}|\alpha\rangle|\beta\rangle\langle\alpha|. \qquad (3.140)$$

$\langle\beta|\hat{O}|\alpha\rangle$ is then the coherent state representation of operator $\hat{O}$. For example, the coherent state representation of the squeezed operator $\hat{S}(r)$ is derived in the last section and is expressed in Eq. (3.139). The connection between the coherent state representation and the number state representation can be easily derived as follows:

$$\langle\beta|\hat{O}|\alpha\rangle = \langle\beta|\Big(\sum_n |n\rangle\langle n|\Big)\hat{O}\Big(\sum_m |m\rangle\langle m|\Big)|\alpha\rangle$$

$$= \sum_{m,n} O_{nm}\langle\beta|n\rangle\langle m|\alpha\rangle$$

$$= e^{-(|\alpha|^2+|\beta|^2)/2}\sum_{m,n}\frac{O_{nm}}{\sqrt{m!n!}}\beta^{*n}\alpha^m$$

$$\equiv e^{-(|\alpha|^2+|\beta|^2)/2}f_O(\beta^*,\alpha), \qquad (3.141)$$

where $O_{nm} \equiv \langle n|\hat{O}|m\rangle$ is the number state representation.

If $\hat{O}$ is a traceable, non-negative definite Hermitian operator, $|O_{nm}|$ is bounded. Then the double infinite series $f_O(\beta^*,\alpha)$ in Eq. (3.141) is absolutely convergent for all β^* and α and is thus an analytic function of the complex variables β^* and α. Indeed, if we write operator $\hat{O}$ in the normally ordered form of $\hat{O}^{(n)}(\hat{a}^\dagger,\hat{a})$, then $\langle\beta|\hat{O}|\alpha\rangle = O^{(n)}(\beta^*,\alpha)\langle\beta|\alpha\rangle$ and $f_O(\beta^*,\alpha) = O^{(n)}(\beta^*,\alpha)e^{\beta^*\alpha}$, which is an analytic function of β^* and α for a well behaved $O^{(n)}(\beta^*,\alpha)$.

According to the theory of complex analysis, the analytic function $f_O(\beta^*,\alpha)$ can be determined everywhere from some arbitrarily small range of β^*,α. Particularly, if we know the value $f_O(\alpha^*,\alpha)$, $f_O(\beta^*,\alpha)$ is then entirely determined. Therefore, $\langle\beta|\hat{O}|\alpha\rangle$ can be uniquely determined by the diagonal elements $\langle\alpha|\hat{O}|\alpha\rangle$.

3.7.1 Glauber-Sudarshan P-Representation of Density Operator

We can apply the above to the density operator for the description of the quantum state of a system. From Section 3.5, we find that the density operator satisfies $\mathrm{Tr}\hat{\rho} = 1$ so it is traceable and it is a non-negative Hermitian from its definition in Eq. (3.102). Therefore, we should be able to write it in the form of the diagonal coherent state representation:

$$\hat{\rho} = \int d^2\alpha P(\alpha)|\alpha\rangle\langle\alpha|, \qquad (3.142)$$

where $P(\alpha)$ is some real function of complex variable α because $\hat{\rho}$ is a Hermitian operator. This is the Glauber-Sudarshan P-Representation first introduced by Glauber [Glauber (1963c)] and Sudarshan [Sudarshan (1963)] independently. It plays a key role in the quantum coherence theory to be discussed in Chapter 5.

The actual form of function $P(\alpha)$ can be derived directly from the density operator $\hat{\rho}$ [Mehta and Sudarshan (1965)]. Consider the function $\langle -\beta|\hat{\rho}|\beta\rangle$ of complex variable β. From Eq. (3.142), we have

$$\langle -\beta|\hat{\rho}|\beta\rangle = \int d^2\alpha P(\alpha)\langle -\beta|\alpha\rangle\langle\alpha|\beta\rangle$$

$$= e^{-|\beta|^2}\int d^2\alpha P(\alpha)e^{-|\alpha|^2}e^{\beta\alpha^*-\beta^*\alpha}, \qquad (3.143)$$

or

$$\langle -\beta|\hat{\rho}|\beta\rangle e^{|\beta|^2} = \int d^2\alpha P(\alpha)e^{-|\alpha|^2}e^{\beta\alpha^*-\beta^*\alpha}. \qquad (3.144)$$

Writing out the complex variables in terms of the real variables: $\beta = u + iv, \alpha = x + iy$, we have $e^{\beta\alpha^*-\beta^*\alpha} = e^{2i(xv-yu)}$ and Eq. (3.144) becomes

$$\langle -\beta|\hat{\rho}|\beta\rangle e^{u^2+v^2} = \int dxdy P(x,y)e^{-(x^2+y^2)}e^{2i(xv-yu)}. \qquad (3.145)$$

So, function $f(u,v) \equiv \langle -\beta|\hat{\rho}|\beta\rangle e^{u^2+v^2}$ is just the Fourier transformation of function $g(x,y) \equiv P(x,y)e^{-(x^2+y^2)}$. The reverse transformation gives

$$g(x,y) = \frac{1}{\pi^2}\int dudv f(u,v)e^{-2i(xv-yu)} \qquad (3.146)$$

or

$$P(\alpha)e^{-|\alpha|^2} = \frac{1}{\pi^2}\int d^2\beta\langle -\beta|\hat{\rho}|\beta\rangle e^{|\beta|^2}e^{\beta^*\alpha-\beta\alpha^*}. \qquad (3.147)$$

We have then expressed $P(\alpha)$ in terms of the elements $\langle -\beta|\hat{\rho}|\beta\rangle$ of the density operator $\hat{\rho}$.

Furthermore, since the average of any normally ordered operator moment $: \hat{a}^n\hat{a}^{\dagger m} : = \hat{a}^{\dagger m}\hat{a}^n$ has the form of

$$\langle: \hat{a}^n\hat{a}^{\dagger m} :\rangle = \langle\hat{a}^{\dagger m}\hat{a}^n\rangle = \text{Tr}\big[\hat{\rho}\hat{a}^{\dagger m}\hat{a}^n\big]$$

$$= \int d^2\alpha P(\alpha)\alpha^{*m}\alpha^n = \langle\alpha^{*m}\alpha^n\rangle_P, \qquad (3.148)$$

we can define the characteristic function $C_N(u,u^*)$ for the normally ordered moment $: \hat{a}^n\hat{a}^{\dagger m} : = \hat{a}^{\dagger m}\hat{a}^n$ as

$$C_N(u,u^*) \equiv \langle: e^{u\hat{a}^\dagger-u^*\hat{a}} :\rangle = \langle e^{u\hat{a}^\dagger}e^{-u^*\hat{a}}\rangle$$

$$= \text{Tr}\big[\hat{\rho}e^{u\hat{a}^\dagger}e^{-u^*\hat{a}}\big] = \int d^2\alpha e^{u\alpha^*-u^*\alpha}P(\alpha). \qquad (3.149)$$

The average of any normally ordered moment $\langle: \hat{a}^n\hat{a}^{\dagger m} :\rangle = \langle\hat{a}^{\dagger m}\hat{a}^n\rangle$ can be expressed as

$$\langle\hat{a}^{\dagger m}\hat{a}^n\rangle = \left[\frac{\partial^{m+n}C_N(u,u^*)}{\partial u^m\partial(-u^*)^n}\right]_{u,u^*=0} \qquad (3.150)$$

where we treat u, u^* as two independent variables. Since $e^{u\alpha^* - u^*\alpha}$ is the kernel of the two-dimensional Fourier transformation, the inverse-Fourier transformation of Eq. (3.149) has the form of

$$P(\alpha) = \frac{1}{\pi^2} \int d^2 u \, e^{-u\alpha^* + u^*\alpha} C_N(u, u^*). \qquad (3.151)$$

It is easier to use Eq. (3.147) for $P(\alpha)$ than Eq. (3.151) because the calculation of $\langle -\beta|\hat{\rho}|\beta\rangle$ is easier than $C_N(u, u^*)$ in most cases.

Let us find $P(\alpha)$ for some commonly encountered states:

(a) Coherent state

For a coherent state of $|\alpha_0\rangle$,
$$\langle -\beta|\hat{\rho}|\beta\rangle = \langle -\beta|\alpha_0\rangle\langle\alpha_0|\beta\rangle = e^{-|\alpha_0|^2 - |\beta|^2} e^{\beta\alpha_0^* - \beta^*\alpha}.$$
From Eq. (3.147), we have

$$P(\alpha)e^{-|\alpha|^2} = \frac{1}{\pi^2} \int d^2\beta \, e^{-|\alpha_0|^2} e^{\beta^*(\alpha - \alpha_0) - \beta(\alpha^* - \alpha_0^*)}$$

$$= e^{-|\alpha_0|^2} \frac{1}{\pi^2} \int d^2\beta \, e^{\beta^*(\alpha - \alpha_0) - \beta(\alpha^* - \alpha_0^*)}$$

$$= e^{-|\alpha_0|^2} \delta^2(\alpha - \alpha_0). \qquad (3.152)$$

So, $P(\alpha) = \delta^2(\alpha - \alpha_0)$ for the coherent state $|\alpha_0\rangle$.

(b) Thermal state

For a thermal state of average photon number $\bar{n}$, $\hat{\rho}_{th} = \sum_n P_n |n\rangle\langle n|$ with $P_n = \bar{n}^n/(\bar{n}+1)^{n+1}$. So, we have

$$\langle -\beta|\hat{\rho}_{th}|\beta\rangle = \sum_n \frac{\bar{n}^n}{(\bar{n}+1)^{n+1}} \frac{(-1)^n}{n!} |\beta|^{2n} e^{-|\beta|^2}$$

$$= \frac{e^{-|\beta|^2}}{\bar{n}+1} \sum_n \frac{(-1)^n}{n!} \left(\frac{\bar{n}|\beta|^2}{\bar{n}+1}\right)^n$$

$$= \frac{e^{-|\beta|^2}}{\bar{n}+1} \exp\left(-\frac{\bar{n}|\beta|^2}{\bar{n}+1}\right). \qquad (3.153)$$

Substituting the above into Eq. (3.147) and carrying out the integration, we obtain

$$P(\alpha) = \frac{1}{\pi\bar{n}} e^{-|\alpha|^2/\bar{n}}. \qquad (3.154)$$

(c) Number states

For the number state of $\hat{\rho} = |n\rangle\langle n|$,

$\langle -\beta|\hat{\rho}|\beta\rangle = \langle -\beta|n\rangle\langle n|\beta\rangle = e^{-|\beta|^2}(-|\beta|^2)^n/n!$.

Substituting it into Eq. (3.147), we obtain

$$P(\alpha)e^{-|\alpha|^2} = \frac{1}{\pi^2 n!} \int d^2\beta (-|\beta|^2)^n e^{\beta^*\alpha - \beta\alpha^*}, \qquad (3.155)$$

which is divergent in the normal sense. But we can express it in terms of the special δ-function as

$$P(\alpha) = \frac{e^{|\alpha|^2}}{n!} \frac{\partial^{2n}}{\partial\alpha^{*n}\partial\alpha^n} \frac{1}{\pi^2} \int e^{\beta^*\alpha - \beta\alpha^*} d^2\beta$$

$$= \frac{e^{|\alpha|^2}}{n!} \frac{\partial^{2n}}{\partial\alpha^{*n}\partial\alpha^n} \delta^{(2)}(\alpha). \qquad (3.156)$$

So, $P(\alpha)$ is the $2n$'th derivative of a δ-function, which is even more singular than a δ-function. This will lead to nonclassical behavior for the number states (see Section 5.2.4).

3.7.2 *Wigner W-Representation of Density Operator*

The Wigner distribution (also called the Wigner function) is a quasi-probability distribution for position and momentum of a particle. It was introduced by Eugene Wigner in 1932 to study quantum corrections to classical statistical mechanics [Wigner (1932)]. Specifically, it is defined for a particle in a pure state of wave function $\psi(x)$ as

$$W(x,p) \equiv \frac{1}{\pi\hbar} \int_{-\infty}^{\infty} dy \psi^*(x+y)\psi(x-y)e^{-2ipy/\hbar}$$

$$= \frac{1}{\pi\hbar} \int_{-\infty}^{\infty} dq \phi^*(p+q)\phi(p-q)e^{-2ixq/\hbar}, \qquad (3.157)$$

where $\phi(p)$ is the Fourier transformation of $\psi(x)$ and x, p are the position and momentum but could be any conjugate variable pair. For a mixed state of $\hat{\rho}$, it becomes

$$W(x,p) = \frac{1}{\pi\hbar} \int_{-\infty}^{\infty} dy \langle x+y|\hat{\rho}|x-y\rangle e^{-2ipy/\hbar}$$

$$= \frac{1}{\pi\hbar} \int_{-\infty}^{\infty} dq \langle p+q|\hat{\rho}|p-q\rangle e^{-2ixq/\hbar}. \qquad (3.158)$$

It is a quasi-probability distribution of the quantum system in $x - p$ phase space in the sense that W is normalized to 1: $\int dx dp W(x,p) = 1$, which can be shown from Eq. (3.158):

$$\int dx dp W(x,p) = \int dx dy \langle x+y|\hat{\rho}|x-y\rangle \frac{1}{\pi \hbar} \int dp e^{-2ipy/\hbar}$$

$$= \int dx dy \langle x+y|\hat{\rho}|x-y\rangle \delta(y)$$

$$= \int dx \langle x|\hat{\rho}|x\rangle = \mathrm{Tr}\hat{\rho} = 1, \qquad (3.159)$$

and the marginal distributions are x and p probability distributions, which can be shown from Eq. (3.157):

$$\int dp W(x,p) = \int dy \psi^*(x+y)\psi(x-y)\delta(y) = |\psi(x)|^2$$

$$\int dx W(x,p) = \int dq \phi^*(p+q)\phi(p-q)\delta(q) = |\phi(p)|^2 \qquad (3.160)$$

for pure states and from Eq. (3.158):

$$\langle x|\rho|x\rangle = \int dp W(x,p), \quad \langle p|\rho|p\rangle = \int dx W(x,p) \qquad (3.161)$$

for mixed states.

There are some interesting properties for the Wigner function. The first one is the overlap of quantum states by the product rule:

$$\mathrm{Tr}(\hat{\rho}_1 \hat{\rho}_2) = 2\pi \hbar \int dx dp W_{\hat{\rho}_1}(x,p) W_{\hat{\rho}_2}(x,p), \qquad (3.162)$$

which can be shown from Eq. (3.158):

$$\mathrm{Tr}(\hat{\rho}_1 \hat{\rho}_2) = 2 \int dx dy_1 dy_2 \langle x+y_1|\hat{\rho}_1|x-y_1\rangle \langle x+y_2|\hat{\rho}_2|x-y_2\rangle$$

$$\times \frac{1}{\pi \hbar} \int dp e^{-2ip(y_1+y_2)/\hbar}$$

$$= 2 \int dx dy_1 dy_2 \langle x+y_1|\hat{\rho}_1|x-y_1\rangle \langle x+y_2|\hat{\rho}_2|x-y_2\rangle \delta(y_1+y_2)$$

$$= 2 \int dx dy_1 \langle x+y_1|\hat{\rho}_1|x-y_1\rangle \langle x-y_1|\hat{\rho}_2|x+y_1\rangle$$

$$= \int dz_1 dz_2 \langle z_1|\hat{\rho}_1|z_2\rangle \langle z_2|\hat{\rho}_2|z_1\rangle$$

$$= \int dz_1 \langle z_1|\hat{\rho}_1 \hat{\rho}_2|z_1\rangle = \mathrm{Tr}(\hat{\rho}_1 \hat{\rho}_2), \qquad (3.163)$$

where we changed integration variables $x+y_1 = z_1, x-y_1 = z_2$ and used relation $\int dz_2 |z_2\rangle \langle z_2| = 1$.

The second one is that the Wigner function can become negative for some values of x, p for some states, which can be shown from Eq. (3.162) for orthogonal states $\text{Tr}(\hat{\rho}_1 \hat{\rho}_2) = 0$:

$$2\pi\hbar \int dx dp W_{\hat{\rho}_1}(x, p) W_{\hat{\rho}_2}(x, p) = \text{Tr}(\hat{\rho}_1 \hat{\rho}_2) = 0, \qquad (3.164)$$

which means either $W_{\hat{\rho}_1}(x, p)$ or $W_{\hat{\rho}_2}(x, p)$ must be negative for some x, p in order for the expression above to stand. So, like the Glauber P-distribution, the Wigner distribution cannot be a true probability distribution.

The third one is that unlike the Glauber P-function, the Wigner function is bounded, which can be shown from Eq. (3.157):

$$
\begin{aligned}
\left| W(x, p) \right| &\le \frac{1}{\pi\hbar} \left| \int dy \left| \psi^*(x+y) \psi(x-y) e^{-2ipy/\hbar} \right| \right. \\
&= \frac{1}{\pi\hbar} \int dy \left| \psi^*(x+y) \psi(x-y) \right| \\
&\le \frac{1}{\pi\hbar} \sqrt{\int dy \left| \psi^*(x+y) \right|^2 \int dy \left| \psi(x-y) \right|^2} \\
&= \frac{1}{\pi\hbar},
\end{aligned}
\qquad (3.165)
$$

where we used the Cauchy-Schwarz inequality for the last inequality above.

The Wigner function has practical significance in quantum optics. As we will show in Section 10.4, homodyne detection technique can be applied for the measurement of the Wigner function of an optical field by the method of quantum state tomography. This allows us to measure the complete quantum state for an optical field.

To prepare for this, let us apply the Wigner function defined in Eq. (3.158) to the virtual harmonic oscillator for a single-mode field. We change x, p to the dimensionless variables $X = x\sqrt{2\omega/m\hbar}$, $Y = p\sqrt{2m/\omega\hbar}$ with $m = 1$. We need to choose the correct coefficient so that W is normalized: $\int dX dY \, W(X, Y) = 1$. The explicit form of the Wigner function for an optical field is then given by

$$W(X, Y) = \frac{1}{2\pi} \int du \langle X + u | \hat{\rho} | X - u \rangle e^{-iuY}, \qquad (3.166)$$

which can be expressed in a more familiar form in quantum optics:

$$W(X, Y) = \frac{1}{(2\pi)^2} \int du dv C_W(u, v) e^{-ivX + iuY} \qquad (3.167)$$

with

$$C_W(u, v) \equiv \text{Tr}\left(\hat{\rho} e^{iv\hat{X} - iu\hat{Y}} \right) = \text{Tr}\left(\hat{\rho} e^{\eta \hat{a}^\dagger - \eta^* \hat{a}} \right), \qquad (3.168)$$

where $\eta = u + iv$ and $\hat{X} = \hat{a} + \hat{a}^\dagger, \hat{Y} = (\hat{a} - \hat{a}^\dagger)/i$.

To prove the expression in Eq. (3.167) is the same as Eq. (3.166), we first express $C_W(u,v)$ in the base of the eigen-states $\{|\xi\rangle\}$ of $\hat{X}$ ($\hat{X}|\xi\rangle = \xi|\xi\rangle$):

$$
\begin{aligned}
C_W(u,v) &= \int d\xi \langle\xi|\hat{\rho}e^{iv\hat{X}-iu\hat{Y}}|\xi\rangle = \int d\xi \langle\xi|\hat{\rho}e^{ivu}e^{-iu\hat{Y}}e^{iv\hat{X}}|\xi\rangle \\
&= \int d\xi \, e^{iv(\xi+u)}\langle\xi|\hat{\rho}e^{-iu\hat{Y}}|\xi\rangle \\
&= \int d\xi \, e^{iv(\xi+u)}\langle\xi|\hat{\rho}|\xi+2u\rangle,
\end{aligned}
\tag{3.169}
$$

where we used Campbell-Baker-Hausdorff theorem in Eq. (3.119) with $[\hat{X}, \hat{Y}] = 2i$, and $e^{iv\hat{X}}|\xi\rangle = e^{iv\xi}|\xi\rangle, e^{-iu\hat{Y}}|\xi\rangle = |\xi + 2u\rangle$. Substituting the above into Eq. (3.167), we have

$$
\begin{aligned}
W(X,Y) &= \frac{1}{(2\pi)^2}\int dudv e^{iuY-ivX}\int d\xi \, e^{iv(\xi+u)}\langle\xi|\hat{\rho}|\xi+2u\rangle \\
&= \frac{1}{2\pi}\int due^{iuY}\int d\xi\langle\xi|\hat{\rho}|\xi+2u\rangle\int dv\frac{e^{iv(\xi+u-X)}}{2\pi} \\
&= \frac{1}{2\pi}\int due^{iuY}\int d\xi\langle\xi|\hat{\rho}|\xi+2u\rangle\delta(\xi+u-X) \\
&= \frac{1}{2\pi}\int du\langle X-u|\hat{\rho}|X+u\rangle e^{iuY},
\end{aligned}
\tag{3.170}
$$

which gives Eq. (3.166) after a variable change of $u \to -u$. With Eq. (3.166), Eq. (3.165) becomes

$$
\left|W(X,Y)\right| \le \frac{1}{2\pi} \quad \text{or} \quad -\frac{1}{2\pi} \le W(X,Y) \le \frac{1}{2\pi}
\tag{3.171}
$$

for pure states.

Now let us find the Wigner function for some known states.

(a) *Coherent state*

For a coherent state $|\alpha\rangle$ with $\alpha = a_1 + ia_2$, we have

$$
\begin{aligned}
C_W(u,v) &= \mathrm{Tr}\left(\hat{\rho}_\alpha e^{\eta\hat{a}^\dagger-\eta^*\hat{a}}\right) = \langle\alpha|e^{-(u^2+v^2)/2}e^{\eta\hat{a}^\dagger}e^{-\eta^*\hat{a}}|\alpha\rangle \\
&= e^{-(u^2+v^2)/2}e^{\eta\alpha^*}e^{-\eta^*\alpha} = e^{i2va_1-i2ua_2-(u^2+v^2)/2}.
\end{aligned}
\tag{3.172}
$$

Taking a Fourier transformation, we have

$$
W(x_1,x_2) = \frac{1}{2\pi}\exp\left[-\frac{1}{2}\left(\bar{x}_1^2 + \bar{x}_2^2\right)\right]
\tag{3.173}
$$

with $\bar{x}_i = x_i - 2a_i$ $(i = 1, 2)$. Here, we changed the variables from X, Y to x_1, x_2. Since $W(x_1, x_2) > 0$, we can treat it as a probability distribution and calculate the variance $\langle \Delta^2 x_i \rangle = \langle \bar{x}_i^2 \rangle = \int d\bar{x}_1 d\bar{x}_2 \bar{x}_i^2 W(\bar{x}_1, \bar{x}_2) = 1$. So, the distribution has a standard deviation of 1 and the contour of one standard deviation for the distribution is $\bar{x}_1^2 + \bar{x}_2^2 = 1$, which is a circle of radius 1 centered at $(2a_1, 2a_2)$. We take this circle as the representation of the size of the quantum noise when depicting the coherent state in X-Y phase space (see Fig. 3.1).

(b) *Coherent squeezed state (css)*

A coherent squeezed state defined in Eq. (3.71) has the form of $|\alpha, r\rangle = \hat{D}(\alpha)\hat{S}(r)|0\rangle = \hat{S}(r)|\alpha_-\rangle$ given by Eq. (3.80) in Section 3.4.3. Here, $\alpha_- = \alpha \cosh r - \alpha^* \sinh r = a_1 e^{-r} + i a_2 e^r$. So, its characteristic function can be calculated as

$$
\begin{aligned}
C_W(u,v) &= \text{Tr}\left(\hat{\rho}_{css} e^{iv\hat{X} - iu\hat{Y}} \right) \\
&= \langle \alpha_- | \hat{S}^\dagger(r) e^{iv\hat{X} - iu\hat{Y}} \hat{S}(r) | \alpha_- \rangle \\
&= \langle \alpha_- | e^{\hat{S}^\dagger(r)(iv\hat{X} - iu\hat{Y})\hat{S}(r)} | \alpha_- \rangle \\
&= \langle \alpha_- | \exp[iv e^r \hat{X} - iu e^{-r} \hat{Y}] | \alpha_- \rangle \\
&= \langle \alpha_- | e^{iv'\hat{X} - iu'\hat{Y}} | \alpha_- \rangle \\
&= e^{-(u'^2 + v'^2)/2} e^{2iv' a_1' - 2iu' a_2'},
\end{aligned}
\tag{3.174}
$$

where we used relations in Eq. (3.61) and $u' \equiv u e^{-r}, v' \equiv v e^r; a_1' = a_1 e^{-r}, a_2' = a_2 e^r$. So, the Wigner function is

$$
\begin{aligned}
W_{css}(x_1, x_2) &= \frac{1}{(2\pi)^2} \int du dv e^{iux_2 - ivx_1} e^{-(u'^2 + v'^2)/2} e^{2iv' a_1' - 2iu' a_2'} \\
&= \frac{1}{(2\pi)^2} \int du' dv' e^{iu' x_2' - iv' x_1'} e^{-(u'^2 + v'^2)/2} e^{2iv' a_1' - 2iu' a_2'} \\
&= \frac{1}{2\pi} \exp\left[-\frac{1}{2}\left(\bar{x}_1'^2 + \bar{x}_2'^2 \right) \right],
\end{aligned}
\tag{3.175}
$$

where we set $x_1' \equiv x_1 e^{-r}, x_2' \equiv x_2 e^r$ and $\bar{x}_1' = x_1' - 2a_1', \bar{x}_2' = x_2' - 2a_2'$. So, the final form is

$$
W_{css}(x_1, x_2) = \frac{1}{2\pi} \exp\left[-\frac{1}{2}\left(\bar{x}_1^2 e^{-2r} + \bar{x}_2^2 e^{2r} \right) \right].
\tag{3.176}
$$

The contour of one standard deviation is $\bar{x}_1^2 e^{-2r} + \bar{x}_2^2 e^{2r} = 1$, which is an ellipse centered at $(2a_1, 2a_2)$ with major and minor axes as e^r and e^{-r}. This ellipse depicts the quantum noise for the coherent squeezed state in X-Y phase space (see Fig. 3.2). Compared with the coherent state, its noise part is squeezed along Y but stretched along X, but the displacement of $(2a_1, 2a_2)$ from the origin is the same.

(c) *Single-photon state* $|1\rangle$ *(sps)*

The characteristic function for a single-photon state is

$$C_W(u,v) = \text{Tr}\big(\hat{\rho}_{sps}e^{\eta\hat{a}^\dagger - \eta^*\hat{a}}\big) = \langle 1|e^{-(u^2+v^2)/2}e^{\eta\hat{a}^\dagger}e^{-\eta^*\hat{a}}|1\rangle$$

$$= e^{-(u^2+v^2)/2}(1 - |\eta|^2) = (1 - u^2 - v^2)e^{-(u^2+v^2)/2}. \quad (3.177)$$

Substituting into Eq. (3.167), we have the Wigner function for a single-photon state:

$$W_{sps}(x_1,x_2) = \frac{1}{2\pi}(x_1^2 + x_2^2 - 1)\exp\left[-\frac{1}{2}(x_1^2 + x_2^2)\right]. \quad (3.178)$$

Note that when $x_1 = 0 = x_2$, $W_{sps}(0,0) = -1/2\pi$, i.e., the minimum value allowed by Eq. (3.171). The negativeness of the Wigner function is an indication of nonclassical behavior of the optical field in the single-photon state.

(d) *Schrödinger cat state*

For the Schrödinger cat state of

$$|\psi\rangle_{cat} = N_r\big(|-ir\rangle + |ir\rangle\big) \quad \text{with} \quad N_r^{-2} = 2\big(1 + e^{-2r^2}\big) \quad (3.179)$$

from Eq. (3.51), the characteristic function is

$$C_W(u,v) = {}_{cat}\langle\psi|e^{-(u^2+v^2)/2}e^{\eta\hat{a}^\dagger}e^{-\eta^*\hat{a}}|\psi\rangle_{cat}$$

$$= N_r^2 e^{-(u^2+v^2)/2}\big(e^{-ir\eta}\langle ir| + e^{ir\eta}\langle -ir|\big)\big(e^{-ir\eta^*}|ir\rangle + e^{ir\eta^*}|-ir\rangle\big)$$

$$= N_r^2 e^{-(u^2+v^2)/2}\big(e^{2iru} + e^{-2iru} + e^{-2rv}e^{-2r^2} + e^{-2rv}e^{-2r^2}\big)$$

$$= 2N_r^2 e^{-(u^2+v^2)/2}\Big[\cos(2ru) + \cosh(2rv)e^{-2r^2}\Big]. \quad (3.180)$$

Then the Wigner function is

$$W_{cat}(x_1,x_2) = \frac{N_r^2}{2\pi}\Big\{e^{-x_1^2/2}\big[e^{-(x_2-2r)^2/2} + e^{-(x_2+2r)^2/2}\big]$$

$$+ e^{-x_2^2/2}e^{-2r^2}\big[e^{-(x_1-2ir)^2/2} + e^{-(x_1+2ir)^2/2}\big]\Big\}$$

$$= \frac{N_r^2}{2\pi}\Big\{e^{-x_1^2/2}\big[e^{-(x_2-2r)^2/2} + e^{-(x_2+2r)^2/2}\big]$$

$$+ 2e^{-(x_1^2+x_2^2)/2}\cos 2rx_1\Big\}. \quad (3.181)$$

It shows a strong oscillation around the origin. The marginal probability $P_{cat} = |\psi_{cat}(x_1)|^2$ is then

$$P_{cat}(x_1) = \int dx_2 W_{cat}(x_1,x_2) = \frac{2N_r^2}{\sqrt{2\pi}}e^{-x_1^2/2}(1 + \cos 2rx_1), \quad (3.182)$$

which is the same as the distribution in Eq. (3.55) but with $x_1 = x\sqrt{2}$.

In Section 10.4, we will discuss the quantum state tomography technique for the measurement of the Wigner function. An essential part of this technique is the measurement of the probability distribution for a rotated quadrature-phase amplitude $\hat{X}_\theta \equiv \hat{a}e^{-i\theta} + \hat{a}^\dagger e^{i\theta} = \hat{X}\cos\theta + \hat{Y}\sin\theta$. Let us now express it in terms of the Wigner function $W(x_1, x_2)$. For this, we need to make a change of variables from (x_1, x_2) to another set of canonical variables of (x_1^θ, x_2^θ):

$$x_1 = x_1^\theta \cos\theta - x_2^\theta \sin\theta, \quad x_2 = x_2^\theta \cos\theta + x_1^\theta \sin\theta. \tag{3.183}$$

Then we have

$$W_\theta(x_1^\theta, x_2^\theta) = W(x_1^\theta \cos\theta - x_2^\theta \sin\theta, x_2^\theta \cos\theta + x_1^\theta \sin\theta). \tag{3.184}$$

Since (x_1^θ, x_2^θ) is a set of canonical variables, the marginal probability $P(x_\theta)$ is then

$$\begin{aligned}
P(x_1^\theta) &= \int dx_2^\theta \, W_\theta(x_1^\theta, x_2^\theta) \\
&= \int dx_2^\theta \, W(x_1^\theta \cos\theta - x_2^\theta \sin\theta, x_2^\theta \cos\theta + x_1^\theta \sin\theta) \\
&= \int dx_1'^\theta dx_2^\theta \, \delta(x_1'^\theta - x_1^\theta) \\
&\quad \times W(x_1'^\theta \cos\theta - x_2^\theta \sin\theta, x_2^\theta \cos\theta + x_1'^\theta \sin\theta). \tag{3.185}
\end{aligned}$$

Making a change of variables from $(x_1'^\theta, x_2^\theta)$ back to (x_1, x_2), we have

$$P(x_1^\theta) = \int dx_1 dx_2 \delta(x_1 \cos\theta + x_2 \sin\theta - x_1^\theta) W(x_1, x_2). \tag{3.186}$$

In Section 10.4, we will make an inverse of the relation above to express the Wigner function in terms of $P(x_1^\theta)$ so that $W(x_1, x_2)$ can be obtained by measuring $P(x_1^\theta)$ with homodyne detection.

3.8 Problems

Problem 3.1 Expressing the number states in terms of the creation operators

(i) Using the commutation relation $[\hat{a}, \hat{a}^\dagger] = 1$ and the method of induction, prove

$$[\hat{a}, \hat{a}^{\dagger n}] = n\hat{a}^{\dagger n-1}, \quad [\hat{a}^n, \hat{a}^\dagger] = n\hat{a}^{n-1}. \tag{3.187}$$

(ii) Using Eq. (3.187) and $\hat{a}|0\rangle = 0$, prove

$$\hat{a}^\dagger \hat{a}(\hat{a}^{\dagger n}|0\rangle) = n\hat{a}^{\dagger n}|0\rangle, \qquad (3.188)$$

that is, $\hat{a}^{\dagger n}|0\rangle$ is the eigen-state of the photon number operator $\hat{n} \equiv \hat{a}^\dagger \hat{a}$ with an eigen-value of n.

(iii) Using Eq. (3.187) and $\hat{a}|0\rangle = 0$, prove

$$\hat{a}^n \hat{a}^{\dagger n}|0\rangle = n\hat{a}^{n-1}\hat{a}^{\dagger n-1}|0\rangle = ... = n!|0\rangle, \qquad (3.189)$$

and prove further $\langle 0|\hat{a}^n \hat{a}^{\dagger n}|0\rangle = n!$ so the normalized form of the number state is

$$|n\rangle = \frac{1}{\sqrt{n!}}\hat{a}^{\dagger n}|0\rangle. \qquad (3.190)$$

Problem 3.2 The field amplitude fluctuation of the number state and Heisenberg uncertainty relation.

Since the photon number state is the eigenstate of the photon number operator $\hat{n}$, the photon number of the optical field is completely determined, that is, $\Delta n \equiv \sqrt{\langle (\hat{n} - \langle \hat{n} \rangle)^2 \rangle} = 0$. But we see from Section 3.1.3 that the corresponding harmonic oscillator's wave function $\psi_n(q)$ has a wide range, which indicates that the position of the harmonic oscillator has a large uncertainty. We will work out its position and momentum uncertainties here.

From Eq. (3.6), we have

$$\hat{Q} = \sqrt{\frac{\hbar}{2\omega}}(\hat{a} + \hat{a}^\dagger), \qquad \hat{P} = \sqrt{\frac{\hbar\omega}{2}}(\hat{a} - \hat{a}^\dagger)/i. \qquad (3.191)$$

For the number state $|n\rangle$, prove

(i) $\langle \hat{Q} \rangle_n = 0, \langle \hat{P} \rangle_n = 0$.

(ii) $\langle \hat{Q}^2 \rangle_n = (n + 1/2)\hbar/\omega, \langle \hat{P}^2 \rangle_n = (n + 1/2)\hbar\omega$.

So, $\Delta Q_n = \sqrt{(n + 1/2)\hbar/\omega}, \Delta P_n = \sqrt{(n + 1/2)\hbar\omega}$ or $\Delta X_n = \Delta Y_n = \sqrt{2n + 1}$. Then, $\Delta Q_n \Delta P_n = (n + 1/2)\hbar > \hbar/2$, or $\Delta X_n \Delta Y_n = 2n + 1 > 1$.

Problem 3.3 The wave functions of the coherent state and the squeezed state.

In Section 3.1.3, we derived the wave functions for the number states. We can use the same method to find the wave functions for the coherent state and squeezed state. ·

(i) Using the definition of the coherent state in Eq. (3.25) and the method in Section 3.1.3, prove the wave function of the coherent state is

$$\psi_\alpha(x) = \frac{1}{(\pi)^{1/4}} e^{-\text{Im}^2(\alpha)-(x-\alpha\sqrt{2})^2/2}, \qquad (3.192)$$

where $x = q/q_0 (q_0 = \sqrt{\hbar/\omega})$.

(ii) Using the definition of the squeezed state in Eq. (3.69) and the method of Section 3.1.3, find the wave function of the squeezed state.

Problem 3.4 The phase uncertainty of the coherent state and the phase-squeezed state.

The uncertainty of the phase of the electric field can be roughly determined from the width of the zero crossing in the electric field function in Fig. 3.4. The width in Fig. 3.4(a) is zero, indicating that the phase is certain at any moment for classical waves. But the widths are finite in Figs. 3.4(b)–(c), indicating the uncertainty in the phase of quantum electromagnetic waves. Using the relation between the phase and the electric field strength in Eq. (3.85) and the uncertainty of the electric field in Eq. (3.86), find the phase uncertainties of the coherent state and the squeezed state.

Problem 3.5 Photon number fluctuation of a phase-squeezed state: $|\alpha, re^{2i\varphi_\alpha}\rangle$.

For the displacement quantity $|\alpha| >>$ the squeezing quantity r, calculate the variance $\langle \Delta^2 n\rangle$ of the phase-squeezed state of $|\alpha, re^{2i\varphi_\alpha}\rangle$ by using Eq. (3.82).

Problem 3.6 Find the photon statistics for the Schrödinger cat state in Eq. (3.51). Calculate the variance $\langle \Delta^2 n\rangle$ for this state.

Problem 3.7 Find the Wigner function for the thermal state.

Problem 3.8 Photon statistics for squeezed vacuum states.

The general formula for the photon statistics of squeezed states is given by Eqs. (3.94) and (3.95) through some rather cumbersome derivation. We will present the outline of a simpler derivation for the special case of squeezed vacuum states.

For the squeezed vacuum state with maximum squeezing in $\hat{X} = \hat{a} + \hat{a}^\dagger$, the state has the form of

$$| - r\rangle = \hat{S}(-r)|0\rangle = \hat{S}^\dagger(r)|0\rangle \quad \text{with} \quad r > 0$$
$$= \sum_n c_n |n\rangle \quad \text{with} \quad c_n \equiv \langle n| - r\rangle. \tag{3.193}$$

(i) By using Eq. (3.62), show

$$(\mu\hat{a} + \nu\hat{a}^\dagger)| - r\rangle = 0, \quad \text{with} \quad \mu = \cosh r, \ \nu = \sinh r. \tag{3.194}$$

(ii) By applying $\langle n|$ to Eq. (3.194) from the left, show the recursive relation:

$$\mu c_{n+1}\sqrt{n+1} + \nu c_{n-1}\sqrt{n} = 0 \tag{3.195}$$

with c_n defined in Eq. (3.193). A special case of $n = 0$ gives $c_1 = 0$.

(iii) By using the result in (ii), show

$$c_n = \begin{cases} 0 & \text{for } n = 2k - 1 \\ (-1)^k c_0 \sqrt{\frac{(2k)!}{(2^k k!)^2}} \left(\frac{\nu}{\mu}\right)^k & \text{for } n = 2k \end{cases} \tag{3.196}$$

where $|c_0|^2 = 1/\mu$ is determined by normalization with the identity

$$\left(1 - \frac{\nu^2}{\mu^2}\right)^{-1/2} = \sum_k \frac{(2k)!}{(2^k k!)^2} \left(\frac{\nu}{\mu}\right)^{2k}. \tag{3.197}$$

$P_n = |c_n|^2$ gives Eq. (3.94) with $\alpha = 0$ for squeezed vacuum state. Based on this result and the principle of multi-photon interference, we will show later in Problem 8.5 that the case of $\alpha \neq 0$ (Eq. (3.94)) is the result of multi-photon interference between a coherent state and a squeezed vacuum state.

Chapter 4

Quantum States of Multi-Mode Fields

When multiple modes of an optical field are excited, we must describe it with a multi-mode quantum state. If all the modes are independent like what is given in the Hamiltonian in Eq. (2.76), we can use single-mode quantum states described in the previous chapter to describe each mode separately. The quantum state of the whole field can be represented by a direct product of all the single-mode states. However, if the modes are correlated, the quantum state for the whole field cannot be written as a direct product of the single-mode states but must be the superposition of the different direct product states. This leads to the concept of entanglement. This is the main emphasis of this chapter.

4.1 Multi-Mode Coherent States of Independent Modes

Let us start with the simplest case of a multi-mode field with all modes independently excited. Since modes are independent, we can use the single-mode quantum states for each mode and the quantum state for the whole is a direct product of these single-mode states:

$$\hat{\rho}_{sys} = \prod_{\lambda} \otimes \hat{\rho}_{\lambda}, \tag{4.1}$$

where $\hat{\rho}_{\lambda}$ is the density operator for describing mode λ. For example, if the optical is in coherent states with its modes excited independently, the quantum state of the field is described by a multi-mode coherent state which is a direct product of the coherent states of each excited mode:

$$|\psi_{sys}\rangle = \prod_{\lambda} \otimes |\alpha_{\lambda}\rangle_{\lambda} \equiv |\{\alpha_{\lambda}\}\rangle. \tag{4.2}$$

For the two-mode case, we have

$$|\psi_{sys}\rangle = |\alpha\rangle_1 \otimes |\beta\rangle_2 \equiv |\alpha, \beta\rangle. \tag{4.3}$$

107

The operators of each mode only act on the state of its own mode:

$$\hat{a}_1|\alpha,\beta\rangle = (\hat{a}_1|\alpha\rangle_1) \otimes |\beta\rangle_2 = (\alpha|\alpha\rangle_1) \otimes |\beta\rangle_2 = \alpha|\alpha,\beta\rangle. \tag{4.4}$$

Similarly,

$$\hat{a}_\mu|\{\alpha_\lambda\}\rangle = \alpha_\mu|\{\alpha_\lambda\}\rangle. \tag{4.5}$$

For a more complicated field operator such as the electric field operator in Eq. (2.90):

$$\hat{\mathbf{E}}(\mathbf{r},t) = \hat{\mathbf{E}}^{(+)}(\mathbf{r},t) + \hat{\mathbf{E}}^{(-)}(\mathbf{r},t), \tag{4.6}$$

where $\left[\hat{\mathbf{E}}^{(-)}(\mathbf{r},t)\right]^\dagger = \hat{\mathbf{E}}^{(+)}(\mathbf{r},t)$ and

$$\hat{\mathbf{E}}^{(+)}(\mathbf{r},t) \equiv i\sqrt{4\pi} \sum_{s=1,2} \int d^3k \sqrt{\frac{\hbar\omega}{2}} \hat{a}_s(\mathbf{k})\hat{\epsilon}_{\mathbf{k},s} \frac{e^{i(\mathbf{k}\cdot\mathbf{r}-\omega t)}}{(2\pi)^{3/2}}, \tag{4.7}$$

we have

$$\hat{\mathbf{E}}^{(+)}(\mathbf{r},t)|\{\alpha_\lambda\}\rangle = i\sqrt{4\pi} \sum_{s=1,2} \int d^3k \sqrt{\frac{\hbar\omega}{2}} \hat{\epsilon}_{\mathbf{k},s} \frac{e^{i(\mathbf{k}\cdot\mathbf{r}-\omega t)}}{(2\pi)^{3/2}} \left(\hat{a}_s(\mathbf{k})|\{\alpha_\lambda\}\rangle\right)$$

$$= \left(i\sqrt{4\pi} \sum_{s=1,2} \int d^3k \sqrt{\frac{\hbar\omega}{2}} \hat{\epsilon}_{\mathbf{k},s}\alpha_s(\mathbf{k}) \frac{e^{i(\mathbf{k}\cdot\mathbf{r}-\omega t)}}{(2\pi)^{3/2}}\right) |\{\alpha_\lambda\}\rangle$$

$$\equiv \vec{\mathcal{E}}(\mathbf{r},t)|\{\alpha_\lambda\}\rangle, \tag{4.8}$$

where

$$\vec{\mathcal{E}}(\mathbf{r},t) \equiv i\sqrt{4\pi} \sum_{s=1,2} \int d^3k \sqrt{\frac{\hbar\omega}{2}} \hat{\epsilon}_{\mathbf{k},s}\alpha_s(\mathbf{k}) \frac{e^{i(\mathbf{k}\cdot\mathbf{r}-\omega t)}}{(2\pi)^{3/2}}. \tag{4.9}$$

The expression above is the same as the classical expression of the electric field obtained from Eqs. (2.14) and (2.66) with $q_{\mathbf{k},s}(0) = \alpha_s(\mathbf{k})\sqrt{\hbar/2\omega}$. This relation for $q_{\mathbf{k},s}$ is the same as Eqs. (2.68) and (2.74) after the quantization. Here we make an exchange of $\hat{a} \leftrightarrow \alpha$.

In the photo-detection theory of Chapter 5, we have the probability of finding photo-electrons in optical detectors proportional to the intensity of the field: $I = \langle\hat{I}\rangle$ with the intensity operator as $\hat{I} \equiv \hat{\mathbf{E}}^{(-)}(\mathbf{r},t) \cdot \hat{\mathbf{E}}^{(+)}(\mathbf{r},t)$. For the multi-mode coherent state, we can easily obtain

$$I = \langle\hat{I}\rangle_\alpha = \langle\{\alpha_\lambda\}|\hat{\mathbf{E}}^{(-)}(\mathbf{r},t) \cdot \hat{\mathbf{E}}^{(+)}(\mathbf{r},t)|\{\alpha_\lambda\}\rangle = \vec{\mathcal{E}}^*(\mathbf{r},t) \cdot \vec{\mathcal{E}}(\mathbf{r},t). \tag{4.10}$$

But in classical wave optics, the intensity I of the field is defined as the absolute square of the amplitude of the electric field. So, $\vec{\mathcal{E}}(\mathbf{r},t)$ corresponds exactly to the amplitude of the electric field in classical optics. Therefore, we also claim that a multi-mode coherent state is a quantum state whose description of the field is closest to the description of classical optics. Next, based on this connection, let us discuss some more about the description of a multi-mode field in classical wave optics.

4.2   Classical Description of Multi-Mode Optical Fields

In Section 1.5, we mentioned that the fluctuations of an electromagnetic field can be represented by a random process $E(\mathbf{r}, t)$ and a random process is a time function of some random variables $\{X_j\}$: $E(\mathbf{r}, t) = f(\{X_j\}, t)$. But no matter what it is, this function must satisfy Maxwell equations for electromagnetic fields. So, we can express it in terms of the modes of the electromagnetic field:

$$\vec{E}(\mathbf{r}, t) = \sum_{\mathbf{k},s} E_{\mathbf{k},s} \hat{\epsilon}_{\mathbf{k},s} u_{\mathbf{k},s}(\mathbf{r}) e^{-i\omega t} + c.c. \qquad (4.11)$$

Compared with Eq. (4.9), $E_{\mathbf{k},s}$ corresponds to the coherent state excitations $\alpha_s(\mathbf{k})$ of the modes. From Eq. (4.11), we find that only $E_{\mathbf{k},s}$ can be arbitrary. If the electromagnetic field is to be described by a random process, only $E_{\mathbf{k},s}$ can be the random variables. Hence, the classical description of a multi-mode field is based on the description of the random variables $E_{\mathbf{k},s}$. This is the starting point of the classical coherence theory.

For simplicity, let us only consider one-dimensional scalar fields with a complex representation:

$$
\begin{aligned}
E(\mathbf{r}, t) &= \frac{1}{\sqrt{2\pi}} \int d\omega E(\omega) e^{i(kz - \omega t)} \\
&= \frac{1}{\sqrt{2\pi}} \int d\omega E(\omega) e^{-i\omega(t - z/c)},
\end{aligned} \qquad (4.12)
$$

where $k = \omega/c$. Defining $\tau \equiv t - z/c$, one-dimensional scalar field is only a function of time:

$$E(\mathbf{r}, t) = E(\tau) = \frac{1}{\sqrt{2\pi}} \int d\omega E(\omega) e^{-i\omega\tau}, \qquad (4.13)$$

where $E(\omega)$ is a random variable similar to $E_{\mathbf{k},s}$ and it can be expressed by $E(\tau)$ as

$$E(\omega) = \frac{1}{\sqrt{2\pi}} \int d\omega E(\tau) e^{i\omega\tau}. \qquad (4.14)$$

$E(\omega)$ is the frequency component of the optical field, whose probability distribution $P(\{E(\omega)\}) = P(E(\omega_1), E(\omega_2), ..., E(\omega_j), ...)$ determines the statistical properties of the field.

4.2.1   *Continuous Waves (CW) and Stationary Processes*

Consider the correlation function of the optical field:

$$\Gamma(t, \tau) \equiv \langle E^*(t) E(t + \tau) \rangle = \int d\{E(\omega)\} P(\{E(\omega)\}) E^*(t) E(t + \tau), \qquad (4.15)$$

Quantum Optics For Experimentalists

where random process $E(t)$ is related to the random variables $\{E(\omega)\}$ through Eq. (4.13). For continuous waves, the starting time is irrelevant, so, $\Gamma(t, \tau) = \Gamma(\tau)$ is independent of time t and continuous waves are usually described by stationary processes. Furthermore, continuous waves are often ergodic, that is, in a sufficiently long time, they include or run through all possible allowed values. Therefore, the average over the probability distribution in Eq. (4.15) can be replaced by an average over time:

$$\Gamma(\tau) \equiv \langle E^*(t)E(t+\tau) \rangle = \lim_{T \to \infty} \frac{1}{T} \int_T dt E^*(t)E(t+\tau). \qquad (4.16)$$

One special case of Eqs. (4.15) and (4.16) is $\tau = 0$: $\Gamma(0) = \langle E^*(t)E(t) \rangle$. This is the intensity I of the optical field. Using Eq. (4.13), we obtain

$$\Gamma(\tau) = \frac{1}{2\pi} \int d\omega_1 d\omega_2 \langle E^*(\omega_1)E(\omega_2) \rangle e^{i(\omega_1 - \omega_2)t} e^{-i\omega_2 \tau}. \qquad (4.17)$$

Since $\Gamma(\tau)$ is unrelated to time t for continuous stationary waves, we must have

$$\langle E^*(\omega_1)E(\omega_2) \rangle = 2\pi S(\omega_1)\delta(\omega_1 - \omega_2), \qquad (4.18)$$

where $S(\omega_1)$ is some function of ω_1. The expression above shows that there is no phase relation between different frequency components of a continuous stationary field since $\langle E^*(\omega_1)E(\omega_2) \rangle = 0$ for $\omega_1 \neq \omega_2$. Hence, Eq. (4.17) changes to

$$\Gamma(\tau) = \int d\omega S(\omega) e^{-i\omega\tau} \qquad (4.19)$$

or

$$S(\omega) = \frac{1}{2\pi} \int d\tau \Gamma(\tau) e^{i\omega\tau}. \qquad (4.20)$$

From Eq. (4.19), we obtain $I = \Gamma(0) = \int d\omega S(\omega)$. So, $S(\omega)$ is regarded as the spectral function of the optical field. The statistical relation in Eq. (4.18) is a signature of continuous stationary fields.

4.2.2 Pulsed Waves and Non-Stationary Processes

For a pulsed field, we can define an instantaneous intensity: $I(t) \equiv E^*(t)E(t)$. But if the detector is not as fast as the pulse, it is impossible to observe the shape of the pulse and all it sees is an integration over time, i.e., the total energy of the pulse:

$$I = \int_{-\infty}^{\infty} dt E^*(t)E(t). \qquad (4.21)$$

Similarly, the correlation function in Eq. (4.16) becomes

$$\Gamma(\tau) = \int dt E^*(t)E(t+\tau)$$

$$= \frac{1}{2\pi} \int dt d\omega_1 d\omega_2 E^*(\omega_1)E(\omega_2)e^{i(\omega_1-\omega_2)t}e^{-i\omega_2\tau}$$

$$= \int d\omega_1 d\omega_2 E^*(\omega_1)E(\omega_2)\delta(\omega_1-\omega_2)e^{-i\omega_2\tau}$$

$$= \int d\omega |E(\omega)|^2 e^{-i\omega\tau}$$

$$= \int d\omega S(\omega)e^{-i\omega\tau}, \tag{4.22}$$

where we used $\frac{1}{2\pi}\int dt\, e^{i\omega t} = \delta(\omega)$. The last line is the same as Eq. (4.19). So, the spectral relation in Eq. (4.20) stands for pulsed light field, too.

4.2.3   Coherence of Optical Fields – Phase Correlations

In the coherence theory, the normalized field correlation function

$$\gamma(\tau) \equiv \Gamma(\tau)/\Gamma(0) = \langle E^*(t)E(t+\tau)\rangle/\langle |E_1(t)|^2\rangle, \tag{4.23}$$

describes the correlation of the phases of the optical field at different times. The degree of correlation of the phases determines the visibility of interference fringes and therefore the coherence of the optical field, as we have seen in Eq. (1.37) of Section 1.5. The coherence time T_c of an optical field is defined as the range within which $\gamma(\tau)$ is appreciably different from zero (see Fig. 4.1).

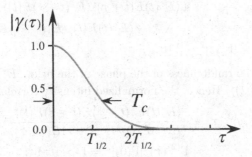

Fig. 4.1   Correlation function $|\gamma(\tau)|$ and the coherence time T_c.

We can also define the cross-correlation function between two optical fields:

$$\Gamma_{12}(\tau) \equiv \langle E_1^*(t)E_2(t+\tau)\rangle, \tag{4.24}$$

and the normalized cross-correlation function:

$$\gamma_{12}(\tau) \equiv \frac{\Gamma_{12}(\tau)}{\sqrt{\Gamma_{11}(0)\Gamma_{22}(0)}} = \frac{\langle E_1^*(t)E_2(t+\tau)\rangle}{\sqrt{\langle |E_1(t)|^2\rangle\langle |E_2(t)|^2\rangle}}. \qquad (4.25)$$

It corresponds to the phase correlation between two optical fields and gives the visibility of the interference fringes for the interference between the two fields. When the two fields correspond to the same optical field but at two different locations, we obtain the spatial coherence function, as we first introduced in Eq. (1.36) of Section 1.5.

4.2.4 Hanbury Brown and Twiss Effect – Intensity Correlations

A more complicated correlation function than Eq. (4.15) is the higher order intensity correlation function of the optical field:

$$\Gamma^{(2)}(\tau) \equiv \langle I(t)I(t+\tau)\rangle = \langle E^*(t)E^*(t+\tau)E(t+\tau)E(t)\rangle. \qquad (4.26)$$

For a continuous stationary optical field, it is independent of time t. It corresponds to the correlation of the intensity of the optical field at different times. For a thermal source with Gaussian statistical distribution, we can apply the Gaussian moment theorem (see Isserlis theorem in Eq. (1.28)):

$$\langle ABCD\rangle = \langle AB\rangle\langle CD\rangle + \langle AC\rangle\langle BD\rangle + \langle AD\rangle\langle BC\rangle. \qquad (4.27)$$

Then Eq. (4.26) becomes

$$\begin{aligned}
\Gamma^{(2)}(\tau) &= \langle E^*(t)E^*(t+\tau)\rangle\langle E(t+\tau)E(t)\rangle \\
&\quad + \langle E^*(t)E(t+\tau)\rangle\langle E^*(t+\tau)E(t)\rangle \\
&\quad\quad + \langle E^*(t)E(t)\rangle\langle E^*(t+\tau)E(t+\tau)\rangle \\
&= |\Gamma(\tau)|^2 + |\Gamma(0)|^2, \qquad\qquad\qquad\qquad (4.28)
\end{aligned}$$

where, due to the randomness of the phase of the field, $\langle E^*(t)E^*(t+\tau)\rangle = 0 = \langle E(t+\tau)E(t)\rangle$. Hence, the normalized intensity correlation function is

$$\begin{aligned}
g^{(2)}(\tau) &\equiv \frac{\langle E^*(t)E^*(t+\tau)E(t+\tau)E(t)\rangle}{\langle E^*(t)E(t)\rangle\langle E^*(t+\tau)E(t+\tau)\rangle} \\
&= \Gamma^{(2)}(\tau)/|\Gamma(0)|^2 = 1 + |\gamma(\tau)|^2, \qquad (4.29)
\end{aligned}$$

where $\gamma(\tau)$ is the temporal coherence function given in Eq. (4.23). This gives the complete explanation of the Hanbury Brown-Twiss photon bunching effect by the classical wave theory, which gives not only $g^{(2)}(0) = 2 > 1$ for photon bunching effect, as we have done in Eq. (1.42) of Section 1.6, but also its temporal dependence on the time delay τ. For the temporal

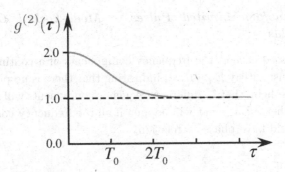

Fig. 4.2 $g^{(2)}(\tau)$ as a function of the time delay τ. It shows the photon bunching effect.

coherence function $|\gamma(\tau)|$ depicted in Fig. 4.1, the expression in Eq. (4.29) gives $g^{(2)}(\tau)$ as a specific function of τ shown in Fig. 4.2, which agrees well with the experimental observation shown in Fig. 1.9.

In general, applying the Cauchy-Schwarz inequality:

$$|\langle AB \rangle|^2 \leq \langle |A|^2 \rangle \langle |B|^2 \rangle \qquad (4.30)$$

to arbitrary optical fields described by classical wave theory, we obtain

$$g^{(2)}(0) \geq g^{(2)}(\tau) \quad \text{or} \quad g^{(2)}(0) \geq g^{(2)}(\infty) = 1. \qquad (4.31)$$

Hence, the classical wave theory always gives rise to the photon bunching effect. In the multi-mode description of a thermal field, which has a Gaussian statistical distribution for all the modes, the size of the photon bunching effect, which is characterized by the excess correlation $g^{(2)} - 1$, is usually related to the average number of modes M of the detected field (see Eq. (4.84) of Problem 4.4).

On the other hand, quantum fields can exhibit photon anti-bunching effect, i.e., $g^{(2)}(0) < 1$, as we have seen in Section 1.7. We will discuss more about this in Section 5.3 of Chapter 5.

For two optical fields, the correlation function between the intensities of the two fields is

$$\Gamma_{12}^{(2)}(\tau) \equiv \langle I_1(t) I_2(t+\tau) \rangle = \langle E_1^*(t) E_2^*(t+\tau) E_2(t+\tau) E_1(t) \rangle. \quad (4.32)$$

The corresponding Cauchy-Schwarz inequality for this quantity is

$$\left[\Gamma_{12}^{(2)}(0) \right]^2 \leq \Gamma_{11}^{(2)}(0) \Gamma_{22}^{(2)}(0). \qquad (4.33)$$

Later in Section 7.4 of Chapter 7, we will show that this inequality is violated by two optical fields with quantum correlation.

4.2.5   Transform-Limited Pulses – Mode-Locked Optical Fields

As we discussed earlier, the frequency components of a continuous stationary field must satisfy Eq. (4.18), indicating that there is no phase correlation between them, that is, different frequency components will not interfere with each other.[1] But what will happen if all the frequency components of an optical field have phase correlation.

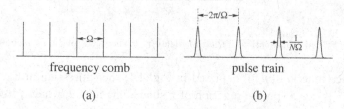

frequency comb   pulse train

(a)   (b)

Fig. 4.3   (a) A comb-like frequency distribution; (b) A pulse train in the time domain.

For simplicity, let us consider a comb-like frequency distribution, as shown in Fig. 4.3(a). Suppose that there are $2N+1$ frequency components of equal height and they are separated by the same quantity Ω. If they all have the same phase, the optical field has the form of

$$E(t) = \sum_{n=-N}^{N} E_0 e^{-i(\omega_0 t + n\Omega)t} = E_0 e^{-i\omega_0 t} \frac{\sin N\Omega t}{\sin \Omega t}. \qquad (4.34)$$

$E_0 = |E_0|e^{i\varphi_0}$ is the complex amplitude of each frequency component. Figure 4.3(b) shows the field function in time domain: it is an infinite pulse train with equal spacing of $2\pi/\Omega$.

For an optical field with a continuous spectrum, if the relationship among all frequency components is fixed, that is, $E(\omega) = A_0 \mathcal{E}(\omega)$ where $\mathcal{E}(\omega)$ is a definite function but A_0 may be a complex random variable, then the optical field has a definite pulse shape $\mathcal{E}(t)$:

$$E(t) = A_0 \int d\omega \mathcal{E}(\omega) e^{-i\omega t} \equiv A_0 \mathcal{E}(t). \qquad (4.35)$$

Its correlation function is

$$\Gamma(t_1, t_2) = \langle E^*(t_1) E(t_2) \rangle = \langle |A_0|^2 \rangle \mathcal{E}^*(t_1) \mathcal{E}(t_2) \qquad (4.36)$$

so the normalized correlation function is

$$\gamma(t_1, t_2) = \Gamma(t_1, t_2)/\sqrt{\Gamma(t_1, t_1)\Gamma(t_2, t_2)} = 1. \qquad (4.37)$$

---

[1]Interference between fields of different frequencies shows up in the form of temporal beating.

Hence, the field within the optical pulse is completely coherent. We find from Eq. (4.35) that the temporal function $\mathcal{E}(t)$ of the field and its spectral function $\mathcal{E}(\omega)$ are a Fourier transformation pair. So, we call this type of field a transform-limited optical pulse, whose spectral width $\Delta\omega$ and pulse width satisfy $\Delta T \Delta\omega = 2\pi$. Therefore, $T_c = 1/\Delta\nu = 2\pi/\Delta\omega = \Delta T$, that is, the coherence time T_c is equal to the pulse length ΔT for a transform-limited pulse. This is consistent with the fact presented in Eq. (4.37) that any two points inside a transform-limited pulse are completely coherent. On the other hand, a non-transform-limited optical pulse has its coherence time T_c always smaller than the pulse length ΔT. Figure 4.4 shows the physical picture for these two cases.

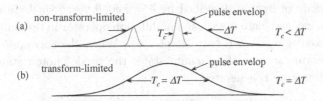

Fig. 4.4 (a) A non-transform-limited pulse with its coherence time $T_c <$ the pulse length ΔT; (b) A transform-limited pulse with its coherence time $T_c =$ the pulse length ΔT.

4.3 Multi-Mode Single-Photon States – Entangled States of Single Photon

Now let us come back to the description of the quantum states of a multi-mode field. We first consider the single-photon case. This photon may be in different modes. Because there is only one photon, this photon must be in a superposition state of the single-photon states for different modes. It has the form of

$$|1\rangle_m = \sum_i c_i |1_i\rangle, \tag{4.38}$$

where $|1_i\rangle \equiv |0\rangle_1...|0\rangle_{i-1}|1\rangle_i|0\rangle_{i+1}...$ is the quantum state with only mode i having one photon excited and other modes in vacuum. For the modes in vacuum, we can omit them in writing the state in Eq. (4.38). Since this state cannot be written as a direct product of the states for each mode, i.e., $\prod_i \otimes |\psi\rangle_i$, it is a single-photon entangled state. We will examine a couple of specific multi-mode single-photon states next.

4.3.1 Two-Mode Single-Photon States

Two-mode single-photon state is the simplest entangled state of photons. There are two modes of the fields being excited but in the meantime there is only one photon. It can be written as

$$|1\rangle_{AB} = c_1 |1\rangle_A |0\rangle_B + c_2 |0\rangle_A |1\rangle_B. \tag{4.39}$$

When a single-photon in a single-mode is incident on a lossless beam splitter, the output state is a two-mode single-photon state (for quantum treatment of a beam splitter, see Section 6.2 in Chapter 6). Another example of two-mode single-photon state is a photon in an arbitrary but defined polarization state. From classical optics, we know that an arbitrary polarization state of light is described by $\hat{\epsilon} = \hat{x}\cos\theta + \hat{y}e^{i\delta}\sin\theta$. In quantum optics, we can associate it with an annihilation operator in the similar form: $\hat{a}_{\hat{\epsilon}} = \hat{a}_x \cos\theta + \hat{a}_y e^{i\delta}\sin\theta$ (see Chapter 6). When the corresponding creation operator acts on the vacuum state, the single-photon state in this polarization state is generated:

$$|1\rangle_{\hat{\epsilon}} = \hat{a}_{\hat{\epsilon}}^\dagger |0\rangle = \cos\theta|1\rangle_x|0\rangle_y + e^{-i\delta}\sin\theta|0\rangle_x|1\rangle_y. \tag{4.40}$$

Setting $\delta = 0$ gives a single-photon state with a linear polarization along θ-direction: $|1\rangle_\theta = \cos\theta|1\rangle_x + \sin\theta|1\rangle_y$. Here, we do not write out the modes in vacuum. $\delta = \pm\pi/2, \theta = \pi/4$ corresponds to a single-photon state in left or right circular polarization $|1\rangle_\pm = (|1\rangle_x \pm i|1\rangle_y)/\sqrt{2}$.

4.3.2 Multi-Frequency Single-Photon States –
Single-Photon Wave Packets

Another multi-mode single-photon state is the multi-frequency single-photon state. Assume all the excited modes have the same polarization and spatial mode function and the only difference is their frequencies. Such a photon state has the form of

$$|1\rangle_{mf} = \sum_i c_i |1_{\omega_i}\rangle = \sum_i c_i \hat{a}_{\omega_i}^\dagger |0\rangle \equiv \hat{A}^\dagger |0\rangle, \tag{4.41}$$

where $\hat{A} \equiv \sum_i c_i^* \hat{a}_{\omega_i}$. Since $|1\rangle_{mf}$ is normalized, i.e., $\sum_i |c_i|^2 = 1$, we have

$$[\hat{A}, \hat{A}^\dagger] = \sum_{i,j} c_i^* c_j [\hat{a}_{\omega_i}, \hat{a}_{\omega_j}^\dagger] = \sum_{i,j} c_i^* c_j \delta_{ij} = \sum_i |c_i|^2 = 1. \tag{4.42}$$

For a continuous spectral distribution, we have $|1\rangle_{mf} = \int d\omega\, c(\omega)\hat{a}^\dagger(\omega)|0\rangle$ with $\hat{a}^\dagger(\omega)$ satisfying commutation relation $[\hat{a}(\omega), \hat{a}^\dagger(\omega')] = \delta(\omega - \omega')$ for the continuous variables (see Section 2.3.5). A special case is $c(\omega) = \phi(\omega)e^{i\omega T}(\phi(\omega) = \text{real})$. It corresponds to a single-photon wave packet:

$$|T\rangle_\phi = \int d\omega \phi(\omega)e^{i\omega T}\hat{a}^\dagger(\omega)|0\rangle \equiv \hat{A}_\phi^\dagger(T)|0\rangle, \qquad (4.43)$$

where $A_\phi(T) = \int d\omega \phi(\omega)e^{-i\omega T}\hat{a}(\omega)$ and satisfies $[A_\phi(T), A_\phi^\dagger(T)] = 1$ due to normalization $\int d\omega |\phi(\omega)|^2 = 1$. The reason that Eq. (4.43) describes a single-photon wave packet is because the probability density of photon detection at time t is (see Section 2.1.5)

$$R(t) = \,_\phi\langle T|\hat{E}^\dagger(t)\hat{E}(t)|T\rangle_\phi = |g(t-T)|^2, \qquad (4.44)$$

where

$$\hat{E}(t) = \frac{1}{\sqrt{2\pi}}\int d\Omega \,\hat{a}(\Omega)e^{-i\Omega t}, \quad g(t) = \frac{1}{\sqrt{2\pi}}\int d\omega \phi(\omega)e^{-i\omega t}. \quad (4.45)$$

For a well-behaved function $\phi(\omega)$, $g(t-T)$ is a wave packet centered at time T (see Fig. 4.5).

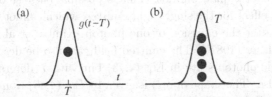

(a) $g(t-T)$ (b) T

Fig. 4.5   (a) Wave packet for a photon in a single temporal mode; (b) N-photon single temporal mode wave packet.

As discussed in Section 2.2.2, a generalized mode is a linear combination of several different modes. Although it consists of multiple modes, its mode function is still a solution of Maxwell equations. Hence, it represents a generalized mode and the annihilation operator $\hat{A}$ or $\hat{A}_\phi(T)$ is the annihilation operator for the photon in this generalized mode after quantization. It gives a quantum description for this generalized mode. The quantum state generated by the action of the corresponding creation operator on vacuum is thus a single-mode single-photon state in this generalized sense. $\hat{A}_\phi(T)$ describes a single temporal mode with a pulse shape of $g(t)$, similar to the operator $\hat{a}_{\hat{e}}$ for the polarization mode with an arbitrary polarization $\hat{e}$. The coherent state corresponding to mode $\hat{A}_\phi^\dagger(T)$ is $|\alpha_\phi(T)\rangle = \hat{D}_{A_\phi}|0\rangle$ $(\hat{D}_{A_\phi} \equiv e^{\alpha\hat{A}_\phi^\dagger(T)-\alpha^*\hat{A}_\phi(T)})$, which describes a laser pulse of shape $g(t - T)$. An N-photon state of single temporal mode is described by the multi-photon state (see more in Section 8.4.1 of Chapter 8):

$$|N, T\rangle_\phi = \frac{1}{\sqrt{N!}} [\hat{A}_\phi^\dagger(T)]^N |0\rangle. \tag{4.46}$$

Because the N photons are in one temporal mode, they are completely indistinguishable and will produce a maximum multi-photon interference effect (see Section 8.4 of Chapter 8). This N-photon state can be produced probabilistically from single-photon states with the help of beam splitters (see Section 8.4.1).

There are two practical ways for the generation of single-photon states. The first one is to use the commonly available two-photon states generated from spontaneous parametric processes. One of the two correlated photons becomes a temporal single-photon state when heralded on the detection of the other photon [Hong and Mandel (1986)]. In this case, we can use the temporal single-photon state in Eq. (4.43) to describe it with T determined by the detection time of the other photon. The second method is to use a short laser pulse and coherently drive a single light emitter such as an atom, an ion, and a quantum dot. This is possible because of the photon anti-bunching effect for the single light emitter, which cannot emit another photon right after the emission of one photon [Kimble et al. (1977); Diedrich and Walther (1987)]. The emitted field can also be described by the temporal single-photon state in Eq. (4.43) but with T determined by the short laser pulse. The shape of $g(t)$ is given by Eq. (4.45) from the spectral function $\phi(\omega)$ of the single light emitter.

4.4   Multi-Mode Two-Photon States – Two-Photon Entangled State

Even though the multi-photon state in Eq. (4.46) is written in the form of multi-frequency mode, it is, as we mentioned earlier, still describes in essence a single-mode field, i.e., a single temporal mode field. In this section, we will discuss some genuine multi-mode multi-photon states, that is, no matter how it is written, the photon state must be described by multiple modes. On the other hand, if we can arrange the mode structure so that each mode is independently excited, we can use single-mode method to treat each mode individually and the whole system is a direct product of all the single-mode states, as we did in Section 4.1. But if a multi-mode state is in some form of superposition, it usually is an entangled state, which cannot be written as a product of the states of independent modes. In this case, the modes are not independent but correlated. We must treat all modes together. So, the multi-mode photon states in this section are usually a superposition state of multi-mode multi-photon states. Let us consider a two-mode case first, say, a two-photon state of polarization.

4.4.1 Two-Photon Polarization States

For the case of two photons in two polarization modes of $\hat{x}, \hat{y}$, we have three base-states: $|2_x, 0_y\rangle, |0_x, 2_y\rangle, |1_x, 1_y\rangle$. The superposition state of the three states with equal weight is

$$|2xy\rangle = \left(|2\rangle_x|0\rangle_y + |1\rangle_x|1\rangle_y + |0\rangle_x|2\rangle_y\right)/\sqrt{3}. \qquad (4.47)$$

This is so-called W-state [Dur (2001)]. Another two-mode two-photon state is the superposition state of maximum photon number:

$$|NOON(2)\rangle = \left(|2\rangle_x|0\rangle_y + |0\rangle_x|2\rangle_y\right)/\sqrt{2}. \qquad (4.48)$$

This is the so-called two-photon NOON state [Ou <u>et al.</u> (1990b); Ou (1997a); Boto <u>et al.</u> (2000)], which has important application in precision phase measurement (see Chapter 11). These two states cannot be written in the form of Eq. (4.46) so they are some sort of entangled states.

4.4.2 Two-Parti Two-Photon States of Polarization – Bell States

Next, we extend the two-photon polarization state to two-parti system of A and B. Now there are four modes: xA, yA, xB, yB. Consider only the cases of one parti each having one photon. We have four possibilities of $|1_x\rangle_A|1_x\rangle_B, |1_y\rangle_A|1_y\rangle_B, |1_x\rangle_A|1_y\rangle_B, |1_y\rangle_A|1_x\rangle_B$. A set of superposition states of the four are:

$$|\Psi_\pm\rangle = \left(|1_x\rangle_A|1_y\rangle_B \pm |1_y\rangle_A|1_x\rangle_B\right)/\sqrt{2}, \qquad (4.49)$$

$$|\Phi_\pm\rangle = \left(|1_x\rangle_A|1_x\rangle_B \pm |1_y\rangle_A|1_y\rangle_B\right)/\sqrt{2}. \qquad (4.50)$$

Here the absolute value of every superposition coefficient is the same. They form the base states for the sub-space of two-parti two-photon polarization states with A and B each having a photon. The four states in Eqs. (4.49) and (4.50) are known as the Bell states. They can violate Bell's inequalities and play an important role in the test of nonlocality of quantum mechanics [Bell (1987)]. The four Bell states are also the basis for the teleportation of a single-photon state of arbitrary polarization [Bennett <u>et al.</u> (1993); Braunstein and Mann (1996); Bouwmeester <u>et al.</u> (1997)]. Problem 8.6 discusses more interesting properties of Bell states (Eq. (8.124)).

4.4.3 Multi-Frequency Two-Photon States – Frequency-Entangled States and Time-Entangled States

Consider now the frequency modes. An interesting one is the two-frequency two-photon state of two-parti system of A and B:

$$|2(\omega_1, \omega_2)\rangle = \left(|1_{\omega_1}\rangle_A |1_{\omega_2}\rangle_B + |1_{\omega_2}\rangle_A |1_{\omega_1}\rangle_B\right)/\sqrt{2}. \tag{4.51}$$

Again, each parti has one photon. This is a frequency-entangled state. When making a direct time-resolved two-photon coincidence measurement, we can observe a two-photon beating effect [Legero et al. (2004)]. Injecting it into a Hong-Ou-Mandel interferometer (Section 8.2.1), we can observe a spatial beating effect [Ou and Mandel (1988); Li et al. (2009)]. For a two-photon state with a continuous spectrum, we have

$$|\Psi_2\rangle = \int_{\Delta\omega} d\omega_1 d\omega_2 \Psi(\omega_1, \omega_2) \hat{a}_A^\dagger(\omega_1) \hat{a}_B^\dagger(\omega_2) |0\rangle, \tag{4.52}$$

which can be obtained from spontaneous parametric down-conversion processes (see Section 6.1.5). When the pump field to the parametric process is a continuous wave with single frequency, we have $\Psi(\omega_1, \omega_2) = V_p \psi(\omega_1) \delta(\omega_1 + \omega_2 - \omega_p)$ (see Section 6.1.5). Hence, Eq. (4.52) is changed to

$$\begin{aligned}|\Psi_2(CW)\rangle &= \int d\omega_1 d\omega_2 \delta(\omega_1 + \omega_2 - \omega_p) \psi(\omega_1) \hat{a}_A^\dagger(\omega_1) \hat{a}_B^\dagger(\omega_2) |0\rangle \\ &= \int d\omega_1 d\omega_2 V_p \delta(\omega_1 + \omega_2 - \omega_p) \\ &\quad \times \sqrt{\psi(\omega_1)}\sqrt{\psi(\omega_p - \omega_2)} \hat{a}_A^\dagger(\omega_1) \hat{a}_B^\dagger(\omega_2) |0\rangle \\ &= C \int d\omega_1 d\omega_2 dT e^{i(\omega_1 + \omega_2 - \omega_p)T} \phi(\omega_1) \varphi(\omega_2) \hat{a}_A^\dagger(\omega_1) \hat{a}_B^\dagger(\omega_2) |0\rangle \\ &= C \int dT e^{-i\omega_p T} |T\rangle_{\phi,A} |T\rangle_{\varphi,B}, \tag{4.53}\end{aligned}$$

where $|T\rangle$ is a single-photon wave packet given by Eq. (4.43). Equation (4.53) indicates that two wave packets are produced simultaneously at time T but with T undetermined. The physical picture of this state is straightforward: the monochromatic continuous pump field means that the pump is an infinitely long wave train with pump photon appearing at any time, but whenever down-converted, it converts to two wave packets simultaneously. Since the time for the pump photon is uncertain, so is the generation time for the two-photon wave packets. The state in Eq. (4.53) is a time-entangled two-photon states and can be used to generate time-bin entangled two-photon states (see Section 8.2.2).

4.5 N-Photon Entangled States

Next, we consider multi-photon states in multiple modes. Here, we assume the total number of photons of the optical field is a fixed number of N. We already encountered the case of $N = 2$ but we will consider the case of $N > 2$.

4.5.1   *Two-Mode N-Photon Entangled States – NOON States*

The simplest is an N-photon state in two modes. Consider the following N-photon state:

$$|NOON\rangle = (|N\rangle_A|0\rangle_B + |0\rangle_A|N\rangle_B)/\sqrt{2}. \qquad (4.54)$$

This state is known as the NOON state [Ou (1997a); Kok et al. (2002)]. In this state, all N photons are either in mode A or mode B. Since all N photons are always together, we can equivalently regard the N photons as one entity with a total energy of $N\hbar\omega_0$, where ω_0 is the frequency of one photon. The equivalent de Broglie wavelength of the combined entity is then λ_0/N. This state can be used in the precision measurement of phase and reach the Heisenberg limit in phase measurement (see Chapter 11) [Ou (1997a)].

4.5.2   *N-Parti Polarization Entangled States – GHZ States and W-States*

Earlier, we mentioned that the Bell states of two-parti system can be used to test nonlocality of quantum mechanics. For a multi-parti system, the nonlocal effect of quantum mechanics becomes more apparent. For a three-parti system, the three-photon Greenberg-Horne-Zeilinger (GHZ) state has the form of

$$|GHZ(3)\rangle = (|1_x\rangle_A|1_x\rangle_B|1_x\rangle_C + |1_y\rangle_A|1_y\rangle_B|1_y\rangle_C)/\sqrt{2}, \qquad (4.55)$$

It was proved that the GHZ state of three-parti system can demonstrate locality violation of quantum mechanics without the need of a Bell-like inequality [Greenberger et al. (1989)]. The GHZ state in Eq. (4.55) is a three-photon entangled state. Moreover, the following W-state has similar property and can be used in quantum information [Dur (2001)]:

$$|W(3)\rangle = (|1_x\rangle_A|1_y\rangle_B|1_y\rangle_C + |1_y\rangle_A|1_x\rangle_B|1_y\rangle_C + |1_y\rangle_A|1_y\rangle_B|1_x\rangle_C)/\sqrt{3}. \qquad (4.56)$$

4.6   Two-Mode Squeezed States – Photon Entangled States of Continuous Variables

The multi-mode states discussed in previous sections are all written in the number state representation. In these states, photon numbers are countable. They are suitable for quantum information applications with discrete variables where information is encoded in each photon. Besides discrete variables, another coding method is to work on the continuous variables. It encodes the information on the amplitude or phase of an optical field. This type of applications of quantum information requires entangled states with continuous variables. The earliest and most commonly used photon entangled states of continuous variables are generated from spontaneous parametric amplifier (see Section 6.1.5 or 6.3.4). They have the form of

$$|\eta_{ab}\rangle = \hat{S}_{ab}(\eta)|0\rangle \qquad (4.57)$$

where the two-mode squeezing operator $\hat{S}_{ab}(\eta)$ is

$$\hat{S}_{ab}(\eta) = e^{\eta \hat{a}^\dagger \hat{b}^\dagger - \eta^* \hat{a}\hat{b}}, \qquad (4.58)$$

which is similar to the squeezing operator $\hat{S}(r)$ in Eq. (3.63): when $a = b$, i.e., the two modes are the same, operator $\hat{S}_{ab}(\eta)$ becomes the single-mode squeezing operator $\hat{S}(r)$. So, the state in Eq. (4.57) is also called two-mode squeezed state. It was first introduced and studied by Caves and Schumaker in 1985 [Caves and Schumaker (1985)]. Operator $\hat{S}_{ab}(\eta)$ has similar properties as $\hat{S}(r)$ (see Problem 4.1):

$$\hat{A} = \hat{S}_{ab}^\dagger \hat{a} \hat{S}_{ab} = G\hat{a} + g\hat{b}^\dagger, \quad \hat{B} = G\hat{b} + g\hat{a}^\dagger \qquad (4.59)$$

where $G \equiv \cosh|\eta|, g \equiv (\eta/|\eta|)\sinh|\eta|$. Next, let us find the properties of the two-mode squeezed state $|\eta_{ab}\rangle$. But the specifics depend on what physical quantity we measure.

4.6.1   *Twin Beams*

When the physical quantity we measure is photon number or intensity, the photon numbers of the two modes are strongly correlated. Define photon number difference operator: $\hat{N}_- \equiv \hat{a}^\dagger \hat{a} - \hat{b}^\dagger \hat{b}$. Then we have in the Schrödinger picture

$$\hat{N}_-|\eta_{ab}\rangle = (\hat{a}^\dagger \hat{a} - \hat{b}^\dagger \hat{b})\hat{S}_{ab}(\eta)|0\rangle = \hat{S}_{ab}\hat{S}_{ab}^\dagger(\hat{a}^\dagger \hat{a} - \hat{b}^\dagger \hat{b})\hat{S}_{ab}(\eta)|0\rangle$$
$$= \hat{S}_{ab}(\hat{A}^\dagger \hat{A} - \hat{B}^\dagger \hat{B})|0\rangle, \qquad (4.60)$$

where we used Eq. (4.59). It can be easily shown from Eq. (4.59) that with $G^2 - g^2 = 1$, $\hat{A}^\dagger \hat{A} - \hat{B}^\dagger \hat{B} = \hat{a}^\dagger \hat{a} - \hat{b}^\dagger \hat{b}$. So, $\hat{N}_-|\eta_{ab}\rangle = \hat{S}_{ab}(\hat{a}^\dagger \hat{a} - \hat{b}^\dagger \hat{b})|0\rangle = 0$.

This shows that the photon numbers of the two modes are exactly the same. Writing $|\eta_{ab}\rangle$ in the number state representation in the general form:

$$|\eta_{ab}\rangle = \sum_{m,n} d_{mn} |m\rangle_a |n\rangle_b, \qquad (4.61)$$

we have straightforwardly

$$\hat{N}_- |\eta_{ab}\rangle = (\hat{a}^\dagger \hat{a} - \hat{b}^\dagger \hat{b}) |\eta_{ab}\rangle = \sum_{m,n} d_{mn} (m-n) |m\rangle_a |n\rangle_b. \qquad (4.62)$$

But $\hat{N}_- |\eta_{ab}\rangle = 0$, so we obtain $d_{mn}(m-n) = 0$, i.e., $d_{mn} = c_m \delta_{mn}$. Hence, Eq. (4.61) becomes

$$|\eta_{ab}\rangle = \sum_m c_m |m\rangle_a |m\rangle_b. \qquad (4.63)$$

From Problem 4.1, we can prove $c_m = g^m / G^{m+1}$ so we obtain the final form of the two-mode squeezed state in number state representation:

$$|\eta_{ab}\rangle = \sum_{m=0} (g^m / G^{m+1}) |m\rangle_a |m\rangle_b. \qquad (4.64)$$

Since each term in the sum has identical photon number for the two modes of a, b, this state is also called "twin beams state".

From the way Eq. (4.64) is written, it is impossible to write the twin beams state in the form of direct product states of any two modes. So, twin beams state is an entangled state. Interestingly, if we only look at one mode, say, mode a, the state is a mixed state with the following density operator:

$$\hat{\rho}_a = \mathrm{Tr}_b |\eta_{ab}\rangle \langle \eta_{ab}|$$
$$= \sum_{m=0} (|g|^{2m} / G^{2m+2}) |m\rangle_a \langle m| = \sum_{m=0} P_m |m\rangle_a \langle m|, \qquad (4.65)$$

where $P_m = |g|^{2m} / G^{2m+2} = \bar{n}^m / (\bar{n}+1)^{m+1}$ ($\bar{n} = |g|^2$). This is the photon statistical distribution of a thermal state (see Section 3.5.3) and the density operator in Eq. (4.65) is exactly the same as that in Eq. (3.108) for a thermal state. Hence, each field of the twin beams, when viewed individually, is in a thermal state. This was first pointed out by Yurke and Potasek [Yurke and Potasek (1987)] and demonstrated by Ou *et al.* who showed $g^{(2)} \to 2$ for only one field [Ou et al. (1999a,b)]. Nowadays, the measurement of $g^{(2)}$ is used to determine the number of modes for the fields out of a parametric amplifier [Liu et al. (2016)].

For the two-mode squeezed state $|\eta_{ab}\rangle$ in Eq. (4.57), although the photon numbers of modes a, b are perfectly correlated, the average photon numbers are $|g|^2 = \sinh^2 |\eta|$ and are relatively small with a finite gain for direct photo-detection. The average photon numbers can be boosted up by acting $\hat{S}_{ab}(\eta)$ on a coherent state at mode a: $\hat{S}_{ab}(\eta)|\alpha\rangle_a$, which has an average photon number of $\langle \hat{N}_A \rangle = G^2|\alpha|^2 + |g|^2 \approx G^2|\alpha|^2, \langle \hat{N}_B \rangle = |g|^2(|\alpha|^2 + 1) \approx |g|^2|\alpha|^2$ for $|\alpha|^2 \gg 1$. On the other hand, since $\hat{A}^\dagger \hat{A} - \hat{B}^\dagger \hat{B} = \hat{a}^\dagger \hat{a} - \hat{b}^\dagger \hat{b}$, we have

$$\langle \Delta^2(\hat{N}_A - \hat{N}_B) \rangle = \langle \Delta^2(\hat{N}_a - \hat{N}_b) \rangle = |\alpha|^2 + 1 \approx |\alpha|^2$$
$$= (\langle \hat{N}_A \rangle + \langle \hat{N}_B \rangle)/(G^2 + |g|^2)$$
$$\ll \langle \hat{N}_A \rangle + \langle \hat{N}_B \rangle \quad \text{for} \quad G^2 \gg 1. \quad\quad (4.66)$$

So, the photon numbers are still correlated. Even for finite G, we have $\langle \Delta^2(\hat{N}_A - \hat{N}_B) \rangle < \langle \hat{N}_A \rangle + \langle \hat{N}_B \rangle = \langle \Delta^2(\hat{N}_A - \hat{N}_B) \rangle_{cs}$ where cs stands for the coherent state average.

4.6.2   Two-Mode Entangled States of Continuous Variables with Einstein-Podolsky-Rosen Correlation

In the previous section, we examined the intensity correlation of the two modes, which corresponds to the amplitude correlation of the two modes. But phase and amplitude are a pair of conjugate observables, so we are also interested in the phase correlation. For this, we use the quadrature-phase amplitudes $\hat{X}, \hat{Y}$ introduced in Section 3.4 for a single mode field. For the two-mode case, we have

$$\hat{X}_a = \hat{a} + \hat{a}^\dagger, \quad \hat{Y}_a = (\hat{a} - \hat{a}^\dagger)/i$$
$$\hat{X}_b = \hat{b} + \hat{b}^\dagger, \quad \hat{Y}_b = (\hat{b} - \hat{b}^\dagger)/i. \quad\quad (4.67)$$

In the previous section, our concern is the quantum states of the field so we used the Schrödinger picture. Now we will use the Heisenberg picture. Then the evolution of the field operators is given in Eq. (4.59). For the evolution of the quadrature-phase amplitudes, we have

$$\hat{X}_A = G\hat{X}_a + g\hat{X}_b, \quad \hat{Y}_A = G\hat{Y}_a - g\hat{Y}_b$$
$$\hat{X}_B = G\hat{X}_b + g\hat{X}_a, \quad \hat{Y}_B = G\hat{Y}_b - g\hat{Y}_a \quad\quad (4.68)$$

with $\hat{X}_A = \hat{A} + \hat{A}^\dagger, \hat{Y}_A = (\hat{A} - \hat{A}^\dagger)/i, \hat{X}_B = \hat{B} + \hat{B}^\dagger, \hat{Y}_B = (\hat{B} - \hat{B}^\dagger)/i$. Here, we assume $g > 0$ for simplicity of argument. From Eq. (4.68), we obtain

$$\hat{X}_A - \hat{X}_B = (G - g)(\hat{X}_a - \hat{X}_b) = (\hat{X}_a - \hat{X}_b)/(G + g)$$
$$\hat{Y}_A + \hat{Y}_B = (G - g)(\hat{Y}_a + \hat{Y}_b) = (\hat{Y}_a + \hat{Y}_b)/(G + g). \quad\quad (4.69)$$

This shows that when $g \to \infty$, $\hat{X}_A - \hat{X}_B \to 0, \hat{Y}_A + \hat{Y}_B \to 0$, that is, $\hat{X}_A = \hat{X}_B, \hat{Y}_A = -\hat{Y}_B$. Therefore, the quadrature-phase amplitudes of the two-mode squeezed state are perfectly correlated and this is true simultaneously for two conjugate physical observables $\hat{X}, \hat{Y}$. This is the famous Einstein-Podolsky-Rosen(EPR) correlation and leads to the EPR paradox for quantum mechanics [Einstein et al. (1935)]. So, the two-mode squeezed state is also called an "EPR-entangled state". The idea of using two-mode squeezed states to demonstrate EPR paradox was first proposed by M. Reid [Reid (1989)] and demonstrated by Ou *et al.* experimentally [Ou et al. (1992b)]. Here the two correlated particles are the two virtual harmonic oscillators for the two modes of the electromagnetic fields, where $\hat{X}, \hat{Y}$ correspond to the position and momentum operators, respectively.

4.6.3   Squeezed State in Multi-Frequency Mode – Spectrum of Squeezing

In the discussion about EPR correlation, we measure separately the quadrature-phase amplitudes of the two modes. But if we measure the two modes all together, we will have completely different outcomes. This is the case when the two modes are two frequency modes. We will obtain the spectrum of squeezing of the optical field.

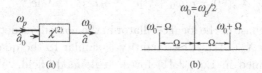

(a)   (b)

Fig. 4.6   (a) Input-output of a degenerate optical parametric oscillator (OPO) below threshold; (b) the correlated frequency components.

Suppose the two modes of the two-mode squeezed state belong to two different frequency modes of a one-dimensional optical field, we may rewrite Eq. (4.59) in terms of the frequency modes as

$$\hat{a}(\omega_0 + \Omega) = G(\Omega)\hat{a}_0(\omega_0 + \Omega) + g(\Omega)\hat{a}_0^\dagger(\omega_0 - \Omega), \qquad (4.70)$$

with $[\hat{a}_0(\omega), \hat{a}_0^\dagger(\omega')] = \delta(\omega - \omega')$. So, the frequency components of $\omega_0 \pm \Omega$ of the field are coupled, as shown in Fig. 4.6. Later in Section 6.3.4, we will prove that the operator evolution in Eq. (4.70) can be achieved in a degenerate optical parametric oscillator (OPO) below threshold, where a pump photon of frequency $\omega_p = 2\omega_0$ is down-converted to two lower frequency photons of frequencies $\omega_0 \pm \Omega$ satisfying the energy conservation $\omega_0 + \Omega + \omega_0 - \Omega = 2\omega_0 = \omega_p$. This conversion couples the two lower frequency photons in a way described by Eq. (4.70). The explicit forms of $G(\Omega), g(\Omega)$ will be derived in Section 6.3.4. But by applying $[\hat{a}(\omega), \hat{a}^\dagger(\omega')] = \delta(\omega - \omega')$ and $[\hat{a}(\omega), \hat{a}(\omega')] = 0$ to Eq. (4.70), we obtain some general properties as follows:

$$|G(\Omega)|^2 - |g(\Omega)|^2 = 1, \quad G(\Omega)g(-\Omega) = G(-\Omega)g(\Omega). \tag{4.71}$$

So, $|G(\Omega)|^2 = 1 + |g(\Omega)|^2 > 1$. With Eq. (4.71), it is straightforward to prove

$$|G(-\Omega)| = |G(\Omega)|, \quad |g(\Omega)| = |g(-\Omega)|,$$
$$\varphi_G^+ + \varphi_g^- = \varphi_G^- + \varphi_g^+, \tag{4.72}$$

where $\varphi_G^\pm = \arg[G(\pm\Omega)], \varphi_g^\pm = \arg[g(\pm\Omega)]$.

In Section 9.5, we will discuss the homodyne detection measurement whose output current spectrum is related to the quantity $S_{X_\varphi}(\Omega)$ defined through the following expression:

$$\langle \hat{X}_\varphi(\Omega)\hat{X}_\varphi^\dagger(\Omega')\rangle \equiv S_{X_\varphi}(\Omega)\delta(\Omega - \Omega'), \tag{4.73}$$

where

$$\hat{X}_\varphi(\Omega) = \hat{a}(\omega_0 + \Omega)e^{-i\varphi} + \hat{a}^\dagger(\omega_0 - \Omega)e^{i\varphi} \tag{4.74}$$

with φ as the phase of the local oscillator in the homodyne detection. Notice that $\hat{X}_\varphi^\dagger(\Omega) = \hat{X}_\varphi(-\Omega)$. This quantity is similar to the quadrature-phase amplitude defined in Eq. (3.41) for a single-mode field. We can regard it as the multi-mode equivalence [Caves and Schumaker (1985)]. Using Eqs. (4.70), (4.71) and (4.72), when $2\varphi = 2\theta_0 \equiv \varphi_G^+ + \varphi_g^- + \pi = \varphi_G^- + \varphi_g^+ + \pi$, we obtain the two conjugate quadrature-phase amplitudes:

$$\hat{X}(\Omega) \equiv \hat{X}_{\theta_0}(\Omega) = (|G(\Omega)| - |g(\Omega)|)\hat{X}_{\theta_0'}^{(0)}(\Omega)e^{i\phi_0},$$
$$\hat{Y}(\Omega) \equiv \hat{X}_{\theta_0 + \pi/2}(\Omega) = (|G(\Omega)| + |g(\Omega)|)\hat{X}_{\theta_0' + \pi/2}^{(0)}(\Omega)e^{i\phi_0} \tag{4.75}$$

with $\theta_0' = \pi/2 + (\varphi_g^- - \varphi_G^-)/2$ and $\phi_0 = (\varphi_G^+ - \varphi_G^-)/2$. $\hat{X}_{\theta_0'}^{(0)}(\Omega)$ is similarly defined in Eq. (4.74) but with $\hat{a}_0(\omega_0 \pm \Omega)$.

From Eq. (4.75), we have

$$\langle \hat{X}(\Omega)\hat{X}^\dagger(\Omega')\rangle = (|G(\Omega)| - |g(\Omega)|)^2\delta(\Omega - \Omega')$$
$$= S_X(\Omega)\delta(\Omega - \Omega')$$
$$\langle \hat{Y}(\Omega)\hat{Y}^\dagger(\Omega')\rangle = (|G(\Omega)| + |g(\Omega)|)^2\delta(\Omega - \Omega')$$
$$= S_Y(\Omega)\delta(\Omega - \Omega'). \tag{4.76}$$

Here, we assume $\hat{a}_0(\omega \pm \Omega)$ are in vacuum so that $\langle \hat{X}_{\theta_0'}^{(0)}(\Omega) \hat{X}_{\theta_0'}^{(0)\dagger}(\Omega') \rangle = \delta(\Omega - \Omega')$. Hence, $S_{X_{\theta_0}}(\Omega) = S_X(\Omega) \equiv (|G(\Omega)| - |g(\Omega)|)^2 = 1/(|G(\Omega)| + |g(\Omega)|)^2 < 1$ and $S_{X_{\theta_0+\pi/2}}(\Omega) = S_Y(\Omega) \equiv (|G(\Omega)| + |g(\Omega)|)^2 > 1$.

Notice that $S_X(\Omega) S_Y(\Omega) = 1$ and for the vacuum state, $g(\Omega) = 0$ and $G(\Omega) = 1$ so that $S_X^{vac}(\Omega) = 1 = S_Y^{vac}(\Omega)$. Therefore, $S_X(\Omega) < S_X^{vac}(\Omega)$ but $S_Y(\Omega) > S_Y^{vac}(\Omega)$. $S_X(\Omega), S_Y(\Omega)$ are similar to $\langle \Delta^2 \hat{X} \rangle, \langle \Delta^2 \hat{Y} \rangle$ of the single-mode field, describing the quadrature-phase amplitude fluctuations of a multi-frequency field. $S_X(\Omega) < S_X^{vac}(\Omega)$ indicates that the squeezed state in multi-frequency mode has one of its quadrature-phase amplitude fluctuations smaller than that of the vacuum state, achieving vacuum quantum noise reduction.

4.7 Problems

Problem 4.1 The derivation of Eqs. (4.59) and (4.64).

(i) Set $\hat{O} \equiv \eta \hat{a}^\dagger \hat{b}^\dagger - \eta^* \hat{a} \hat{b}$, prove that $[\hat{a}, \hat{O}] = \eta \hat{b}^\dagger$, $[\hat{b}^\dagger, \hat{O}] = \eta^* \hat{a}$.

(ii) Using the result of (i) and Eq. (3.121), prove Eq. (4.59) and the following

$$\hat{S}_{ab} \hat{a} \hat{S}_{ab}^\dagger = G\hat{a} - g\hat{b}^\dagger. \tag{4.77}$$

(iii) Using the left-hand side of Eq. (4.77) and definition of $|\eta_{ab}\rangle$ in Eq. (4.57), prove

$$(G\hat{a} - g\hat{b}^\dagger)|\eta_{ab}\rangle = 0. \tag{4.78}$$

(iv) Substitute $|\eta_{ab}\rangle$ in Eq. (4.63) into Eq. (4.78) to show Eq. (4.64), i.e., the final form of $|\eta_{ab}\rangle$.

Problem 4.2 Heralded single-photon state

For the multi-frequency two-photon state $|\Psi_2\rangle$ in Eq. (4.53), the detection of a photon at time $t = T_0$ in field B will project the state $|\Psi_2\rangle$ into $|\psi_A\rangle \equiv {}_B\langle T_0|\Psi_2\rangle$ for field A where

$$|T_0\rangle_B = \hat{E}_B^\dagger(T_0)|vac\rangle \quad \text{with} \quad \hat{E}_B(T_0) = \frac{1}{\sqrt{2\pi}} \int d\omega e^{-i\omega T_0} \hat{a}_B(\omega). \tag{4.79}$$

Show that the projected state $|\psi_A\rangle$ is a single-photon state of the form in Eq. (4.43). Find T in term of T_0.

Problem 4.3 Using Eq. (4.78) and the operator algebra in Section 3.6, prove that the coherent state representation of the twin beams state is given by

$$\langle \alpha, \beta | \eta_{ab} \rangle = \frac{1}{G} \exp\left[\frac{g\alpha^* \beta^*}{G} - \frac{1}{2}(|\alpha|^2 + |\beta|^2) \right]. \tag{4.80}$$

Problem 4.4 Photon bunching effect for pulsed multi-mode thermal state.

Consider a pulsed thermal field described by M temporal modes $\{f_j(t)\}(j = 1, 2, ..., M)$:

$$E(t) = \sum_{j=1}^{M} E_j f_j(t), \tag{4.81}$$

where $\{E_j\}(j = 1, 2, ..., M)$ are independent complex Gaussian random variables with the same average absolute square: $\langle |E_j|^2 \rangle = I_0$, which are similar to the complex random variable introduced in Section 1.6, and the mode functions $\{f_j(t)\}(j = 1, 2, ..., M)$ satisfy the orthonormal relation:

$$\int dt f_j^*(t) f_k(t) = \delta_{jk}. \tag{4.82}$$

For fast pulse detection, the pulses are usually much faster than the response of the detectors so that the observed intensity is a time integral:

$$I = \int dt |E(t)|^2. \tag{4.83}$$

(i) Evaluate $\langle I \rangle$ and $\langle I^2 \rangle$.
(ii) Show that the normalized intensity correlation function

$$g^{(2)} \equiv \frac{\langle I^2 \rangle}{\langle I \rangle^2} = 1 + \frac{1}{M}. \tag{4.84}$$

This is generally true for arbitrary thermal fields with Gaussian statistics and M is the average number of modes of the field [Goodman (2015); de Riedmatten et al. (2004)]. We will come back to this when treating photon bunching effect in spontaneous parametric processes in Problem 7.1 of Chapter 7 (Eq. (7.64)).

Chapter 5

Theory of Photo-detection and Quantum Theory of Coherence

Now we have learned how to describe a complicated system quantum mechanically in terms of quantum states. This description is developed from the fundamental theory of quantum mechanics and is thus logically self-consistent. However, how do we know this description is correct? In physics, we know the ultimate test is the experiments. As we will see in this chapter, the photo-detection theory is the bridge between the theory and experiments in quantum optics. It connects the experimentally measurable quantities with the quantum states that we use to describe the system. These measurable quantities are the subjects of the coherence theory of light, which is about the fluctuations of optical fields in both space and time. The coherence theory uses statistical methods to describe an optical field and this description is closely related to experimentally measurable quantities through the photo-detection theory. We will examine these in this chapter.

5.1 Classical Theory of Coherence and Semi-Classical Theory of Photo-Detection

5.1.1 *Classical Coherence Theory*

The classical theory of optical coherence was developed around the end of 1940's by Emil Wolf and was first documented in his classic textbook "Principles of Optics" with M. Born [Born and Wolf (1999)]. It mostly deals with optical interference effect. It is based completely on the wave theory of light and treated electromagnetic fields as random variables. The time evolution of the fields is thus described by stochastic processes. In this aspect, coherence theory of light is also known as statistical optics in analogy to statistical mechanics. In comparison, classical Maxwell electromagnetic theory of light is equivalent to classical Newtonian mechanics and is deterministic. It explains well the phenomena such as interference and diffraction of light waves. Similar to statistical mechanics, coherence theory of light uses statistical methods to study the fluctuations of electromagnetic fields through such quantities as correlation functions.

Consider the general solution in Eq. (2.22) for the Maxwell wave equation with $\mathbf{A}_\lambda(\mathbf{r})$ in Eq. (2.44):

$$\mathbf{A}(\mathbf{r},t) = \sum_{\mathbf{k},s} q_{\mathbf{k},s}(0)\hat{\epsilon}_{\mathbf{k},s} u_{\mathbf{k},s}(\mathbf{r})e^{-i\omega t} + c.c., \tag{5.1}$$

where $u_{\mathbf{k},s}(\mathbf{r})$ is the spatial mode function, $\hat{\epsilon}_{\mathbf{k},s}$ describes the polarization mode, and $\omega = |\mathbf{k}|c$. $q_{\mathbf{k},s}(0)$ gives the mode excitation. The electric field can be obtained from Eq. (2.14) (we will not give the magnetic field because intensity is related to the electric field only):

$$\mathbf{E}(\mathbf{r},t) = \sum_{\mathbf{k},s} E_{\mathbf{k},s}\hat{\epsilon}_{\mathbf{k},s} u_{\mathbf{k},s}(\mathbf{r})e^{-i\omega t} + c.c., \tag{5.2}$$

where $E_{\mathbf{k},s} \equiv ikq_{\mathbf{k},s}(0)$ describes the electric field excitation for mode $\mathbf{k}, s$.

In coherence theory of light, $\{E_{\mathbf{k},s}\}$ are a set of random variables with a known probability density $P(\{E_{\mathbf{k},s}\})$. So, the electric field is a random process. The measurable quantities (also known as observables) are the correlation functions:

$$\Gamma^{(N,M)}(\mathbf{r}_1,t_1;...;\mathbf{r}_N,t_N;\mathbf{r}_{N+1},t_{N+1};...;\mathbf{r}_{N+M},t_{N+M})$$
$$= \langle E_{j_1}^*(\mathbf{r}_1,t_1)...E_{j_N}^*(\mathbf{r}_N,t_N)E_{j_{N+1}}(\mathbf{r}_{N+1},t_{N+1})...E_{j_{N+M}}(\mathbf{r}_{N+M},t_{N+M})\rangle_P \tag{5.3}$$

Here,

$$E_j(\mathbf{r},t) \equiv \sum_{\mathbf{k},s} E_{\mathbf{k},s}[\hat{\epsilon}_{\mathbf{k},s}]_j u_{\mathbf{k},s}(\mathbf{r})e^{-i\omega t} \tag{5.4}$$

is the j-th component of the positive frequency part of the electric field in Eq. (5.2) and the average in Eq. (5.3) is with respect to the probability distribution $P(\{E_{\mathbf{k},s}\})$. The most commonly used correlation functions are the second-order field correlation function:

$$\Gamma_{ij}^{(1,1)}(\mathbf{r}_1,t_1;\mathbf{r}_2,t_2) = \langle E_i^*(\mathbf{r}_1,t_1)E_j(\mathbf{r}_2,t_2)\rangle_P \tag{5.5}$$

and the fourth-order field correlation or the intensity correlation function:

$$\Gamma^{(2,2)}(\mathbf{r}_1,t_1;\mathbf{r}_2,t_2) = \langle |\mathbf{E}(\mathbf{r}_1,t_1)|^2|\mathbf{E}(\mathbf{r}_2,t_2)|^2\rangle_P = \langle I(\mathbf{r}_1,t_1)I(\mathbf{r}_2,t_2)\rangle_P, \tag{5.6}$$

where $I(\mathbf{r}, t) \equiv |\mathbf{E}(\mathbf{r}, t)|^2$ is proportional to the intensity of the field. If we write $E = |E|e^{i\varphi}$, then Eq. (5.5) becomes

$$\Gamma_{ij}^{(1,1)}(\mathbf{r}_1, t_1; \mathbf{r}_2, t_2) = \langle |E_i(\mathbf{r}_1, t_1) E_j(\mathbf{r}_2, t_2)| e^{i[\varphi(\mathbf{r}_2, t_2) - \varphi(\mathbf{r}_1, t_1)]} \rangle_P. \quad (5.7)$$

Since $e^{i\varphi}$ is a fast varying function of φ, $\Gamma_{ij}^{(1,1)}$ is more about phase correlation whereas $\Gamma^{(2,2)}$ is only related to intensity correlation.

Another quantity deals with fourth-order coherence and is associated with two-photon interference phenomena:

$$\tilde{\Gamma}_{ij}^{(2,2)}(\mathbf{r}_1, t_1; \mathbf{r}_2, t_2) = \langle E_i^*(\mathbf{r}_1, t_1) E_j^*(\mathbf{r}_2, t_2) E_i(\mathbf{r}_2, t_2) E_j(\mathbf{r}_1, t_1) \rangle_P. \quad (5.8)$$

5.1.2 *Semi-Classical Theory of Photo-Detection*

To see how the correlation functions are related to what we can measure in the lab, we look into the detail of the photo-detection process.

In experiments, we measure optical fields with photo-detectors via the photoelectric effect where light fields interact with photo-sensitive media to produce electric currents and optical energy is converted into electric energy. However, as is well-known, classical wave theory has difficulty in explaining photoelectric effect and the concept of photon was introduced by Einstein to understand it [Einstein (1905)]. On the other hand, Mandel, Sudarshan and Wolf in 1964 introduced a semiclassical theory of photo-detection [Mandel et al. (1964)] and successfully explained with classical waves of light basically all aspects of photoelectric effect including the cut-off frequency $\hbar\omega > \hbar\omega_c = W_0$ (W_0 is the work function of the medium).

In the semiclassical theory, even though light fields are described as electromagnetic waves, atoms of the medium are treated quantum mechanically. The photoelectric process is modeled through the interaction between atoms and light waves via the interaction Hamiltonian

$$\hat{H}_I(t) = -\frac{e}{m}\hat{\mathbf{p}}(t) \cdot \mathbf{A}(\mathbf{r}, t), \quad (5.9)$$

where $\hat{\mathbf{p}}(t)$ is the momentum operator for electrons in the atoms and $\mathbf{A}(\mathbf{r}, t)$ is the vector potential in Eq. (5.1) for the electromagnetic waves. Using the perturbation theory in quantum mechanics, we can calculate the probability of ejecting (ionizing) an electron, which is a photoelectric event that gives rise to the photoelectric current for measurement. In this calculation, we have to first assume the field is deterministic or it is one of the realizations in the statistical ensemble of its stochastic process. With some reasonable assumption about the bandwidth of the field and the time response of the medium, we may find that the differential probability ΔP_1 of obtaining one photoelectron within a relatively small time interval Δt is given by

$$\Delta P_1 = \eta I(\mathbf{r}, t)\Delta t, \qquad (5.10)$$

where $I(\mathbf{r}, t) = |\mathbf{E}(\mathbf{r}, t)|^2$ is proportional to the intensity of the field and η is some proportional constant related to the atoms.

We can also find the joint differential probability $\Delta P_2(\mathbf{r}, t; \mathbf{r}', t')$ for two photoelectric events to occur. Assuming the two events are independent if atoms are localized and separated, the joint probability is the product of the probabilities of the two separate events at $\mathbf{r}, t; \mathbf{r}', t'$:

$$\Delta P_2(\mathbf{r}, t; \mathbf{r}', t') = \eta\eta' I(\mathbf{r}, t)I(\mathbf{r}', t')\Delta t\Delta t'. \qquad (5.11)$$

This can be extended to N multiple photoelectric events:

$$\Delta P_N(\mathbf{r}_1, t_1; ...; \mathbf{r}_N, t_N) = \eta_1...\eta_N I(\mathbf{r}_1, t_1)...I(\mathbf{r}_N, t_N)\Delta t_1...\Delta t_N. \qquad (5.12)$$

If atoms are separated within a distance that is much smaller than that over which the field changes, the formulae above are correct even for the same location: $\mathbf{r}_1 = ... = \mathbf{r}_N \equiv \mathbf{r}$.

So far, we only dealt with photoelectric events in one or more differential time interval Δt and we find the differential probability of finding one photoelectron in Δt is given in Eq. (5.10), which is much smaller than unity due to short time interval. Then, this differential quantity is also the expected average number of photoelectron within Δt.[1] So, for a finite interval T, the total average photoelectrons is simply the sum of all the events in the interval:

$$\langle n \rangle_T = \eta \int_T I(\mathbf{r}, t)dt. \qquad (5.13)$$

If all the photoelectric events are independent of each other, it is a good approximation that they follow the Poissonian distribution with an average number $\langle n \rangle_T$ in Eq. (5.13). Then we obtain the probability $P(n, T)$ of n events in a finite interval T:

$$P(n, T) = \frac{(\langle n \rangle_T)^n}{n!}e^{-\langle n \rangle_T}$$

$$= \frac{1}{n!}\left[\eta \int_T I(\mathbf{r}, t)dt\right]^n \exp\left[-\eta \int_T I(\mathbf{r}, t)dt\right]. \qquad (5.14)$$

Notice that the formula above for Poissonian distribution is true only if the field is deterministic and the randomness is purely from the probabilistic nature of quantum mechanics for the photoelectric events.

---

[1]The expected value for the number of photoelectrons is $\sum_n nP_n \approx P_1$ because higher order probabilities such as $P_2 \sim (\Delta t)^2$ is much smaller than $P_1 \sim \Delta t$ for short Δt.

For a fluctuating field, we must treat the field quantity $\mathbf{E}(\mathbf{r}, t)$ as a random variable and make an ensemble average of all the expressions above over the probability distribution $P(\{E_{\mathbf{k},s}\})$. We then obtain probability densities:

$$p_1(\mathbf{r}, t) \equiv \Delta P_1 / \Delta t = \eta \langle I(\mathbf{r}, t) \rangle_P, \tag{5.15}$$

$$p_2(\mathbf{r}, t; \mathbf{r}', t') \equiv \Delta P_2(\mathbf{r}, t; \mathbf{r}', t') / \Delta t \Delta t' = \eta \eta' \langle I(\mathbf{r}, t) I(\mathbf{r}', t') \rangle_P, \tag{5.16}$$

$$p_N(\mathbf{r}_1, t_1; ...; \mathbf{r}_N, t_N) = \eta_1 ... \eta_N \langle I(\mathbf{r}_1, t_1) ... I(\mathbf{r}_N, t_N) \rangle_P. \tag{5.17}$$

The same is true for the probability $P(n, T)$ of n events in a finite time interval T:

$$P(n, T) = \frac{1}{n!} \left\langle \left[\eta \int_T I(\mathbf{r}, t) dt \right]^n \exp \left[- \eta \int_T I(\mathbf{r}, t) dt \right] \right\rangle_P. \tag{5.18}$$

Later on, we will present a similar expression for quantized fields from Glauber's full quantum photon-detection theory. But before going to quantized fields, let us first look at the application of the classical coherence theory to Hanbury Brown-Twiss effect.

5.1.3 *More on Classical Explanation of Hanbury Brown-Twiss Effect*

In Section 1.5, we discussed in detail the second-order coherence function and its relation to the phase correlation of optical fields. We will not repeat it here except that we need to point out that the basis for that discussion is Eq. (5.15) for photoelectric events in a single detector. In Sections 1.6 and 4.2.4, we discussed the intensity fluctuations of the optical field with an implicit assumption that detectors measure directly the intensity of the optical fields. Now with the semiclassical theory of photo-detection in Eqs. (5.15)–(5.17), we can explore what the experimental observations will be and how they are connected to various correlation functions of the optical fields. Without the involvement of quantum theory of light, Hanbury Brown-Twiss experiment is the perfect platform to discuss the semiclassical theory of photo-detection and most of the discussion can be applied to the quantum theory of photo-detection.

In experiments, intensity correlation can be measured by two detectors through coincidence measurement (see Chapter 7 for more). We will count coincidence of two photoelectric events within a time interval of ΔT for a time period of T. Take $t' = t + \tau$ in Eq. (5.16). Then $p_2(\mathbf{r}, t; \mathbf{r}', t + \tau) \Delta t \Delta \tau$ is the joint probability that one photoelectron appears at time t within Δt and another one at time $t + \tau$ within $\Delta \tau$. Similar to Eq. (5.13), we find the average rate R_c of coincidence count within a finite coincidence window ΔT to be

$$R_c(t) = \frac{\langle n_c(\Delta T)\rangle_{\Delta t}}{\Delta t} = \int_{-\frac{\Delta T}{2}}^{\frac{\Delta T}{2}} d\tau p_2(\mathbf{r}, t; \mathbf{r}', t+\tau)$$

$$= \eta\eta' \int_{-\frac{\Delta T}{2}}^{\frac{\Delta T}{2}} d\tau \langle I(\mathbf{r},t)I(\mathbf{r}',t+\tau)\rangle. \quad (5.19)$$

Normally, photo-detectors and the subsequent circuit have some intrinsic time resolution T_R. Then $\Delta T \geq T_R$. Equation (5.19) connects the intensity correlation function with the measurable quantity R_c.

Let us write

$$\langle I(\mathbf{r},t)I(\mathbf{r}',t+\tau)\rangle \equiv [1 + \lambda(\mathbf{r},t;\mathbf{r}',t+\tau)]\langle I(\mathbf{r},t)\rangle\langle I(\mathbf{r}',t+\tau)\rangle, \quad (5.20)$$

where quantity λ characterizes the correlation in intensity fluctuations of the fields. Then Eq. (5.19) becomes

$$R_c = R_1 R_2 T_R \left[1 + \frac{1}{T_R} \int_{-\frac{T_R}{2}}^{\frac{T_R}{2}} d\tau \lambda(\mathbf{r}, t; \mathbf{r}', t+\tau) \right], \quad (5.21)$$

where, according to Eq. (5.13), $R_1 = \eta\langle I(\mathbf{r},t)\rangle, R_2 = \eta'\langle I(\mathbf{r}',t+\tau)\rangle$ are the average rates at which photoelectric events arrive at the two detectors, respectively. Here we take $\Delta T = T_R$ and assume R_1, R_2 do not change significantly within T_R so that we can pull them out of the integral. The first term in Eq. (5.21) stems from purely accidental coincidence event from the two detectors as if the fields are uncorrelated. The second term is from the excess contribution due to correlation in intensity fluctuations.

Hanbury Brown-Twiss Effect

For a thermal field, the probability distribution $P(\{E_{\mathbf{k},s}\})$ for the field in Eq. (5.4) is Gaussian. We then use Isserlis Theorem in Eq. (1.28) for multivariate Gaussian distribution to obtain for polarized field

$$\langle I(\mathbf{r}_1,t_1)I(\mathbf{r}_2,t_2)\rangle_P = \langle |E(\mathbf{r}_1,t_1)|^2 |E(\mathbf{r}_2,t_2)|^2 \rangle_P$$

$$= \langle |E(\mathbf{r}_1,t_1)|^2\rangle_P \langle |E(\mathbf{r}_2,t_2)|^2\rangle_P$$

$$+ \langle E(\mathbf{r}_1,t_1)E^*(\mathbf{r}_2,t_2)\rangle_P \langle E^*(\mathbf{r}_1,t_1)E(\mathbf{r}_2,t_2)\rangle_P$$

$$+ \langle E^*(\mathbf{r}_1,t_1)E^*(\mathbf{r}_2,t_2)\rangle_P \langle E(\mathbf{r}_1,t_1)E(\mathbf{r}_2,t_2)\rangle_P$$

$$= \langle I(\mathbf{r}_1,t_1)\rangle\langle I(\mathbf{r}_2,t_2)\rangle \left[1 + |\gamma(\mathbf{r}_1,t_1;\mathbf{r}_2,t_2)|^2 \right], \quad (5.22)$$

where the last term in the second equation is zero because of random phase fluctuations. The equation above is the same as Eq. (4.29) in Section 4.2.4

for time variable only. But here we include spatial variables. So, the excess intensity correlation function $\lambda(r_1, t_1; r_2, t_2)$ is directly related to the second-order coherence function: $\lambda(r_1, t_1; r_2, t_2) = |\gamma(r_1, t_1; r_2, t_2)|^2$. This explains well the HBT effect (Section 1.2): whenever r_1, r_2 are within coherence area of the field, there is an excess intensity fluctuation, otherwise, there is none. It is interesting to note that although γ is best to characterize phase correlation, it nevertheless is related to excess intensity correlation λ, which has nothing to do with phase. This is all because of the Gaussian nature of the probability distribution.

For the HBT experiment with a beam splitter and time delay, we take $r = r'$ in Eq. (5.21) and assume the field is stationary. Then we can write

$$R_c = R_1 R_2 T_R \left[1 + \frac{1}{T_R} \int_{-\frac{T_R}{2}}^{\frac{T_R}{2}} d\tau \lambda(\tau) \right], \tag{5.23}$$

where $\lambda(\tau)$ is related to the excess intensity fluctuations of the incoming field and $\lambda(\tau) = |\gamma(\tau)|^2$ for the thermal light. When the resolution time T_R of the detection system is much larger than the coherence time T_c of the field, $|\gamma(\tau)| \sim 0$ for $\tau = T_R/2 \gg T_c$ so that we can replace the integration limits with $\pm\infty$:

$$\int_{-\frac{T_R}{2}}^{\frac{T_R}{2}} d\tau |\gamma(\tau)|^2 \approx \int_{-\infty}^{\infty} d\tau |\gamma(\tau)|^2 = T_c. \tag{5.24}$$

So, Eq. (5.23) becomes

$$R_c = R_1 R_2 T_R \left[1 + \frac{T_c}{T_R} \right]. \tag{5.25}$$

Experimentally, we can define a measurable quantity $g^{(2)} \equiv R_c/R_1 R_2 T_R$. Then we have $g^{(2)} = 1 + T_c/T_R = 1 + 1/M$ with $M \equiv T_R/T_c$. In the language of temporal mode, T_c is roughly the size of a single-temporal mode. Then $M = T_R/T_c$ is the number of distinguishable or non-overlapping (orthogonal) temporal modes detected by the detectors within the detector's resolution time T_R. This relation is the same as that in Eq. (4.84) of Problem 4.4 for pulsed fields. Here, we proved it for the CW field.

The result above can be applied to stellar intensity interferometry where intensity correlation at two different locations is measured for the light from a distant star to determine $|\gamma_{12}|$ via $\lambda(r_1, r_2)$. We learned from Eq. (1.45) of Problem 1.2 in Chapter 1 that the spatial coherence function γ_{12} can be used to measure the size of a star by the method of stellar interferometry [Michelson (1890, 1920); Michelson and Pease (1921)]. But here, we can measure γ_{12} without observing interference fringes as in the method of Michelson, which can be very unstable due to atmospherical fluctuations. This is a new type of stellar interferometry based on intensity correlation technique [Hanbury Brown and Twiss (1956b)]. Stellar intensity interferometry was the first application of the intensity correlation technique immediately after it was invented in the famous Hanbury Brown-Twiss photon bunching experiment [Hanbury Brown and Twiss (1956a)].

HBT effect with thermal light can also be applied to optical imaging based on intensity correlation through a strange "ghost image" phenomenon [Pittman et al. (1995); Bennink et al. (2002)], which is a kind of nonlocal fourth-order classical wave interference effect [Gatti et al. (2004)].

5.2　Glauber's Photo-detection Theory and Quantum Theory of Coherence

5.2.1　*Photo-Electric Measurement and Normal Ordering*

In parallel with the work by Mandel, Sudarshan, and Wolf [Mandel et al. (1964)] on the semiclassical theory of photo-detection, Glauber [Glauber (1964)] developed a full quantum theory of photo-detection, in which both atomic media and optical fields are described quantum mechanically. The result is very similar to the semiclassical theory. In particular, Glauber evaluated the differential probability of one photoelectric event in a relatively small time interval Δt as

$$\Delta P_1(\mathbf{r}, t) \propto \langle \hat{\mathbf{E}}^{(-)}(\mathbf{r}, t) \cdot \hat{\mathbf{E}}^{(+)}(\mathbf{r}, t)\rangle_\psi \Delta t, \tag{5.26}$$

where the average is over the quantum state ψ of the optical field and

$$\hat{\mathbf{E}}^{(+)}(\mathbf{r}, t) = \sum_{\mathbf{k},s} l(\mathbf{k})\hat{a}_{\mathbf{k},s}\hat{\epsilon}_{\mathbf{k},s}u_{\mathbf{k},s}(\mathbf{r})e^{-i\omega t} = [\hat{\mathbf{E}}^{(-)}(\mathbf{r}, t)]^\dagger \tag{5.27}$$

is the positive frequency part of the electric field operator in Eq. (2.82) or Eq. (2.90) in Section 2.3.3. The joint differential probability of two photoelectric events is given by

$$\Delta P_2(\mathbf{r}_1, t_1; \mathbf{r}_2, t_2)$$

$$\propto \sum_{i,j} \langle \hat{E}_i^{(-)}(\mathbf{r}_1, t_1)\hat{E}_j^{(-)}(\mathbf{r}_2, t_2)\hat{E}_j^{(+)}(\mathbf{r}_2, t_2)\hat{E}_i^{(+)}(\mathbf{r}_1, t_1)\rangle_\psi \Delta t_1 \Delta t_2$$

$$= \langle : \hat{I}(\mathbf{r}_1, t_1)\hat{I}(\mathbf{r}_2, t_2) : \rangle_\psi \Delta t_1 \Delta t_2, \tag{5.28}$$

where :: denote normal ordering of the creation and annihilation operators $\hat{a}^\dagger$, $\hat{a}$ and

$$\hat{I}(\mathbf{r}, t) \equiv \hat{\mathbf{E}}^{(-)}(\mathbf{r}, t) \cdot \hat{\mathbf{E}}^{(+)}(\mathbf{r}, t) \tag{5.29}$$

is the intensity operator of the optical field. More generally for N photoelectric events, we have

$$\Delta P_N(\mathbf{r}_1, t_1; ...; \mathbf{r}_N, t_N) \propto \langle : \hat{I}(\mathbf{r}_1, t_1)...\hat{I}(\mathbf{r}_N, t_N) : \rangle_\psi \Delta t_1...\Delta t_N. \quad (5.30)$$

Equations (5.26), (5.28) and (5.30) are similar to the semiclassical equivalents in Eqs. (5.15)–(5.17) except that the average is now over the quantum state ψ.

Notice that Eqs. (5.26), (5.28) and (5.30) are for differential probabilities valid only for infinitesimal time intervals and cannot be normalized by time integration. Based on Glauber's quantum theory of photo-detection, Kelley and Kleiner [Kelley and Kleiner (1964)] derived the probability of counting n photoelectrons in a finite time interval T. The result is

$$P(n, T) = \frac{1}{n!} \left\langle : \left[\eta \int_T \hat{I}(\mathbf{r}, t) dt \right]^n \exp\left[-\eta \int_T \hat{I}(\mathbf{r}, t) dt \right] : \right\rangle_\psi. \quad (5.31)$$

This formula is similar to Eq. (5.18) for the classical wave theory.

5.2.2 *Glauber's Quantum Theory of Coherence*

From the results of the quantum theory of photo-detection, Glauber then defines the general quantum correlation functions [Glauber (1963a)]:

$$G^{(N,M)}(\mathbf{x}_1; \mathbf{x}_2...; \mathbf{x}_{N+M})$$
$$\equiv \langle : \hat{E}^{(-)}(\mathbf{x}_1)...\hat{E}^{(-)}(\mathbf{x}_N)\hat{E}^{(+)}(\mathbf{x}_{N+1})...\hat{E}^{(+)}(\mathbf{x}_{N+M}) : \rangle_\psi \quad (5.32)$$

with $\mathbf{x}_i \equiv \mathbf{r}_i, t_i$. These quantities are directly related to the measurement of photoelectric events and relevant coherence phenomena. For example, the second-order quantity

$$G^{(1,1)}(\mathbf{x}_1; \mathbf{x}_2) = \langle \hat{E}^{(-)}(\mathbf{x}_1)\hat{E}^{(+)}(\mathbf{x}_2) \rangle_\psi \quad (5.33)$$

has its normalized quantity

$$g_1 \equiv G^{(1,1)}(\mathbf{x}_1; \mathbf{x}_2) / \sqrt{G^{(1,1)}(\mathbf{x}_1; \mathbf{x}_1) G^{(1,1)}(\mathbf{x}_2; \mathbf{x}_2)}$$

directly related to the visibility of interference fringe and describes phase correlation of the optical fields. The fourth-order quantity

$$G^{(2,2)}(\mathbf{x}_1; \mathbf{x}_2) = \langle \hat{E}^{(-)}(\mathbf{x}_1)\hat{E}^{(-)}(\mathbf{x}_2)\hat{E}^{(+)}(\mathbf{x}_2)\hat{E}^{(+)}(\mathbf{x}_1) \rangle_\psi \quad (5.34)$$

is related to the intensity correlation and coincidence measurement (see more in Chapter 7).

For a multi-mode coherent state given by

$$|\psi\rangle = \prod_{\mathbf{k},s} |\{\alpha_{\mathbf{k},s}\}\rangle, \quad (5.35)$$

we have

$$\hat{E}_j^{(+)}(\mathbf{r},t)|\psi\rangle = \mathcal{E}_j(\mathbf{r},t)|\psi\rangle \tag{5.36}$$

with $\vec{\mathcal{E}}(\mathbf{r},t) \equiv \sum_{\mathbf{k},s} l(\mathbf{k})\alpha_{\mathbf{k},s}\hat{\epsilon}_{\mathbf{k},s}e^{i\mathbf{k}\cdot\mathbf{r}-i\omega t}$ being a complex number and normal ordering leads to

$$\hat{E}^{(+)}(\mathbf{x}_{N+1})...\hat{E}^{(+)}(\mathbf{x}_{N+M})|\psi\rangle = [\mathcal{E}(\mathbf{x}_{N+1})...\mathcal{E}(\mathbf{x}_{N+M})]|\psi\rangle \tag{5.37}$$

and

$$\langle\psi|\hat{E}^{(-)}(\mathbf{x}_1)...\hat{E}^{(-)}(\mathbf{x}_N) = \langle\psi|[\mathcal{E}^*(\mathbf{x}_1)...\mathcal{E}^*(\mathbf{x}_N)] \tag{5.38}$$

For the multi-mode coherent state of Eq. (5.35), the correlation function in Eq. (5.32) then becomes

$$G^{(N,M)}(\mathbf{x}_1;\mathbf{x}_2...;\mathbf{x}_{N+M}) = \mathcal{E}^*(\mathbf{x}_1)...\mathcal{E}^*(\mathbf{x}_N)\mathcal{E}(\mathbf{x}_{N+1})...\mathcal{E}(\mathbf{x}_{N+M}). \tag{5.39}$$

If we let $\{\alpha_{\mathbf{k},s}\}$ fluctuate and become random variables, then $\mathcal{E}(\mathbf{r},t) = \mathcal{E}(\{\alpha_{\mathbf{k},s}\},t)$ is a random process. Suppose $\{\alpha_{\mathbf{k},s}\}$ are described by a probability distribution: $P(\{\alpha_{\mathbf{k},s}\}) = P(\alpha_1,\alpha_2,...)$, which is the same as $P(\{E_{\mathbf{k},s}\})$ given earlier in classical coherence theory of light. Then we need to average Eq. (5.39) over this probability distribution and Eq. (5.39) becomes

$$G^{(N,M)}(\mathbf{x}_1;\mathbf{x}_2...;\mathbf{x}_{N+M}) = \langle\mathcal{E}^*(\mathbf{x}_1)...\mathcal{E}^*(\mathbf{x}_N)\mathcal{E}(\mathbf{x}_{N+1})...\mathcal{E}(\mathbf{x}_{N+M})\rangle_P, \tag{5.40}$$

which is in the same form as the quantity $\Gamma^{(N,M)}$ in Eq. (5.3) for classical coherence theory with $E(\mathbf{x})$ replaced by $\mathcal{E}(\mathbf{x})$.

5.2.3   *Connection between Quantum and Classical Theory and Optical Equivalence Theorem*

Notice that when $\{\alpha_{\mathbf{k},s}\}$ is a set of random variables, we are not certain about which coherent states the system is in and the uncertainty in determining the quantum state leads to a statistical description of quantum state for the system. In this case, we use a statistical mixture of coherent states or a mixed state given by the density operator:

$$\hat{\rho}_{cl} = \int \prod_{\mathbf{k},s} d^2\{\alpha_{\mathbf{k},s}\}|\{\alpha_{\mathbf{k},s}\}\rangle\langle\{\alpha_{\mathbf{k},s}\}|P_{cl}(\{\alpha_{\mathbf{k},s}\}) \tag{5.41}$$

for the quantum state of the system. The subscript "cl" means that this quantum mechanical description by the density operator $\hat{\rho}_{cl}$ gives exactly the same result as the classical wave theory, i.e., Eq. (5.40) for the quantum description is identical to Eq. (5.3) of the classical description.

On the other hand, we learned from Section 3.7.1 that any arbitrary quantum state of light can be described by a density operator which has the Glauber-Sudarshan P-representation or the coherent representation [Glauber (1963b); Sudarshan (1963)]:

$$\hat{\rho} = \int d^2\alpha |\alpha\rangle\langle\alpha| P_G(\alpha) \qquad (5.42)$$

for a single-mode field or

$$\hat{\rho} = \int \prod_{\mathbf{k},s} d^2\{\alpha_{\mathbf{k},s}\} |\{\alpha_{\mathbf{k},s}\}\rangle\langle\{\alpha_{\mathbf{k},s}\}| P_G(\{\alpha_{\mathbf{k},s}\}) \qquad (5.43)$$

for a multi-mode field. Here functions $P_G(\alpha)$ and $P_G(\{\alpha_{\mathbf{k},s}\})$ are some function of α's with normalization:

$$\int d^2\alpha P_G(\alpha) = 1 \quad \text{and} \quad \int \prod_{\mathbf{k},s} d^2\{\alpha_{\mathbf{k},s}\} P_G(\{\alpha_{\mathbf{k},s}\}) = 1. \qquad (5.44)$$

Comparing Eqs. (5.41) and (5.43), we find that if $P_G = P_{cl}$, then quantum and classical theories of light are equivalent. This is the so-called Optical Equivalence Theorem [Sudarshan (1963)]. So, it seems that there is no need for a quantum theory of light and the classical wave theory can cover the quantum theory if $P_G = P_{cl}$. However, this is a big "if" and this is where the difference lies between the classical wave theory and the quantum theory of light.

5.2.4 *Classical and Non-Classical States of Light*

Let us examine the difference between the two distributions of P_G and P_{cl}. We start with P_{cl}, which we know better. It is a probability distribution so it is normalized to one: $\int \prod_{\mathbf{k},s} d^2\{\alpha_{\mathbf{k},s}\} P_{cl}(\{\alpha_{\mathbf{k},s}\}) = 1$ and P_{cl} must not be negative: $P_{cl} \geq 0$. Now let us look at P_G: it satisfies the normalization equation in Eq. (5.44) just like P_{cl}. But does it satisfy the non-negative condition like P_{cl} for an arbitrary quantum state? It turns out that the answer is "NO".

To prove it, we just need to present a counter-example. Consider the single-photon state $|1\rangle$ for one mode. We showed in Section 3.7.1 that it has a Glauber P-representation of

$$P_G^{|1\rangle}(\alpha, \alpha^*) = e^{\alpha\alpha^*} \frac{\partial^2}{\partial\alpha\partial\alpha^*} \delta^{(2)}(\alpha). \qquad (5.45)$$

In terms of real variables x, y with $\alpha = x + iy, \alpha^* = x - iy$, we have the change of variables:

$$\frac{\partial}{\partial\alpha} = \frac{\partial}{\partial x}\frac{\partial x}{\partial\alpha} + \frac{\partial}{\partial y}\frac{\partial y}{\partial\alpha} = \frac{1}{2}\left(\frac{\partial}{\partial x} - i\frac{\partial}{\partial y}\right), \quad \frac{\partial}{\partial\alpha^*} = \frac{1}{2}\left(\frac{\partial}{\partial x} + i\frac{\partial}{\partial y}\right). \qquad (5.46)$$

Then Eq. (5.45) becomes

$$P_G^{|1\rangle}(x,y) = \frac{e^{x^2+y^2}}{4}\left(\frac{\partial^2}{\partial x^2} + \frac{\partial^2}{\partial y^2}\right)\delta(x)\delta(y). \tag{5.47}$$

Although $P_G^{|1\rangle}$ is written in terms of the special functions $\delta(x), \delta(y)$, we can take it as a limiting case of a Gaussian function:

$$\delta(x) = \lim_{\sigma\to 0}\frac{1}{\sigma\sqrt{2\pi}}e^{-x^2/2\sigma^2} \tag{5.48}$$

and

$$\frac{d^2}{dx^2}\delta(x) = \lim_{\sigma\to 0}\frac{1}{\sigma^5\sqrt{2\pi}}(x^2-\sigma^2)e^{-x^2/2\sigma^2}. \tag{5.49}$$

In Fig. 5.1, we plot a Gaussian function of width σ and its second derivative. We find that the second derivative is negative for $|x| < \sigma$. So, $P_G^{|1\rangle}$ is not non-negative for all values of α. This is not a surprise since a single-photon state describes the optical field as a particle of photon which does not have a counterpart in wave theory.

In fact, if two states are orthogonal with $\mathrm{Tr}(\hat{\rho}_1\hat{\rho}_2) = 0$, then we have

$$\mathrm{Tr}(\hat{\rho}_1\hat{\rho}_2) = \mathrm{Tr}\left[\int d^2\alpha P_G^{(1)}(\alpha)|\alpha\rangle\langle\alpha|\int d^2\beta P_G^{(2)}(\beta)|\beta\rangle\langle\beta|\right]$$

$$= \int d^2\alpha d^2\beta P_G^{(1)}(\alpha)P_G^{(2)}(\beta)|\langle\alpha|\beta\rangle|^2 = 0. \tag{5.50}$$

Since $|\langle\alpha|\beta\rangle|^2 > 0$, one of $P_G^{(1)}(\alpha)$ and $P_G^{(2)}(\beta)$ must be less than zero in order for the equation above to stand. Especially if a state $\hat{\rho} = \int d^2\beta P_G(\beta)|\beta\rangle\langle\beta|$ is orthogonal to a coherent state $|\alpha_0\rangle$ with $\langle\alpha_0|\hat{\rho}|\alpha_0\rangle = 0$, Eq. (5.50) becomes

$$\int d^2\beta|\langle\alpha_0|\beta\rangle|^2 P_G(\beta) = 0, \tag{5.51}$$

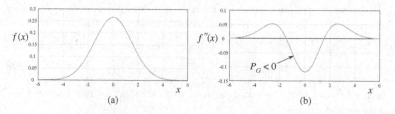

(a)   (b)

Fig. 5.1   (a) Gaussian function of width $\sigma = 1.5$ as a limit for a δ-function; (b) its second derivative. Note that there is a region where $f''(x) < 0$, leading to $P_G < 0$.

and we must have $P_G(\beta) < 0$ for some β. It is easy to show that the Schrödinger cat state in Eq. (3.51) satisfies this.

Since P_G is not always non-negative, in the quantum description of light, we can categorize the quantum states of an optical field into two types: (1) those with $P_G \geq 0$ for all and (2) those with $P_G < 0$ for some values of α. The former has a classical wave correspondence of $P_{cl} = P_G$ and can be described by the classical wave theory whereas the latter does not have such a correspondence and cannot be described classically. In this sense, we call the former "classical states" and the latter "non-classical".

An example of classical states is a thermal state whose Glauber-Sudarshan P-representation is derived in Section 3.7.1 and has following form

$$P_G^{th}(\alpha, \alpha^*) = \frac{1}{\pi \bar{n}} e^{-|\alpha|^2/\bar{n}} \tag{5.52}$$

with $\bar{n}$ as the mean photon number. This function is always positive and can serve as a true probability distribution P_{cl} for the classical wave description. So, the quantum explanation of the HBT experiment is exactly the same as the classical explanation. Besides thermal states, coherent states are classical state. So, the light fields from lasers can be described classically.

As shown earlier, an example of non-classical states is a single-photon state. All photon number states with a non-zero number of photon are non-classical states since their P-functions are higher order derivatives of δ-function. On the other hand, there are many other non-classical states which may not have an explicit form of Glauber P-function. But how do we know these states are non-classical? As we will see, there are some non-classical phenomena that we can observe to find out if a light field is non-classical.

5.3 Anti-bunching Effect

The first observed non-classical phenomenon is the photon anti-bunching effect of light [Kimble et al. (1977)].

Consider intensity correlation function

$$\begin{aligned}
\Gamma^{(2,2)}(t, t+\tau) &= \langle : \hat{I}(t+\tau)\hat{I}(t) : \rangle_{\hat{\rho}} \\
&= \langle \hat{E}^{(-)}(t)\hat{E}^{(-)}(t+\tau)\hat{E}^{(+)}(t+\tau)\hat{E}^{(+)}(t) \rangle_{\hat{\rho}} \\
&= \langle \mathcal{E}^*(t)\mathcal{E}^*(t+\tau)\mathcal{E}(t+\tau)\mathcal{E}(t) \rangle_{P_G} \\
&= \langle I(t+\tau)I(t) \rangle_{P_G}, \tag{5.53}
\end{aligned}$$

where $I(t) \equiv |\mathcal{E}(t)|^2$. If P_G is a true probability distribution, then $I(t), I(t+\tau)$ must satisfy the Cauchy-Schwarz inequality:

$$|\langle I(t+\tau)I(t)\rangle|^2 \leq \langle I^2(t+\tau)\rangle\langle I^2(t)\rangle. \qquad (5.54)$$

For a stationary process, we have $\langle I^2(t+\tau)\rangle = \langle I^2(t)\rangle = \langle I^2(0)\rangle$. Then Eq. (5.54) becomes

$$|\langle I(t+\tau)I(t)\rangle| \leq \langle I^2(0)\rangle \quad \text{or} \quad g^{(2)}(\tau) \leq g^{(2)}(0) \qquad (5.55)$$

after dividing both sides by $\langle I(0)\rangle^2$. Since $g^{(2)}(\tau) \to 1$ for $\tau \to \infty$, we also have $g^{(2)}(0) \geq 1$. Therefore, all classical states of light must satisfy $g^{(2)}(0) \geq 1$ and $g^{(2)}(\tau) \leq g^{(2)}(0)$. We learned that for a thermal state, $g_{th}^{(2)}(0) = 2 > 1$ and this is the so-called photon bunching effect.

On the other hand, if P_G is not a true probability distribution, the Cauchy-Schwarz inequality may be violated and we may not have the photon bunching effect. Therefore, a violation of either of the inequalities $g^{(2)}(0) \geq 1$ and $g^{(2)}(\tau) \leq g^{(2)}(0)$ means that the optical field is in a non-classical state of light. For example, a single-photon state will have zero intensity correlation function:

$$\langle : \hat{I}(t+\tau)\hat{I}(t) : \rangle_{|1\rangle} = \langle 1|\hat{E}^{(-)}(t)\hat{E}^{(-)}(t+\tau)\hat{E}^{(+)}(t+\tau)\hat{E}^{(+)}(t)|1\rangle$$
$$= 0 \qquad (5.56)$$

because there are two annihilation operators in $\hat{E}^{(+)}(t+\tau)\hat{E}^{(+)}(t)$. So, we have $g_{|1\rangle}^{(2)}(0) = 0 = g_{|1\rangle}^{(2)}(\tau)$ for any value of τ. This leads to the photon anti-bunching effect $g_{|1\rangle}^{(2)}(0) < 1$.

The photon anti-bunching effect was first observed by Kimble, Dagenais, and Mandel in the resonant fluorescence from single atoms of Sodium [Kimble et al. (1977)] (see Section 7.3.2) and later by Diedrich and Walther from single ions [Diedrich and Walther (1987)].

Another violation of Cauchy-Schwarz inequality is by the intensity cross-correlation function

$$\Gamma_{12}^{(2,2)} = \langle : \hat{I}_1(t)\hat{I}_2(t) : \rangle_{\hat{\rho}} = \langle I_1(t)I_2(t)\rangle_{P_G}, \qquad (5.57)$$

which, for classical states with $P_G \geq 0$, satisfies

$$|\langle I_1(t)I_2(t)\rangle|^2 \leq \langle I_1^2(t)\rangle\langle I_2^2(t)\rangle \quad \text{or} \quad [g_{12}^{(2)}]^2 \leq g_{11}^{(2)}g_{22}^{(2)}. \qquad (5.58)$$

This inequality is violated by the two-photon state from parametric down-conversion. We will discuss this non-classical phenomenon in Section 7.4.

5.4 Photon Statistics and Photon Correlations

Another nonclassical phenomenon occurs in photon statistics. As we will see in the following, classical states and some nonclassical states will give rise to completely different statistical behavior in the number of photons in those states.

Given an arbitrary state of light, what is the probability of finding the system in $\{|n_i\rangle\}$? Or we make photon number measurements of all the modes $\hat{N}_1, \hat{N}_2, ...$, what is the probability of finding outcome $n_1, n_2, ...(\equiv \{n\})$?

For a pure state $|\psi\rangle = \sum_{\{n\}} c_{\{n\}}|\{n\}\rangle$, we have

$$P(\{n\}) = |c_{\{n\}}|^2 = |\langle\{n\}|\psi\rangle|^2 = \langle\{n\}|\psi\rangle\langle\psi|\{n\}\rangle$$
$$= \left\langle\{n\}\big|\hat{\rho}_\psi\big|\{n\}\right\rangle. \tag{5.59}$$

For a mixed state of $\hat{\rho} = \sum_i p_i|\psi_i\rangle\langle\psi_i|$,

$$P(\{n\}) = \sum_i p_i|\langle\{n\}|\psi_i\rangle|^2$$
$$= \sum_i p_i\langle\{n\}|\psi_i\rangle\langle\psi_i|\{n\}\rangle$$
$$= \left\langle\{n\}\big|\hat{\rho}\big|\{n\}\right\rangle = \mathrm{Tr}\left(\hat{\rho}\big|\{n\}\right\rangle\left\langle\{n\}\big|\right). \tag{5.60}$$

Let us write $\hat{\rho}$ in Glauber-Sudarshan P-representation: $\hat{\rho} = \int d^2\{\alpha\}P_G(\{\alpha\})|\{\alpha\}\rangle\langle\{\alpha\}|$. Then we have after substituting into Eq. (5.60)

$$P(\{n\}) = \mathrm{Tr}\left(\hat{\rho}\big|\{n\}\right\rangle\left\langle\{n\}\big|\right)$$
$$= \int d^2\{\alpha\}P_G(\{\alpha\})\mathrm{Tr}\left(\big|\{\alpha\}\right\rangle\left\langle\{\alpha\}\big|\{n\}\right\rangle\left\langle\{n\}\big|\right)$$
$$= \int d^2\{\alpha\}P_G(\{\alpha\})\prod_i \frac{|\alpha_i|^{2n_i}}{n_i!}e^{-|\alpha_i|^2}. \tag{5.61}$$

Next we use the property of normal ordering $: \hat{N}^m := \hat{a}^{\dagger m}\hat{a}^m$. This gives $\langle\alpha| : \hat{N}^m : |\alpha\rangle = \langle\alpha|\hat{a}^{\dagger m}\hat{a}^m|\alpha\rangle = \alpha^{*m}\alpha^m$ and $\langle\alpha| : e^{\beta\hat{N}} : |\alpha\rangle = e^{\beta|\alpha|^2}$, which leads to $(|\alpha|^{2m}/m!)e^{-|\alpha|^2} = \langle\alpha| : (\hat{N}^m/m!)e^{-\hat{N}} : |\alpha\rangle$. Then we can rewrite Eq. (5.61) as

$$P(\{n\}) = \int d^2\{\alpha\} P_G(\{\alpha\}) \prod_i \frac{|\alpha_i|^{2n_i}}{n_i!} e^{-|\alpha_i|^2}$$

$$= \int d^2\{\alpha\} P_G(\{\alpha\}) \prod_i \left\langle \alpha_i \left| : \frac{\hat{N}_i^{n_i}}{n_i!} e^{-\hat{N}_i} : \right| \alpha_i \right\rangle$$

$$= \left\langle : \prod_i \frac{\hat{N}_i^{n_i}}{n_i!} e^{-\hat{N}_i} : \right\rangle_{\hat{\rho}}. \tag{5.62}$$

Here $\{n\} = \{n_1, n_2, ...\}$.

In practice, it is hard to count photon number in each individual mode. But we can easily measure the total number of photons in the whole field: $N_{tot} = \sum_i n_i$. We will then calculate the probability $P(n)$ of finding the total photon number $N_{tot} = n$:

$$P(n) = \sum_{\{n\}} P(\{n\}) \delta_{N_{tot},n} \quad (N_{tot} = \sum_i n_i)$$

$$= \left\langle : \sum_{\{n\}} \prod_i \frac{\hat{N}_i^{n_i}}{n_i!} e^{-\hat{N}_i} \delta_{N_{tot},n} : \right\rangle_{\hat{\rho}}$$

$$= \frac{1}{n!} \left\langle : \left(\sum_i \hat{N}_i \right)^n e^{-\sum_i \hat{N}_i} : \right\rangle_{\hat{\rho}}. \tag{5.63}$$

Here we used the multinomial expansion identity:

$$\left(\sum_i \hat{N}_i \right)^n = \sum_{n_1+n_2+...=n} n! \frac{\hat{N}_1^{n_1} \hat{N}_2^{n_2}...\hat{N}_i^{n_i}...}{n_1! n_2!...n_i!...}. \tag{5.64}$$

Defining the total number operator $\hat{N} = \sum_i \hat{N}_i$, we have

$$P(n) = \left\langle : \frac{\hat{N}^n}{n!} e^{-\hat{N}} : \right\rangle_{\hat{\rho}}. \tag{5.65}$$

For a multi-mode coherent state $|\psi\rangle = |\{\alpha\}\rangle$ such as the state from a laser, we obtain the Poissonian distribution:

$$P(n) = \frac{U^n}{n!} e^{-U}, \tag{5.66}$$

where $U \equiv \sum_i |\alpha_i|^2 = \langle n \rangle$ is the average total photon number. From this, we have the photon number variance $\langle \Delta^2 n \rangle = U = \langle n \rangle$, which is typical for a Poissonian distribution.

For an arbitrary state described by a density operator $\hat{\rho}$, we can once again write it in the Glauber-Sudarshan P-representation: $\hat{\rho} = \int d^2\{\alpha\} P_G(\{\alpha\}) |\{\alpha\}\rangle\langle\{\alpha\}|$ and the photon number probability is

$$P(n) = \int d^2\{\alpha\} P_G(\{\alpha\})\Big\langle\{\alpha\}\Big| : \frac{\hat{N}^n}{n!}e^{-\hat{N}} : \Big|\{\alpha\}\Big\rangle$$

$$= \int d^2\{\alpha\} P_G(\{\alpha\})\frac{U^n}{n!}e^{-U} \quad (U \equiv \sum_i |\alpha_i|^2)$$

$$= \Big\langle \frac{U^n}{n!}e^{-U} \Big\rangle_{P_G}. \tag{5.67}$$

It is straightforward to see that $\langle n\rangle = \sum n P(n) = \langle U\rangle_{P_G}$ and $\langle n(n-1)\rangle = \sum n(n-1)P(n) = \langle U^2\rangle_{P_G}$. Thus, we have $\langle n^2\rangle = \langle U(U+1)\rangle_{P_G}$ and the variance of photon number

$$\langle \Delta^2 n\rangle = \langle n^2\rangle - \langle n\rangle^2$$

$$= \langle U\rangle_{P_G} + \langle U^2\rangle_{P_G} - \langle U\rangle_{P_G}^2$$

$$= \langle n\rangle + \langle \Delta^2 U\rangle_{P_G}. \tag{5.68}$$

The second term in the last line of the equation above gives the extra photon fluctuations that deviate from a Poissonian distribution. For a classical state of light, P_G is a true probability distribution and $P_G \geq 0$. So, we have

$$\langle \Delta^2 U\rangle_{P_G} = \Big\langle \big(U - \langle U\rangle\big)^2 \Big\rangle_{P_G}$$

$$= \int d^2\{\alpha\} P_G(\{\alpha\})(U - \langle U\rangle)^2 \geq 0. \tag{5.69}$$

Hence, we have $\langle \Delta^2 n\rangle \geq \langle n\rangle$ and this leads to a super-Poissonian or Poissonian distribution for photon number. Therefore, all classical states always have super-Poissonian or Poissonian photon number statistics.

On the other hand, nonclassical states have $P_G < 0$ for some values of α's and this may lead to $\langle \Delta^2 U\rangle_{P_G} < 0$ and $\langle \Delta^2 n\rangle < \langle n\rangle$, which gives the sub-Poissonian distribution. So, a sub-Poissonian photon distribution always means nonclassical for the state of the optical field and this type of field cannot be described by classical wave theory. A simple example is the number state $|n\rangle$ for which $\langle \Delta^2 n\rangle = 0 < n = \langle n\rangle$. The amplitude-squeezed state depicted in Fig. 3.5(a) also exhibits sub-Poisson photon statistics (see Eq. (9.80) in Problem 9.2)

Although the photon concept is quantum mechanical and requires the quantization of the optical field, the photoelectrons produced in photo-detection process can be described by both the semi-classical wave theory and the quantum theory of light as seen in Sections 5.2 and 5.3. So, experimentally we can count photoelectrons and measure their statistics totally independent of how we describe the optical field. The counting probability distribution for photoelectrons is given in Eq. (5.18) by the semiclassical theory of photo-detection, and in Eq. (5.31) for the full quantum theory. It is straightforward to see that these two formulae are the same if $P(\{E_{\mathbf{k},s}\}) = P_G(\{\alpha_{\mathbf{k},s}\})$, i.e., the light field is a classical field. Furthermore, we find Eqs. (5.31) and (5.67) are the same if we set $U = \eta \int_T dt I(t)$ with $I(t) = |\vec{\mathcal{E}}(t)|^2$ and $\vec{\mathcal{E}}(t)$ given in Eq. (5.36). So, the variance in counting photoelectrons is the same as that in Eq. (5.68) and the conclusion about photon statistics is applicable to the statistics of photoelectrons that is observable experimentally.

For example, for thermal fields, which can be described as classical waves, from what we discuss in this section, we should observe a super-Poissonian statistics for counting the photoelectrons. The effect of sub-Poissonian photon statistics in counting photoelectrons was first observed by Short and Mandel for the light from resonant fluorescence of Sodium [Short and Mandel (1983)].

5.5   Quantum Noise and Its Reduction by Squeezed States and Twin Beams

The phenomena of quantum noise reduction by the squeezed states and twin beams are also non-classical phenomena that cannot be explained by the classical wave theory. We will prove their non-classicality in the following. For simplicity of argument, we will only consider single-mode cases. The multi-mode cases are discussed in Problem 5.1.

In the homodyne detection of optical fields (see Section 9.3), the output photo-current fluctuation $\langle \Delta^2 i \rangle$ is directly proportional to the quantity $\langle \Delta^2 \hat{X}_\varphi \rangle = 1 + \langle : \Delta^2 \hat{X}_\varphi : \rangle$ with $\hat{X}_\varphi = \hat{a}e^{-i\varphi} + \hat{a}^\dagger e^{i\varphi}$. Here, the first term of 1 corresponds to the shot noise contribution in photo-detection. With Glauber-Sudashan P-representation and $x_\varphi \equiv \alpha e^{-i\varphi} + \alpha^* e^{i\varphi}$, we have

$$\langle : \Delta^2 \hat{X}_\varphi : \rangle = \langle \Delta^2 x_\varphi \rangle_{P_G} = \langle \big(x_\varphi - \langle x_\varphi \rangle\big)^2 \rangle_{P_G}$$
$$= \int d^2\{\alpha\} P_G(\alpha, \alpha^*)\big(x_\varphi - \langle x_\varphi \rangle\big)^2 \geq 0 \quad \text{if } P_G \geq 0. \quad (5.70)$$

So, classical states always have $\langle \Delta^2 \hat{X}_\varphi \rangle_{cl} \geq 1$, or homodyne detection of classical fields have the shot noise as the lower limit in the photo-current fluctuation. On the other hand, we know from Section 3.4 that squeezed states have $\langle \Delta^2 \hat{X}_\varphi \rangle < 1$ for some φ. For the vacuum state, $\langle \Delta^2 \hat{X}_\varphi^{vac} \rangle = 1$. Thus, squeezed states are non-classical states of light that have noise below the vacuum quantum noise level. We will discuss more about the detection of the squeezed states in Sections 9.5 and 10.1.2.

Violation of Cauchy-Schwarz inequality in intensity cross-correlation function in Eq. (5.58) is an indication of non-classical correlation between two optical fields. We will show in Section 7.4 that a two-photon state from spontaneous parametric processes at low average photon level can give rise to the violation. However, when the light intensity is high, the equality in Eq. (5.58) is approached, leading to small or no violation. In this case, the nonclassical behavior is exhibited in intensity difference between two beams.

From the photo-detection theory in Section 9.2, the fluctuation in the difference of the photo-currents of two detectors is proportional to the intensity difference between fields being detected:

$$\langle \Delta^2 i_- \rangle \propto \langle \Delta^2 (\hat{I}_1 - \hat{I}_2) \rangle$$
$$= \langle \hat{I}_1 \rangle + \langle \hat{I}_2 \rangle + \langle : \Delta^2 (\hat{I}_1 - \hat{I}_2) : \rangle, \tag{5.71}$$

where $\hat{I}_j \equiv \hat{a}_j^\dagger \hat{a}_j$ ($j = 1, 2$). The first two terms are again the contributions from the shot noise of the two detectors, which are uncorrelated. For classical fields with $P_G \geq 0$, it is straightforward to show $\langle : \Delta^2 (\hat{I}_1 - \hat{I}_2) : \rangle \geq 0$. We thus have the shot noise as the limit for intensity difference measurement of two classical fields. But for twin beams with coherent state boost discussed in Section 4.6, we have from Eq. (4.66) $\langle \Delta^2 (\hat{I}_1 - \hat{I}_2) \rangle = (\langle \hat{I}_1 \rangle + \langle \hat{I}_2 \rangle)/(G^2 + |g|^2) < (\langle \hat{I}_1 \rangle + \langle \hat{I}_2 \rangle) = \langle \Delta^2 (\hat{I}_1 - \hat{I}_2) \rangle_{cs}$ ("cs" denotes coherent state). So, the non-classical correlation between the intensities of the twin beams gives rise to noise reduction in intensity difference that leads to detected noise below the classical shot noise limit. We will discuss more about noise reduction in twin beams in Section 10.1.4.

5.6 Remarks about Normal Ordering and Its Relation with Classical and Nonclassical Phenomena

In most of the discussions about the difference between classical and quantum theories, the emphasis is on the non-commutation of operators in quantum theory. However, as can be seen in our discussions in the previous sections about the nonclassical phenomena, we almost always end up with some normally ordered quantities in our calculation. This reveals directly the similarity and the subtle difference between the classical and quantum theory of light. The normal ordering seems to eliminate the difference of ordering of the operators and thus leads to the equivalence between the two theories, that is, the classical and quantum theories provide the same explanations for a large class of optical phenomena,' including the well-known Hanbury Brown-Twiss photon bunching effect.

On the other hand, there indeed exists a difference between classical and quantum theories, which lies in the calculation of these normally ordered quantities with the employment of the Glauber-Sudashan P-representation for taking average. The possible negativeness of the P-function rules out its interpretation as a classical probability density and leads to no classical equivalence. Since the P-function is a quasi-probability density, which is normalized to 1, most of its values must be positive and negativeness of the P-function is thus rare. This means that the nonclassical states of light are not common in optics. As we will show in the next chapter, nonclassical states cannot be produced from classical states by linear interaction (see Section 6.2.4). Nonlinear optical processes must be involved for the generation of nonclassical states.

5.7   Problems

Problem 5.1 Consider a multi-mode operator

$$\hat{d} = \sum_{\lambda} (\beta_\lambda a_\lambda + \beta_\lambda^* a_\lambda^\dagger). \qquad (5.72)$$

Given the multi-mode density matrix

$$\hat{\rho} = \int d^2\{\alpha_\lambda\}\, |\{\alpha_\lambda\}\rangle\langle\{\alpha_\lambda\}|\, P_G(\{\alpha_\lambda\}) \quad \text{with} \quad \int d^2\{\alpha_\lambda\} P_G(\{\alpha_\lambda\}) = 1, \qquad (5.73)$$

(i) Calculate $\langle \hat{d} \rangle$ and $\langle \hat{d}^2 \rangle$.
(ii) Prove that

$$\langle \Delta \hat{d}^2 \rangle \geq \sum_{\lambda} \beta_\lambda \beta_\lambda^* \qquad (5.74)$$

for any classical field with $P(\{\alpha_\lambda\}) \geq 0$ at any value of $\{\alpha_\lambda\}$. (Hint: try to write $\langle \Delta \hat{d}^2 \rangle$ in terms of the quantity (not operator) $d = \sum_\lambda (\beta_\lambda \alpha_\lambda + \beta_\lambda^* \alpha_\lambda^*)$ and its fluctuation average $\langle \Delta d^2 \rangle_P$ and prove $\langle \Delta d^2 \rangle_P > 0$ for the P-distribution given).
(iii) Calculate $\langle \Delta \hat{d}^2 \rangle$ for thermal state with

$$P(\{\alpha_\lambda\}) = \prod_{\lambda} P_{th}(\alpha_\lambda, \bar{n}_\lambda), \qquad (5.75)$$

where $P_{th}(\alpha_\lambda, \bar{n}_\lambda)$ is given in Eq. (5.52) with $\bar{n}_\lambda$ as the average photon number in each mode.

(iv) Assume $\{\beta_\lambda\}$ have a common phase φ for all λ. Find $\langle \Delta \hat{d}^2 \rangle$ for the multi-mode squeezed state:

$$|\psi\rangle = \prod_\lambda \otimes |\psi_\lambda\rangle \qquad (5.76)$$

where $|\psi_\lambda\rangle$ is given in Eq. (3.64) of Section 3.4.2 with $r = r_\lambda$ for each mode. Find the minimum of $\langle \Delta \hat{d}^2 \rangle$ as the phase φ is changed. Compare it with RHS of Eq. (5.74) and discuss.

Problem 5.2 The photon number uncertainty of the amplitude-squeezed state.

Prove, when the displacement quantity $|\alpha| \gg$ the squeezing quantity r, the amplitude-squeezed state $|\alpha, -re^{2i\varphi_\alpha}\rangle$ has an uncertainty in photon number as $\Delta n \equiv \sqrt{\langle \Delta^2 \hat{n} \rangle}$ with $\Delta n = e^{-r}\sqrt{\langle \hat{n} \rangle} < \sqrt{\langle \hat{n} \rangle}$, that is, the amplitude-squeezed state has a sub-Poissonian photon number distribution.

Problem 5.3 $g^{(2)}$ for the modified coherent state.

A modified coherent state is a coherent squeezed state $|\alpha, \xi\rangle$ with small excitations, that is, $|\alpha|, |\xi| \ll 1$.

(i) Calculate $g^{(2)} \equiv \langle : \hat{n}^2 : \rangle / \langle \hat{n} \rangle^2$ and expand it in the first two orders of $|\alpha|, |\xi|$.

(ii) Find the condition for which $g^{(2)}$ is minimum. What is the minimum value of $g^{(2)}$?

Problem 5.4 $g^{(2)}$ for the number state.

Prove that $g^{(2)} \equiv \langle : \hat{n}^2 : \rangle / \langle \hat{n} \rangle^2$ is $1 - 1/N$ for the number state $|N\rangle$.

This result shows that the criterion for claiming that a state contains a single-photon state is $g^{(2)} < 1/2$.

Problem 5.5 Another Cauchy-Schwarz inequality for classical fields.

We have seen in Section 5.3 how photon bunching effect for classical fields arises from the Cauchy-Schwarz inequality: $|\langle AB \rangle|^2 \le \langle A^2 \rangle \langle B^2 \rangle$ when we set $A = I(t), B = I(t+\tau)$. Now by setting $A = \Delta I(t) \equiv I(t) - \langle I(t) \rangle, B = \Delta I(t+\tau) \equiv I(t+\tau) - \langle I(t+\tau) \rangle$, show the following inequality

$$|g^{(2)}(0) - 1| \ge |g^{(2)}(\tau) - 1| \qquad (5.77)$$

for stationary classical fields [Rice and Carmichael (1988)], with $g^{(2)}(\tau) \equiv \langle I(t+\tau)I(t)\rangle/\langle I(t+\tau)\rangle\langle I(t)\rangle = \langle I(\tau)I(0)\rangle/\langle I(0)\rangle^2$. This inequality was violated by the fields emitted from cavity QED system [Foster et al. (2000)] and optical parametric oscillator with a coherent state injection [Lu and Ou (2002)], demonstrating another nonclassical effect.

Chapter 6

Generation and Transformation of Quantum States

The quantum states discussed in Chapters 3 and 4 are generated by nonlinear coupling and linear transformation of different modes of optical fields. Optical resonators (cavities) are special linear devices which are often used to enhance nonlinear interaction between different modes of the fields. They are also spectral filters to shape the frequency modes of the optical fields. The combination of linear and nonlinear devices can produce some quite exotic quantum states with interesting properties. These are the topics of this chapter.

6.1 Generation of Quantum States: Nonlinear Interactions between Light Fields

The interesting quantum states encountered in the previous Chapters are described in the bases of the eigen-states of the free field Hamiltonian in Eq. (2.76). However, they cannot be generated from that Hamiltonian. To produce those quantum states, we need nonlinear interactions, which lead to coupling between different modes of the optical fields.

The most commonly encountered optical fields in the lab are those in the coherent states, which are produced from lasers, and those in the thermal states, which can be obtained from blackbody radiators or discharged atomic vapor gas. Because of the favorable properties of lasers, they have become popular sources of light to replace the thermal sources ever since their invention in the 1960s. In this section, we will see how to produce interesting quantum states from the coherent states via nonlinear optical processes, which are the subjects of nonlinear optics.

6.1.1 A Brief Introduction to Nonlinear Optics: Three-Wave Mixing and Four-Wave Mixing

According to the electromagnetic theory, the total energy of an optical field in a medium is given by[1]

$$U = \frac{1}{8\pi} \int d^3\mathbf{r}[\mathbf{D}(\mathbf{r},t) \cdot \mathbf{E}(\mathbf{r},t) + \mathbf{B}(\mathbf{r},t) \cdot \mathbf{H}(\mathbf{r},t)]. \qquad (6.1)$$

Most of the optical media are not magnetic but are dielectric, so that $\mathbf{B} = \mathbf{H}$ and $\mathbf{D} = \mathbf{E} + 4\pi\mathbf{P}$ with the electric polarization $\mathbf{P}$ given by

$$\mathbf{P} = \vec{\chi}^{(1)} : \mathbf{E} + \vec{\chi}^{(2)} : \mathbf{EE} + \vec{\chi}^{(3)} : \mathbf{EEE} + ... \qquad (6.2)$$

Here, $\vec{\chi}^{(1)}, \vec{\chi}^{(2)}, \vec{\chi}^{(3)}$, etc. are tensors and : denotes the tensor product. The first term of the expression above leads to linear optics best described by the indices of refraction for isotropic media or birefringence for anisotropic media such as optical crystals. It can be treated as a free field in the formalism of Chapter 2 with some modifications.

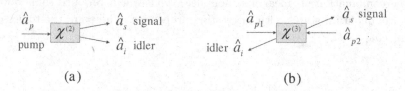

(a)   (b)

Fig. 6.1   (a) Three-wave mixing; (b) Four-wave mixing.

The higher order terms in Eq. (6.2) are responsible for the phenomena in nonlinear optics. The $\chi^{(2)}$-term corresponds to three-wave mixing (Fig. 6.1(a)): two waves come from each factor of the product of $\mathbf{EE}$ and the third is the generated wave, or the reverse process. Likewise, the $\chi^{(3)}$-term leads to four-wave mixing (Fig. 6.1(b)). $\chi^{(2)}$-nonlinearity normally dominates in bulk solid materials such as crystals whereas $\chi^{(3)}$-term is the leading term in gaseous atomic and molecular media because $\chi^{(2)}$ is zero due to central symmetry of randomly oriented atoms and molecules. These nonlinear wave mixing processes give rise to coupling and interaction among the modes of the optical fields. The interaction Hamiltonian stems from these interactions has the form of

$$\hat{H}_{int} = \frac{1}{2} \int \hat{\mathbf{P}}^{NL}(\mathbf{r},t) \cdot \hat{\mathbf{E}}(\mathbf{r},t)d^3\mathbf{r}, \qquad (6.3)$$

---

[1]We are using cgs unit here.

where

$$\hat{\mathbf{P}}^{NL} \equiv \chi^{(2)} : \hat{\mathbf{E}}\hat{\mathbf{E}} + \chi^{(3)} : \hat{\mathbf{E}}\hat{\mathbf{E}}\hat{\mathbf{E}} + \dots \qquad (6.4)$$

In the interaction picture, the quantum state of the system evolves under the unitary operator

$$\hat{\mathcal{U}} = \exp\left(\frac{1}{i\hbar} \int dt \hat{H}_{int}\right). \qquad (6.5)$$

The time integral in Eq. (6.5) and the spatial integral in Eq. (6.1) will give rise to energy and momentum conservation, respectively. The latter is also known as phase matching in nonlinear optics. These conservation laws will restrict the number of modes that can effectively couple to each other so that we only need to consider a few modes of the optical fields. Let us start with the simplest case of three modes for three-wave mixing (TWM). The most general form is

$$\hat{H}_{int}^{TWM} = i\hbar\eta \hat{a}_1^\dagger \hat{a}_2^\dagger \hat{a}_3 + h.c., \qquad (6.6)$$

where $h.c.$ denotes the Hermitian conjugate. Energy conservation requires $\omega_3 = \omega_1 + \omega_2$. Likewise, the simplest form for four-wave mixing (FWM) is

$$\hat{H}_{int}^{FWM} = i\hbar\eta \hat{a}_1^\dagger \hat{a}_2^\dagger \hat{a}_3 \hat{a}_4 + h.c. \qquad (6.7)$$

with $\omega_1 + \omega_2 = \omega_3 + \omega_4$. Note that the following form of interaction

$$\hat{H}_{int}^{FWM'} = i\hbar\eta \hat{a}_1^\dagger \hat{a}_2^\dagger \hat{a}_3^\dagger \hat{a}_4 + h.c. \qquad (6.8)$$

is also possible from four-wave mixing but we do not consider it because in practice, energy conservation requires $\omega_4 = \omega_1 + \omega_2 + \omega_3$ so that $\hat{a}_4$ field has a much higher frequency. Furthermore, its effect can also be too weak to implement experimentally (see the end of Section 6.1.4).

6.1.2 *Two-Photon Processes: Parametric Processes*

Normally, the nonlinear coefficients $\chi^{(2)}$ and $\chi^{(3)}$ are very small, requiring strong coupling fields to produce significant nonlinear effects. So far, at least one of the waves in three-wave mixing and two in four-wave mixing must be strong to have an observable quantum or classical effect in the lab. Let us make the $\hat{a}_3$ field in Eq. (6.6) for three-wave mixing or $\hat{a}_3, \hat{a}_4$ in Eq. (6.7) for four-wave mixing strong so that we can replace them with c-numbers A_3, A_4, respectively. Then we obtain the Hamiltonian for the parametric processes:

$$\hat{H}_{int}^{P} = i\hbar(\zeta \hat{a}_1^\dagger \hat{a}_2^\dagger - \zeta^* \hat{a}_1 \hat{a}_2) \qquad (6.9)$$

where $\zeta = \eta A_3$ for three-wave mixing or $\zeta = \eta A_3 A_4$ for four-wave mixing.[2] This is a two-photon process since two photons in $\hat{a}_1, \hat{a}_2$ are simultaneously annihilated or created. This Hamiltonian will lead to an evolution operator for the state of the fields:

$$\hat{U}^P = e^{-i\hat{H}_{int}^P t/\hbar} = e^{\xi \hat{a}_1 \hat{a}_2 - \xi^* \hat{a}_1^\dagger \hat{a}_2^\dagger} \tag{6.10}$$

with $\xi = -\zeta^* t$. If initially only $\hat{a}_3$, $\hat{a}_4$ are excited with lasers and $\hat{a}_1$, $\hat{a}_2$ are in vacuum, the state of the output field is

$$|\Psi\rangle = \hat{S}_{12}(\xi)|vac\rangle = e^{\xi \hat{a}_1 \hat{a}_2 - \xi^* \hat{a}_1^\dagger \hat{a}_2^\dagger}|vac\rangle. \tag{6.11}$$

This state is exactly the two-mode squeezed state discussed in Section 4.6.

In some cases, it is more convenient to consider the evolution of the operators defined as

$$\hat{O}(t) = \hat{U}^{P\dagger} \hat{O} \hat{U}^P. \tag{6.12}$$

This is similar to the Heisenberg picture but $\hat{U}$ is only related to the interaction Hamiltonian $\hat{H}_{int}$. On the other hand, with the final state $|\Psi_f\rangle$ related to the initial state $|\Psi_i\rangle$ by $|\Psi_f\rangle = \hat{U}^P|\Psi_i\rangle$, we have the expectation value $\langle \hat{O}\rangle(t) = \langle \Psi_f|\hat{O}|\Psi_f\rangle = \langle \Psi_i|\hat{U}^{P\dagger}\hat{O}\hat{U}^P|\Psi_i\rangle = \langle \Psi_i|\hat{O}(t)|\Psi_i\rangle$. So, it is meaningful to evaluate $\hat{O}(t)$.

For $\hat{O} = \hat{a}_1, \hat{a}_2$, we obtain (Section 4.6)

$$\begin{cases} \hat{A}_1 \equiv \hat{U}^{P\dagger} \hat{a}_1 \hat{U}^P = G\hat{a}_1 + g e^{i\delta} \hat{a}_2^\dagger \\ \hat{A}_2 \equiv \hat{U}^{P\dagger} \hat{a}_2 \hat{U}^P = G\hat{a}_2 + g e^{i\delta} \hat{a}_1^\dagger \end{cases} \tag{6.13}$$

with $G \equiv \cosh|\xi|$, $g \equiv \sinh|\xi|$ and $e^{i\delta} = \xi/|\xi|$. This is also the evolution equation for a parametric amplifier.

6.1.2.1   *Generation of two-photon and four-photon states*

For $|\xi| \ll 1$, we make an expansion of the exponential in Eq. (6.11) and obtain the state in number state base as

$$|\Psi\rangle \approx |vac\rangle - \xi^*|1\rangle_1|1\rangle_2 + \xi^{*2}|2\rangle_1|2\rangle_2 + \dots. \tag{6.14}$$

With ξ small, the second term dominates and this is a two-photon state that is widely used in two-photon interference. See Section 8.2 for more. For four-photon coincidence measurement, the first two terms have no contribution and the dominating term is the four-photon state $|2\rangle_1|2\rangle_2$ with two photons in each mode of 1 and 2.

Other states such as entangled states can be produced by using beam splitters on the two-photon and four-photon states in Eq. (6.14). See Section 6.2 for more.

---

[2]$H_{int}^{FWM'}$ in Eq. (6.8) can also give the same Hamiltonian but requires much higher frequency field in $\hat{a}_4$ than all other fields, which is not practical.

6.1.2.2 Generation of squeezed states

When the two quantum fields become the same field, i.e., $\hat{a}_1 = \hat{a}_2 \equiv \hat{a}$, the interaction Hamiltonian in Eq. (6.9) changes to

$$\hat{H}_{int}^S = i\hbar(\zeta \hat{a}^{\dagger 2} - \zeta^* \hat{a}^2). \tag{6.15}$$

This is the reverse process of second harmonic generation. The state of the system evolves from vacuum to

$$|\xi\rangle = \hat{S}(\xi)|vac\rangle = e^{\xi \hat{a}^2 - \xi^* \hat{a}^{\dagger 2}}|vac\rangle \qquad (\xi = -\zeta^* t), \tag{6.16}$$

which is the squeezed vacuum state discussed in Section 3.4.

Other states such as coherent squeezed states can be produced by superposing a coherent state and a squeezed state with a beam splitter. See Section 6.2 for more.

6.1.3 One-Photon Process: Frequency Conversion

When $\hat{a}_1$ field in Eq. (6.6) and $\hat{a}_1$, $\hat{a}_4$ fields in Eq. (6.7) are strong and can be replaced by c-numbers, we obtain the following Hamiltonian:

$$\hat{H}_{int}^F = i\hbar(\zeta \hat{a}_3 \hat{a}_2^{\dagger} - \zeta^* \hat{a}_3^{\dagger} \hat{a}_2) \tag{6.17}$$

where $\zeta = \eta A_1^*$ for three-wave mixing and $\zeta = \eta A_1^* A_4$ for four-wave mixing. This is a one-photon process in which only one photon is annihilated or created at a time. It annihilates one photon in mode $\hat{a}_3$ but in the meantime creates another in mode $\hat{a}_2$ or vice versa. If modes $\hat{a}_2$, $\hat{a}_3$ have different frequencies, this process realizes a frequency conversion of photons. The first frequency up-conversion of a quantum field was realized experimentally by Huang and Kumar [Huang and Kumar (1992)] for a squeezed state. It was also applied to convert photons at optcom wavelength of 1550nm to photons of 800 nm at which the photon counting technique is more mature [Vandevender and Kwiat (2004)] or the reverse process in which a photon at the atomic transition wavelength around 800 nm to optcom wavelength of 1550nm for quantum communication [Ding and Ou (2010); Takesue (2010)].

As we will see in Section 6.2, the Hamiltonian in Eq. (6.17) gives rise to an evolution operator of $\mathcal{U}^F = \exp(\xi \hat{a}_3^{\dagger} \hat{a}_2 - \xi^* \hat{a}_3 \hat{a}_2^{\dagger})$, which is for a linear lossless beam splitter with amplitude transmissivity $t = \cos|\xi|$ and reflectivity $r = (\xi/|\xi|)\sin|\xi|$.

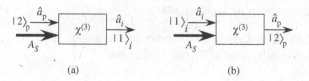

Fig. 6.2   A photon number converter.

6.1.4   A Photon Number Doubler

In the Hamiltonian for four-wave mixing in Eq. (6.7), if we only let one of the fields, say $\hat{a}_4$ be strong and replace it with a c-number A_4 and furthermore, we let $\hat{a}_1$, $\hat{a}_2$ be the same field: $\hat{a}_1 = \hat{a}_2$, we obtain the Hamiltonian for a photon number converter (Fig. 6.2):

$$\hat{H}_{int}^{N} = i\hbar(\zeta \hat{a}_3 \hat{a}_2^{\dagger 2} - \zeta^* \hat{a}_3^{\dagger} \hat{a}_2^2) \tag{6.18}$$

with $\zeta = \eta A_4$. The time evolution operator has the form of

$$\hat{U}^{(N)}(\xi) = e^{\xi \hat{a}_3^{\dagger} \hat{a}_2^2 - \xi^* \hat{a}_3 \hat{a}_2^{\dagger 2}} \tag{6.19}$$

with $\xi = -\zeta^* t$. This process converts one photon in $\hat{a}_3$ to two photons in $\hat{a}_2$. This is somewhat similar to the parametric process in Eq. (6.9) but can do it in an efficient way if $\xi \sim 1$, as we will see in the following.

Consider the case when $\hat{a}_3$ is in a single-photon state and $\hat{a}_2$ is in vacuum: $|\Psi\rangle_{in} = |1\rangle_3 |0\rangle_2$. Defining $\hat{A} \equiv \xi \hat{a}_3^{\dagger} \hat{a}_2^2 - \xi^* \hat{a}_3 \hat{a}_2^{\dagger 2}$, it is straightforward to show that $\hat{A}|1\rangle_3 |0\rangle_2 = -\xi^* \sqrt{2}|0\rangle_3 |2\rangle_2$ and $\hat{A}|0\rangle_3 |2\rangle_2 = \xi\sqrt{2}|1\rangle_3 |0\rangle_2$. So, using these two identities and after expanding the exponential in Eq. (6.19) in an infinite series, we obtain the output state as

$$|\Psi\rangle_{out} = \hat{U}^{(N)}(\xi)|\Psi\rangle_{in} = \cos\theta|1\rangle_3 |0\rangle_2 - e^{-j\delta} \sin\theta|0\rangle_3 |2\rangle_2 \tag{6.20}$$

with $\theta \equiv \sqrt{2}|\xi|$ and $e^{j\delta} = \xi/|\xi|$. If $\theta = \pi/2$, we can achieve 100% conversion from $|1\rangle_3$ to $|2\rangle_2$. The process is reversed if the initial state is $|0\rangle_3 |2\rangle_2$:

$$\hat{U}^{(N)}(\xi)|0\rangle_3 |2\rangle_2 = \cos\theta|0\rangle_3 |2\rangle_2 - e^{j\delta} \sin\theta|1\rangle_3 |0\rangle_2. \tag{6.21}$$

This reverse process with $\theta = \pi/2$ can be used to take out the two-photon state in a weak coherent state $(|\alpha\rangle_2 \approx |0\rangle + \alpha|1\rangle_2 + (\alpha^2/2)|2\rangle_2 + ...)$:

$$\hat{U}^{(N)}(\theta = \pi/2)|0\rangle_3 |\alpha\rangle_2$$
$$\approx |0\rangle_3 |0\rangle_2 + \alpha|0\rangle_3 |1\rangle_2 - e^{j\varphi}(\alpha^2/2)|1\rangle_3 |0\rangle_2 + ... \tag{6.22}$$

Note that for the α^2 term, $\hat{a}_2$-field has no two-photon state, leading to photon anti-bunching.[3]

---

[3]A two-photon destructive interference effect can also take out the two-photon state from a weak coherent state for photon anti-bunching effect. See Problem 6.2.

On the other hand, the condition of $\theta = \pi/2$ or $\xi = \pi/2\sqrt{2}$ is harder to achieve than the parametric process in Eq. (6.9) from four-wave mixing because $\xi \propto \eta A_4$ so that much higher power at A_4 is needed.

6.1.5 Multi-mode Parametric Process

In previous sections, we treated the nonlinear processes with few-mode models, placing each wave for wave mixing in single mode. Though simple and straightforward in demonstrating the physics, it is far from experimental reality where multi-mode excitations are common. On the other hand, it is also not practical to consider all the modes in the optical fields. In experiments, the spatial modes are usually well-defined because most lasers are from optical cavities which produce Gaussian spatial modes as we discussed in Section 2.2.1. So, it is a reasonable assumption that all the fields involved in wave mixing are in single spatial modes. Hence, we can use one-dimensional approximation for each field, which only involves frequency and temporal modes. Furthermore, phase matching conditions and energy conservation in nonlinear wave mixing [Boyd (2003)] place restriction on the polarization states of the optical fields so that only one specific polarization state is allowed for each wave. This means that we can treat the optical fields as scalar fields. Optical filtering, either spectral or spatial, can further restrict the directions of propagation and the frequency bands of the optical fields.

Since both three-wave mixing and four-wave mixing lead to the same Hamiltonian for parametric interaction, we will only consider three-wave mixing due to its simplicity. Because of the restrictions above, we can write the electric field operator in Eq. (6.4) in terms of separate waves:

$$\hat{\mathbf{E}} = \hat{\mathbf{E}}_1 + \hat{\mathbf{E}}_2 + \hat{\mathbf{E}}_3 \qquad (6.23)$$

for three-wave mixing. Here $\hat{\mathbf{E}}_i = \hat{\mathbf{E}}_i^{(-)} + \hat{\mathbf{E}}_i^{(+)}$ for $i = 1, 2, 3$ with

$$\left[\hat{\mathbf{E}}_i^{(-)}\right]^{\dagger} = \hat{\mathbf{E}}_i^{(+)} = \frac{\hat{e}_i}{\sqrt{2\pi}} \int_{\Delta\omega_i} d\omega \hat{a}_i(\omega) e^{i(k_i z - \omega t)}, \qquad (6.24)$$

where $k_i = n_i \omega / c$ and n_i is the index of refraction which is a function of ω due to dispersion. Here we take the field operator in the wave propagation form of Eq. (2.105) in the one-dimensional and quasi-monochromatic approximation with a fixed polarization state.

Substituting Eq. (6.23) into Eq. (6.3) and considering only the relevant terms,[4] we arrive at

$$\hat{H}_{int}^{M} = i\hbar\eta \int_{0}^{L} dz \hat{E}_{1}^{(-)} \hat{E}_{2}^{(-)} \hat{E}_{3}^{(+)} + h.c. \tag{6.25}$$

with $\eta = 6\hat{e}_{3} \cdot (\bar{\chi}^{(2)} : \hat{\epsilon}_{1}^{*}\hat{\epsilon}_{2}^{*})/i\hbar$ and L as the interaction length. The letter M denotes multi-mode here. Substituting Eq. (6.24) into the expression above and carrying out the spatial integral and time integral, we have

$$\frac{1}{i\hbar} \int_{-\infty}^{\infty} dt \hat{H}_{int}^{M}$$

$$= \int d\omega_{1} d\omega_{2} \varphi(\omega_{1}, \omega_{2}) \hat{a}_{1}^{\dagger}(\omega_{1}) \hat{a}_{2}^{\dagger}(\omega_{2}) \hat{a}_{3}(\omega_{1} + \omega_{2}) + h.c., \tag{6.26}$$

where we used $\int e^{i\omega t} dt = 2\pi\delta(\omega)$ and

$$\varphi(\omega_{1}, \omega_{2}) \equiv \frac{\eta L}{\sqrt{2\pi}} \frac{\sin\beta}{\beta} e^{-i\beta} \tag{6.27}$$

with $\beta \equiv \Delta k_{|\omega_{3}=\omega_{1}+\omega_{2}} L/2$ and $\Delta k \equiv k_{1} + k_{2} - k_{3}$ as the phase mismatch.

When the $\hat{a}_{3}$-field is in a coherent state from a strong laser (usually known as the pump field), we can replace the operator with a c-number: $\hat{a}_{3}(\omega_{1} + \omega_{2}) \rightarrow \alpha_{p}(\omega_{1} + \omega_{2})$ where $\alpha_{p}(\omega)$ is the spectral profile of the strong pump laser. Then Eq. (6.26) becomes

$$\frac{1}{i\hbar} \int_{-\infty}^{\infty} dt \hat{H}_{int}^{M} = \xi \int d\omega_{1} d\omega_{2} \Phi(\omega_{1}, \omega_{2}) \hat{a}_{1}^{\dagger}(\omega_{1}) \hat{a}_{2}^{\dagger}(\omega_{2}) + h.c. \tag{6.28}$$

with

$$\Phi(\omega_{1}, \omega_{2}) \equiv \frac{\eta L}{\xi\sqrt{2\pi}} \frac{\sin\beta}{\beta} e^{-i\beta} \alpha_{p}(\omega_{1} + \omega_{2}), \tag{6.29}$$

where ξ is such that $\Phi(\omega_{1}, \omega_{2})$ satisfies the normalization condition

$$\int d\omega_{1} d\omega_{2} |\Phi(\omega_{1}, \omega_{2})|^{2} = 1. \tag{6.30}$$

ξ is usually proportional to $L\sqrt{P}$ with P as the peak power of the strong laser. $\Phi(\omega_{1}, \omega_{2})$ is also known as the joint two-photon spectral function.

When the power of the pump laser is relatively small, ξ is small and we can expand the exponential in the unitary operator in Eq. (6.5) in an infinite series and take the first few terms:

$$\hat{U} = \exp\left(\int_{-\infty}^{\infty} dt \frac{\hat{H}_{int}^{M}}{i\hbar}\right) \approx 1 + \int_{-\infty}^{\infty} dt \frac{\hat{H}_{int}^{M}}{i\hbar} + \frac{1}{2}\left[\int_{-\infty}^{\infty} dt \frac{\hat{H}_{int}^{M}}{i\hbar}\right]^{2}. \tag{6.31}$$

---

[4]Terms such as $\hat{E}_{3}^{(+)} \hat{E}_{2}^{(+)} \hat{E}_{1}^{(-)}$ are equivalent to $\hat{E}_{1}^{(+)} \hat{E}_{2}^{(+)} \hat{E}_{3}^{(-)}$ if we interchange the indices. Terms such as $\hat{E}_{1}^{(+)} \hat{E}_{1}^{(+)} \hat{E}_{3}^{(-)}$ are special cases of $\hat{E}_{1}^{(+)} \hat{E}_{2}^{(+)} \hat{E}_{3}^{(-)}$ by setting 1=2.

For spontaneous processes, we have the output state

$$|\Phi\rangle \approx \left(1 - \frac{|\xi|^2}{2}\right)|vac\rangle + \xi \int d\omega_1 d\omega_2 \Phi(\omega_1, \omega_2)|\omega_1\rangle_1|\omega_2\rangle_2$$

$$+\frac{\xi^2}{2} \int d\omega_1 d\omega_2 d\omega_1' d\omega_2' \Phi(\omega_1, \omega_2)\Phi(\omega_1', \omega_2')|\omega_1, \omega_1'\rangle_1|\omega_2, \omega_2'\rangle_2. \quad (6.32)$$

Here $|\omega_1\rangle_1|\omega_2\rangle_2 \equiv \hat{a}_1^\dagger(\omega_1)\hat{a}_2^\dagger(\omega_2)|vac\rangle$ is a two-photon state and the second term in the expression above is the multi-frequency two-photon state in Eq. (4.52) that we first encountered in Section 4.4.3. The A, B systems over there correspond to fields 1 and 2 here. $|\omega_1, \omega_1'\rangle_1 |\omega_2, \omega_2'\rangle_2 \equiv \hat{a}_1^\dagger(\omega_1)\hat{a}_1^\dagger(\omega_1')\hat{a}_2^\dagger(\omega_2)\hat{a}_2^\dagger(\omega_2')|vac\rangle$ is a four-photon state. We will discuss more of the four-photon state in the third term above in Section 8.3.

When the peak power of the pump laser is high, as usually in the case of pumping by short pulses, ξ becomes large and we cannot make an expansion of the exponential and write explicitly the output state. In this case, it is easy to work with the evolution of the operators as in Eq. (6.13).

In general, the evolution is very complicated and we have

$$\hat{b}_1(\omega) = \hat{\mathcal{U}}^\dagger \hat{a}_1(\omega)\hat{\mathcal{U}} = \int_I G_1(\omega, \omega')\hat{a}_1(\omega')d\omega' + \int_{II} g_1(\omega, \omega')\hat{a}_2^\dagger(\omega')d\omega'$$
$$(6.33a)$$

$$\hat{b}_2(\omega) = \hat{\mathcal{U}}^\dagger \hat{a}_2(\omega)\hat{\mathcal{U}} = \int_{II} G_2(\omega, \omega')\hat{a}_2(\omega')d\omega' + \int_I g_2(\omega, \omega')\hat{a}_1^\dagger(\omega')d\omega',$$
$$(6.33b)$$

where I, II represent the bands of the two fields of $\hat{a}_1, \hat{a}_2$, respectively and $G_{1,2}, g_{1,2}$ have very complicated dependence on ξ and $\Phi(\omega_1, \omega_2)$. Notice that $\hat{b}_{1,2}(\omega)$ is related to all frequency components of $\hat{a}_{1,2}(\omega')$ in general.

On the other hand, when the pump field is of single frequency at ω_p, that is, $\alpha_p(\omega) = \alpha_0\delta(\omega - \omega_p)$, Eq. (6.28) becomes

$$\frac{1}{i\hbar} \int_{-\infty}^{\infty} dt \hat{H}_{int}^M = \xi \int d\omega_1 f(\omega_1)\hat{a}_1^\dagger(\omega_1)\hat{a}_2^\dagger(\omega_p - \omega_1) + h.c. \quad (6.34)$$

where $f(\omega_1)$ is some spectral function centered around ω_{10} at the center of band I of the $\hat{a}_1$-field. The frequency component of the $\hat{a}_2$-field is completely correlated with the $\hat{a}_1$-field by $\omega_2 = \omega_p - \omega_1$. Notice that in this case, the joint spectral function Φ is not normalizable because of the single frequency nature of $\alpha_p(\omega)$ but we can always choose ξ so that $\int d\omega_1 |f(\omega_1)|^2 = 1$.

Now unlike Eq. (6.33), it is straightforward to find that $\hat{b}_{1,2}(\omega)$ is only related to the frequency components of $\hat{a}_{1,2}(\omega)$ and $\hat{a}_{2,1}(\omega_p - \omega)$:

$$\hat{b}_1(\omega) = G(\omega)\hat{a}_1(\omega) + g(\omega)\hat{a}_2^\dagger(\omega_p - \omega)$$
$$\hat{b}_2(\omega) = G(\omega)\hat{a}_2(\omega) + g(\omega)\hat{a}_1^\dagger(\omega_p - \omega) \quad (6.35)$$

with $G(\omega) = \cosh[\xi|f(\omega)|]$, $g(\omega) = \sinh[\xi|f(\omega)|]$. This is the relation for a parametric amplifier with ξ^2 as the gain parameter and $f(\omega)$ as the gain spectral profile. When the two fields of $\hat{a}_1$, $\hat{a}_2$ become one field: $\hat{a}_1 = \hat{a}_2 = \hat{a}_0$, we arrive at the evolution equations in Eq. (4.70) in Section 4.6.3 for the squeezed state in multi-frequency mode with a spectrum of squeezing in Eq. (4.76).

Single-frequency pumping corresponds to continuous-wave (cw) operation of lasers. However, cw lasers usually have a finite bandwidth $\delta\omega_p$. But as long as $\delta\omega_p$ is much smaller than the bandwidth of $\sin\beta/\beta$ in Eq. (6.27), our treatment here is still valid (see [Ou (2007)] for details).

Normal modes and complete temporal modes

For pumping by ultra-short pulses, we have a large $\delta\omega_p$ that is comparable to the bandwidth of $\sin\beta/\beta$ in Eq. (6.27). However, instead of working with the complicated expressions in Eq. (6.33), we resort to a technique called singular value decomposition [Gentle (1998)] to rewrite the joint two-photon spectral function $\Phi(\omega_1, \omega_2)$ as

$$\Phi(\omega_1, \omega_2) = \sum_k r_k \phi_k(\omega_1)\psi_k(\omega_2) \qquad (k = 1, 2, ...), \qquad (6.36)$$

where $\phi_k(\omega_1)$ and $\psi_k(\omega_2)$ are the complex functions which satisfy the ortho-normal conditions $\int \phi_{k1}^*(\omega)\phi_{k2}(\omega)d\omega = \delta_{k1,k2}$ and $\int \psi_{k1}^*(\omega)\psi_{k2}(\omega)d\omega = \delta_{k1,k2}$, respectively and the parameters $r_k(k = 1, 2, 3, ...)$ are real and non-negative: $r_k \geq 0$. From the normalization relation in Eq. (6.30) for $\Phi(\omega_1, \omega_2)$ and ortho-normal relations for $\phi_k(\omega_1), \psi_k(\omega_2)$, we find $\{r_k\}(k = 1, 2, 3, ...)$ satisfy the normalization condition $\sum_k r_k^2 = 1$. $\{r_k\}(k = 1, 2, 3, ...)$ are referred to as the mode amplitudes. For the sake of clarity, the mode index k are arranged in a descending order, so that $r_k \geq r_{k+1}$ for $k \geq 1$. For this arrangement, the mode functions $\phi_1(\omega_s)$ and $\psi_1(\omega_i)$ are referred to as the fundamental mode.

With the mode decomposition in Eq. (6.36), we can rewrite Eq. (6.28) as

$$\frac{1}{i\hbar}\int_{-\infty}^{\infty} dt\hat{H}_{int}^M = \xi\sum_k r_k \int d\omega_1 d\omega_2 \phi_k(\omega_1)\psi_k(\omega_2)\hat{a}_1^\dagger(\omega_1)\hat{a}_2^\dagger(\omega_2) + h.c.$$

$$= \xi\sum_k r_k \hat{A}_k^\dagger \hat{B}_k^\dagger + h.c., \qquad (6.37)$$

where

$$\hat{A}_k^\dagger \equiv \int d\omega_1 \phi_k(\omega_1)\hat{a}_1^\dagger(\omega_1), \quad \hat{B}_k^\dagger \equiv \int d\omega_2 \psi_k(\omega_2)\hat{a}_2^\dagger(\omega_2). \qquad (6.38)$$

Because of the ortho-normal relations for $\phi_k(\omega_1), \psi_k(\omega_2)$ and the commutation relations $[\hat{a}_i(\omega), \hat{a}_j^\dagger(\omega')] = \delta_{ij}\delta(\omega - \omega')$, $\hat{A}_k, \hat{B}_k$ satisfy

$$[\hat{A}_k, \hat{A}_{k'}^\dagger] = \delta_{k,k'}, \quad [\hat{B}_k, \hat{B}_{k'}^\dagger] = \delta_{k,k'}, \quad [\hat{A}_k, \hat{B}_{k'}^\dagger] = 0, \qquad (6.39)$$

which means that $\{\hat{A}_k, \hat{B}_k\}$ are the annihilation operators for two sets of orthogonal modes. As a matter of fact, from Sections 2.2.2 and 4.3.2 on temporal modes, we find that $\{\hat{A}_k, \hat{B}_k\}$ in Eq. (6.38) define two sets of orthogonal temporal modes. Moreover, these are two sets of normal modes for H_{int}^M because they have a one-to-one coupling between the two sets but are decoupled within each set.

With the interaction Hamiltonian in the decoupled form of Eq. (6.37), we can rewrite the evolution operator as

$$\hat{\mathcal{U}} = \exp\left(\xi \sum_k r_k \hat{A}_k^\dagger \hat{B}_k^\dagger + h.c.\right) \qquad (6.40)$$

and the evolution equations as

$$\hat{C}_k = \cosh(r_k\xi)\hat{A}_k + \sinh(r_k\xi)\hat{B}_k^\dagger,$$
$$\hat{D}_k = \cosh(r_k\xi)\hat{B}_k + \sinh(r_k\xi)\hat{A}_k^\dagger. \qquad (6.41)$$

So, each pair of $\{\hat{A}_k, \hat{B}_k\}$ follows the single-mode model of the parametric process discussed in Section 6.1.2 and produces a two-mode squeezed state.

The materials covered in this section will be the basis for the experimental study of pulsed entangled quantum states in Part 2 of this book.

6.2 Linear Transformation: Beam Splitters

6.2.1 *General Formalism*

Optical beam splitters are important devices in optical interference: they are usually used to split the amplitude of an incoming wave or combine two waves for interference. Therefore, there are four ports: 2 input ports and 2 output ports and a beam splitter then couples these four modes. The wave behavior of beam splitters is well-documented in the classical electromagnetic wave theory [Born and Wolf (1999)]. Quantum mechanically, the relationship between the four input and output operators has been studied by a number of researchers [Zeilinger (1981); Yurke et al. (1986); Prasad et al. (1987); Ou et al. (1987); Fearn and Loudon (1987, 1989); Campos et al. (1989)] and there is a simple relation for the field operators of the four modes in Heisenberg picture:

$$\begin{cases} \hat{b}_1 = t\hat{a}_1 + r\hat{a}_2, \\ \hat{b}_2 = t'\hat{a}_2 + r'\hat{a}_1. \end{cases} \qquad (6.42)$$

t, r, t', r' are the complex amplitude transmissivity and reflectivity of the beam splitter, respectively. From the commutation relations

$$[\hat{b}_k, \hat{b}_l^\dagger] = \delta_{kl}, \quad (k, l = 1, 2) \tag{6.43}$$

we obtain $|t|^2 + |r|^2 = 1$, $|t'|^2 + |r'|^2 = 1$, and $t^*r' + r^*t' = 0$, which leads to $|t| = |t'|$, $|r| = |r'|$, and

$$\varphi_t - \varphi_r + \varphi_{t'} - \varphi_{r'} = \pi. \tag{6.44}$$

The phase relation above is universal and independent of the specifics of the beam splitter. This relation can also be derived via input-output energy conservation from classical wave theory [Ou and Mandel (1989); Smiles-Mascarenhas (1991)].

In general, t and r are complex numbers. However, by carefully choosing the reference point, we may arbitrarily change φ_t, φ_r, $\varphi_{t'}$, $\varphi_{r'}$ within the restriction in Eq. (6.44). For simplicity, we choose φ_t, φ_r, $\varphi_{t'}$ to be zero then $\varphi_{r'} = -\pi$ according to Eq. (6.44). Therefore, Eq. (6.42) becomes:

$$\begin{cases} \hat{b}_1 = t\hat{a}_1 + r\hat{a}_2, \\ \hat{b}_2 = t\hat{a}_2 - r\hat{a}_1. \end{cases} \quad (t, r > 0) \tag{6.45}$$

6.2.2 State Transformation of Number States through a Beam Splitter

The operator relationship in Eq. (6.45), together with the input state, is usually enough to determine the properties at the output ports. However, the approach above with operators lacks the visual connections to such interesting phenomena in quantum information as quantum entanglement and other nonclassical effects in the output ports. So, to see what emerges from the beam splitter, it is better to work in the Schrödinger picture and find the output state of the beam splitter. For this purpose, a unitary evolution operator is needed to connect the input and output states.

We start in the Heisenberg picture, in which the output operators are connected to the input operators by a unitary transformation:

$$\begin{cases} \hat{b}_1 = \hat{U}^\dagger \hat{a}_1 \hat{U} = t\hat{a}_1 + r\hat{a}_2, \\ \hat{b}_2 = \hat{U}^\dagger \hat{a}_2 \hat{U} = t\hat{a}_2 - r\hat{a}_1. \end{cases} \tag{6.46}$$

Here, $\hat{U}$ is a function of $\hat{a}_1$ and $\hat{a}_2$. (In Appendix A, we will discuss a simple derivation of the explicit form of $\hat{U}$ using the operator algebra presented in Section 3.6.) The state is unchanged and is the same as the input state in this picture:

$$|\Psi\rangle = |\phi\rangle_{in}. \tag{6.47}$$

All the properties at the output can be calculated by averaging the operators $\hat{b}_1, \hat{b}_2$ in Eq. (6.46) over the state in Eq. (6.47).

On the other hand, all these properties can also be equivalently calculated in the Schrödinger picture in which the output operators $\hat{b}_1, \hat{b}_2$ are the same as the input operators $\hat{a}_1, \hat{a}_2$, but the output state is connected to the input by

$$|\Psi\rangle_{out} = \hat{U}|\phi\rangle_{in}. \tag{6.48}$$

Generally, in order to find the output state, we need the explicit form of $\hat{U}$ (see Appendix A). In many cases, however, this is not necessary, as illustrated below. For simplicity, let us first consider a single photon state input at port 1:

$$|\phi\rangle_{in} = |1\rangle_1 \otimes |0\rangle_2 = (\hat{a}_1^\dagger|0\rangle_1) \otimes |0\rangle_2. \tag{6.49}$$

The output state then becomes:

$$|\Psi\rangle_{out} = \hat{U}\hat{a}_1^\dagger|0\rangle_1 \otimes |0\rangle_2 = \hat{U}\hat{a}_1^\dagger\hat{U}^\dagger\hat{U}|0\rangle_1 \otimes |0\rangle_2, \tag{6.50}$$

where we insert the unitary relation $\hat{U}^\dagger\hat{U} = 1$ between $\hat{a}_1$ and $|0\rangle_1$.

It is easy to see that $\hat{U}|0\rangle_1 \otimes |0\rangle_2 = |0\rangle_1 \otimes |0\rangle_2$, that is, vacuum input gives vacuum output for a beam splitter. (This can also be easily confirmed directly from the explicit form of $\hat{U}$ in Appendix A.) To find $\hat{U}\hat{a}_1\hat{U}^\dagger$, we invert Eq. (6.45) to relate $\hat{a}_1, \hat{a}_2$ in terms of $\hat{b}_1, \hat{b}_2$:

$$\hat{a}_1 = \hat{U}\hat{b}_1\hat{U}^\dagger = t\hat{b}_1 - r\hat{b}_2,$$
$$\hat{a}_2 = \hat{U}\hat{b}_2\hat{U}^\dagger = t\hat{b}_2 + r\hat{b}_1. \tag{6.51}$$

So, the principle of reversibility gives:

$$\begin{cases} \hat{U}\hat{a}_1\hat{U}^\dagger = t\hat{a}_1 - r\hat{a}_2, \\ \hat{U}\hat{a}_2\hat{U}^\dagger = t\hat{a}_2 + r\hat{a}_1. \end{cases} \tag{6.52}$$

The relation above can also be derived directly from the explicit form of $\hat{U}$ in Appendix A. Therefore, we have, for the output state:

$$|\Psi\rangle_{out} = (t\hat{b}_1^\dagger - r\hat{b}_2^\dagger)|0\rangle_1 \otimes |0\rangle_2 = t|1,0\rangle - r|0,1\rangle. \tag{6.53}$$

Note that we replaced the input operators $\hat{a}_1, \hat{a}_2$ by the same output operators $\hat{b}_1, \hat{b}_2$ in the Schrödinger picture.

Likewise, we can find the output state of an input state of $|1,1\rangle$ in the Hong-Ou-Mandel interferometer [Hong et al. (1987)] (see Section 8.2.1):

$$\begin{aligned} |\Psi\rangle_{out} &= \hat{U}|1\rangle_1|1\rangle_2 = \hat{U}\hat{a}_1^\dagger|0\rangle_1\hat{a}_2^\dagger|0\rangle_2 = \hat{U}\hat{a}_1^\dagger\hat{a}_2^\dagger|0\rangle \\ &= \hat{U}\hat{a}_1^\dagger\hat{U}^\dagger\hat{U}\hat{a}_2^\dagger\hat{U}^\dagger\hat{U}|0\rangle = (\hat{U}\hat{a}_1^\dagger\hat{U}^\dagger)(\hat{U}\hat{a}_2^\dagger\hat{U}^\dagger)|0\rangle \\ &= (t\hat{a}_1^\dagger - r\hat{a}_2^\dagger)(t\hat{a}_2^\dagger + r\hat{a}_1^\dagger)|0\rangle \\ &= (t^2 - r^2)\hat{a}_1^\dagger\hat{a}_2^\dagger|0\rangle + tr(\hat{a}_2^{\dagger2} - \hat{a}_1^{\dagger2})|0\rangle \\ &= (t^2 - r^2)|1,1\rangle + \sqrt{2}tr(|2,0\rangle - |0,2\rangle), \end{aligned} \tag{6.54}$$

and for a 50:50 beam splitter, we obtain the two-photon NOON state:

$$|\Psi\rangle_{out} = (|2,0\rangle - |0,2\rangle)/\sqrt{2}. \tag{6.55}$$

The disappearance of the $1,1\rangle$ state above is the result of two-photon Hong-Ou-Mandel interference (Section 8.2.1).

For a general input state of $|M, N\rangle$, we find the output state as

$$|\Psi\rangle_{out} = \hat{U}|M\rangle_1|N\rangle_2 = \frac{1}{\sqrt{M!N!}}\hat{U}\hat{a}_1^{\dagger M}|0\rangle_1\hat{a}_2^{\dagger N}|0\rangle_2$$

$$= \frac{1}{\sqrt{M!N!}}\hat{U}\hat{a}_1^{\dagger M}\hat{U}^{\dagger}\hat{U}\hat{a}_2^{\dagger N}\hat{U}^{\dagger}\hat{U}|0\rangle$$

$$= \frac{1}{\sqrt{M!N!}}(\hat{U}\hat{a}_1^{\dagger}\hat{U}^{\dagger})^M(\hat{U}\hat{a}_2^{\dagger}\hat{U}^{\dagger})^N|0\rangle$$

$$= \frac{1}{\sqrt{M!N!}}(t\hat{a}_1^{\dagger} - r\hat{a}_2^{\dagger})^M(t\hat{a}_2^{\dagger} + r\hat{a}_1^{\dagger})^N|0\rangle$$

$$= \sum_{m=0}^{M}\sum_{n=0}^{N}\frac{(-1)^m\sqrt{M!N!}t^{M+N-m-n}r^{m+n}}{(M-m)!m!(N-n)!n!}\hat{a}_1^{\dagger M-m+n}\hat{a}_2^{\dagger N-n+m}|vac\rangle$$

$$= \sum_{m=0}^{M}\sum_{n=0}^{N}\frac{(-1)^m\sqrt{M!N!(M-m+n)!(N-n+m)!}}{(M-m)!m!(N-n)!n!}$$

$$\times t^{M+N-m-n}r^{m+n}|M-m+n, N-n+m\rangle, \tag{6.56}$$

which can be regrouped as

$$|\Psi\rangle_{out} = \sum_{k=0}^{M+N}c_k|k, M+N-k\rangle, \tag{6.57}$$

where c_k collects the coefficients of the common terms of $|k, M+N-k\rangle$ in Eq. (6.56) and is in a very complicated form for the general case. But for some special cases, we can derive its explicit form. For example, for a 50:50 beam splitter with $M = N$, the above is simplified as

$$|\Psi\rangle_{out} = \frac{1}{N!2^N}(\hat{a}_1^{\dagger 2} - \hat{a}_2^{\dagger 2})^N|0\rangle$$

$$= \frac{1}{2^N}\sum_{k=0}^{N}(-1)^{N-k}\frac{\hat{a}_1^{\dagger 2k}\hat{a}_2^{\dagger 2(N-k)}}{k!(N-k)!}|0\rangle$$

$$= \frac{1}{2^N}\sum_{k=0}^{N}(-1)^{N-k}\frac{\sqrt{(2k)!(2N-2k)!}}{k!(N-k)!}|2k, 2N-2k\rangle. \tag{6.58}$$

Another special case is when $M = 1$ with $t = r = 1/\sqrt{2}$. In this case, the coefficient in Eq. (6.57) has an explicit form of

$$c_k = \frac{2k-N-1}{2^{N+1}}\sqrt{\frac{N!}{k!(N-k+1)!}} \quad \text{for } 1 \le k \le N \tag{6.59}$$

and $c_0 = -\sqrt{(N+1)/2^{N+1}} = -c_{N+1}$. Taking $n_1 = k, n_2 = N + 1 - k$ as the photon numbers of the two outputs, the output state becomes

$$|\Psi\rangle_{out} = \sum_{n_1=0,n_2=N+1}^{N+1,0} \frac{n_1 - n_2}{2^{N+1}} \sqrt{\frac{N!}{n_1!n_2!}} |n_1, n_2\rangle. \qquad (6.60)$$

It is interesting to note that $c_k = 0$ for $k = (N+1)/2$ or $n_1 = n_2$ when $N = $ odd. This is a multi-photon destructive interference effect which is a generalization of the two-photon Hong-Ou-Mandel effect [Ou (1996)]. A multi-photon bunching effect shows up in the probability of finding $|N + 1, 0\rangle$ at the outputs due to multi-photon construction interference. We will discuss more about this in Section 8.3.

6.2.3 *State Transformation of an Arbitrary State*

The technique used in Section 6.2.2 can also be applied to an arbitrary input state in the Glauber-Sudarshan P-representation [Glauber (1963b); Sudarshan (1963)] to derive a general relation between the input and the output states. In the Glauber-Sudarshan P-representation, the input and output states are described by the density operators as

$$\hat{\rho}_{in} = \int d^2\alpha_1 d^2\alpha_2 P_{in}(\alpha_1, \alpha_2) |\alpha_1, \alpha_2\rangle\langle\alpha_1, \alpha_2|, \qquad (6.61)$$

$$\hat{\rho}_{out} = \int d^2\alpha_1 d^2\alpha_2 P_{out}(\alpha_1, \alpha_2) |\alpha_1, \alpha_2\rangle\langle\alpha_1, \alpha_2|, \qquad (6.62)$$

where $P_{in/out}(\alpha_1, \alpha_2)$ is a quasi-probability distribution and can completely describe the incoming/outgoing fields of the beam splitter. $|\alpha_1, \alpha_2\rangle$ is the coherent state base. Our goal is to find the connection between P_{in} and P_{out}. From Eq. (6.48), the output density operator is given by

$$\hat{\rho}_{out} = \hat{U}\hat{\rho}_{in}\hat{U}^\dagger$$
$$= \int d^2\alpha_1 d^2\alpha_2 P_{in}(\alpha_1, \alpha_2) \hat{U}|\alpha_1, \alpha_2\rangle\langle\alpha_1, \alpha_2|\hat{U}^\dagger. \qquad (6.63)$$

Obviously, $\hat{U}|\alpha_1, \alpha_2\rangle$ is the output state corresponding to a coherent state input state of $|\alpha_1, \alpha_2\rangle$, and, from classical optics and Eq. (6.45), we know the output is also a coherent state of the form:

$$\hat{U}|\alpha_1, \alpha_2\rangle = |\beta_1, \beta_2\rangle, \qquad (6.64)$$

with

$$\begin{cases} \beta_1 = t\alpha_1 + r\alpha_2 \\ \beta_2 = t\alpha_2 - r\alpha_1. \end{cases} \qquad (6.65)$$

The relation above can also be derived by using the method discussed in Section 6.2.2 and writing the coherent state in the form of the displacement operator (see Eq. (3.26)):

$$|\alpha\rangle = \hat{D}(\alpha)|0\rangle, \tag{6.66}$$

with

$$\hat{D}(\alpha) = \exp(\alpha\hat{a} - \alpha^*\hat{a}^\dagger). \tag{6.67}$$

Then, we have:

$$\hat{U}|\alpha_1, \alpha_2\rangle = \hat{U}\hat{D}_1(\alpha_1)\hat{D}_2(\alpha_2)|0, 0\rangle$$
$$= \hat{U}\hat{D}_1(\alpha_1)\hat{D}_2(\alpha_2)\hat{U}^\dagger|0, 0\rangle. \tag{6.68}$$

But, with

$$\hat{U}\hat{D}_1(\alpha_1)\,\hat{D}_2\,(\alpha_2)\hat{U}^\dagger$$
$$= \hat{U}\exp(\alpha_1\hat{a}_1 - \alpha_1^*\hat{a}_1^\dagger + \alpha_2\hat{a}_2 - \alpha_2^*\hat{a}_2^\dagger)\hat{U}^\dagger$$
$$= \exp(\alpha_1\hat{U}\hat{a}_1\hat{U}^\dagger - \alpha_1^*\hat{U}\hat{a}_1^\dagger\hat{U}^\dagger + \alpha_2\hat{U}\hat{a}_2\hat{U}^\dagger - \alpha_2^*\hat{U}\hat{a}_2^\dagger\hat{U}^\dagger)$$
$$= \exp[\alpha_1(t\hat{a}_1 - r\hat{a}_2) - \alpha_1^*(t\hat{a}_1^\dagger - r\hat{a}_2^\dagger)$$
$$\qquad + \alpha_2(t\hat{a}_2 + r\hat{a}_1) - \alpha_2^*(t\hat{a}_2^\dagger + r\hat{a}_1^\dagger)]$$
$$= \exp[(t\alpha_1 + r\alpha_2)\hat{a}_1 - (t\alpha_1^* + r\alpha_2^*)\hat{a}_1^\dagger$$
$$\qquad + (t\alpha_2 - r\alpha_1)\hat{a}_2 - (t\alpha_2^* - r\alpha_1^*)\hat{a}_2^\dagger]$$
$$= \hat{D}_1(\beta_1)\hat{D}_2(\beta_2), \tag{6.69}$$

we have Eqs. (6.64) and (6.65).

Substituting Eq. (6.64) into Eq. (6.63) and making a change of variables from α to β by Eq. (6.65), we find the output state as

$$\hat{\rho}_{out} = \int d^2\beta_1 d^2\beta_2 |\beta_1, \beta_2\rangle\langle\beta_1, \beta_2| P_{in}(t\beta_1 - r\beta_2, t\beta_2 + r\beta_1). \tag{6.70}$$

Therefore, we have:

$$P_{out}(\beta_1, \beta_2) = P_{in}(t\beta_1 - r\beta_2, t\beta_2 + r\beta_1). \tag{6.71}$$

This relation between input and output P-functions was first derived in [Ou et al. (1987)]. It shows that if the input state is a classical state with $P_{in} \geq 0$, the output state must also retain this property. Thus, it is impossible to generate nonclassical states from classical states by linear transformations. On the other hand, linear transformation will keep the nonclassical properties of the input states. In other words, if certain non-classical properties are not exhibited for some nonclassical states, they may show up after some linear transformations on these states. We will see this in the next section.

The format discussed here for P-representation can be applied to other phase space function such as Wigner function. We will leave the derivation to Problem 6.4 and present the result as follows. For an input Wigner function $W_{in}(X_1, Y_1; X_2, Y_2)$, the output Wigner function is

$$W_{out}(X_1, Y_1; X_2, Y_2)$$
$$= W_{in}(tX_1 - rX_2, tY_1 - rY_2; tX_1 + rX_2, tY_1 + rY_2), \quad (6.72)$$

which is similar to Eq. (6.71). Here t, r must be real.

6.2.4 State Transformation of Squeezed States

To further demonstrate the usefulness of the simple technique in Section 6.2.2, we consider the homodyne (mixing) of a coherent state and a squeezed vacuum state by a beam splitter. From Sections 3.2.2 and 3.4.2, we find that the coherent state and the squeezed vacuum state can be expressed as

$$|\alpha\rangle_1 = \hat{D}_1(\alpha)|0\rangle$$
$$|\zeta\rangle_2 = \hat{S}_2(\zeta)|0\rangle, \quad (6.73)$$

where

$$\hat{D}_1(\alpha) = \exp(\alpha\hat{a}_1 - \alpha^*\hat{a}_1^\dagger)$$
$$\hat{S}_2(\zeta) = \exp(\zeta\hat{a}_2^2 - \zeta^*\hat{a}_2^{\dagger 2}). \quad (6.74)$$

The output state is then

$$|\Psi\rangle_{out} = \hat{U}\hat{D}_1(\alpha)\hat{S}_2(\zeta)|0\rangle$$
$$= \hat{U}\hat{D}_1(\alpha)\hat{S}_2(\zeta)\hat{U}^\dagger\hat{U}|0\rangle$$
$$= (\hat{U}\hat{D}_1(\alpha)\hat{U}^\dagger)(\hat{U}\hat{S}_2(\zeta)\hat{U}^\dagger)|0\rangle. \quad (6.75)$$

It is easy to show that

$$\hat{U}\hat{D}_1(\alpha)\hat{U}^\dagger = \exp(\alpha\hat{U}\hat{a}_1\hat{U}^\dagger - \alpha^*\hat{U}\hat{a}_1^\dagger\hat{U}^\dagger)$$
$$\hat{U}\hat{S}_2(\zeta)\hat{U}^\dagger = \exp(\zeta\hat{U}\hat{a}_2^2\hat{U}^\dagger - \zeta^*\hat{U}\hat{a}_2^{\dagger 2}\hat{U}^\dagger). \quad (6.76)$$

From Eq. (6.69) by setting $\alpha_1 = \alpha, \alpha_2 = 0$, we have

$$\hat{U}\hat{D}_1(\alpha)\hat{U}^\dagger = \exp(t\alpha\hat{a}_1 - t^*\alpha^*\hat{a}_1^\dagger - r\alpha\hat{a}_2 + r^*\alpha^*\hat{a}_2^\dagger)$$
$$= \hat{D}_1(t\alpha)\hat{D}_2(-r\alpha). \quad (6.77)$$

So the output state becomes

$$|\Psi\rangle_{out} = \hat{D}_1(t\alpha)\hat{D}_2(-r\alpha)[\hat{U}\hat{S}_2(\zeta)\hat{U}^\dagger]|0\rangle. \quad (6.78)$$

If we are only interested in the output state of one output port, say, port 2, we may trace out the state in port 1:

$$\hat{\rho}_2 = \text{Tr}_1 |\Psi\rangle_{out}\langle\Psi|_{out}$$
$$= \hat{D}_2(-r\alpha)\hat{\rho}_2^s\hat{D}_2^\dagger(-r\alpha), \tag{6.79}$$

where

$$\hat{\rho}_2^s = \text{Tr}_1[\hat{U}\hat{S}_2(\zeta)\hat{U}^\dagger|0\rangle\langle 0|\hat{U}\hat{S}_2^\dagger(\zeta)\hat{U}^\dagger] \tag{6.80}$$

is the output state at port 2 with only squeezed state input. $\hat{D}_1$ is traced out in Eq. (6.79). Furthermore,

$$\hat{U}\hat{S}_2(\zeta)\hat{U}^\dagger = \exp[\zeta(t\hat{a}_2 + r\hat{a}_1)^2 - \zeta^*(t^*\hat{a}_2^\dagger + r^*\hat{a}_1^\dagger)^2]$$
$$\approx \hat{S}_2(\zeta) \quad \text{for} \quad t \to 1, \ r \to 0. \tag{6.81}$$

So in the limit of transparent beam splitter ($t \to 1, r \to 0$) but with $\beta \equiv -r\alpha = $ finite, the output state in port 2 is

$$\hat{\rho}_2 \approx \hat{D}_2(\beta)\hat{S}_2(\zeta)|0\rangle\langle 0|\hat{S}_2^\dagger(\zeta)\hat{D}_2^\dagger(\beta), \tag{6.82}$$

or a pure state of

$$|\psi_2\rangle_{out} = \hat{D}_2(\beta)\hat{S}_2(\zeta)|0\rangle, \tag{6.83}$$

which is a coherent squeezed state discussed in Section 3.4.3. We have shown in Eq. (3.77) of Section 3.4.3 that

$$\hat{S}_2^\dagger(\zeta)\hat{D}_2(\beta)\hat{S}_2(\zeta) = \hat{D}_2(\beta\mu - \beta^*\nu^*), \tag{6.84}$$

where

$$\nu \equiv (\zeta/|\zeta|)\sinh|\zeta|, \quad \mu \equiv \sqrt{1 + |\nu|^2} = \cosh|\zeta|.$$

So the output state is

$$|\psi_2\rangle_{out} = \hat{S}_2(\zeta)\hat{D}_2(\beta\mu - \beta^*\nu^*)|0\rangle, \tag{6.85}$$

which is the squeezed coherent state discussed in Section 3.4.3 and can also be produced by directly injecting a coherent state into a degenerate parametric amplifier [Yuen (1976)]. Experimentally, the technique of using a highly transmissive beam splitter to combine a squeezed vacuum with a strong coherent state is quite popular for the generation of a coherent squeezed state.

Next we look into the case of two squeezed states input to the beam splitter with one from each side. We will consider a special case when the beam splitter is 50:50 and the squeezed states have equal strength but are 180 degree out of phase. From Eq. (6.73), the input state is simply

$$|\Psi\rangle_{in} = \hat{S}_1(-\zeta)\hat{S}_2(\zeta)|0\rangle \tag{6.86}$$

and the output state is

$$|\Psi\rangle_{out} = \hat{U}\hat{S}_1(-\zeta)\hat{S}_2(\zeta)|0\rangle$$
$$= \hat{U}\hat{S}_1(-\zeta)\hat{S}_2(\zeta)\hat{U}^\dagger\hat{U}|0\rangle$$
$$= (\hat{U}\hat{S}_1(-\zeta)\hat{U}^\dagger)(\hat{U}\hat{S}_2(\zeta)\hat{U}^\dagger)|0\rangle. \tag{6.87}$$

From the first part of Eq. (6.81), we have

$$\hat{U}\hat{S}_2(\zeta)\hat{U}^\dagger = \exp\left[\frac{\zeta}{2}\left(\hat{a}_2 + \hat{a}_1\right)^2 - \frac{\zeta^*}{2}\left(\hat{a}_2^\dagger + \hat{a}_1^\dagger\right)^2\right]$$
$$= \exp\left[\frac{\zeta}{2}\left(\hat{a}_1^2 + \hat{a}_2^2\right) - \frac{\zeta^*}{2}\left(\hat{a}_1^{\dagger 2} + \hat{a}_2^{\dagger 2}\right) + \zeta\hat{a}_1\hat{a}_2 - \zeta^*\hat{a}_1^\dagger\hat{a}_2^\dagger\right], \tag{6.88}$$

and

$$\hat{U}\hat{S}_1(-\zeta)\hat{U}^\dagger = \exp\left[-\frac{\zeta}{2}\left(\hat{a}_2 - \hat{a}_1\right)^2 + \frac{\zeta^*}{2}\left(\hat{a}_2^\dagger - \hat{a}_1^\dagger\right)^2\right]$$
$$= \exp\left[\frac{\zeta^*}{2}\left(\hat{a}_1^{\dagger 2} + \hat{a}_2^{\dagger 2}\right) - \frac{\zeta}{2}\left(\hat{a}_1^2 + \hat{a}_2^2\right) + \zeta\hat{a}_1\hat{a}_2 - \zeta^*\hat{a}_1^\dagger\hat{a}_2^\dagger\right], \tag{6.89}$$

where we set $t = r = 1/\sqrt{2}$. It is straightforward to show that if $\hat{A} \equiv \zeta^*(\hat{a}_1^{\dagger 2} + \hat{a}_2^{\dagger 2})/2 - \zeta(\hat{a}_1^2 + \hat{a}_2^2)/2$ and $\hat{B} \equiv \zeta\hat{a}_1\hat{a}_2 - \zeta^*\hat{a}_1^\dagger\hat{a}_2^\dagger$, then

$$[\hat{A}, \hat{B}] = 0,$$

so that we can write $\exp(\hat{B} \pm \hat{A}) = \exp(\pm\hat{A})\exp(\hat{B})$ in Eqs. (6.88), (6.89). The output state then becomes

$$|\Psi\rangle_{out} = \exp[2\zeta\hat{a}_1\hat{a}_2 - 2\zeta^*\hat{a}_1^\dagger\hat{a}_2^\dagger], \tag{6.90}$$

which is a two-mode squeezed state or a twin-beams state that can be produced from a nondegenerate parametric amplifier (see Sections 4.6 and 6.1.2).

6.3 Optical Resonators: Input-Output Theory of an Open Quantum System and Model of Decoherence

Optical resonators or cavities are commonly used optical devices that can clean up the spatial modes, act as bandpass filters, and enhance the field strength for nonlinear optical interaction. As we have seen in Section 2.2.1, the field inside an optical resonator has a discrete set of modes. Outside of the resonator, the spatial modes are matched through Gaussian waves. But since there is no resonator outside, the frequency is continuous. So, how do a discrete set of modes and a continuous set of modes couple together? Since optical resonators are often used as optical filters, this question also covers the mode transformation of an optical filter.

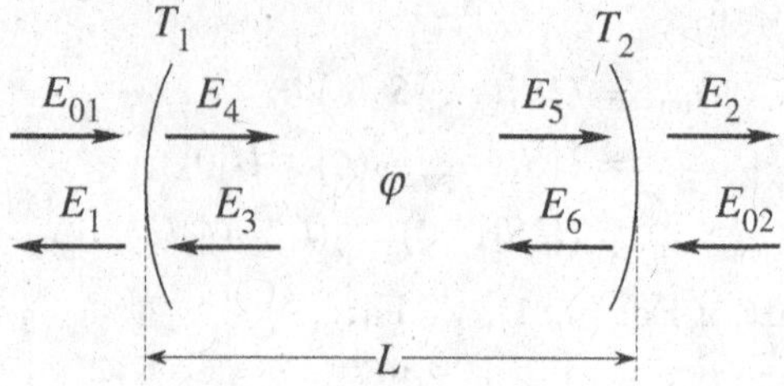

Fig. 6.3   Input and output waves coupled to the waves inside a cavity of transmissivity of T_1, T_2, respectively.

6.3.1   *Classical Wave Model*

Since spatial modes are matched, we will only consider the simplest case when a monochromatic plane wave of frequency ω enters a Fabry-Perot cavity with two mirrors of transmissivity $T_1, T_2 \sim 1$, respectively, as shown in Fig. 6.3.

Referring to Fig. 6.3, we denote the input fields as E_{01}, E_{02} and outgoing fields as E_1, E_2 and the fields inside as E_3, E_4, E_5, E_6. Treating each mirror as a beam splitter, we have

$$E_1 = E_3\sqrt{T_1} + E_{01}\sqrt{1 - T_1}, \quad E_4 = E_{01}\sqrt{T_1} - E_3\sqrt{1 - T_1}$$
$$E_2 = E_5\sqrt{T_2} - E_{02}\sqrt{1 - T_2}, \quad E_6 = E_{02}\sqrt{T_2} + E_5\sqrt{1 - T_2}. \quad (6.91)$$

Fields E_3, E_5 are respectively related to E_6, E_4 by $E_3 = E_6 e^{i\varphi}, E_5 = E_4 e^{i\varphi}$ with $\varphi = kL = nL\omega/c$ as the phase due to propagation (L is the length of the cavity and n is the index of refraction). Solving first the fields inside the cavity in terms of E_{01}, E_{02}, we obtain

$$E_6 = \frac{E_{01}e^{i\varphi}\sqrt{(1 - T_2)T_1} + E_{02}\sqrt{T_2}}{1 + e^{2i\varphi}\sqrt{(1 - T_1)(1 - T_2)}}. \quad (6.92)$$

Substituting back for the output fields, we have

$$E_1 = \frac{E_{01}(\sqrt{1 - T_1} + e^{2i\varphi}\sqrt{1 - T_2}) + E_{02}e^{i\varphi}\sqrt{T_1 T_2}}{1 + e^{2i\varphi}\sqrt{(1 - T_1)(1 - T_2)}} \quad (6.93)$$

$$E_2 = \frac{E_{01}e^{i\varphi}\sqrt{T_1 T_2} - E_{02}(\sqrt{1 - T_2} + e^{2i\varphi}\sqrt{1 - T_1})}{1 + e^{2i\varphi}\sqrt{(1 - T_1)(1 - T_2)}}. \quad (6.94)$$

Due to a π-phase shift at one of the two reflecting surfaces, the total round trip phase is $\pi + 2\varphi$. We then obtain the resonance condition:

$$2\varphi + \pi = 2N\pi \quad \text{or} \quad \omega_0 = (2N - 1)\pi c/2nL. \quad (6.95)$$

Here ω_0 is the resonance frequency and we have $e^{2i\varphi} = -\exp[i2nL(\omega - \omega_0)/c] \equiv -e^{i\delta}$ with $\delta \equiv 2\pi(\omega - \omega_0)/\Omega_{FSR}$ ($\Omega_{FSR} = 2\pi c/2nL$ is the free spectral range). If the input fields have frequency near resonance, we have $\delta \ll 1$ and $e^{2i\varphi} = -e^{i\delta} \approx -(1 + i\delta)$. Furthermore, for high finesse cavity, we have $T_1 \ll 1$, $T_2 \ll 1$. With these approximations, Eqs. (6.92)–(6.94) become

$$E_6 = \frac{E_{01}e^{i\varphi}\sqrt{T_1} + E_{02}\sqrt{T_2}}{(T_1 + T_2)/2 - i\delta}$$

$$E_1 = \frac{E_{02}e^{i\varphi}\sqrt{T_1 T_2} - E_{01}[(T_1 - T_2)/2 + i\delta]}{(T_1 + T_2)/2 - i\delta}$$

$$E_2 = \frac{E_{01}e^{i\varphi}\sqrt{T_1 T_2} + E_{02}[(T_2 - T_1)/2 + i\delta]}{(T_1 + T_2)/2 - i\delta}. \tag{6.96}$$

If we only have one input field, say E_{01}, both the transmitted field and the field inside have a Lorentzian line shape:

$$E_2 = \frac{2E_{01}e^{i\varphi}\sqrt{T_1 T_2}}{T_1 + T_2}\frac{\Delta\Omega/2}{\Delta\Omega/2 - i(\omega - \omega_0)},$$

$$E_6 = \frac{2E_{01}e^{i\varphi}\sqrt{T_1}}{T_1 + T_2}\frac{\Delta\Omega/2}{\Delta\Omega/2 - i(\omega - \omega_0)} \tag{6.97}$$

where the full line width at half maximum is

$$\Delta\Omega \equiv \Omega_{FSR}(T_1 + T_2)/2\pi = \Omega_{FSR}/\mathcal{F} \tag{6.98}$$

with $\mathcal{F} \equiv 2\pi/(T_1 + T_2)$ defined as the finesse of the cavity and Ω_{FSR} as the free spectral range. Equation (6.98) is often used for determining the finesse of the cavity: $\mathcal{F} = \Omega_{FSR}/\Delta\Omega$, with Ω_{FSR}, $\Delta\Omega$ measured by recording the transmitted intensity in a scan over one full spectral range of the cavity.

Notice that when the input field is on resonance, the intensity of the inside field is $|E_6|^2 = 4|E_{01}|^2 T_1/(T_1 + T_2)^2$, or the enhancement factor is $B = 4T_1/(T_1 + T_2)^2$, which can be quite large if T_1, T_2 are small. For example, let $T_1 = 0.02$, $T_2 = 0.01$. Then we have $B = 89$: the power inside the cavity is nearly 90 times of the input power.

6.3.2 Further Reading: Intra-cavity Second Harmonic Generation

In quantum optics, many of the interesting non-classical states are generated through nonlinear interaction such as three- and four-wave mixing discussed in Section 6.1.1. Three-wave mixing, because of the high nonlinear coefficients in some nonlinear crystals such as $LiNbO_3$ and $KNbO_3$, has become popular in quantum optical experiments for generating two-photon states and squeezed states. It requires a high power pump field at twice of the frequency of the quantum fields. This field is usually obtained through second harmonic generation (SHG) via nonlinear crystals, which requires high power for good conversion efficiency. For pulsed fields, this is not a problem since short optical pulses have extremely high peak power. For cw fields, however, this becomes a challenge.

Optical cavities were first proposed and tested for the enhancement of the nonlinear conversion [Ashkin et al. (1966)]. The large enhancement factor for the field inside cavity is important for intra-cavity second harmonic generation [Kozlovsky et al. (1988); Polzik and Kimble (1991)], where high power is a must for efficient nonlinear frequency conversion. We will consider this technique in this section.

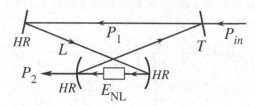

Fig. 6.4   Cavity enhancement of second harmonic generation in a bow-tie cavity configuration. HR: high reflector for the fundamental field.

Figure 6.4 shows a schematic for intra-cavity second harmonic generation with a bow-tie shaped enhancement cavity for the fundamental field at ω_0. A second harmonic field at $2\omega_0$ is generated inside the nonlinear crystal in a single-pass fashion.[5] Suppose the cavity has an input coupler of transmissivity T and an intra-cavity loss of L, which can be modeled as a two-ended cavity with $T_1 = T$ and $T_2 = L$. Moreover, there is a nonlinear loss $L_n \equiv P_2/P_1$ due to nonlinear conversion of the intra-cavity power P_1 to second harmonic power P_2. This loss is proportional to intra-cavity fundamental power P_1 as

$$L_n = P_2/P_1 = E_{NL}P_1, \tag{6.99}$$

where effective nonlinear coefficient E_{NL} is defined as the single-pass SHG conversion efficiency:

$$P_2 = E_{NL}P_1^2 \tag{6.100}$$

---

[5]Traveling wave cavities are preferred over standing wave cavities because of the simplicity in SHG geometry.

for $E_{NL}P_1 \ll 1$. E_{NL} can be obtained directly by measuring the single-pass non-saturated P_2 and the input P_1 and is a fixed number for a certain focussing geometry and crystal length (Typically, $E_{NL} \sim 10^{-4}W^{-1}$ for 1-cm long LiNbO$_3$ and $\sim 10^{-3}W^{-1}$ for 1-cm long KNbO$_3$).

Referring to Eq. (6.97), we have E_6 as the fundamental field inside the cavity and $P_1 = |E_6|^2$, $T_1 = T$ as the input coupling at one end of the cavity and $T_2 = L + L_n$ at the other end but with nonlinear loss included. The input fundamental field is on resonance and the input power is related to E_{01}: $P_{in} = |E_{01}|^2$. With these, Eq. (6.97) can be re-written as

$$4TP_{in} = (T + L + E_{NL}P_1)^2 P_1. \qquad (6.101)$$

The overall SHG conversion efficiency is defined as $\eta \equiv P_2/P_{in} = E_{NL}P_1^2/P_{in}$. Then Eq. (6.101) can be re-written in terms of η as

$$4T\sqrt{P_{in}E_{NL}} = \sqrt{\eta}(T + L + \sqrt{\eta E_{NL}P_{in}})^2, \qquad (6.102)$$

or

$$4(T/L)\sqrt{P_{in}/P_0} = \sqrt{\eta}(T/L + 1 + \sqrt{\eta}\sqrt{P_{in}/P_0})^2, \qquad (6.103)$$

where $P_0 \equiv L^2/E_{NL}$ is a characteristic power for the device defined by the round-trip intra-cavity loss L and nonlinear conversion coefficient E_{NL}. This is a cubic equation for η. Figure 6.5(a) shows η as a function of the dimensionless quantity $\sqrt{P_{in}/P_0}$ for a set of parameters of T/L. We find that there is an optimum input power for the conversion efficiency η at each value of T/L. Or, in other words, with each input power P_{in}, the input coupler coefficient T can be optimized for impedance matching $T = L + L_n$

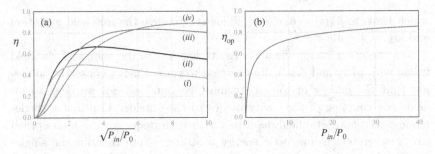

Fig. 6.5   (a) Intra-cavity second harmonic conversion efficiency η as a function of $\sqrt{P_{in}/P_0}$ for various values of T/L: $(i)1; (ii)2; (iii)5; (iv)8$.   (b) Optimized η_{op} as a function of P_{in}/P_0. $P_0 \equiv L^2/E_{NL}$.

to reach the maximum conversion efficiency, with

$$T_{op} = \frac{L}{2}\left(1 + \sqrt{1 + 4P_{in}/P_0}\right)$$

$$\eta_{op} = \frac{4P_{in}/P_0}{(1 + \sqrt{1 + 4P_{in}/P_0})^2}. \qquad (6.104)$$

In Fig. 6.5(b), we plot the optimized η_{op} versus the dimensionless quantity P_{in}/P_0 and it shows a quick increase up to somewhere around $P_{in}/P_0 = 1$ and a slower increase towards unit conversion after that.

6.3.3 Quantum Treatment of Cavity Loss: Decoherence of an Open Quantum System

As we discussed in Section 2.2.1, an optical resonator (cavity) defines a discrete set of modes for the optical waves inside it. If the cavity is ideal in the sense that there is no loss and interaction, the system is isolated and each mode will evolve independently. Of course, there does not exist such an ideal system in reality. For simplicity, let us ignore the interaction between different modes[6] and concentrate on the losses, which couple the system to the outside fields and make it an open system.

After the quantization of the cavity modes, we end up with the Hamiltonian of the whole system as:

$$\hat{H}_{cav} = \sum_\lambda \hat{H}_\lambda \quad \text{with} \quad \hat{H}_\lambda = \hbar\omega_\lambda(\hat{a}_\lambda^\dagger \hat{a}_\lambda + 1/2). \qquad (6.105)$$

Here each individual mode $\hat{a}_\lambda$ evolves independently in the Heisenberg picture:

$$\frac{d\hat{a}_\lambda}{dt} = \frac{1}{j\hbar}[\hat{a}_\lambda, \hat{H}_{cav}] = -j\omega_\lambda \hat{a}_\lambda, \qquad (6.106)$$

which leads to $\hat{a}_\lambda(t) = \hat{a}_\lambda e^{-j\omega_\lambda t}$. This is of course the free field evolution and $[\hat{a}_\lambda, \hat{a}_{\lambda'}^\dagger] = \delta_{\lambda\lambda'}$.

On the other hand, when the cavity has losses, the energy of the field inside will decay and gradually decrease to zero. Energy conservation means that the energy of the cavity must go outside. So, we need to include the coupling of the cavity modes to the outside. Consider a single-mode electromagnetic field described by $\hat{a}$ with frequency ω_0 in an optical cavity where electromagnetic energy is stored. Note that for the single-mode field, we have $[\hat{a}, \hat{a}^\dagger] = 1$. The loss of the cavity can be modeled

---

[6]Interaction between the modes will lead to a new set of eigen-modes of the system so we can relabel them and treat the new ones as independent modes. See Problem 6.1.

as an interaction of the single-mode field with a continuous multi-mode field outside the cavity described by $\{\hat{b}(\omega)\}$ with a free-field Hamiltonian $\hat{H}_{out} = \int d\omega \hbar \omega [\hat{b}^\dagger(\omega)\hat{b}(\omega) + 1/2]$ and $[\hat{b}(\omega), \hat{b}^\dagger(\omega')] = \delta(\omega - \omega')$. Notice that for different cavity modes, we assume that the eigen-frequencies of the cavity modes are well separated beyond their linewidths, so the modes can be thought of as independent and coupled to independent groups of outside frequency modes that are only within the linewidth of each cavity mode. We will show at the end that this assumption holds. Here, we also assume that the spatial mode of the cavity matches that of the outside fields. With these two assumptions, we can use one-dimensional quasi-monochromatic approximation (Sections 2.3.4, 2.3.5) for the fields outside the cavity.

The coupling Hamiltonian between the cavity mode and the field outside has the form of

$$\hat{H}_c = j\hbar \int d\omega \kappa(\omega) [\hat{b}^\dagger(\omega)\hat{a} - \hat{a}^\dagger \hat{b}(\omega)]. \tag{6.107}$$

This interaction is similar to the beam splitter Hamiltonian discussed in Section 6.2 and corresponds to linear coupling.

Gardiner and Collett worked out a detailed formalism to connect the field inside the cavity with the fields outside the cavity, after making some reasonable Markovian approximation by assuming all outside modes are coupled to the cavity mode with equal strength: $\kappa(\omega) = \sqrt{\gamma/2\pi}$ [Gardiner and Collett (1985)]. They derived an equation of motion for $\hat{a}$, the field inside the cavity, as

$$\frac{d\hat{a}}{dt} = -(\gamma/2 + j\omega_0)\hat{a} - \sqrt{\gamma}\hat{b}_{in}, \tag{6.108}$$

where

$$\hat{b}_{in} = \frac{1}{\sqrt{2\pi}} \int d\omega b_{in}(\omega)e^{-j\omega t} \tag{6.109}$$

is the input field to the cavity with $[\hat{b}_{in}(\omega), \hat{b}^\dagger_{in}(\omega')] = \delta(\omega - \omega')$. This is the one-dimensional quasi-monochromatic field with its spatial mode matched to the cavity mode. So, the evolution equation above shows that the loss constant γ is related to the coupling coefficient κ, and the outside field comes into the cavity as a noise term, which preserves the commutation relation for $\hat{a}$: $[\hat{a}, \hat{a}^\dagger] = 1$. With a time reversal of the evolution equation in Eq. (6.108), the coupling in Eq. (6.107) can also give rise to the following equation for the output field $\hat{b}_{out}$ [Gardiner and Collett (1985)]:

$$-\frac{d\hat{a}}{dt} = -(\gamma/2 - j\omega_0)\hat{a} + \sqrt{\gamma}\hat{b}_{out}. \tag{6.110}$$

Combining Eqs. (6.108) and (6.110), we obtain the input-output relation as related to the inside field: [Gardiner and Collett (1985)]

$$\hat{b}_{out} - \hat{b}_{in} = \sqrt{\gamma}\hat{a}. \tag{6.111}$$

The evolution equations in Eqs. (6.108) and (6.110) and the input-output relation in Eq. (6.111) complete the input-output formalism that relates the discrete mode $\hat{a}$ inside the cavity to the outside traveling continuous waves $\hat{b}_{in}, \hat{b}_{out}$.

With the evolution equation in Eq. (6.108) and the boundary condition in Eq. (6.111), we can find the output field $\hat{b}_{out}$ in terms of the input field $\hat{b}_{in}$. For this, let us write $\hat{a} = (1/\sqrt{2\pi}) \int d\omega \hat{a}(\omega)e^{-j\omega t}$ and substitute it into Eq. (6.108). We have

$$\frac{1}{\sqrt{2\pi}} \int d\omega(-j\omega)\hat{a}(\omega)e^{-j\omega t} = -(\gamma/2 + j\omega_0)\frac{1}{\sqrt{2\pi}} \int d\omega \hat{a}(\omega)e^{-j\omega t}$$
$$- \sqrt{\gamma}\frac{1}{\sqrt{2\pi}} \int d\omega \hat{b}_{in}(\omega)e^{-j\omega t}, \tag{6.112}$$

or

$$\frac{1}{\sqrt{2\pi}} \int d\omega \left[(j\omega - \gamma/2 - j\omega_0)\hat{a}(\omega)e^{-j\omega t} - \sqrt{\gamma}\hat{b}(\omega)\right]e^{-j\omega t} = 0. \tag{6.113}$$

This gives

$$\hat{a}(\omega) = \frac{\sqrt{\gamma}\hat{b}_{in}(\omega)}{j(\omega - \omega_0) - \gamma/2}. \tag{6.114}$$

Substituting into Eq. (6.111), we obtain

$$\hat{b}_{out}(\omega) = \hat{b}_{in}(\omega)\frac{\omega - \omega_0 - j\gamma/2}{\omega - \omega_0 + j\gamma/2}. \tag{6.115}$$

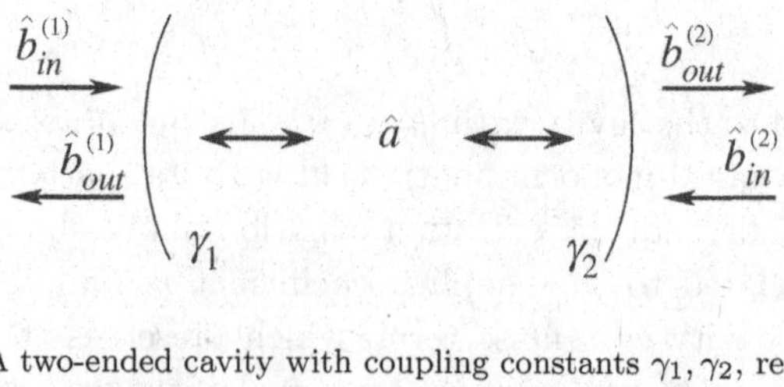

Fig. 6.6   A two-ended cavity with coupling constants γ_1, γ_2, respectively.

For a two-ended cavity in Fig. 6.6, we denote the two coupling coefficients as γ_1, γ_2 and the equation of motion for the cavity mode $\hat{a}$ is

$$\frac{d\hat{a}}{dt} = -[(\gamma_1 + \gamma_2)/2 + j\omega_0]\hat{a} - \sqrt{\gamma_1}\hat{b}_{in}^{(1)} - \sqrt{\gamma_2}\hat{b}_{in}^{(2)} \tag{6.116}$$

and each set of input/output fields satisfies a boundary condition as in Eq. (6.111):

$$\hat{b}_{out}^{(1)} - \hat{b}_{in}^{(1)} = \sqrt{\gamma_1}\hat{a}, \quad \hat{b}_{out}^{(2)} - \hat{b}_{in}^{(2)} = \sqrt{\gamma_2}\hat{a}. \tag{6.117}$$

Similarly we can find

$$\hat{a}(\omega) = \frac{\sqrt{\gamma_1}\hat{b}_{in}^{(1)}(\omega) + \sqrt{\gamma_2}\hat{b}_{in}^{(2)}(\omega)}{j(\omega - \omega_0) - (\gamma_1 + \gamma_2)/2} \tag{6.118}$$

and

$$\hat{b}_{out}^{(1)}(\omega) = \frac{\hat{b}_{in}^{(1)}(\omega)[\omega - \omega_0 - j(\gamma_1 - \gamma_2)/2] - j\hat{b}_{in}^{(2)}(\omega)\sqrt{\gamma_1\gamma_2}}{\omega - \omega_0 + j(\gamma_1 + \gamma_2)/2}$$

$$\hat{b}_{out}^{(2)}(\omega) = \frac{\hat{b}_{in}^{(2)}(\omega)[\omega - \omega_0 - j(\gamma_2 - \gamma_1)/2] - j\hat{b}_{in}^{(1)}(\omega)\sqrt{\gamma_1\gamma_2}}{\omega - \omega_0 + j(\gamma_1 + \gamma_2)/2}. \tag{6.119}$$

With $[\hat{b}_{in}^{(k)}(\omega), \hat{b}_{in}^{(l)\dagger}(\omega')] = \delta_{kl}\delta(\omega - \omega')$ $(k, l = 1, 2)$ for the input fields, it can be easily checked that $[\hat{b}_{out}^{(k)}(\omega), \hat{b}_{out}^{(l)\dagger}(\omega')] = \delta_{kl}\delta(\omega - \omega')$ for the output fields. Notice that the above expressions are similar to those in Eq. (6.96) with a proper phase φ and the coefficients satisfy the conditions for a lossless beam splitter discussed in Section 6.2. So, a cavity acts as a filter that can be modeled by a lossless beam splitter with frequency dependent transmissive and reflective coefficients. Notice that the effect of the cavity is only to the frequency components within the linewidth: $|\omega - \omega_0| \sim \gamma_1, \gamma_2$ so the assumption of independence about different cavity modes holds.

Comparing Eq. (6.119) with Eq. (6.97), we can relate the decay constants γ_1, γ_2 to the parameters of the cavity:

$$\gamma_i = \Omega_{FSR}T_i/2\pi \quad (i = 1, 2), \text{ and } \gamma_1 + \gamma_2 = \Omega_{FSR}/\mathcal{F}, \tag{6.120}$$

where the free spectral range $\Omega_{FSR} \equiv 2\pi/t_r = 2\pi c/2nL$ with t_r and $2nL$ as the cavity round-trip time and round-trip optical length, respectively.

We can now answer the question about what happens to a photon when it is filtered by an optical cavity. As we discussed in Section 2.5, we cannot talk about a photon without its mode. For a field with well-defined spatial modes, we can describe it by one-dimensional approximation with frequency modes (Section 2.3.5). So, the single photon that enters an optical cavity filter is a wave packet in a temporal mode described in Eq. (4.43) of Section 4.3.2. The cavity then acts as a frequency-dependent beam splitter with relations in Eq. (6.119) and splits the single-photon wave packet into two output ports, similar to the case discussed in Eq. (6.53) of Section 6.2.2:

$$|T\rangle_{out} = t|T\rangle_1|0\rangle_2 - r|0\rangle_1|T\rangle_2 \tag{6.121}$$

with

$$|T\rangle_k \equiv c_k \int d\omega f_k(\omega)\phi(\omega)e^{j\omega T}\hat{a}_k^\dagger(\omega)|0\rangle_k \quad (k = 1, 2) \qquad (6.122)$$

and $f_1(\omega) = [(\omega - \omega_0) - j(\gamma_1 - \gamma_2)/2]/[(\omega - \omega_0) - j(\gamma_1 + \gamma_2)/2]$, $f_2(\omega) = -j\sqrt{\gamma_1\gamma_2}/[(\omega - \omega_0) - j(\gamma_1 + \gamma_2)/2]$. $c_k^{-2} \equiv \int d\omega|f_k\phi|^2$ is for normalization. The transmitted and reflected fields are still in single-photon wave packets $|T\rangle_1, |T\rangle_2$ but their shapes are modified by the filter cavity. $t^2 \equiv 1/c_1^2, r^2 \equiv 1/c_2^2$ give the probabilities of transmission and reflection, respectively.

6.3.4 Optical Resonators with Nonlinear Interactions

A filter is still a linear device that cannot produce non-classical states from classical states. So, we need nonlinear interaction between different modes of the cavity. For this, we consider a three-wave mixing described by Hamiltonian in Eq. (6.6). If field $\hat{a}_3$ is a strong pump field at ω_p so that it can be replaced by a field amplitude $\alpha_p e^{-j\omega_p t}$ and the other fields are the two cavity modes denoted by $\hat{a}, \hat{b}$, the interaction Hamiltonian becomes

$$\hat{H}_I = j\hbar\zeta(\hat{b}^\dagger\hat{a}^\dagger e^{-j\omega_p t} - \hat{a}\hat{b}e^{j\omega_p t}), \qquad (6.123)$$

with $\zeta \equiv \eta\alpha_p$. Here η includes the nonlinear coefficient and spatial mode integral in Eq. (6.3), and other physical constants so that $|\alpha_p|^2 = \langle\hat{a}_3^\dagger\hat{a}_3\rangle$ is the photon number of the pump field $\hat{a}_3$. The Hamiltonian in the equation above is that for the parametric interaction and the device with the cavity is known as the optical parametric oscillator (OPO). Now we need to add $\hat{H}_I$ to $\hat{H}_{cav}$ for the evolution equation:

$$\frac{d\hat{a}}{dt} = \frac{1}{j\hbar}[\hat{a}, \hat{H}_{cav} + \hat{H}_I] - \frac{\alpha}{2}\hat{a} - \sqrt{\alpha}\hat{a}_{in},$$

$$\frac{d\hat{b}}{dt} = \frac{1}{j\hbar}[\hat{b}, \hat{H}_{cav} + \hat{H}_I] - \frac{\beta}{2}\hat{b} - \sqrt{\beta}\hat{b}_{in} \qquad (6.124)$$

with output coupling decay constants α, β for the two cavity modes. Here, for simplicity, we do not include intra-cavity loss other than the output coupling. With the Hamiltonian in Eq. (6.123), we have

$$\frac{d\hat{a}}{dt} = -\left(j\omega_a + \frac{\alpha}{2}\right)\hat{a} + \zeta\hat{b}^\dagger e^{-j\omega_p t} - \sqrt{\alpha}\hat{a}_{in},$$

$$\frac{d\hat{b}}{dt} = -\left(j\omega_b + \frac{\beta}{2}\right)\hat{b} + \zeta\hat{a}^\dagger e^{-j\omega_p t} - \sqrt{\beta}\hat{b}_{in}. \qquad (6.125)$$

Normally in an experiment, there are some frequency bands that are detected with center frequencies as ω_{a0}, ω_{b0} for $\hat{a}, \hat{b}$ fields. These center frequencies are, for example, from local oscillators of homodyne detection (see Section 9.3 for detail). Now let us go to the rotating frame:

$\hat{a} = \hat{A}(t)e^{-j\omega_{a0}t}, \hat{b} = \hat{B}(t)e^{-j\omega_{b0}t}$ with $\omega_{a0} + \omega_{b0} = \omega_p$ and write further: $\hat{A}(t) = (1/\sqrt{2\pi}) \int d\Omega \hat{A}(\Omega)e^{-j\Omega t}$, etc. and similar to Eq. (6.114), we can change Eq. (6.125) to

$$\left[-j(\Omega + \Delta_a) + \frac{\alpha}{2} \right] \hat{A}(\Omega) = \zeta \hat{B}^\dagger(-\Omega) - \sqrt{\alpha}\hat{A}_{in}(\Omega),$$

$$\left[-j(\Omega + \Delta_b) + \frac{\beta}{2} \right] \hat{B}(\Omega) = \zeta \hat{A}^\dagger(-\Omega) - \sqrt{\beta}\hat{B}_{in}(\Omega), \qquad (6.126)$$

where $\Delta_a \equiv \omega_{a0} - \omega_a, \Delta_b \equiv \omega_{b0} - \omega_b$ are detuning of the frequency components of our interest $(\omega_{a0}, \omega_{b0})$ from the cavity resonance frequencies (ω_a, ω_b). Notice that Ω is the frequency off-set from the frequencies of our interest: ω_{a0}, ω_{b0} that are determined by the detection system (say, homodyne detection). Solving jointly $\hat{A}(\Omega)$, $\hat{B}(\Omega)$ with $\hat{A}^\dagger(-\Omega)$, $\hat{B}^\dagger(-\Omega)$, we have

$$\hat{A}(\Omega) = \frac{\sqrt{\alpha}\hat{A}_{in}(\Omega)\left[\beta/2 + j(\Delta_b - \Omega)\right] + \zeta\sqrt{\beta}\hat{B}^\dagger_{in}(-\Omega)}{\zeta^2 - \left[\alpha/2 - j(\Delta_a + \Omega)\right]\left[\beta/2 + j(\Delta_b - \Omega)\right]},$$

$$\hat{B}(\Omega) = \frac{\sqrt{\beta}\hat{B}_{in}(\Omega)\left[\alpha/2 + j(\Delta_a - \Omega)\right] + \zeta\sqrt{\alpha}\hat{A}^\dagger_{in}(-\Omega)}{\zeta^2 - \left[\beta/2 - j(\Delta_b + \Omega)\right]\left[\alpha/2 + j(\Delta_a - \Omega)\right]}. \qquad (6.127)$$

From the input-output relation in Eq. (6.111), we have

$$\hat{A}_{out}(\Omega) = G(\Omega)\hat{A}_{in}(\Omega) + g(\Omega)\hat{B}^\dagger_{in}(-\Omega),$$

$$\hat{B}_{out}(\Omega) = e^{j\theta}\left[G(-\Omega)\hat{B}_{in}(\Omega) + g(-\Omega)\hat{A}^\dagger_{in}(-\Omega)\right] \qquad (6.128)$$

with

$$G(\Omega) \equiv \frac{\zeta^2 + \left[\alpha/2 + j(\Delta_a + \Omega)\right]\left[\beta/2 + j(\Delta_b - \Omega)\right]}{\zeta^2 - \left[\alpha/2 - j(\Delta_a + \Omega)\right]\left[\beta/2 + j(\Delta_b - \Omega)\right]},$$

$$g(\Omega) \equiv \frac{\zeta\sqrt{\alpha\beta}}{\zeta^2 - \left[\beta/2 + j(\Delta_b - \Omega)\right]\left[\alpha/2 - j(\Delta_a + \Omega)\right]},$$

$$e^{j\theta} \equiv \frac{\zeta^2 - \left[\beta/2 + j(\Delta_b + \Omega)\right]\left[\alpha/2 - j(\Delta_a - \Omega)\right]}{\zeta^2 - \left[\beta/2 - j(\Delta_b + \Omega)\right]\left[\alpha/2 + j(\Delta_a - \Omega)\right]}. \qquad (6.129)$$

With the commutation relations $[\hat{A}_{in}(\Omega), \hat{A}^\dagger_{in}(\Omega')] = \delta(\Omega - \Omega') = [\hat{B}_{in}(\Omega), \hat{B}^\dagger_{in}(\Omega')]$ and $[\hat{A}_{in}(\Omega), \hat{B}_{in}(\Omega')] = 0$ for the input fields, it can be easily checked that the same relations are satisfied by the output fields $\hat{A}_{out}(\Omega)$, $\hat{B}_{out}(\Omega)$.

The input-output relations in Eq. (6.128) are exactly those for the two-mode squeezed states discussed in Section 4.6 and are similar to those in Eq. (6.35) in Section 6.1.5 for the case of single-frequency pumped traveling wave three-wave mixing. Notice that the gain parameters in Eq. (6.129) depend on the coupling decay constants which are important in the case of optical cavities. The gain parameters are the highest when resonance conditions $\Delta_a = 0 = \Delta_b$ are met and we have

$$G(\Omega) = \frac{\zeta^2 + (\alpha/2 + j\Omega)(\beta/2 - j\Omega)}{\zeta^2 - (\alpha/2 - j\Omega)(\beta/2 - j\Omega)},$$

$$g(\Omega) = \frac{\zeta\sqrt{\alpha\beta}}{\zeta^2 - (\alpha/2 - j\Omega)(\beta/2 - j\Omega)}. \tag{6.130}$$

Note that $G(0), g(0)$ become infinity when $\zeta^2 = \alpha\beta/4$. This means that a threshold is reached for laser-like oscillation which is why the device is named as optical parametric oscillator (OPO). The quantum states to our interest are produced below threshold. The situation above threshold is not covered in the treatment here. For more, refer to [Heidmann et al. (1987)].

In the case when the two cavity modes are the same: $\hat{a} = \hat{b}$, the device becomes a degenerate OPO with $\omega_a = \omega_b$ and we choose the detection frequency at $\omega_0 = \omega_p/2$. We can simply replace b with a in Eq. (6.123) and the Hamiltonian for degenerate OPO is then

$$\hat{H}_D = j\hbar\frac{\zeta}{2}(\hat{a}^{\dagger 2}e^{-j\omega_p t} - \hat{a}^2 e^{j\omega_p t}), \tag{6.131}$$

where we replaced ζ with $\zeta/2$ because the nonlinear coefficient for the degenerate case is half of that for the non-degenerate case.[7] $\hat{H}_D$ is a special case of the Hamiltonian for second harmonic generation:

$$\hat{H}_{SHG} = j\hbar\frac{\eta}{2}(\hat{b}\hat{a}^{\dagger 2} - \hat{b}^\dagger\hat{a}^2) \tag{6.132}$$

when the harmonic field has a strong excitation so that we can replace $\hat{b}$ with its average $\langle\hat{b}\rangle = \alpha_p e^{-j\omega_p t}$ where $|\alpha_p|^2$ is the photon number of the pump field and $\zeta = \eta\alpha_p$. The evolution equation for the field in OPO is changed from Eq. (6.125) to

$$\frac{d\hat{a}}{dt} = -\left(j\omega_a + \frac{\beta}{2}\right)\hat{a} + \zeta\hat{a}^\dagger e^{-2j\omega_0 t} - \sqrt{\beta}\hat{a}_{in}, \tag{6.133}$$

where β is the output coupling constant for the OPO field. Now let us include the effect of loss which can be modeled as the coupling to another outside field $\hat{c}_{in}$ with a coupling coefficient γ. Then Eq. (6.133) becomes

$$\frac{d\hat{a}}{dt} = -\left(j\omega_a + \frac{\beta}{2} + \frac{\gamma}{2}\right)\hat{a} + \zeta\hat{a}^\dagger e^{-2j\omega_0 t} - \sqrt{\beta}\hat{a}_{in} - \sqrt{\gamma}\hat{c}_{in}. \tag{6.134}$$

---

[7]This point can be seen from the $\chi^{(2)}$-nonlinear term in Eq. (6.4) where if the E-field has two modes: $E = E_1 + E_2$, then $P_{NL}^{(2)} = \chi^{(2)}E_1^2 + \chi^{(2)}E_2^2 + 2\chi^{(2)}E_1 E_2$. So, the nonlinear coefficients for the degenerate cases (first and second terms) is half of that for the non-degenerate case (third term).

Using the same technique as in solving Eq. (6.125) and $\hat{A}(t) = \hat{a}(t)e^{j\omega_0 t}$, we find the output operator as

$$\hat{A}_{out}(\Omega) = G_D(\Omega)\hat{A}_{in}(\Omega) + g_D(\Omega)\hat{A}_{in}^\dagger(-\Omega)$$
$$+ G_L(\Omega)\hat{C}_{in}(\Omega) + g_L(\Omega)\hat{C}_{in}^\dagger(-\Omega) \qquad (6.135)$$

with

$$G_D(\Omega) \equiv \frac{\zeta^2 + \left[(\beta+\gamma)/2 + j(\Delta-\Omega)\right]\left[(\beta-\gamma)/2 + j(\Delta+\Omega)\right]}{\zeta^2 - \left[(\beta+\gamma)/2 + j(\Delta-\Omega)\right]\left[(\beta+\gamma)/2 - j(\Delta+\Omega)\right]}$$

$$g_D(\Omega) \equiv \frac{\zeta\beta}{\zeta^2 - \left[(\beta+\gamma)/2 + j(\Delta-\Omega)\right]\left[(\beta+\gamma)/2 - j(\Delta+\Omega)\right]},$$

$$G_L(\Omega) \equiv \frac{\sqrt{\beta\gamma}\left[(\beta+\gamma)/2 + j(\Delta-\Omega)\right]}{\zeta^2 - \left[(\beta+\gamma)/2 + j(\Delta-\Omega)\right]\left[(\beta+\gamma)/2 - j(\Delta+\Omega)\right]}$$

$$g_L(\Omega) \equiv \frac{\zeta\sqrt{\beta\gamma}}{\zeta^2 - \left[(\beta+\gamma)/2 + j(\Delta-\Omega)\right]\left[(\beta+\gamma)/2 - j(\Delta+\Omega)\right]}. \qquad (6.136)$$

Here $\Delta \equiv \omega_0 - \omega_a$ is the cavity detuning. The threshold of the degenerate OPO is $|\zeta_{th}| = (\beta+\gamma)/2$, which is the same as the non-degenerate case from Eq. (6.130). When there is no loss, or, $\gamma = 0$, we have $G_L = 0 = g_L$ and Eq. (6.136) becomes

$$G_D(\Omega) = G(\Omega) = \frac{\zeta^2 + \left[\beta/2 + j(\Delta-\Omega)\right]\left[\beta/2 + j(\Delta+\Omega)\right]}{\zeta^2 - \left[\beta/2 + j(\Delta-\Omega)\right]\left[\beta/2 - j(\Delta+\Omega)\right]}$$

$$g_D(\Omega) = g(\Omega) = \frac{\zeta\beta}{\zeta^2 - \left[\beta/2 + j(\Delta-\Omega)\right]\left[\beta/2 - j(\Delta+\Omega)\right]}. \qquad (6.137)$$

Then, Eq. (6.135) is exactly in the form of Eq. (4.70) in Section 4.6.3 for the squeezed states in multi-frequency mode. It can be easily checked that $G(\Omega), g(\Omega)$ satisfy Eq. (4.71). From Eq. (4.76) of Section 4.6.3 and Eq. (6.137), we can find the spectrum of squeezing $S_X(\Omega) = (|G(\Omega)| - |g(\Omega)|)^2 = 1/(|G(\Omega)| + |g(\Omega)|)^2$. Figure 6.7 shows $S_X(\Omega)$ as a function of $2\Omega/\beta$ for various parameters of ζ, Δ. The optimum squeezing of $S_X(\Omega) = 0$ occurs at threshold $\zeta = \beta/2$ with $\Delta = 0$ and $\Omega = 0$.

In practice, there will always be some losses inside the cavity. But we can lock the cavity on resonance with ω_0 so that $\Delta = 0$. Then Eq. (6.136)

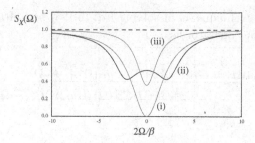

Fig. 6.7 Spectrum of squeezing $S_X(\Omega)$ for various values of ζ, Δ: (i) $\zeta = \beta/2, \Delta = 0$; (ii) $\zeta = \beta/2, \Delta = 5\beta/4$; (iii) $\zeta = \beta/8, \Delta = 0$.

becomes

$$G_D(\Omega) = \frac{\zeta^2 + \left[(\beta + \gamma)/2 - j\Omega\right]\left[(\beta - \gamma)/2 + j\Omega\right]}{\zeta^2 - \left[(\beta + \gamma)/2 - j\Omega\right]^2}$$

$$g_D(\Omega) = \frac{\zeta\beta}{\zeta^2 - \left[(\beta + \gamma)/2 - j\Omega\right]^2},$$

$$G_L(\Omega) = \frac{\sqrt{\beta\gamma}\left[(\beta + \gamma)/2 - j\Omega\right]}{\zeta^2 - \left[(\beta + \gamma)/2 - j\Omega\right]^2}$$

$$g_L(\Omega) = \frac{\zeta\sqrt{\beta\gamma}}{\zeta^2 - \left[(\beta + \gamma)/2 - j\Omega\right]^2}. \tag{6.138}$$

For $\Omega \ll (\beta + \gamma)/2$ and near threshold operation, the quantities above have a common phase determined by the common denominator. Then it is straightforward to show the minimum value of the spectrum is

$$S_X(\Omega) = (|G_D(\Omega)| - |g_D(\Omega)|)^2 + (|G_L(\Omega)| - |g_L(\Omega)|)^2. \tag{6.139}$$

With $\Omega \ll (\beta + \gamma)/2$ or $\Omega \sim 0$, we have

$$S_X(0) \approx \frac{[\zeta - (\beta - \gamma)/2]^2 + \beta\gamma}{[\zeta + (\beta + \gamma)/2]^2}. \tag{6.140}$$

When the threshold is approached: $\zeta \to \zeta_{th} \equiv (\beta + \gamma)/2$, we obtain

$$S_X(0) \approx \frac{\gamma}{\beta + \gamma} = \frac{L}{T + L}, \tag{6.141}$$

where we used Eq. (6.120) with T as the transmissivity of the coupling mirror and L as the round-trip loss excluding T. This is the best squeezing that can be achieved for the field generated from the OPO under threshold. Equation (6.141) has a straightforward explanation: the optimum noise level out of an OPO is just the fraction of photons lost among the overall output photons during one round trip. Since squeezing is due to photon pair correlation between the field components, it is highly sensitive to loss. The lost photons become uncorrelated. So, the fractional uncorrelated photons set the limit of noise level out of OPO.

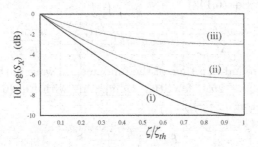

Fig. 6.8   The amount of squeezing $S_X(0)$ in log-scale as a function of ζ/ζ_{th} for three values of γ/β: (i)$\gamma/\beta = 0.11$; (ii) $\gamma/\beta = 0.3$; (iii) $\gamma/\beta = 1$.

In Fig. 6.8, we plot S_X from Eq. (6.140) in log-scale as a function of ζ/ζ_{th} for three values of $\gamma/\beta = L/T$. We find that pumping around 90% of threshold should be enough to obtain the maximum squeezing given in Eq. (6.141). Thus, the threshold value is an important parameter in the experiment for generating squeezed state. We estimate it in the following.

From Eq. (6.138), the threshold is reached when $\zeta = \zeta_{th} = (\beta + \gamma)/2$. From Eq. (6.120), we find β and γ are related to experimentally measurable quantities so we need to do the same for ζ. From Eq. (6.132) for degenerate case ($\hat{a} = \hat{b}$), we find $\zeta = \eta\alpha_p$ with $|\alpha_p|^2$ as the photon number of the pump field. But η is still undetermined since it depends on the nonlinear coefficient of the nonlinear medium and the mode overlap of the three waves. In fact, η is related to the effective nonlinear coefficient E_{NL} that can be measured with single-pass un-saturated second harmonic generation (SHG) (see Section 6.3.2). To find the connection, we use the same OPO cavity and nonlinear medium for SHG. From the Hamiltonian in Eq. (6.132) for SHG, we find the evolution equation of the harmonic field $\hat{b}$ as

$$\frac{d\langle\hat{b}\rangle}{dt} = \frac{\eta}{2}\langle\hat{a}\rangle^2, \tag{6.142}$$

where we assume the fields are all in coherent states and take average of the fields. For single-pass of the harmonic field, we can approximate

$$\langle\hat{b}\rangle \approx \eta\Delta t\langle\hat{a}\rangle^2/2 \tag{6.143}$$

with Δt as the transient time for the harmonic wave passing through the nonlinear medium. Then, with the power P_2 of the generated SHG and the power P_1 of the fundamental wave defined as

$$P_2 = \frac{\hbar\omega_2|\langle \hat{b}\rangle|^2}{\Delta t}, \quad P_1 = \frac{\hbar\omega_1|\langle \hat{a}\rangle|^2}{t_r}, \tag{6.144}$$

we obtain from Eq. (6.143)

$$P_2 = \frac{\hbar\omega_2|\langle \hat{b}\rangle|^2}{\Delta t} = \frac{1}{4}\hbar\omega_2\eta^2\Delta t\left(\frac{P_1 t_r}{\hbar\omega_1}\right)^2 = \frac{\eta^2 t_r^2\Delta t}{\hbar\omega_2}P_1^2. \tag{6.145}$$

Here, t_r is the round-trip time of the fundamental wave in the OPO cavity and we used $\hbar\omega_2 = 2\hbar\omega_1$ for SHG. Comparing with Eq. (6.100), we have (see also [Wu et al. (1987)])

$$E_{NL} = \frac{\eta^2 t_r^2\Delta t}{\hbar\omega_2}. \tag{6.146}$$

Now come back to the OPO. From the threshold condition of $\zeta_{th} = (\beta+\gamma)/2$ with $\zeta = \eta\alpha_p$ and $\beta+\gamma = 2\pi/t_r\mathcal{F}$ (Eq. (6.120)) and Eq. (6.146), we find the threshold power as

$$P_2^{th} = \frac{\hbar\omega_2|\alpha_p|^2}{\Delta t} = \frac{\hbar\omega_2|\zeta_{th}|^2}{\eta^2\Delta t} = \frac{\pi^2}{E_{NL}\mathcal{F}^2}. \tag{6.147}$$

Here, $\mathcal{F} = 2\pi/(T+L)$ is the finesse of the OPO cavity at fundamental frequency $\omega_1 = \omega_0$.

Experimental detection by homodyne measurement scheme of the squeezed states in multi-frequency mode leads to observation of quantum noise reduction in a wide spectral range and will be covered in Sections 9.5 and 10.1.2.

6.4 Problems

Problem 6.1 Coupling of two cavity modes: frequency splitting.

Consider a cavity with a birefringent crystal inside. It can be described by two polarization modes (e-ray and o-ray) of the optical field. The polarizations of the two modes are orthogonal to each other and labeled as x, y. The frequencies of the two modes are ω_1, ω_2, respectively, which correspond to resonant frequencies of the two polarizations, that is, only when the frequency of the incident field to the cavity is equal to these eigen-frequencies, there are outputs from the other side. The free Hamiltonian for the two modes is

$$\hat{H}_0 = \hbar\omega_1\hat{a}_x^\dagger\hat{a}_x + \hbar\omega_2\hat{a}_y^\dagger\hat{a}_y. \tag{6.148}$$

For a real birefringent crystal, due to imperfectness in growing process, the optical axis rotates by a very small amount. However, such a rotation will produce a coupling between the two polarization modes of x and y. The coupling has the form of

$$\hat{H}_I = \hbar g(\hat{a}_x^\dagger \hat{a}_y + h.c.). \tag{6.149}$$

(i) Make a mode transformation

$$\hat{A} = \hat{a}_x \cos\theta + \hat{a}_y \sin\theta,$$
$$\hat{B} = -\hat{a}_x \sin\theta + \hat{a}_y \cos\theta, \tag{6.150}$$

which corresponds to a polarization rotation. Find the angle θ so that $\hat{H} = \hat{H}_0 + \hat{H}_I$ can be written in the free form of $\hat{H}_0$ and compare the eigenfrequencies of $\hat{H}$ with those of $\hat{H}_0$ for different values of $\delta = \omega_1 - \omega_2$. Prove that operators $\hat{A}$, $\hat{B}$ are annihilation operators by showing that they satisfy the commutation relation.

(ii) Consider the state $|\psi_A\rangle = |\alpha_A\rangle_A|0\rangle_B$ where mode $\hat{A}$ is in a coherent state while mode $\hat{B}$ is in vacuum. Prove that $|\psi_A\rangle$ is an eigenstate of both $\hat{a}_x$, $\hat{a}_y$ and therefore $|\psi_A\rangle = |\alpha_x^A\rangle_x|\alpha_y^A\rangle_y$. Find their eigenvalues α_x^A, α_y^A (Hint: use $\hat{A}|\psi_A\rangle = \alpha_A|\psi_A\rangle$, $\hat{B}|\psi_A\rangle = 0$).

(iii) Do the same for the state $|\psi_B\rangle = |0\rangle_A|\alpha_B\rangle_B$.

(iv) Assume that a light field in coherent state with frequency ω and polarization x is incident into the cavity. What does the output (frequency and polarization) look like as frequency ω is scanned?

Problem 6.2 Anti-bunching by mixing of a coherent state and a two-photon state [Shafiei et al. (2004)].

Consider the following input states for a 50:50 beam splitter. The input mode a is in a two-photon state

$$|\eta\rangle_a \approx |0\rangle_a + \eta|2\rangle_a \quad (|\eta| \ll 1), \tag{6.151}$$

and the b mode is in a coherent state

$$|\alpha\rangle_b \approx |0\rangle_b + \alpha|1\rangle_b + \frac{\alpha^2}{\sqrt{2}}|2\rangle_b + \frac{\alpha^3}{\sqrt{6}}|3\rangle_b \quad (|\alpha| \ll 1). \tag{6.152}$$

(i) Find the output state of the beam splitter. Keep only up to the 3-photon terms.

(ii) Calculate up to the second non-zero order the normalized intensity correlation function $g^{(2)} = \langle \hat{n}^2 \rangle / \langle \hat{n} \rangle^2$ for each output field of the BS and find the condition for anti-bunching: $g^{(2)} < 1$.

Problem 6.3 Three-photon NOON state by mixing of a coherent state and a two-photon state [Shafiei et al. (2004)].

Use the output state in Problem 6.2 to obtain the condition for which the three-photon part of the state is a three-photon NOON state (Section 4.5.1).

Problem 6.4 State transformation of the Wigner function for a beam splitter.

Use the characteristic function in Eq. (3.168) for the Wigner function and the operator transformation in Eq. (6.52) for a lossless beam splitter to prove the transformation of the Wigner function in Eq. (6.72).

Problem 6.5 Conversion between 1-photon and 3-photon for the photon doubler in Section 6.1.4.

For the interaction in Eq. (6.18), if the input state is $|\Psi_{in}\rangle = |1\rangle_3 |1\rangle_2$, find the output state. What if the input state is $|\Psi_{in}\rangle = |0\rangle_3 |3\rangle_2$?

Problem 6.6 Non-degenerate optical parametric amplifier with losses.

When a non-degenerate optical parametric oscillator (NOPO) is operated below threshold, it becomes a non-degenerate parametric amplifier (NOPA). We discussed the case without loss in Section 6.3.4. Now let us include the intra-cavity losses by modeling the losses as the coupling of the two cavity modes to other outside fields, just like the degenerate case in Eq. (6.134) but now for two non-degenerate modes. For simplicity, we assume the two cavity modes have the same output coupling coefficient of $\alpha = \beta$ and the same loss coefficient of γ.

(i) Using the method in Section 6.3.4 to prove that for doubly resonant condition of $\Delta_a = 0 = \Delta_b$, the output fields from the NOPO have the frequency component at $\Omega = 0$ in the form of [Ou et al. (1992a)]

$$\hat{a}_{out} = \bar{G}\hat{a}_{in} + \bar{g}\hat{b}^{\dagger}_{in} + \bar{G}'\hat{a}_0 + \bar{g}'\hat{b}^{\dagger}_0,$$
$$\hat{b}_{out} = \bar{G}\hat{b}_{in} + \bar{g}\hat{a}^{\dagger}_{in} + \bar{G}'\hat{b}_0 + \bar{g}'\hat{a}^{\dagger}_0, \qquad (6.153)$$

with $\bar{G}^2 - \bar{g}^2 + \bar{G}'^2 - \bar{g}'^2 = 1$ and

$$\bar{G} = [(\beta^2 - \gamma^2)/4 + |\zeta|^2]/M, \qquad \bar{g} = \zeta\beta/M$$

$$\bar{G}' = \sqrt{\beta\gamma}(\gamma + \beta)/2M, \qquad \bar{g}' = \zeta\sqrt{\gamma\beta}/M, \qquad (6.154)$$

where ζ is proportional to the pump amplitude and β, γ proportional to the cavity round-trip output coupling T and loss L, respectively, and $M \equiv (\gamma + \beta)^2/4 - |\zeta|^2$. $\hat{a}_0$, $\hat{b}_0$ denote the vacuum modes coupled in through the losses.

(ii) Show that the EPR correlation between the two outputs (see Section 4.6.2), i.e., $\langle\Delta^2(\hat{X}_{a_{out}} - \hat{X}_{b_{out}})\rangle$, is $2\gamma/(\beta+\gamma)$ at best when threshold is reached: $\zeta = \zeta_{th} \equiv (\beta + \gamma)/2$, similar to Eq. (6.141).

(iii) Show that the threshold power for the non-degenerate OPO discussed here is the same as that in Eq. (6.147) for a degenerate OPO.

Problem 6.7 Intra-cavity second harmonic generation with double resonance condition [Collett and Walls (1985); Ou and Kimble (1993)].

In Section 6.3.2, we considered the system of intra-cavity second harmonic generation (SHG) in a single-pass configuration for the harmonic wave by using classical wave arguments. We will treat the system with both waves on resonance and with quantum arguments here using the method in Section 6.3.4. We start with the Hamiltonian for SHG in Eq. (6.132) which is re-written as

$$\hat{H}_{SHG} = j\hbar\frac{\eta}{2}(\hat{b}\hat{a}^{\dagger 2} - \hat{b}^{\dagger}\hat{a}^2) \qquad (6.155)$$

Denoting the decay constants as γ_1 and γ_2 for the fundamental and harmonic fields, respectively, which are related to the corresponding coupling coefficients T_1, T_2 by Eq. (6.120), we have the equations of motion as

$$\frac{d\hat{a}}{dt} = -\left(j\omega_a + \frac{\gamma_1}{2}\right)\hat{a} + \eta\hat{b}\hat{a}^{\dagger} - \sqrt{\gamma_1}\hat{a}_{in},$$

$$\frac{d\hat{b}}{dt} = -\left(j\omega_b + \frac{\gamma_2}{2}\right)\hat{b} - \frac{1}{2}\eta\hat{a}^2 - \sqrt{\gamma_2}\hat{b}_{in}. \qquad (6.156)$$

For simplicity, we will assume double resonance condition $\omega_a - \omega_0 = 0$ and $\omega_b - 2\omega_0 = 0$ with ω_0 as the frequency of the fundamental input field and no intra-cavity losses for both waves. The equations above are coupled nonlinear differential operator equations that are impossible to solve. However, both the harmonic and fundamental fields are strong in SHG so we can write $\hat{b} = \langle\hat{b}\rangle + \Delta\hat{b}$ and $\hat{a} = \langle\hat{a}\rangle + \Delta\hat{a}$ with $|\langle\hat{b}\rangle|^2 \gg \langle\Delta\hat{b}^{\dagger}\Delta\hat{b}\rangle$ and

$|\langle\hat{a}\rangle|^2 \gg \langle\Delta\hat{a}^\dagger\Delta\hat{a}\rangle$. Substituting these to Eq. (6.156) and keeping only up to the linear terms in $\Delta\hat{b}$ and $\Delta\hat{a}$, we can find the linearized equations for $\Delta\hat{b}$ and $\Delta\hat{a}$.

(i) Use the method outlined above to derive equations for $\langle\hat{b}\rangle$, $\langle\hat{a}\rangle$ (zeroth order) and for $\Delta\hat{b}$, $\Delta\hat{a}$ (first order).

(ii) For second harmonic generation, the fundamental wave has a non-zero coherent state input but the harmonic wave has vacuum input. Find the steady state solutions for $\langle\hat{b}\rangle$, $\langle\hat{a}\rangle$ and $\langle\hat{b}_{out}\rangle$.

(iii) Using the method in deriving Eq. (6.147), show that the conversion efficiency from input fundamental to output harmonic waves follows a cubic equation similar to Eq. (6.102) in Section 6.3.2 but with the effective $T^{\text{eff}} = T_1$ and $E_{NL}^{\text{eff}} = 4E_{NL}/T_2$ for this cubic equation. Here, E_{NL} is the single-pass non-saturated nonlinear conversion coefficient given in Eqs. (6.100) and (6.146). So, the effective nonlinear conversion coefficient is enhanced by a factor of $4/T_2$ due to the harmonic resonance cavity.

(iv) Use the method in Section 6.3.4 to solve the linearized equations for $\Delta\hat{b}$, $\Delta\hat{a}$ and find the spectra of squeezing for both the fundamental and the harmonic waves.

Problem 6.8 Generalized Hong-Ou-Mandel effects for three and four photons.

Hong-Ou-Mandel two-photon interference effect is for two photons entering on a lossless beam splitter with one from each side [see Eqs. (6.54) and (6.55) in Section 6.2.2 and more in Section 8.2.1]. Now, consider three or four photons entering a lossless beam splitter of transmissivity T and reflectivity R.

(i) Show that the output state for $|2\rangle_a|1\rangle_b$ input state is

$$|\Psi_3\rangle_{out} = \sqrt{3T^2R}|3\rangle_A|0\rangle_B + \sqrt{3TR^2}|0\rangle_A|3\rangle_B$$
$$+(T-2R)\sqrt{T}|2\rangle_A|1\rangle_B + (R-2T)\sqrt{R}|1\rangle_A|2\rangle_B, \quad (6.157)$$

where a, b are the input modes and A, B are the output modes of the BS. If we make a three-photon coincidence measurement with detection of two photons at A and one photon at B, then the coincidence will be zero when $T = 2R = 2/3$. This is a three-photon destructive interference effect [Sanaka et al. (2006)], similar to the Hong-Ou-Mandel effect for two photons.

(ii) Show that the output state for $|2\rangle_a|2\rangle_b$ input state is

$$|\Psi_4\rangle_{out} = = TR\sqrt{6}\big(|4\rangle_A|0\rangle_B + |0\rangle_A|4\rangle_B\big) + \big(T^2 + R^2 - 4TR\big)|2\rangle_A|2\rangle_B$$
$$+(T-R)\sqrt{6TR}\big(|3\rangle_A|1\rangle_B + |1\rangle_A|3\rangle_B\big). \qquad (6.158)$$

Similar to (i), if we make a four-photon coincidence measurement with detection of two photons at both A and B, then the coincidence will be zero when $T = (3 \pm \sqrt{6})/6$, $R = (3 \mp \sqrt{6})/6$. This is a four-photon destructive interference effect [Liu et al. (2007)].

Problem 6.9 Schrödinger cat state with a loss [Walls and Milburn (1985)]

We can model the loss with a beam splitter of transmissivity $T = 1 - \gamma$. Use the result of Problem 6.4 and the Wigner function in Eq. (3.178) for the Schrödinger cat state in Eq. (3.51) of Section 3.3 to find the Wigner function for the Schrödinger cat state with a loss of γ. Show the result in Eq. (3.56). You need to trace out one of the outputs of the BS.

PART 2

Experimental Techniques in Quantum Optics and Their Applications

Chapter 7

Experimental Techniques of Quantum Optics I: Photon Counting Technique

In the classical world where we are, it is usually difficult to observe quantum phenomena. What we measure in experiments, such as current, voltage, etc., are all macroscopic classical quantities. Then, how can we observe the quantum behavior of optical fields from these measured classical quantities? It turns out that, as we discussed in Chapter 5, the quantum behaviors of light are revealed in the statistical correlations among the measured quantities. The experimental techniques to be discussed in this chapter and Chapter 9 will be some special experimental methods for observing these statistical correlations that will show quantum phenomena in optics. To understand these, we must connect what we learned in the previous chapters about the quantum theory of light with the physical observables measured in the experiments. In this way, we will use the quantum theory of light to explain the phenomena observed in experiments and furthermore to guide the experiments in search of new quantum phenomena. In this chapter, we will discuss the photon counting technique, one of the two techniques commonly used in the experimental study of quantum optics.

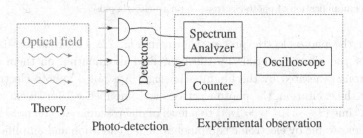

Fig. 7.1   The measurement process in optics experiment: from incoming optical field to recorded outputs of the measurable quantities. Photo-detection process bridges between theoretical description of the optical fields and the experimental observations.

7.1   The Process of Photo-Detection

In any one of the optical experiments, one device is essential, that is, the photo-detector. Although we can see light with our bare eyes, they are incredibly crude and inaccurate devices. Photo-detectors used in the lab are usually made of photo-sensitive materials. The common materials are semiconductors such as silicon, germanium, and gallium arsenide, etc. They are made into photo-electric diodes. As shown in Fig. 7.1, photo-detectors are the bridges between optical fields and the electric currents and voltages we measure. For optical fields, we describe them with the quantum theory we established in the previous chapters. The electric currents and voltages are important parts of the experimental phenomena we can measure in the lab. The theory of photo-detection connects the theory with the experiments.

In order to observe quantum behavior of light, the photo-detection process must be a quantum process. Fortunately, this process exhibits the very first observed effect correctly explained by the quantum theory of light, that is, the famous photo-electric effect. Einstein was the first to understand this effect when he proposed the concept of photon. The process of photo-detection is also a quantum measurement process. So, the investigation of this process will help us understand the essence of quantum mechanics.

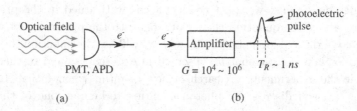

Fig. 7.2   The process of photo-electric detection: (a) the generation of photo-electrons; (b) the amplification of photo-electrons to photo-electric pulses.

In the photo-electric process shown in Fig. 7.2, electrons are excited out of the photo-materials by the light field illuminating on them. The electrons generated by the illuminating light are called "photo-electrons". After the excitation, the photo-electrons are amplified within an extremely short time $(T_R \sim 10^{-9}s = 1ns)$ to a level of macroscopic electric pulses that are measurable by electronic equipments. The generation and amplification are usually done in one single device such as photo-multiplier tubes (PMT), avalanche photo-diodes (APD).

When the illuminating light is not so strong, the photo-electric pulses are separated from each other. We can then work on each individual pulses by first performing pulse height discrimination to eliminate the dark counts and then pulse shaping for further operation. During this process, the detector is quenched and cannot generate another photo-electric pulse and so, it is not responsive to light. This time period ($\sim 10^{-6}$ s $= 1$ μs) is known as the "dead time" of the detector. This limits the maximum count rate to around 10^6 counts per second (cps). For this type of photo-detector, the light intensity must be low enough so that the photo-electric pulses are resolved in time by the detector. Then, we can count each photo-electric pulse. This is photon counting technique, as shown in Fig. 7.3.

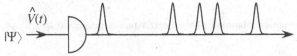

Fig. 7.3 Photon counting technique for weak optical fields.

On the other hand, when the light intensity is large and there is no pulse processing so that there is no dead time limit for the detector, the photo-electric pulses from different photo-electrons will overlap to form a continuous photo-electric current. Since the generation of photo-electrons is random due to its probabilistic nature, the photo-current is a random process and will fluctuate with time, as shown in Fig. 7.4.

Fig. 7.4 Continuous photo-current under relatively strong light intensity.

The experimental technique discussed in this chapter and its applications in next chapter will deal with the case of weak light intensity where we can count photo-electric pulses.

7.2 Detection Probabilities of Photo-Electrons

When the optical field is weak, the photo-electric events are rare. So, within the duration of one photo-electric pulse ($\Delta t \sim T_R \sim 1$ ns), the probability of generating two photo-electrons is very small: $P_2 \ll 1$, that is, the photoelectric pulses are separated and do not overlap, as shown in Fig. 7.5. In this case, the photo-electric pulses are countable. We then use digital counters to register the number of the photo-electric pulses and make a counting measurement.

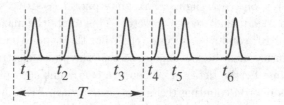

Fig. 7.5 The non-overlapping photo-electric pulses generated by an optical field with low intensity.

The theory of photo-electric detection was established around 1963 by Glauber and Mandel independently. Mandel's theory is a detection theory with quantized atoms but for classical optical fields·(waves) while Glauber's theory is all quantum mechanical including the treatment of the optical fields. Their results are similar and are all about the probability of detecting photo-electrons. In Glauber's theory of photo-detection, when an optical field in the quantum state of $|\Psi\rangle$ illuminates on a photo-detector, the probability of producing a photo-electron in the infinitesimal time interval of t to $t + dt$ is given by

$$p_1 dt = \alpha \langle \hat{\mathbf{E}}^{(-)}(\mathbf{r}, t) \cdot \hat{\mathbf{E}}^{(+)}(\mathbf{r}, t) \rangle_\Psi dt. \qquad (7.1)$$

Here, α is a constant related to the detector but independent of the quantum state $|\Psi\rangle$. $\hat{\mathbf{E}}^{(+)}(\mathbf{r}, t)$ and $\hat{\mathbf{E}}^{(-)}(\mathbf{r}, t)$ are respectively the positive and negative frequency parts of the electric field operators of the optical field:

$$\hat{\mathbf{E}}^{(-)\dagger}(\mathbf{r}, t) = \hat{\mathbf{E}}^{(+)}(\mathbf{r}, t) = i\sqrt{4\pi} \sum_{s=1,2} \int d^3\mathbf{k} \sqrt{\frac{\hbar\omega}{2}} \, \hat{a}_s(\mathbf{k}) \hat{\epsilon}_{\mathbf{k},s} \frac{e^{i(\mathbf{k}\cdot\mathbf{r}-\omega t)}}{(2\pi)^{3/2}}. \qquad (7.2)$$

For an optical field with only one polarization, under the one-dimensional quasi-monochromatic approximation (see Sections 2.3.4 and 2.3.5), Eq. (7.1) becomes

$$p_1 dt = \alpha' \langle \hat{E}^{(-)}(t) \hat{E}^{(+)}(t) \rangle_\Psi dt, \qquad (7.3)$$

where

$$\left[\hat{E}^{(-)}(t)\right]^\dagger = \hat{E}^{(+)}(t) = \frac{1}{\sqrt{2\pi}} \int d\omega \hat{a}(\omega) e^{-i\omega t}. \qquad (7.4)$$

Hence, during a finite time interval of T, the probability of finding a photo-electron is

$$P_1 = \int_T p_1 dt = \alpha' \int_T dt \langle \hat{E}^{(-)}(t) \hat{E}^{(+)}(t) \rangle_\Psi. \qquad (7.5)$$

By the definition of probability, we know that the number of photo-electrons registered during time T is $N_e \propto P_1$, that is,

$$N_e = C_0 P_1 = \eta \int_T dt \langle \hat{E}^{(-)}(t) \hat{E}^{(+)}(t) \rangle_\Psi = \int_T dt R_e(t). \qquad (7.6)$$

C_0, η are all some constants. In the expression above, the counting rate of photo-electric pulses is then $R_e(t) = \eta \langle \hat{E}^{(-)}(t) \hat{E}^{(+)}(t) \rangle_\Psi$. From Sections 2.3.4 and 2.3.5, we learned that $\langle \hat{E}^{(-)}(t) \hat{E}^{(+)}(t) \rangle_\Psi = R_p(t)$ is the photon number rate of the optical field in state $|\Psi\rangle$. So, we have $R_e(t) = \eta R_p(t)$ or $\eta = R_e(t)/R_p(t)$, that is, η is the ratio between the counted number of photo-electrons and the arriving number of photons. So, the physical meaning of η is the quantum efficiency of the photo-detector, i.e., the percentage of photons converted to photo-electrons.

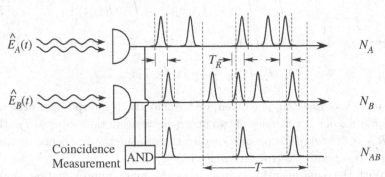

Fig. 7.6   Coincidence measurement between two detectors: only when two pulses overlap within T_R, is a coincidence pulse generated.

If we place two detectors at two separate locations denoted as $\mathbf{r}'$ and $\mathbf{r}''$ shown in Fig. 7.6, we can measure the correlation of the optical field at these two locations. Actually, what we can really measure is the statistical correlation between the photo-electrons generated by the two detectors. The simplest method is to use an "AND" gate to measure the coincidence count N_c of simultaneously registering two photo-electric pulses within the opening time T_R of the "AND" gate, as shown in Fig. 7.6. The connection between this measurement and the optical field is given from Glauber's photo-detection theory: for an optical field in quantum state $|\Psi\rangle$, the joint probability of producing two photo-electrons, one by detector A at $\mathbf{r}'$ during time interval t' to $t' + dt'$ and the other by detector B at $\mathbf{r}''$ during time interval t'' to $t'' + dt''$, is

$$p_2(\mathbf{r}'t', \mathbf{r}''t'')dt'dt'' = \alpha'\alpha'' \langle \hat{\mathbf{E}}_B^{(-)}(\mathbf{r}'', t'') \hat{\mathbf{E}}_A^{(-)}(\mathbf{r}', t')$$
$$\times \hat{\mathbf{E}}_A^{(+)}(\mathbf{r}', t') \hat{\mathbf{E}}_B^{(+)}(\mathbf{r}'', t'') \rangle_\Psi dt'dt''. \quad (7.7)$$

Integrating over time, we obtain the joint detection probability of two photo-electrons in a finite time interval T:

$$P_2 = \int_T p_2(t', t'')dt'dt''. \quad (7.8)$$

As shown in Fig. 7.6, we usually use the "AND" gate with a time window of T_R in the experiment to record the joint probability of simultaneous arrival of the two photo-electric pulses. To interpret the physical meaning of Eq. (7.8) as related to the experimental observations, we rewrite the time integral in Eq. (7.8) as

$$P_2 = \int_T dt \int_{T_R} p_2(t, t + \tau)d\tau \equiv \int_T dt R_{p2}(t), \quad (7.9)$$

where we make changes of variables: $t', t'' \to t, t + \tau$, and define

$$R_{p2}(t) \equiv \int_{T_R} p_2(t, t + \tau)d\tau. \quad (7.10)$$

From the way Eq. (7.9) is written, we see that R_{p2} is interpreted as the rate of simultaneous appearance of two photo-electrons in time interval T_R, that is, $R_{p2} = dP_2/dt$. It is proportional to the measured coincidence counting rate of two photo-electrons in time window T_R.

With the one-dimensional quasi-monochromatic approximation, similar to Eq. (7.3), Eq. (7.7) is changed to

$$p_2(t', t'')dt'dt'' = \alpha'\alpha'' \langle \hat{E}_B^{(-)}(t'') \hat{E}_A^{(-)}(t') \hat{E}_A^{(+)}(t') \hat{E}_B^{(+)}(t'') \rangle_\Psi dt'dt''. \quad (7.11)$$

Here, $\hat{E}_{1,2}^{(\pm)}(t)$ is given in Eq. (7.4). So, similar to Eq. (7.6), the coincidence count N_c of two photo-electrons in time T is proportional to P_2 and Eq. (7.9) becomes

$$N_c = \int_T dt R_2(t), \quad (7.12)$$

where $R_2(t) \propto R_{p2}(t)$ and has the form of

$$R_2(t) \equiv \int_{T_R} \eta_A \eta_B \langle \hat{E}_B^{(-)}(t + \tau) \hat{E}_A^{(-)}(t) \hat{E}_A^{(+)}(t) \hat{E}_B^{(+)}(t + \tau) \rangle_\Psi d\tau. \quad (7.13)$$

Here, η_A, η_B are the quantum efficiencies of detectors A and B, respectively. R_2 is the coincidence counting rate for simultaneous appearance of two photo-electrons in time T_R measured in the experiment. As mentioned before, this rate can be measured in the experiment with the "AND" gate method for the coincidence of two photo-electric pulses (see Fig. 7.6).

For a continuous optical field, the change of the field with time is a stationary process. Then, $R_e(t)$ of Eq. (7.6) and $R_2(t)$ of Eq. (7.13) are independent of t and become constant. The total count of photo-electrons and the total coincidence count within time T are $N_e = R_e T$ and $N_c = R_2 T$, respectively.

For a non-continuous pulsed optical field (multi-frequency), we need to integrate over the response time of the detector to obtain the overall probability of photo-electrons. When the pulse duration is much shorter than the response time of the detector, which is usually the case for ultra-short pulses, the time integral will cover the whole pulse and is equivalent to integration over the whole length of the pulse. In this case, we have the probability of generating a photo-electron in one single pulse:

$$P_1 = \int_{-\infty}^{\infty} p_1(t)dt = \alpha \int_{-\infty}^{\infty} dt \langle \hat{\mathbf{E}}^{(-)}(t) \cdot \hat{\mathbf{E}}^{(+)}(t) \rangle_\Psi. \qquad (7.14)$$

In the one-dimensional quasi-monochromatic approximation, the equation above changes to

$$P_1 = \eta \int_{-\infty}^{\infty} dt \langle \hat{E}^{(-)}(t) \hat{E}^{(+)}(t) \rangle_\Psi, \qquad (7.15)$$

where η is the quantum efficiency of the detector. Similarly, the probability for generating two photo-electrons in one pulse is

$$P_2 = \eta_A \eta_B \int_{-\infty}^{\infty} dt_1 dt_2 \langle \hat{E}_B^{(-)}(t_2) \hat{E}_A^{(-)}(t_1) \hat{E}_A^{(+)}(t_1) \hat{E}_B^{(+)}(t_2) \rangle_\Psi. \qquad (7.16)$$

If the repetition rate or the number of pulses per second is R_p, the count rate of photo-electric pulses is then $R_1 = P_1 R_p$, and the coincidence rate for two photo-electric pulses is $R_2 = P_2 R_p$.

7.3 Photon Counting

In quantum optics experiments, when the light field is weak, we can use the photon counting techniques to study the behavior of the optical field, so as to obtain the photon statistics, photon correlation, etc. of the optical field. These measurements will reveal the quantum behaviors of the optical

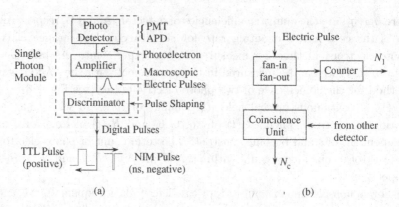

Fig. 7.7   (a) Commonly used conversion process from photons to standardized digital electronic pulses in photon counting technique. (b) Digital electronic pulse counting and coincidence counting.

field and exhibit quantum phenomena, such as quantum interference and quantum entanglement.

The centerpiece of the photon counting technique is the conversion from optical signal to standardized digital electronic pulses. Once we obtain these digital pulses, we can apply the mature digital techniques to process and analyze the data in these pulses to obtain various statistical properties of the optical field. Figure 7.7(a) shows a typical conversion process from light to standardized digital electronic pulses. After the illumination of the detector by the optical field, the generated photo-electron undergoes an initial amplification. This function is usually built in the detectors. Photo-multiplier tubes (PMT) were popular devices in the early days but were replaced by avalanche photo-diodes (APD), which output an electric pulse triggered by one photo-electron. The gain of the initial amplification is around 10^{3-4}. The electric pulse after the initial amplification is still weak ($i_e \sim 10^4 e/1ns = 1.6 \times 10^{-6} Amp$) and requires further amplification to become a macroscopic electric pulse. The gain of the initial amplification by the PMTs or APDs usually depends on many uncontrollable factors so it fluctuates widely, leading to a large variation of the size of the electric pulse after the initial amplification. Furthermore, there also exist some thermally excited electrons in the initial amplification stage. They give rise to electric pulses of relatively small size and are named "dark pulses" because they are there even without the optical field. To eliminate the dark pulses and produce a uniform electric pulse, a discriminator is used to set up certain threshold for the electric pulses. This will cut most of the dark pulses and some pulses with small size. Another role of the discriminator is to shape the pulses into some standardized digital pulses. Nowadays, this part of technology is rather mature and the whole process is integrated into a commercially available single-photon module which takes in a photon through fiber coupling and directly converts it into a standardized digital pulse for output. There are two types of commonly used digital pulses: one is the TTL positive pulse with an adjustable pulse width of 100 ns; the other is the ultrashort NIM negative pulse with a pulse width of a few ns. The TTL pulses are used for registering the number whereas the NIM pulses are for time-related measurement. As shown in Fig. 7.7(b), the standardized pulse from the output of the single-photon module is input into a "fan-in fan-out" device and is duplicated to a number of standardized pulses for further processing such as simple number counting and coincidence measurement with pulses from other detectors.

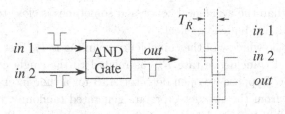

Fig. 7.8   Coincidence measurement with an "AND" gate.

7.3.1   *Coincidence Measurement with an "AND" Gate*

The simplest method to make a coincidence measurement is to use an "AND" gate. It can be generalized to the coincidence of multiple pulses from multiple detectors for the measurement of multi-photon correlation effects. As shown in Fig. 7.8, a coincidence pulse is generated only when the two or more input pulses have some overlap in the "AND" gate. This requires the difference in arrival times of all the pulses to be smaller than the pulse width. The coincidence counting rate is given in Eq. (7.13) where T_R is determined by the width of the longer pulse of the two input pulses.

With coincidence counts and the single counts from each detector, we can analyze the data to obtain the statistical properties of the optical fields. In general, the higher the coincidence count is, the stronger the correlation is between the optical fields measured by the detectors. However, we need a reference to judge how strong the correlation is. This reference is the accidental coincidence. In the following, we will discuss how to define and measure the accidental coincidence in different circumstances.

Accidental coincidence count is the coincidence count from two randomly generated pulses. It corresponds to the accidental events between two photo-detection events of completely uncorrelated optical fields. So, if the measured coincidence count is higher than the accidental coincidence count, this indicates a positive correlation between the fields, otherwise, it is a negative correlation. For example, as we will see later, for optical fields in a two-photon state, the probability is very high for detecting two photo-electric pulses simultaneously by photo-detectors. In this case, the measured coincidence count is much higher than the accidental count. This is the photon bunching effect. On the other hand, for a field in a single-photon state, since there is only one photon at a time, two detectors can only generate one photo-electric pulse. The coincidence count in this case is much lower than the accidental count and sometimes is close to zero. This is the photon anti-bunching effect.

Suppose the window of the coincidence counting is T_R. Let us calculate the accidental coincidence rate R_{ac} as related to the single count rate of each detector. Suppose we open detector A, B for a time interval $T \gg T_R$. If the pulses from the two detectors are generated randomly, the probability density for the two detectors generating the pulses is the same, i.e., $p_A = p_B = 1/T$. Then, the probability of generating a pulse by each of the two detectors in the coincidence window T_R is $P_A = p_A T_R = T_R/T = P_B$. Since the pulses from the two detectors are random, the joint probability of generating two pulses, one from each detector, is $P_{AB} = P_A P_B = (T_R/T)^2$ (see Section 1.4.3). If we register N_A, N_B counts in time period T from the two detectors, respectively, that is, the sample size is N_A, N_B, respectively, the counts in time period T_R from each of the two detectors are $n_A = N_A P_A = N_A T_R/T = R_A T_R, n_B = N_B P_B = N_B T_R/T = R_B T_R$, respectively, where $R_A = N_A/T, R_B = N_B/T$ are the counting rate for the two detectors, respectively. The above is for one random variable. With two together, the sample size becomes $N_{AB} = N_A N_B$ (see Section 1.4.3 for joint probability). So, the coincidence count in the time window T_R is $n_{AB} = N_{AB} P_{AB} = N_A N_B (T_R/T)^2$. Writing in terms of rates, we have the coincidence rate $R_{AB} = n_{AB}/T_R = R_A R_B T_R$. Since we are discussing random events, this is the accidental coincidence rate and is written as

$$R_{ac} = R_A R_B T_R. \tag{7.17}$$

The expression above is for stationary fields because $p_A = p_B = 1/T =$

constant. It can also be obtained from Eqs. (7.13) and (7.16). When A, B are two independent optical fields, we have

$$\langle \hat{E}_B^{(-)}(t+\tau)\hat{E}_A^{(-)}(t)\hat{E}_A^{(+)}(t)\hat{E}_B^{(+)}(t+\tau)\rangle_\Psi$$
$$= \langle \hat{E}_A^{(-)}(t)\hat{E}_A^{(+)}(t)\rangle\langle \hat{E}_B^{(-)}(t+\tau)\hat{E}_B^{(+)}(t+\tau)\rangle. \quad (7.18)$$

Using Eqs. (7.6) and (7.13), we obtain

$$R_{ac} = \int_{T_R} R_A(t)R_B(t+\tau)d\tau. \quad (7.19)$$

For stationary fields, $R_A(t), R_B(t+\tau)$ are independent of time and they can be pulled out of the integration in the expression above. Then, we arrive at Eq. (7.17). Sometimes, due to the circumstance of the experiment, we need to close and open the detectors intermittently. If the intermittent time is much larger than the characteristic time such as correlation time and coherence time of the optical fields, we can still treat them as stationary fields. But Eq. (7.17) needs modification since it only stands for the period when the detectors are open. If the ratio between open and close times is r, the actually measured count rates are then $R'_A = rR_A/(r+1), R'_B = rR_B/(r+1), R'_{AB} = rR_{AB}/(r+1)$. Using Eq. (7.17), we obtain the relation between all the count rates as

$$R'_{ac} = R'_A R'_B T_R (r+1)/r. \quad (7.20)$$

For non-stationary pulsed optical fields, using Eq. (7.18) for independent fields and Eq. (7.16) for joint probability in the pulsed case, we have

$$P_2 = \eta_A\eta_B \int_{-\infty}^{\infty} dt_1 dt_2 \langle \hat{E}_B^{(-)}(t_2)\hat{E}_B^{(+)}(t_2)\rangle\langle \hat{E}_A^{(-)}(t_1)\hat{E}_A^{(+)}(t_1)\rangle$$
$$= \eta_A \int_{-\infty}^{\infty} dt_1 \langle \hat{E}_A^{(-)}(t_1)\hat{E}_A^{(+)}(t_1)\rangle\eta_B \int_{-\infty}^{\infty} dt_2 \langle \hat{E}_B^{(-)}(t_2)\hat{E}_B^{(+)}(t_2)\rangle$$
$$= P_A P_B. \quad (7.21)$$

Here, we used Eq. (7.15) for pulsed fields. If the repetition rate of the optical pulses is R_p, we obtain the accidental coincidence rate as

$$R_{ac} = P_2 R_p = P_A P_B R_p = R_A R_B/R_p, \quad (7.22)$$

where $P_A = R_A/R_P$, $P_B = R_B/R_P$. For intermittent pulsed fields, Eq. (7.22) needs to be multiplied by a factor of $(r+1)/r$, similar to Eq. (7.20).

In the experiment, the repetition rate of optical pulses from pulsed optical fields is easy to measure and it is straightforward to find the accidental coincidence rate from Eq. (7.22) after we measure the count rate from

each detector. For continuous optical fields, however, we need to measure
the coincidence window time T_R. This can be done by using background
light in the lab, which is usually uncorrelated white light, for the illumination of the detectors. We register simultaneously the three counting rates:
R_A, R_B, R_{AB} for this light. Then the coincidence window time T_R can be
calculated from Eq. (7.17). T_R is usually quite stable for a coincidence device so we only need to measure it once. Then we can use it to calculate the
accidental coincidence rate from Eq. (7.17) for any light in the experiment.

After finding the accidental coincidence rate, we can compare it with the
coincidence rate measured from the optical fields of our concern to find the
correlation properties of the fields. For this, we go back to the coincidence
rate in Eq. (7.13) and rewrite it as

$$R_{AB}(t) = \int_{T_R} d\tau R_A(t) R_B(t+\tau) \frac{\langle \hat{E}_B^{(-)}(t+\tau)\hat{E}_A^{(-)}(t)\hat{E}_A^{(+)}(t)\hat{E}_B^{(+)}(t+\tau)\rangle_\Psi}{\langle \hat{E}_A^{(-)}(t)\hat{E}_A^{(+)}(t)\rangle\langle \hat{E}_B^{(-)}(t+\tau)\hat{E}_B^{(+)}(t+\tau)\rangle}$$

$$= \int_{T_R} d\tau R_A(t) R_B(t+\tau) g_{AB}^{(2)}(t, t+\tau), \qquad (7.23)$$

where

$$g_{AB}^{(2)}(t, t+\tau) \equiv \frac{\langle \hat{E}_B^{(-)}(t+\tau)\hat{E}_A^{(-)}(t)\hat{E}_A^{(+)}(t)\hat{E}_B^{(+)}(t+\tau)\rangle_\Psi}{\langle \hat{E}_A^{(-)}(t)\hat{E}_A^{(+)}(t)\rangle\langle \hat{E}_B^{(-)}(t+\tau)\hat{E}_B^{(+)}(t+\tau)\rangle} \qquad (7.24)$$

is the normalized second-order intensity correlation function. For statio-
nary optical fields, $R_A(t), R_B(t+\tau)$, and $g_{AB}^{(2)}(t, t+\tau) = g_{AB}^{(2)}(\tau)$ are all
independent of t. Equation (7.23) then becomes

$$R_{AB}(t) = R_A R_B \int_{T_R} d\tau g_{AB}^{(2)}(\tau), \qquad (7.25)$$

which is also independent of t. Using $R_{ac} = R_A R_B T_R$, Eq. (7.25) can be
rewritten as

$$\frac{R_{AB}}{R_{ac}} = \frac{1}{T_R} \int_{T_R} d\tau g_{AB}^{(2)}(\tau). \qquad (7.26)$$

Therefore, the ratio between the measured coincidence rate and the accidental coincidence rate is just the average of the normalized second-order
intensity correlation function over the coincidence window T_R. When the
coincidence window T_R is much smaller than the characteristic time of the
optical fields, such as the coherence time, the optical fields change very
little within T_R and $g_{AB}^{(2)}(\tau) \approx g_{AB}^{(2)}(0)$. Then Eq. (7.26) becomes

$$R_{AB}/R_{ac} \approx g_{AB}^{(2)}(0), \qquad (7.27)$$

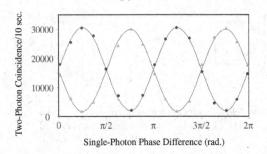

Fig. 7.9   Application of the simple coincidence counting method: two-photon interference fringes as phase of some field is scanned. Reproduced from [Sun et al. (2006)]

that is, the ratio of measured coincidence to the accidental coincidence is the normalized intensity correlation function at zero time delay.

In the experiment, we usually change some physical parameters and observe how the coincidence counting rate varies with these parameters. A typical example is shown in Fig. 7.9 where the two-photon coincidence counts show interference fringes as the phase of some field is scanned. Notice that the period is π instead of 2π, which is typical of two-photon interference fringes. In Chapter 8, we will discuss more about two-photon interference as well as multi-photon interference phenomena with more than two photons, where multi-photon coincidence counts change with phases, forming multi-photon interference fringe pattern.

7.3.2   Time-Resolved Coincidence Measurement

In most multi-photon measurement experiments, the simple coincidence counting method in Fig. 7.8 is enough. But for more complicated optical fields, we need to measure the time correlation between photons. It is related to the normalized second-order intensity correlation function $g_{AB}^{(2)}(\tau)$. The measurement of this quantity requires the time-delayed coincidence measurement method by introducing a time delay T, as shown in Fig. 7.10(a). In this case, the measured coincidence rate in Eq. (7.25) changes to

$$R_{AB}(t) = R_A R_B \int_{T_R} d\tau g_{AB}^{(2)}(T + \tau). \tag{7.28}$$

When the coincidence window T_R is much smaller than the characteristic time of the optical fields, $g_{AB}^{(2)}(T + \tau) \approx g_{AB}^{(2)}(T)$ for τ within T_R. Then, from Eq. (7.28), Eq. (7.27) changes to

$$R_{AB} \approx R_{ac} g_{AB}^{(2)}(T). \tag{7.29}$$

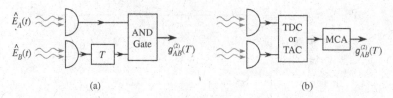

(a)   (b)

Fig. 7.10   (a) Time-delayed coincidence measurement. (b) Measurement of $g_{AB}^{(2)}(\tau)$ by time-to-digital converter (TDC) or time-to-analog converter (TAC) and multi-channel analyzer (MCA).

Hence, as the time delay T is scanned, we obtain the normalized second-order intensity correlation function $g_{AB}^{(2)}(\tau)$. The non-zero range of $g_{AB}^{(2)}(\tau)$ gives the intensity correlation time of field A and field B.

The method above for scanning time delay T is done one step at a time and is quite time-consuming. A more direct method is to use a time-to-digital converter (TDC) or a time-to-analog converter (TAC) and a multi-channel analyzer (MCA), as shown in Fig. 7.10(b). It can measure $g_{AB}^{(2)}(\tau)$ in a single run. An example of time-delayed coincidence measurement is the Hanbury Brown-Twiss (HBT) experiment showing the photon bunching effect, where the time delay is achieved by changing the position of one of the detectors (see Fig. 1.9 in Section 1.6).

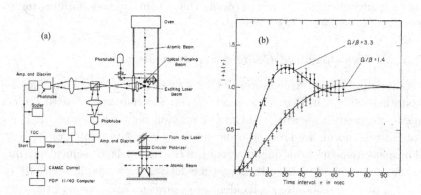

Fig. 7.11   Application of time-delayed coincidence counting technique for the measurement of $g_{AB}^{(2)}(\tau)$. (a) Experimental setup. (b) Exhibition of the photon anti-bunching effect $g^{(2)}(0) < g^{(2)}(\tau)$. Adapted from [Dagenais and Mandel (1978)].

A typical example of applying modern digital technology for the measurement is the observation of the photon anti-bunching effect shown in Fig. 7.11, where the fluorescence from single atoms driven by a laser resonant to the atomic transition line is detected by two phototubes after split by a beam splitter and the time-delayed coincidences $N_c(\tau)$ are measured by a time-to-digital converter (TDC) (Fig. 7.11(a)). The normalized intensity correlation function $g^{(2)}(\tau) = 1 + \lambda(\tau)$ can be extracted from $N_c(\tau)$ by the formula $g^{(2)}(\tau) = N_c(\tau)/N_c(\infty)$, which exhibits the photon anti-bunching behavior of $g^{(2)}(0) < g^{(2)}(\tau)$ (Fig. 7.11(b)).

7.4   Theoretical Description of Experiments

After discussing the experimental techniques, we investigate in this section how we can use the multi-mode description of optical fields discussed in Chapter 4 to explain the experimental observations and make prediction for further experiments. The combination of experiment and theory is the emphasis of this book. From this section on, we will see how this is done in detail.

We start with the two-photon correlation measurement experiment for the simple process of spontaneous parametric down-conversion (SPDC), where a photon of higher energy is split into a pair of lower energy photons known as "signal" and "idler" via interaction with a $\chi^{(2)}$-nonlinear crystal. The first description of this experiment was provided by Burnham and Weinberg in 1970 [Burnham and Weinberg (1970)] who observed a correlation time of 4 ns. In 1985, Friberg *et al.* measured the time correlation function between the two down-converted photons with faster modern digital electronic devices and obtained a correlation time of 100 ps [Friberg et al. (1985a)]. However, because the correlation time between the two down-converted photons ($\sim 100fs$) is much shorter than the response time of the detectors ($\sim 100ps$), the measured time correlation function is simply the time response function of the digital electronic devices. Because of the large bandwidth from SPDC sources, no progress was made until 1999 when Ou and Lu made the first measurement of the true time correlation function from a narrow band spontaneous parametric down-conversion inside an optical cavity [Ou and Lu (1999)].

From Chapters 4 and 6, we find the two-photon state from the SPDC process has the form of

$$|\Psi_2\rangle = \int d\omega_1 d\omega_2 \Psi(\omega_1, \omega_2) \hat{a}_s^\dagger(\omega_1) \hat{a}_i^\dagger(\omega_2)|0\rangle, \qquad (7.30)$$

where "s, i" represent the signal and idler fields, respectively. For a SPDC process pumped by a CW field, we have

$$\Psi(\omega_1, \omega_2) = \psi(\omega_1)\delta(\omega_1 + \omega_2 - \omega_p). \qquad (7.31)$$

Here, ω_p is the angular frequency of the pump field, $\psi(\omega_1)$ is the spectral function for the down-converted photons. For type-I down-conversion processes near frequency degeneracy, $\psi(\omega_1)$ is a symmetric function with respect to $\omega_0 \equiv \omega_p/2$: $\psi(\omega_1) = \psi(\omega_p - \omega_1)$.

Using photo-detectors to make a direct detection of the down-converted fields, we first register single detector count rate. From Eq. (7.6), we have the detected count rate for the signal field as

$$R_{1s} = \eta_s \langle \hat{E}_s^{(-)}(t) \hat{E}_s^{(+)}(t) \rangle_{\Psi_2}, \tag{7.32}$$

where the field operator has the following form under the one-dimensional quasi-monochromatic approximation:

$$\left[\hat{E}_s^{(-)}(t)\right]^\dagger = \hat{E}_s^{(+)}(t) = \frac{1}{\sqrt{2\pi}} \int d\omega \hat{a}_s(\omega) e^{-i\omega t}. \tag{7.33}$$

To evaluate Eq. (7.32), we first perform the following calculation:

$$\hat{E}_s^{(+)}(t)|\Psi_2\rangle = \frac{1}{\sqrt{2\pi}} \int d\omega \hat{a}_s(\omega) e^{-i\omega t}$$

$$\times \int d\omega_1 d\omega_2 \Psi(\omega_1, \omega_2) \hat{a}_s^\dagger(\omega_1) \hat{a}_i^\dagger(\omega_2)|0\rangle$$

$$= \frac{1}{\sqrt{2\pi}} \int d\omega_1 d\omega_2 \Psi(\omega_1, \omega_2) e^{-i\omega_1 t} \hat{a}_i^\dagger(\omega_2)|0\rangle, \tag{7.34}$$

where we used the commutation relation $[\hat{a}_s(\omega), \hat{a}_s^\dagger(\omega_1)] = \delta(\omega - \omega_1)$. Hence,

$$\langle \hat{E}_s^{(-)}(t) \hat{E}_s^{(+)}(t) \rangle_{\Psi_2} = \frac{1}{2\pi} \int d\omega_1 d\omega_2 d\omega_1' d\omega_2' e^{-i(\omega_1 - \omega_1')t}$$

$$\times \Psi(\omega_1, \omega_2) \Psi(\omega_1', \omega_2') \langle 0|a_i(\omega_2') a_i^\dagger(\omega_2)|0\rangle$$

$$= \frac{1}{2\pi} \int d\omega_1 d\omega_2 d\omega_1' e^{-i(\omega_1 - \omega_1')t} \Psi(\omega_1, \omega_2) \Psi(\omega_1', \omega_2)$$

$$= \frac{1}{2\pi} \int d\omega_1 |\psi(\omega_1)|^2, \tag{7.35}$$

where we used Eq. (7.31). Finally,

$$R_{1s} = \eta_s \frac{1}{2\pi} \int d\omega_1 |\psi(\omega_1)|^2. \tag{7.36}$$

Similarly, we can find the single detector count rate for the idler field as

$$R_{1i} = \eta_i \frac{1}{2\pi} \int d\omega_1 |\psi(\omega_1)|^2. \tag{7.37}$$

In order to find the correlation time between the signal and idler fields, we need to calculate the intensity correlation function

$$\Gamma^{(2)}(t, t + \tau) = \langle \hat{E}_s^{(-)}(t) \hat{E}_i^{(-)}(t + \tau) \hat{E}_i^{(+)}(t + \tau) \hat{E}_s^{(+)}(t) \rangle_{\Psi_2}. \tag{7.38}$$

For this, we first calculate

$$\hat{E}_i^{(+)}(t+\tau)\hat{E}_s^{(+)}(t)|\Psi_2\rangle = \frac{1}{2\pi}\int d\omega_s d\omega_i \hat{a}_s(\omega_s)e^{-i\omega_s t}a_i(\omega_i)e^{-i\omega_i(t+\tau)}$$

$$\times \int d\omega_1 d\omega_2 \Psi(\omega_1,\omega_2)\hat{a}_s^\dagger(\omega_1)\hat{a}_i^\dagger(\omega_2)|0\rangle$$

$$= G(t,t+\tau)|0\rangle \tag{7.39}$$

with

$$G(t,t+\tau) \equiv \frac{1}{2\pi}\int d\omega_1 d\omega_2 \Psi(\omega_1,\omega_2)e^{-i\omega_1 t}e^{-i\omega_2(t+\tau)} \tag{7.40}$$

being the two-photon wave function of SPDC. For the SPDC process pumped by a CW field, $\Psi(\omega_1,\omega_2)$ is given in Eq. (7.31). So,

$$G(t,t+\tau) = \frac{e^{-i\omega_p t}}{2\pi}\int d\omega_2 \psi(\omega_p-\omega_2)e^{-i\omega_2\tau} \equiv e^{-i\omega_p(t+\tau/2)}g(\tau) \tag{7.41}$$

with

$$g(\tau) \equiv \frac{1}{2\pi}\int d\Omega\, \psi(\omega_p/2-\Omega)e^{-i\Omega\tau}. \tag{7.42}$$

Hence, we obtain from Eq. (7.13) the coincidence count rate between the signal and idler fields as

$$R_2 = \eta_s\eta_i\int_{T_R} d\tau |g(\tau)|^2. \tag{7.43}$$

For the single-pass SPDC process in a $\chi^{(2)}$-nonlinear medium, we usually have $T_R \gg T_c = 1/\Delta\omega_{PDC}$ where $\Delta\omega_{PDC}$ is the spectral width of the SPDC fields, i.e., the width of $\psi(\omega)$ so T_c is the width of $g(\tau)$. Hence, we can take the integration range in Eq. (7.43) as $(-\infty,+\infty)$. Moreover, since

$$\int_{-\infty}^{+\infty} d\tau |g(\tau)|^2 = \frac{1}{4\pi^2}\int_{-\infty}^{+\infty} d\tau d\Omega d\Omega' \psi(\omega_p/2-\Omega)e^{-i\Omega\tau}\psi^*(\omega_p/2-\Omega')e^{i\Omega'\tau}$$

$$= \frac{1}{2\pi}\int_{-\infty}^{+\infty} d\Omega d\Omega' \psi(\omega_p/2-\Omega)\psi^*(\omega_p/2-\Omega')\delta(\Omega-\Omega')$$

$$= \frac{1}{2\pi}\int d\Omega |\psi(\omega_p/2-\Omega)|^2, \tag{7.44}$$

we obtain the coincidence rate as

$$R_2 = \eta_s\eta_i\frac{1}{2\pi}\int d\Omega |\psi(\omega_p/2-\Omega)|^2. \tag{7.45}$$

Here, we used the relation

$$\frac{1}{2\pi}\int_{-\infty}^{+\infty} d\tau e^{i\Omega'\tau}e^{-i\Omega\tau} = \delta(\Omega-\Omega'). \tag{7.46}$$

Comparing with Eqs. (7.36) and (7.37), we have

$$R_2 = \eta_i R_{1s} = \eta_s R_{1i}. \qquad (7.47)$$

Especially in the ideal case when the quantum efficiency of the detectors is 100%, i.e., $\eta_s = \eta_i = 1$, we arrive at

$$R_2 = R_{1s} = R_{1i}. \qquad (7.48)$$

The physical meaning of the expression above is straightforward: the probability of two-photon detection is the same as that of single-photon detection, that is, whenever we detect one photon, we also detect two photons. So, the detected field must be in a two-photon state. Notice that even in the non-ideal case of η_s, $\eta_i < 1$ due to the existence of losses, Eq. (7.47) indicates that the coincidence count rate is proportional to the single-detector count rate. This is the characteristic behavior of detecting a two-photon-state. Figure 7.12 shows this proportional relationship, which was measured on the SPDC fields with resonant enhancement [Ou and Lu (1999)].

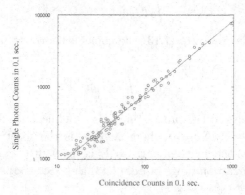

Fig. 7.12 The characteristic property of a two-photon state: the linear relationship between the coincidence count rate and the single-detector count rate. Reproduced from [Lu and Ou (2000)].

In addition to the proportional characteristic property demonstrated in Fig. 7.12, another characteristic of the SPDC fields is the high photon correlation between the two down-converted fields. This can be seen from the ratio between the coincidence count rate to the accidental coincidence rate:

$$R_2/R_{ac} = \eta_i R_{1s}/R_{1s}R_{1i}T_R = \eta_i/R_{1i}T_R = \eta_i/N_i(T_R) \qquad (7.49)$$

with $N_i(T_R) \equiv R_{1i}T_R$ as the average number of counts in time interval T_R for the idler field. When the coincidence window T_R is small (typically

$\sim 1ns$) so that the average count for idler field is much smaller than 1, i.e., $N_i(T_R) = R_{1i}T_R \ll 1$, we find from Eq. (7.49) that $R_2/R_{ac} \gg 1$ and it increases as R_{1i} decreases. This phenomenon was first observed by Friberg, Hong and Mandel [Friberg et al. (1985b)]. The large coincidence-to-accidental ratio (CAR) indicates a strong photon bunching effect. So, $CAR \equiv R_2/R_{ac} \gg 1$ is another two-photon characteristic of the SPDC source. This property was used to confirm the two-photon characteristic for the light source from a four-wave mixing process in optical fibers [Li et al. (2004); Fan et al. (2005)].

The linear relationship in Eq. (7.47) does not describe completely the true situations in the experiment. In fact, because of the existence of accidental coincidences, there should be a quadratic term in Eq. (7.47). For SPDC processes, the quadratic term originates from the accidental coincidence of two photons from two random pairs, one from each pair. The generation of two pairs of photons cannot be described by the state in Eq. (7.30). We must consider higher-order terms from parametric down-conversion. Problem 7.1 discusses this case and gives such higher-order term in Eq. (7.57), from which we can find the extra quadratic term (see Problem 7.3) added to Eq. (7.47) for modification. The final result is

$$R_2 = \eta_i R_{1s} + \beta R_{1s}^2, \qquad (7.50)$$

where $\beta \propto T_R$.

The optical fields from SPDC processes are of strong nonclassical nature. This is exhibited mainly in the violation of Cauchy-Schwarz inequality that stands for the classical sources. From the non-negative P-distribution for classical fields, we can derive Cauchy-Schwarz inequality (see Eq. (5.58) in Section 5.3):

$$\langle I_1 I_2 \rangle \leq \sqrt{\langle I_1^2 \rangle \langle I_2^2 \rangle}, \qquad (7.51)$$

which shows that the cross-correlation between the intensities of two fields is always smaller than the geometric average of the intensity auto-correlations of two fields. The intensity cross-correlation of two optical fields can be measured directly by registering the coincidence counts between the detections of the two fields. The auto-correlation of an optical field can be measured by splitting it into two with a 50:50 beam splitter and then making a coincidence measurement between the two split fields. Since a beam splitter reduces light intensity, we need to normalize the inequality in Eq. (7.51). Dividing two sides of Eq. (7.51) by $\langle I_1 \rangle \langle I_2 \rangle$, we obtain

$$g_{12}^{(2)}(0) \leq \sqrt{g_{11}^{(2)}(0) g_{22}^{(2)}(0)}, \qquad (7.52)$$

where $g_{12}^{(2)}(0) = \langle I_1 I_2 \rangle / \langle I_1 \rangle \langle I_2 \rangle$, $g_{11}^{(2)}(0) = \langle I_1^2 \rangle / \langle I_1 \rangle^2$, $g_{22}^{(2)}(0) = \langle I_2^2 \rangle / \langle I_2 \rangle^2$. The average here is for the probability distribution of the classical fields. For quantum states, we use Eqs. (7.24) and (7.27) for cross-correlation and the beam splitter method discussed above for intensity auto-correlation.

For the state of SPDC in Eqs. (7.30) and (7.31), we have already obtained $g_{12}^{(2)}(0) = \Gamma^{(2)} / \langle I_s \rangle \langle I_i \rangle \propto |g(0)|^2 \neq 0$ from Eqs. (7.38), (7.40) and (7.42). It is also straightforward to show that $g_{11}^{(2)}(0) = 0 = g_{22}^{(2)}(0)$, with the subscripts $s = 1, i = 2$, because the signal field and idler field each have only one photon, as shown in Eq. (7.30). With these values for $g_{12}^{(2)}, g_{11}^{(2)}, g_{22}^{(2)}$, it is obvious that the inequality in Eq. (7.52) is violated. In the actual experiment, however, because of the existence of the accidental coincidences, $g_{11}^{(2)}(0), g_{22}^{(2)}(0) \neq 0$. These accidental coincidences stem from random two-pair events in SPDC. This situation cannot be described by the quantum state in Eq. (7.38) since it neglects the contribution from higher-order terms (see its derivation in Section 6.1.5). In Problems 7.1 and 7.3, we will consider the case of two-pair generation in SPDC and calculate the coincidence count rate from each field.

Nevertheless, we can approach this from the experimental point of view. In the experiment, we usually measure $g_{12}^{(2)}(\tau)$. For this, the Cauchy-Schwarz inequality in Eq. (7.51) becomes

$$\langle I_1(t) I_2(t + \tau) \rangle \leq \sqrt{\langle I_1^2(t) \rangle \langle I_2^2(t + \tau) \rangle} \tag{7.53}$$

$$g_{12}^{(2)}(\tau) \leq \sqrt{g_{11}^{(2)}(0) g_{22}^{(2)}(0)}. \tag{7.54}$$

Integrating Eq. (7.54) and using Eq. (7.26), we have

$$R_{si}/R_{ac} \leq \sqrt{g_{ss}^{(2)}(0) g_{ii}^{(2)}(0)}. \tag{7.55}$$

Here, $s = 1, i = 2$. R_{si}/R_{ac} is exactly the coincidence-to-accidental ratio (CAR) measured in the experiment. From Problem 7.1, we have $g_{ss}^{(2)}(0) = 2 = g_{ii}^{(2)}(0)$. Then, Eq. (7.55) becomes

$$R_{si}/R_{ac} \leq 2. \tag{7.56}$$

As we discussed about Eq. (7.49) for SPDC, R_{si}/R_{ac} can become very large for small T_R. In this case, the Cauchy-Schwarz inequality in Eq. (7.56) is violated for the optical fields from SPDC. This demonstrates the nonclassical property for the optical fields generated in SPDC. Experimentally, Clauser first demonstrated the violation of Eq. (7.52) with atomic cascade emission [Clauser (1974)]. Zou *et al.* first observed the violation of Eq. (7.55) for SPDC [Zou et al. (1991a)].

7.5   Problems

Problem 7.1 Photon bunching effect for the signal or idler field alone in SPDC.

After we consider the higher-order term, the quantum state of SPDC in Eq. (7.30) changes to (see Section 6.1.5)

$$|\Psi_2\rangle = \int d\omega_1 d\omega_2 \Psi(\omega_1,\omega_2)\hat{a}_s^\dagger(\omega_1)\hat{a}_i^\dagger(\omega_2)|0\rangle$$

$$+\frac{1}{2}\int d\omega_1 d\omega_2 d\omega_1' d\omega_2' \Psi(\omega_1,\omega_2)\Psi(\omega_1',\omega_2')$$

$$\times \hat{a}_s^\dagger(\omega_1)\hat{a}_i^\dagger(\omega_2)\hat{a}_s^\dagger(\omega_1')\hat{a}_i^\dagger(\omega_2')|0\rangle. \qquad (7.57)$$

we will work on the signal field only.

(i) For the SPDC process pumped by a CW field, Eq. (7.31) gives $\Psi(\omega_1,\omega_2) = \psi(\omega_1)\delta(\omega_1+\omega_2-\omega_p)$. Use this and the state in Eq. (7.57) to prove

$$\Gamma_{ss}^{(2)}(t,t+\tau) \equiv \langle \hat{E}_s^{(-)}(t)\hat{E}_s^{(-)}(t+\tau)\hat{E}_s^{(+)}(t+\tau)\hat{E}_s^{(+)}(t)\rangle_{\Psi_2}$$

$$= H^2(0) + H(\tau)H^*(\tau), \qquad (7.58)$$

where $H(\tau) \equiv (1/2\pi)\int d\omega_1|\psi(\omega_1)|^2 e^{-i\omega_1\tau}$. From Eq. (7.35), we have $H(0) = \langle \hat{E}_s^{(-)}(t)\hat{E}_s^{(+)}(t)\rangle_{\Psi_2}$. From this, prove $g_{ss}^{(2)}(\tau) = 1 + |\gamma(\tau)|^2$ with $\gamma(\tau) \equiv H(\tau)/H(0)$. This result is the same as that for a thermal source (see Eq. (4.29) in Section 4.2.4).

(ii) For the SPDC process pumped by a pulsed field, $\Psi(\omega_1,\omega_2)$ does not have the frequency correlation in Eq. (7.31). We can calculate with Eqs. (7.15) and (7.16). Prove the single-photon counting probability is

$$P_s \equiv \eta \int_{-\infty}^{\infty} dt \langle \hat{E}_s^{(-)}(t)\hat{E}_s^{(+)}(t)\rangle_{\Psi_2}$$

$$= \eta \int d\omega_1 d\omega_2 |\Psi(\omega_1,\omega_2)|^2, \qquad (7.59)$$

and the two-photon counting probability for the signal field is

$$P_{ss} \equiv \eta^2 \int_{-\infty}^{\infty} dt_1 dt_2 \langle \hat{E}_s^{(-)}(t_2)\hat{E}_s^{(-)}(t_1)\hat{E}_s^{(+)}(t_1)\hat{E}_s^{(+)}(t_2)\rangle_\Psi$$

$$= \eta^2(\mathcal{A}+\mathcal{E}), \qquad (7.60)$$

where

$$\mathcal{A} \equiv \int d\omega_1 d\omega_2 d\omega_1' d\omega_2' |\Psi(\omega_1,\omega_2)\Psi(\omega_1',\omega_2')|^2 = P_s^2 \qquad (7.61)$$

$$\mathcal{E} \equiv \int d\omega_1 d\omega_2 d\omega_1' d\omega_2' \Psi(\omega_1,\omega_2)\Psi(\omega_1',\omega_2')\Psi^*(\omega_1,\omega_2')\Psi^*(\omega_1',\omega_2). \qquad (7.62)$$

Hence,

$$g_{ss}^{(2)} \equiv P_{ss}/P_s^2 = 1 + \mathcal{E}/\mathcal{A}. \tag{7.63}$$

Notice that $\mathcal{E} \leq \mathcal{A}$ so that $g_{ss}^{(2)} \leq 2$. When $\mathcal{E} = \mathcal{A}$, we have $g_{ss}^{(2)} = 2$, or exactly the same as a single-mode thermal source. The calculation for the idler field is identical.

(iii) For the pulse spontaneous parametric process described by Eq. (6.36) in Section 6.1.4, use Eqs. (7.61)–(7.63) to prove

$$g_{ss}^{(2)} = 1 + \sum_j \lambda_j^4 \tag{7.64}$$

with

$$\lambda_j \equiv \frac{\sinh r_j \xi}{\sqrt{\sum_j \sinh^2 r_j \xi}}. \tag{7.65}$$

For M modes with equal $r_j = r$, we have $g_{ss}^{(2)} = 1 + 1/M$. This recovers the result in Eq. (4.84) of Chapter 4, which is based on classical wave theory.

Problem 7.2 Intensity correlation between the signal and idler fields of pulse-pumped SPDC.

For the SPDC process pumped by a pulsed field, we have calculated the single and two-photon detection probabilities for each field of SPDC in the previous problem. We now continue to calculate the intensity correlation between the two fields from SPDC. Use Eq. (7.16) for pulsed fields to prove the coincidence counting probability for the two-photon detection of the signal and the idler fields is

$$P_{si} \equiv \eta^2 \int_{-\infty}^{\infty} dt_1 dt_2 \langle \hat{E}_i^{(-)}(t_2)\hat{E}_s^{(-)}(t_1)\hat{E}_s^{(+)}(t_1)\hat{E}_i^{(+)}(t_2) \rangle_\Psi$$

$$= \eta_s \eta_i \int d\omega_1 d\omega_2 |\Psi(\omega_1, \omega_2)|^2 = \eta_i P_s = \eta_s P_i. \tag{7.66}$$

This result is exactly the same as Eq. (7.47) for the case of CW pumping.

Problem 7.3 The contribution to photon coincidence counting from randomly generated two pairs of photons in SPDC.

Using the second term in Eq. (7.57), we can calculate its contribution to photon coincidence counting. Similar to Problem 7.1, the calculation result depends on the bandwidth of the pump field for SPDC.

(i) For the SPDC process pumped by a pulsed field, we need to use Eq. (7.16) to calculate the two-photon coincidence counting probability P_{si}. In Problem 7.2, we have calculated the contribution from the first term in Eq. (7.57). Now we can use similar method to calculate the contribution from the second term. Prove the contribution to P_{si} from the second term of Eq. (7.57) is

$$P_{si}^{(2)} \equiv \eta_s \eta_i \int_{-\infty}^{\infty} dt_1 dt_2 \langle \hat{E}_i^{(-)}(t_2) \hat{E}_s^{(-)}(t_1) \hat{E}_s^{(+)}(t_1) \hat{E}_i^{(+)}(t_2) \rangle_{\Psi_2^{(2)}}$$

$$= C \eta_s \eta_i (\mathcal{A} + \mathcal{E})^2 = C(1 + \mathcal{E}/\mathcal{A})^2 P_s P_i \qquad (7.67)$$

with C as a constant. Find the constant C.

(ii) Do the same as (i) but for the SPDC process pumped by a CW field.

Chapter 8

Applications of Photon Counting Techniques: Multi-Photon Interference and Entanglement

Photon counting techniques are the most commonly used experimental methods for the observation of multi-photon interference effects, which rely on multi-photon coincidence measurement. There are other experimental methods developed for coincidence measurement, e.g., frequency up-conversion [Dayan et al. (2005); Lukens et al. (2013)]. But because of the easiness of operation and good efficiency for coincidence measurement, photon counting techniques are the favorite of experimentalists for observing some quantum optical and multi-photon interference phenomena. There is a monograph by the current author with some detailed coverage on various multi-photon interference effects [Ou (2007)]. Ten years have passed since the publication of the monograph and we also introduced the mode theory for quantum optics in this book. So, in this chapter we will treat some of the old multi-photon interference effects from a new perspective of mode theory to establish a good conceptual understanding and then we will discuss some of the recent developments.

8.1 Multi-Photon Interference in General

8.1.1 Single-Photon Interference and Two-Photon Interference

When we talk about single-photon and multi-photon effects, one immediately considers the average photon number, i.e., $\langle n \rangle$: single-photon case corresponds to $\langle n \rangle \ll 1$ whereas multi-photon case to $\langle n \rangle \gg 1$. These are only superficial but not fundamental difference. Consider the famous Young's double slit and its variations such as Michelson and Mach-Zehnder interferometers. The interference fringe patterns do not depend on how many photons there are as long as the exposure time is long enough to es-

217

tablish the fringes, even for the case when there is only one photon between the source and the observation screen [Taylor (1909)]. This prompted Dirac to make the following statement about photon interference [Dirac (1930)]:

Each photon only interferes with itself.

Different photons never interfere.

This suggests that it is the single-photon effect that is responsible for interference phenomena even at large average photon number: $\langle n \rangle \gg 1$. However, the situation changed after intensity correlation technique was invented by Hanbury Brown and Twiss. In 1967, Pfleegor and Mandel demonstrated interference phenomenon in intensity correlation [Pfleegor and Mandel (1967a,b)]. They observed an interference pattern in coincidence measurement between two detectors in the low photon number limit when there is no more than one photon at a time between the source and the detectors.

There is one thing in Pfleegor-Mandel experiment that is fundamentally different from traditional interference: intensity correlation. From what we learned in Chapter 7, both detectors must register a photoelectron respectively in order to obtain a signal. This means that Pfleegor-Mandel interference effect is a two-photon effect. To see more clearly about this, let us examine Pfleegor-Mandel experiment in more detail as follows.

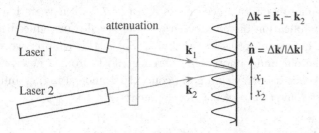

Fig. 8.1   Pfleegor-Mandel two-photon interference experiment with two independent but attenuated lasers.

In Pfleegor-Mandel experiment, two optical fields from two independent but heavily attenuated lasers are allowed to superimpose in an area where two photo-detectors are located at $\mathbf{r}_1, \mathbf{r}_2$ (x_1, x_2 in Fig. 8.1). Let us write the field operator as

$$\hat{E}(\mathbf{r}) = \hat{a}_1 e^{i\mathbf{k}_1 \cdot \mathbf{r}} + \hat{a}_2 e^{i\mathbf{k}_2 \cdot \mathbf{r}} \tag{8.1}$$

Here we assume plane wave mode for the fields. Let the fields be respectively in the coherent states: $|\psi\rangle = |Ae^{i\varphi_{10}}\rangle \otimes |Ae^{i\varphi_{20}}\rangle$ with the same amplitude

A but different phases $\varphi_{10}, \varphi_{20}$. By Glauber's photo-detection theory, the two-photon coincidence probability is then proportional to

$$P_2(\mathbf{r}_1, \mathbf{r}_2) \propto \langle \hat{E}^\dagger(\mathbf{r}_1)\hat{E}^\dagger(\mathbf{r}_2)\hat{E}(\mathbf{r}_1)\hat{E}(\mathbf{r}_2)\rangle_\psi$$
$$= ||\hat{E}(\mathbf{r}_1)\hat{E}(\mathbf{r}_2)|\psi\rangle||^2$$
$$= ||\Phi_2(\mathbf{r}_1, \mathbf{r}_2)|\psi\rangle||^2 = |\Phi_2(\mathbf{r}_1, \mathbf{r}_2)|^2 \qquad (8.2)$$

with

$$\Phi_2(\mathbf{r}_1, \mathbf{r}_2) \equiv A^2 \Big[\big(e^{i\mathbf{k}_1\cdot\mathbf{r}_1}e^{i\mathbf{k}_2\cdot\mathbf{r}_2} + e^{i\mathbf{k}_1\cdot\mathbf{r}_2}e^{i\mathbf{k}_2\cdot\mathbf{r}_1}\big)e^{i(\varphi_{10}+\varphi_{20})}$$
$$+ e^{i\mathbf{k}_1\cdot(\mathbf{r}_1+\mathbf{r}_2)}e^{2i\varphi_{10}} + e^{i\mathbf{k}_2\cdot(\mathbf{r}_1+\mathbf{r}_2)}e^{2i\varphi_{20}} \Big]. \qquad (8.3)$$

Since the two lasers are independent so that $\varphi_{10} - \varphi_{20}$ is random, P_2 has contributions only from the absolute values of the three terms in Eq. (8.3), where the last two terms exhibit no interference and only the first term contains the addition of two quantities and gives rise to interference effect:

$$\left| A^2\big(e^{i\mathbf{k}_1\cdot\mathbf{r}_1}e^{i\mathbf{k}_2\cdot\mathbf{r}_2} + e^{i\mathbf{k}_1\cdot\mathbf{r}_2}e^{i\mathbf{k}_2\cdot\mathbf{r}_1}\big)e^{i(\varphi_{10}+\varphi_{20})} \right|^2$$
$$= 2A^4[1 + \cos\Delta\mathbf{k}\cdot(\mathbf{r}_1 - \mathbf{r}_2)] = 2A^4[1 + \cos 2\pi(x_1 - x_2)/L], \qquad (8.4)$$

where $\Delta\mathbf{k} \equiv \mathbf{k}_1 - \mathbf{k}_2 \approx \Delta\theta|\mathbf{k}_1|\hat{\mathbf{n}}$ for a small angle $\Delta\theta$ between $\mathbf{k}_1$ and $\mathbf{k}_2$ with $|\mathbf{k}_1| = |\mathbf{k}_2| = 2\pi/\lambda$ for wavelength λ. $x_j(j = 1, 2) \equiv \mathbf{r}_j \cdot \hat{\mathbf{n}}$ is the location of detector j along the fringe direction $\hat{\mathbf{n}} \equiv \Delta\mathbf{k}/|\Delta\mathbf{k}|$ and $L \equiv 2\pi/|\Delta\mathbf{k}| \approx \lambda/\Delta\theta$ is the fringe spacing. Combining the three terms, we have

$$P_2(\mathbf{r}_1, \mathbf{r}_2) \propto 2A^4 + 2A^4[1 + \cos 2\pi(x_1 - x_2)/L]$$
$$= 4A^4[1 + 0.5\cos 2\pi(x_1 - x_2)/L]. \qquad (8.5)$$

This is consistent with the result of Pfleegor-Mandel experiment [Pfleegor and Mandel (1967a,b)]. If we discard the parenthesis of the first term in Eq. (8.3), there are totally four terms in the expression. Figure 8.2 graphically depicts these four situations. Figures 8.2(a) and (b) correspond to the first two terms in which the two detected photons are respectively from two different lasers whereas Figs. 8.2(c) and (d) correspond to the last two terms in which the two detected photons are from the same laser. Since coincidence measurement cannot distinguish the cases in Figs. 8.2(a) and (b), they give rise to interference effect shown in Eq. (8.4). No interference occurs for the cases in Figs. 8.2(c) and (d) because they are from different lasers and distinguishable.[1]

---

[1] Interference could arise from these two terms if there is a fixed phase difference between them, which requires phase correlation between two lasers and erases distinguishability.

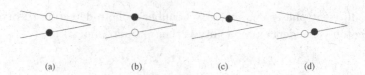

(a)   (b)   (c)   (d)

Fig. 8.2 Four possibilities for the two-photon coincidence detected in Pfleegor-Mandel two-photon interference experiment.

Since interference effect exhibited in coincidence measurement involves two photons, it is called two-photon interference. From what we see in Figs. 8.2(a) and (b), the two detected photons act as one entity and in the spirit of Dirac, the two-photon entity interferes with the entity itself. This consideration leads to the concept of two-photon wave function.

8.1.2 Two-Photon Wave Function

From Eq. (8.2), we see that the two-photon detection probability is the absolute square of the function $\Phi(\mathbf{r}_1, \mathbf{r}_2)$, which consists of four terms corresponding to the four situations depicted in Fig. 8.2. Concentrating on the first two terms, we find that it is the superposition of these two terms that gives rise to the two-photon interference effect. We can further write them explicitly as follows:

$$A^2 e^{i\mathbf{k}_1 \cdot \mathbf{r}_1} e^{i\mathbf{k}_2 \cdot \mathbf{r}_2} e^{i(\varphi_{10}+\varphi_{20})} = A e^{i\mathbf{k}_1 \cdot \mathbf{r}_1} e^{i\varphi_{10}} A e^{i\mathbf{k}_2 \cdot \mathbf{r}_2} e^{i\varphi_{20}} = \phi_1(\mathbf{r}_1)\phi_2(\mathbf{r}_2),$$
$$A^2 e^{i\mathbf{k}_1 \cdot \mathbf{r}_2} e^{i\mathbf{k}_2 \cdot \mathbf{r}_1} e^{i(\varphi_{10}+\varphi_{20})} = A e^{i\mathbf{k}_1 \cdot \mathbf{r}_2} e^{i\varphi_{10}} A e^{i\mathbf{k}_2 \cdot \mathbf{r}_1} e^{i\varphi_{20}} = \phi_1(\mathbf{r}_2)\phi_2(\mathbf{r}_1).$$
$$(8.6)$$

Here $\phi_j(\mathbf{r}) \equiv A e^{i\mathbf{k}_j \cdot \mathbf{r}} e^{i\varphi_{j0}} (j = 1, 2)$ is the one-photon wave function, whose absolute value square is the single-photon detection probability. Hence, the two superposing amplitudes are each the product of two one-photon wave functions, one for each photon. So, in two-photon interference, we need to consider the wave functions of the two photons together and obtain a new wave function which we call "two-photon wave function". The absolute value square of the two-photon wave function gives the two-photon detection probability in exactly the same way as the one-photon wave function.

Notice that there is an important feature in two-photon interference that is different from single-photon interference: the interference pattern does not depend on the phase difference between the two superimposed fields. This is because the two-photon phase is the sum of the phases of the two fields: $\varphi_{2p} = \varphi_{10} + \varphi_{20}$ and appears in both superposing terms and eventually cancels in the final result. Because of this, two-photon interference is phase-insensitive.[2]

The concept of two-photon wave function can be generalized to arbitrary N-photon cases when an N-photon coincidence measurement is performed. We will discuss three- and four-photon interference later in Section 8.3.

8.2 Various Two-Photon Interference Effects

In Pfleegor-Mandel experiment, which was done with classical sources of lasers, we find from Eq. (8.5) that the visibility of interference is only 50%. It turns out that this is not uniquely limited to Pfleegor-Mandel experiment. Mandel was the first to point out that the visibility of two-photon interference with classical sources cannot exceed 50% [Mandel (1983)], which was later generally proved [Ou (1988)]. Richter in 1977 and Mandel in 1983 proposed to use nonclassical fields for two-photon interference and showed that it is possible to achieve a visibility of 100% in two-photon interference [Richter (1977); Mandel (1983)]. A quick examination of Eqs. (8.3) and (8.5) and the related Fig. 8.2 reveals that the reason for 50% visibility is the contributions of the two cases where two photons come from the same laser (Figs. 8.2(c) and (d)) and this contribution always exists for classical fields.

To reach a visibility higher than 50%, we need to reduce this type of contribution and this leads to quantum sources with photon anti-bunching [Ou (1988)]. A single-photon state is the most anti-bunched photon source and gives no two-photon event and thus no contribution from the cases in Figs. 8.2(c) and (d). Indeed, for a quantum state of $|\psi\rangle = |1\rangle_1 \otimes |1\rangle_2$, where each side has only one photon, we can easily check that the visibility of two-photon interference is 100%:

$$P_2(\mathbf{r}_1, \mathbf{r}_2) \propto 2[1 + \cos 2\pi(x_1 - x_2)/L]. \tag{8.7}$$

The first two-photon interference experiment with a two-photon quantum state was performed by Ghosh and Mandel, who demonstrated a two-photon interference fringe is consistent with Eq. (8.7) [Ghosh and Mandel (1987)]. After that, various two-photon interference phenomena were discovered with two-photon sources [Mandel (1999)]. In the following, we will discuss a number of typical ones.

---

[2]Phase-sensitive two-photon interference may occur if the two photons together follow separate paths for interference. See Section 8.2.2.

8.2.1   *Hong-Ou-Mandel Interference*

Perhaps the most well-known two-photon interference phenomenon is the Hong-Ou-Mandel effect [Hong et al. (1987)]. Its simplicity in both geometrical structure and physical picture makes it a favorite example in quantum optics textbooks. It has become a standard technique for testing particle indistinguishability. Furthermore, it is also the basis for linear optical quantum computing [Knill et al. (2001)] and a number of protocols in quantum information [Zeilinger (1999)]. As we will see in the following, it clearly illustrates the importance of mode concept in understanding quantum optical phenomena.

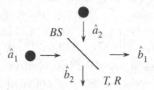

Fig. 8.3   Hong-Ou-Mandel interferometer with two photons.

The Hong-Ou-Mandel interferometer involves simply a 50:50 beam splitter and two photons input with one respectively from each side, as shown in Fig. 8.3. With an input state of $|1\rangle_1|1\rangle_2$ to the beam splitter of amplitude transmissivity t and reflectivity r, the output state in single-mode case was derived in Section 6.2.2 as

$$|\Psi\rangle_{out} = \sqrt{2}tr\big(|2\rangle_1|0\rangle_2 - |0\rangle_1|2\rangle_2\big) + (t^2 - r^2)|1\rangle_1|1\rangle_2$$
$$= \frac{1}{\sqrt{2}}\big(|2\rangle_1|0\rangle_2 - |0\rangle_1|2\rangle_2\big) \quad \text{for } t = r = \frac{1}{\sqrt{2}}. \qquad (8.8)$$

The lack of state $|1,1\rangle$ in the output state $|\Psi\rangle_{out}$ when $t = r = 1/\sqrt{2}$ is a result of destructive two-photon interference. The multi-mode treatment for a two-photon state from spontaneous parametric down-conversion can be found in the monograph by the current author [Ou (2007)]. But here, we will take a different approach from the point of view of two-photon wave function discussed earlier in Section 8.1.2. This will illustrate further its difference from the traditional single-photon interference.

Taking the same notations in Section 8.1.2, we assume the wave functions for the two input photons are $\phi_1 = Ae^{i\varphi_{10}}u_1(1), \phi_2 = Ae^{i\varphi_{20}}u_2(2)$ with the same amplitude of $A(> 0)$. u_1, u_2 are their mode functions and the numbers in the parenthesis denote input ports. So, the two-photon wave function for the input fields is simply

$$\Psi_{in}^{(2)} = \phi_{10}\phi_{20} = A^2 e^{i(\varphi_{10}+\varphi_{20})} u_1(1)u_2(2). \tag{8.9}$$

After the beam splitter, each wave is split into two:

$$\phi_{10} \rightarrow tAe^{i\varphi_{10}}u_1(1) + rAe^{i\varphi_{10}}u_1(2),$$
$$\phi_{20} \rightarrow t'Ae^{i\varphi_{20}}u_2(2) + r'Ae^{i\varphi_{20}}u_2(1), \tag{8.10}$$

where t, r, t', r' are the complex amplitude transmissivity and reflectivity for the two sides of the beam splitter, respectively. The two-photon wave function for the output is then

$$\begin{aligned}
\Psi_{out}^{(2)} &= [tAe^{i\varphi_{10}}u_1(1) + rAe^{i\varphi_{10}}u_1(2)][t'Ae^{i\varphi_{20}}u_2(2) + r'Ae^{i\varphi_{20}}u_2(1)] \\
&= tr'A^2 e^{i(\varphi_{10}+\varphi_{20})}u_1(1)u_2(1) + rt'A^2 e^{i(\varphi_{10}+\varphi_{20})}u_1(2)u_2(2) \\
&\quad + tt'A^2 e^{i(\varphi_{10}+\varphi_{20})}u_1(1)u_2(2) + rr'A^2 e^{i(\varphi_{10}+\varphi_{20})}u_1(2)u_2(1) \\
&= A^2 e^{i(\varphi_{10}+\varphi_{20})}\big[tr'u_1(1)u_2(1) + rt'u_1(2)u_2(2) \\
&\quad + tt'u_1(1)u_2(2) + rr'u_1(2)u_2(1)\big] \\
&= A^2 e^{i(\varphi_{10}+\varphi_{20})}\big[tr'u_1(1)u_2(1) + rt'u_1(2)u_2(2) \\
&\quad + (tt' + rr')u(1)u(2)\big] \quad \text{if } u_1 = u_2 \equiv u. \tag{8.11}
\end{aligned}$$

Pictorially, the four terms in Eq. (8.11) correspond to the four possibilities shown in Fig. 8.4. The first two terms (Figs. 8.4(a) and (b)) are distinguishable whereas the last two terms (Figs. 8.4(c) and (d)) are indistinguishable if $u_1 = u_2 \equiv u$ and are summed together in the last line of Eq. (8.11). Notice that the last two terms in Eq. (8.11), which correspond to two photons going to separate ports, can be written as

$$A^2 e^{i(\varphi_{10}+\varphi_{20})}(tt' + rr')u(1)u(2) = (\Psi_{out_c}^{(2)} + \Psi_{out_d}^{(2)})u(1)u(2), \tag{8.12}$$

where $\Psi_{out_c}^{(2)} = tt'A^2 e^{i(\varphi_{10}+\varphi_{20})}$, $\Psi_{out_d}^{(2)} = rr'A^2 e^{i(\varphi_{10}+\varphi_{20})}$ are the two-photon wave functions without the indistinguishable mode functions $u(1)u(2)$ for the cases (c) and (d) in Fig. 8.4, respectively. The spatial indistinguishability leads to the superposition of the two-photon wave functions.

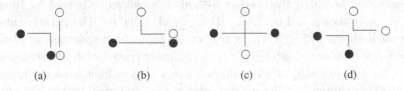

(a)   (b)   (c)   (d)

Fig. 8.4   Four possibilities for the two photons in Hong-Ou-Mandel interferometer.

For a lossless beam splitter, the phases of t, r, t', r' must satisfy $\varphi_t + \varphi_{t'} - \varphi_r - \varphi_{r'} = \pi$ due to energy conservation from input to output (see Section 6.2.1 and Appendix A). Then, for a 50:50 beam splitter, we have

$$\Psi^{(2)}_{out_c} + \Psi^{(2)}_{out_d} \propto (tt' + rr')$$

$$= e^{i(\varphi_r + \varphi_{r'})}[e^{i(\varphi_t + \varphi_{t'} - \varphi_r - \varphi_{r'})} + 1]/\sqrt{2}$$

$$= e^{i(\varphi_r + \varphi_{r'})}[e^{i\pi} + 1]/\sqrt{2}$$

$$= 0. \qquad (8.13)$$

So, the last two superposition terms in Eq. (8.11) add to zero due to destructive interference, leading to no appearance for the two photons at separate ports. In this case, the result based on the two-photon wave function in Eq. (8.11) is the same as that based on the quantum state in Eq. (8.8). Notice that the interference discussed here is between two-photon wave functions $\Psi^{(2)}_{out_c}$, $\Psi^{(2)}_{out_d}$ so it depends on the phase difference of the two wave functions, which is the phase difference of the phase sum of each photon, i.e., the difference between $\varphi_t + \varphi_{t'}$ for case (c) and $\varphi_r + \varphi_{r'}$ for case (d) in Fig. 8.4. The initial phase $\varphi_{10} + \varphi_{20}$ is canceled out in this way. Notice that, unlike the single-photon interference, the two-photon interference does not depend on the phase difference $\varphi_{10} - \varphi_{20}$ between two input phases. Another thing is that the complete cancelation depends on the complete overlap of the mode functions $u_1 = u_2$. This means that photon indistinguishability relies on the mode of the photons (see Section 8.4).

Experimentally, if we make a coincidence measurement between the two output fields of the beam splitter, we should expect no coincidence count except accidental count according to Eq. (8.8) or the discussion above. However, these arguments only apply to the single-mode case. In the experiment, we usually have multi-frequency modes and the two-photon wave functions become two-photon wave packets. Only when the two wave packets overlap at the beam splitter, which correspond to $u_1 = u_2 \equiv u$ in Eq. (8.11), a complete destructive interference can occur, leading to no coincidence count. Otherwise, we have a non-zero coincidence count depending on how well the overlap is. This is what was observed by Hong *et al.* and shown in Fig. 8.5(b) [Hong et al. (1987)]. The experimental layout is shown in Fig. 8.5(a), where two photons from spontaneous parametric down-conversion (SPDC) in a nonlinear crystal of KDP are input to a 50:50 beam splitter after filtering out scattered light from the pump field. Two-photon coincidence measurement is performed on the two output fields with the photon counting techniques discussed in Chapter 7. As

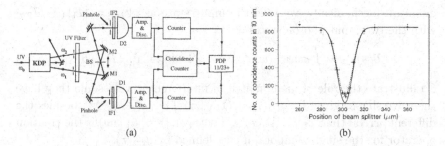

(a)   (b)

Fig. 8.5 Hong-Ou-Mandel experiment. (a) Experimental layout; (b) Two-photon coincidence dip as the position of the beam splitter is scanned. IF: interference filter; BS: beam splitter. Reproduced from [Hong et al. (1987)].

the position of the beam splitter is scanned, which changes the overlap of the two input photons, a dip occurs in the coincidence count at optimum overlap for destructive interference.

Hong-Ou-Mandel experiment with two independent photons

In the two-photon wave function discussion above, we take the two-photon wave function as the product of the single-photon wave functions ϕ_{10}, ϕ_{20}. We can do this for two single-mode photons. But if the two input photons are of multi-mode nature (frequency mode), this is true only when the two input photons are independent of each other and each photon corresponds to a single-photon wave packet in a single-temporal mode, as discussed in Section 4.3.2. Although the original experiment by Hong et al. was performed with correlated photons from SPDC, the argument above shows that the two-photon destructive interference effect should occur even for two independent photons. Indeed, the Hong-Ou-Mandel effect was observed with photons from a variety of independent single-photon sources such as the heralded single photon from parametric down-conversion [Wang and Rhee (1999); de Riedmatten et al. (2003)], single quantum dots [Santori et al. (2002)], single atoms [Beugnon et al. (2006)], single ions [Maunz et al. (2007)]. In the following, we will discuss this situation.

In writing Eq. (8.10), we already assumed the mode functions are different for two input photons but we set them equal in Eq. (8.11) for complete destructive interference and only qualitatively discussed the situation when they are different. For a quantitative discussion, we need to specify the mode functions u_1, u_2. For independent photons, we can describe them with wave packets of different shapes and arrival times. To treat this case quantum mechanically, we take the input state as $|\Psi\rangle_{in} = |T_1\rangle_1 \otimes |T_2\rangle_2$ with the two input photon states as

$$|T_{1,2}\rangle_{1,2} = \int d\omega \phi_{1,2}(\omega) e^{i\omega T_{1,2}} \hat{a}_{1,2}^\dagger(\omega)|0\rangle \equiv \hat{A}_{1,2}^\dagger(T_{1,2})|0\rangle. \quad (8.14)$$

To illustrate the role of mode match in interference, we assume they have the wave shapes given by $g_{1,2}(\tau) = (1/\sqrt{2\pi}) \int d\omega \phi_{1,2}(\omega) e^{-i\omega\tau}$ besides the different arrival time $T_{1,2}$. $\hat{A}_{1,2}^\dagger(\tau) = \int d\omega \phi_{1,2}(\omega) e^{i\omega\tau} \hat{a}_{1,2}^\dagger(\omega)$ is the creation operator for the single temporal mode defined by $g_{1,2}(\tau)$.

Taking the one-dimensional quasi-monochromatic approximation, we have the output field operators of the beam splitter as

$$\hat{E}_1^{(o)}(t) = \sqrt{T}\hat{E}_1(t) + \sqrt{R}\hat{E}_2(t - D/c),$$
$$\hat{E}_2^{(o)}(t) = \sqrt{T}\hat{E}_2(t) - \sqrt{R}\hat{E}_1(t + D/c), \quad (8.15)$$

where T, R are the transmissivity and reflectivity of the beam splitter, D is the displacement of the beam splitter relative to some symmetric position, and input field operator $\hat{E}_{1,2}(t) = (1/\sqrt{2\pi}) \int d\omega \hat{a}_{1,2}(\omega) e^{-i\omega t}$.

To find the coincidence counting probability in detecting two photons at two output ports, we first calculate the following quantity:

$$\hat{E}_1^{(o)}(t_1)\hat{E}_2^{(o)}(t_2)|T_1\rangle_1|T_2\rangle_2$$
$$= \left[\sqrt{T}\hat{E}_1(t_1) + \sqrt{R}\hat{E}_2(t_1 - D/c)\right]$$
$$\times \left[\sqrt{T}\hat{E}_2(t_2) - \sqrt{R}\hat{E}_1(t_2 + D/c)\right]|T_1\rangle_1|T_2\rangle_2$$
$$= \left[T\hat{E}_1(t_1)\hat{E}_2(t_2) - R\hat{E}_2(t_1 - D/c)\hat{E}_1(t_2 + D/c)\right]|T_1\rangle_1|T_2\rangle_2$$
$$= \left[Tg_1(t_1 - T_1)g_2(t_2 - T_2)\right.$$
$$\left. - Rg_2(t_1 - T_2 - D/c)g_1(t_2 - T_1 + D/c)\right]|0\rangle, \quad (8.16)$$

where $\hat{E}_j\hat{E}_j|T_1\rangle_1|T_2\rangle_2 = 0$ $(j = 1, 2)$. So, the probability of two-photon coincidence detection at the two outputs is proportional to

$$\langle T_1|\langle T_2|\hat{E}_2^{(o)\dagger}(t_2)\hat{E}_1^{(o)\dagger}(t_1)\hat{E}_1^{(o)}(t_1)\hat{E}_2^{(o)}(t_2)|T_1\rangle_1|T_2\rangle_2$$
$$= \left|Tg_1(t_1 - T_1)g_2(t_2 - T_2)\right.$$
$$\left. - Rg_2(t_1 - T_2 - D/c)g_1(t_2 - T_1 + D/c)\right|^2. \quad (8.17)$$

The total two-photon coincidence detection probability is an integration of t_1, t_2 over the whole photon wave packet:

$$P_2 \propto \int dt_1 dt_2 \langle T_1|\langle T_2|\hat{E}_2^{(o)\dagger}(t_2)\hat{E}_1^{(o)\dagger}(t_1)\hat{E}_1^{(o)}(t_1)\hat{E}_2^{(o)}(t_2)|T_1\rangle_1|T_2\rangle_2$$
$$\propto 1 - \frac{2TR}{T^2 + R^2}\mathcal{V}_{12}(T_1 - T_2 - D/c) \quad (8.18)$$

with the visibility of interference defined as

$$\mathcal{V}_{12}(T_1 - T_2 - D/c) \equiv \int \frac{dt_1 dt_2}{2\pi} g_1^*(t_1 - T_1) g_2^*(t_2 - T_2)$$
$$\times g_2(t_1 - T_2 - D/c) g_1(t_2 - T_1 + D/c)$$
$$= \left| \int d\omega \phi_1^*(\omega) \phi_2(\omega) e^{i\omega(T_1 - T_2 - D/c)} \right|^2. \tag{8.19}$$

Here, we used $\int dt |g_1(t)|^2 = 2\pi = \int dt |g_2(t)|^2$ for single-photon states. When the paths are balanced with $D/c = T_1 - T_2$, the visibility $\mathcal{V}_{12}$ depends on the mode overlap:

$$\mathcal{V}_{12}(0) = \left| \int d\omega \phi_1^*(\omega) \phi_2(\omega) \right|^2 \le 1. \tag{8.20}$$

The inequality above comes from the Cauchy-Schwarz inequality and $\int d\omega |\phi_1(\omega)|^2 = 1 = \int d\omega |\phi_2(\omega)|^2$ for single-photon states. The equal sign stands when $\phi_1(\omega) = \phi_2(\omega)$ or complete mode match between the two input photons. Set $P_2(\infty) = P_{20}$ and $T = R$. As the beam splitter position D is scanned through $c(T_1 - T_2)$, P_2/P_{20} shows an interference dip from 1 down to $1 - \mathcal{V}_{12}(0)$, i.e., the Hong-Ou-Mandel interference dip.

8.2.2 Time-bin Entanglement and Franson Interferometer

In both Pfleegor-Mandel and Hong-Ou-Mandel interference experiment, the interference effects are independent of phases of individual fields. But this is not a characteristic of two-photon interference. In this section, we will study a type of two-photon interference experiment in which the measured two-photon coincidence counts are a function of the phase difference between optical fields involved in the interference.

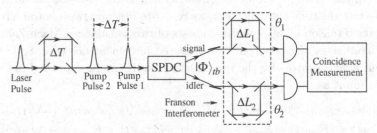

Fig. 8.6 Schematic for producing a time-bin entangled state and its detection with a Franson interferometer.

Consider a spontaneous parametric down-conversion process pumped by two coherent pulses with a delay of ΔT, as shown in Fig. 8.6. Each pump pulse creates a pulsed two-photon state of the form

$$|\Phi\rangle = \int d\omega_1 d\omega_2 \Phi(\omega_1, \omega_2)|\omega_1\rangle_1 |\omega_2\rangle_2, \qquad (8.21)$$

which was obtained from Eq. (6.32) in Section 6.1.5. The time delay ΔT between the pump pulses can be introduced as a phase factor of $e^{i\omega_p \Delta T}$ in the pump profile $\alpha_p(\omega_p)$ in $\Phi(\omega_1, \omega_2)$ from Eq. (6.29). Then the quantum state for the time-bin entangled state is

$$|\Phi\rangle_{tb} = [|\Phi(0)\rangle + |\Phi(\Delta T)\rangle]/\sqrt{2}$$

$$= \frac{1}{\sqrt{2}} \int d\omega_1 d\omega_2 \Phi(\omega_1, \omega_2)[1 + e^{i(\omega_1 + \omega_2)\Delta T}|\omega_1\rangle_1 |\omega_2\rangle_2. \qquad (8.22)$$

Here, we assume ΔT is much larger than the inverse of the spectral width of $\Phi(\omega_1, \omega_2)$ so that the two pulses are not overlapping:

$$\langle \Phi(0)|\Phi(\Delta T)\rangle = \int d\omega_1 d\omega_2 |\Phi(\omega_1, \omega_2)|^2 e^{i(\omega_1 + \omega_2)\Delta T} \approx 0. \qquad (8.23)$$

Since the two pulses are coherent to each other, the two-photon state in Eq. (8.22) is a time-entangled state similar to that in Eq. (4.53) of Section 4.4.3. But different from the state in Eq. (4.53), the state here in Eq. (8.22) is a pulsed two-photon state. In fact, if the two-photon spectral function $\Phi(\omega_1, \omega_2)$ is factorable: $\Phi(\omega_1, \omega_2) = \phi(\omega_1)\psi(\omega_2)$, which corresponds to a transform-limited two-photon state, the two-photon state in Eq. (8.22) can be rewritten as

$$|\Phi\rangle_{tb} = [|\phi_t(0)\rangle_1 |\psi_t(0)\rangle_2 + |\phi_t(\Delta T)\rangle_1 |\psi_t(\Delta T)\rangle_2]/\sqrt{2}, \qquad (8.24)$$

where $|\phi_t(\tau)\rangle_1 \equiv \int d\omega_1 e^{i\omega_1 \tau}\phi(\omega_1)|\omega_1\rangle_1, |\psi_t(\tau)\rangle_2 \equiv \int d\omega_2 e^{i\omega_2 \tau}\psi(\omega_2)|\omega_2\rangle_2$ are the single-temporal mode single-photon states defined in Section 4.3.2. The expression above shows the entanglement of the two photons in time.

To test the time entanglement, we resort to Franson two-photon interferometer [Franson (1989)], which consists of two unbalanced Mach-Zehnder interferometers whose outputs are detected in coincidence (Fig. 8.6). The output field operators for the two interferometers can be expressed in terms of the input as

$$\hat{E}_{out1}(t) = [\hat{E}_{in1}(t) + \hat{E}_{in1}(t + \Delta L_1/c) + \hat{E}_{in01}(t) - \hat{E}_{in01}(t + \Delta L_1/c)]/2$$

$$\hat{E}_{out2}(t) = [\hat{E}_{in2}(t) + \hat{E}_{in2}(t + \Delta L_2/c) + \hat{E}_{in02}(t) - \hat{E}_{in02}(t + \Delta L_2/c)]/2, \qquad (8.25)$$

where $\hat{E}_{in\{1,2\}}(t) = (1/\sqrt{2\pi})\int d\omega e^{-i\omega t}\hat{a}_{\{1,2\}}(\omega)$ and $\Delta L_{1,2}$ are the path difference for the two unbalanced MZ interferometers. $\hat{E}_{in01}, \hat{E}_{in02}$ are the vacuum inputs in the unused input ports of the two interferometers.

To calculate the two-photon coincidence detection probability, we first evaluate the following

$$\hat{E}_{out1}(t+\tau)\hat{E}_{out2}(t)|\Phi\rangle_{tb}$$

$$= \Big[\hat{E}_{in1}(t+\tau)\hat{E}_{in2}(t) + \hat{E}_{in1}(t+\tau)\hat{E}_{in2}(t+\Delta L_2/c)$$

$$+\hat{E}_{in1}(t+\tau+\Delta L_1/c)\hat{E}_{in2}(t+\Delta L_2/c)$$

$$+\hat{E}_{in1}(t+\tau+\Delta L_1/c)\hat{E}_{in2}(t)\Big]|\Phi\rangle_{tb}/2. \qquad (8.26)$$

Here the contributions from the unused vacuum inputs are zero. The above can be calculated from

$$\hat{E}_{in1}(t')\hat{E}_{in2}(t)|\Phi\rangle_{tb} = \int \frac{d\omega_1 d\omega_2}{2\pi}\Phi(\omega_1,\omega_2)e^{-i(\omega_1 t'+\omega_2 t)}[1 + e^{i(\omega_1+\omega_2)\Delta T}]|0\rangle$$

$$= [F(t',t) + F(t'-\Delta T, t-\Delta T)]|0\rangle, \qquad (8.27)$$

with a temporal two-photon wave function

$$F(t',t) \equiv \frac{1}{2\pi}\int d\omega_1 d\omega_2 \Phi(\omega_1,\omega_2)e^{-i(\omega_1 t'+\omega_2 t)}. \qquad (8.28)$$

Since the two-photon spectral function $\Phi(\omega_1,\omega_2)$ has a wide spectrum due to the broad band nature of the down-conversion and the pump pulse, the two-photon wave function $F(t',t)$ has a narrow range around t', $t = 0$ in the order of the reciprocal bandwidth of $\Phi(\omega_1,\omega_2)$. Substituting Eq. (8.27) into Eq. (8.26), we have a total of eight terms:

$$\hat{E}_{out1}(t+\tau)\hat{E}_{out2}(t)|\Phi\rangle_{tb}$$

$$= [F(t+\tau,t) + F(t+\tau+\Delta L_1/c, t+\Delta L_2/c) + F(t+\tau+\Delta L_1/c, t)$$

$$+F(t+\tau, t+\Delta L_2/c) + F(t+\tau-\Delta T+\Delta L_1/c, t-\Delta T)$$

$$+F(t+\tau-\Delta T, t-\Delta T) + F(t+\tau-\Delta T, t-\Delta T+\Delta L_2/c)$$

$$+F(t+\tau-\Delta T+\Delta L_1/c, t-\Delta T+\Delta L_2/c)]|0\rangle. \qquad (8.29)$$

In Franson interferometer, we set the path difference $\Delta L_1, \Delta L_2 \sim c\Delta T$ and the coincidence window is much shorter than ΔT so that $t, \tau \ll \Delta T$ but longer than the reciprocal bandwidth of $\Phi(\omega_1,\omega_2)$. Notice that the domain of $F(t+\tau,t)$ in which $F(t+\tau,t) \neq 0$ is much smaller than ΔT because of Eq. (8.23). Then only two terms in Eq. (8.29) are non-zero:

$$\hat{E}_{out1}(t+\tau)\hat{E}_{out2}(t)|\Phi\rangle_{tb}$$

$$= [F(t+\tau,t) + F(t+\tau-\Delta T+\Delta L_1/c, t-\Delta T+\Delta L_2/c)]|0\rangle. \qquad (8.30)$$

These two surviving terms correspond to the two photons from the first pulse going through the long path and from the second pulse but going through the short path, respectively. If the delay between the pump pulses matches the path delays, i.e., ΔL_1, $\Delta L_2 \sim c\Delta T$, these two possibilities are indistinguishable leading to two-photon interference.

The two-photon coincidence probability for the two outputs of the interferometers is then

$$P_2 \propto \int dt d\tau \, _{tb}\langle\Phi|\hat{E}_{out2}^{\dagger}(t)\hat{E}_{out1}^{\dagger}(t+\tau)\hat{E}_{out1}(t+\tau)\hat{E}_{out2}(t)|\Phi\rangle_{tb}$$

$$= \int dt d\tau \|\hat{E}_{out1}(t+\tau)\hat{E}_{out2}(t)|\Phi\rangle_{tb}\|^2$$

$$= \int dt d\tau \left| F(t+\tau,t) + F(t+\tau-\Delta T+\Delta L_1/c, t-\Delta T+\Delta L_2/c)\right|^2$$

$$= 1 + |\mathcal{V}|\cos[\omega_{10}(\Delta T - \Delta L_1/c) + \omega_{20}(\Delta T - \Delta L_2/c) - \epsilon_0]$$

$$= 1 + |\mathcal{V}|\cos(\theta_1 + \theta_2 - \epsilon_0), \tag{8.31}$$

where we used the normalization: $\int d\omega_1 d\omega_2|\Phi(\omega_1,\omega_2)|^2 = 1$ and

$$\mathcal{V} \equiv \int d\Omega_1 d\Omega_2|\Phi(\omega_{10}+\Omega_1,\omega_{20}+\Omega_2)|^2$$

$$\times e^{i\Omega_1(\Delta T - \Delta L_1/c)+i\Omega_2(\Delta T - \Delta L_2/c)}$$

$$\equiv |\mathcal{V}|e^{i\epsilon_0}. \tag{8.32}$$

So the two-photon coincidence shows an interference fringe as a function of the phases $\theta_j \equiv \omega_{j0}(\Delta T - \Delta L_j/c)$ $(j = 1,2)$ of the two MZ interferometers. The visibility of the interference fringe depends on how well the path differences ΔL_1, ΔL_2 match the delay ΔT between the pump pulses.

Experimentally, time entanglement of two photons was first observed in Franson interferometer by Ou et al. [Ou et al. (1990a)] and Kwiat et al. [Kwiat et al. (1990)] independently. The pulsed time-bin entangled state was realized by Brendel et al. [Brendel et al. (1999)].

The Franson interference effect shown with a time-bin entangled two-photon state is a nonlocal effect in the sense that the two photons from SPDC can be spatially separated and analyzed locally by two separate MZ interferometers, respectively. This is similar to a polarization entangled two-photon state and can be used to demonstrate the violation of Bell's inequality [Franson (1989)]. Because of this, it has applications in quantum cryptography [Ekert (1991)].

8.2.3   Distinguishability in Two-Photon Interference and Quantum Erasers

As demonstrated in both Hong-Ou-Mandel and Franson interferometers, two-photon interference stems from indistinguishability in the paths taken by two photons. However, because of the involvement of two photons, there usually is no one-photon interference which shows in single detector counts. This point can be easily checked for both phase independent Hong-Ou-Mandel interference and phase-sensitive Franson interference effects. In both cases, single detector counts are constant, exhibiting no interference effect.

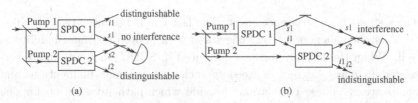

Fig. 8.7 Schematics for one-photon interference with two correlated photons. (a) Disappearance of one-photon interference due to distinguishability between $i1$ and $i2$. (b) Restoration of one-photon interference due to indistinguishability between $i1$ and $i2$. Adapted from [Zou et al. (1991b)].

To further illustrate this, let us consider a simple scheme shown in Fig. 8.7(a), where the pump field for SPDC is split into two for pumping two SPDC processes. We start with the pump field in a coherent state $|\alpha_p\rangle$. After the first 50:50 beam splitter, the state becomes $|\alpha_p/\sqrt{2}\rangle_1|\alpha_p/\sqrt{2}\rangle_2$. If we superpose the two output fields with another beam splitter, interference effect will occur. But now let the pump fields be down-converted to signal and idler fields, respectively, by interacting them with two nonlinear crystals, as shown in Fig. 8.7(a). For the two SPDC processes together, the Hamiltonian of the system is

$$H_{2SPDC} = \chi \hat{a}_1 \hat{a}_{s1}^\dagger \hat{a}_{i1}^\dagger + \chi \hat{a}_2 \hat{a}_{s2}^\dagger \hat{a}_{i2}^\dagger + h.c. \qquad (8.33)$$

For simplicity, we use single-mode treatment here. From Chapter 6, we find the state after the SPDC is

$$|\Psi_2\rangle = |vac\rangle + \eta(|1\rangle_{s1}|1\rangle_{i1} + e^{i\varphi}|1\rangle_{s2}|1\rangle_{i2}), \qquad (8.34)$$

where η is proportional to χ, α_p and $|\eta|^2$ is related to the down-conversion probability. We now superpose two signal fields, as shown in Fig. 8.7(a). It is straightforward to find the intensity of the superposed field $\hat{a}_s = (\hat{a}_{s1} +$

$\hat{a}_{s2})/\sqrt{2}$ as

$$\langle\Psi_2|\hat{a}_s^\dagger\hat{a}_s|\Psi_2\rangle = \left(\langle\Psi_2|\hat{a}_{s1}^\dagger\hat{a}_{s1}|\Psi_2\rangle + \langle\Psi_2|\hat{a}_{s2}^\dagger\hat{a}_{s2}|\Psi_2\rangle\right.$$
$$\left. + \langle\Psi_2|\hat{a}_{s1}^\dagger\hat{a}_{s2}|\Psi_2\rangle + \langle\Psi_2|\hat{a}_{s2}^\dagger\hat{a}_{s1}|\Psi_2\rangle\right)\Big/2$$
$$= |\eta|^2 + |\eta|^2|\langle1_{i2}|1_{i1}\rangle|\cos\varphi$$
$$= |\eta|^2, \tag{8.35}$$

which shows no interference effect. The last line is because $\langle1_{i2}|1_{i1}\rangle = 0$.

This phenomenon can be understood with photon distinguishability as well. When the two photons are produced in SPDC, they are highly correlated in many degrees of freedom such as frequency, momentum, and polarization. These correlations provide which-path information for one of the two correlated photons from the other, thus leading to no interference effect due to the complementarity principle of quantum mechanics. This can be seen from the disappearance of the interference term $\langle\Psi_2|\hat{a}_{s2}^\dagger\hat{a}_{s1}|\Psi_2\rangle = |\eta|^2 e^{i\varphi}\langle1_{i2}|1_{i1}\rangle = 0$ because $\langle1_{i2}|1_{i1}\rangle$ is zero due to distinguishability between $i1, i2$.

Compared with other interference experiments demonstrating the complementarity principle, where measurement is usually performed on the particle participating in the interference experiment to gain the which-path information, the scheme here does not make such a measurement and the which-path information is obtained through quantum correlation. But this also tells us how to restore the one-photon interference effect: making $i1, i2$ photons indistinguishable. This can be achieved by injecting $i1$ field into the second SPDC process and aligning it with $i2$ field, as shown in Fig. 8.7(b). If the two modes are completely overlapping, there is no way to distinguish from which SPDC process the idler photon is generated, i.e., $\langle1_{i2}|1_{i1}\rangle = 1$. This leads to coherence and one-photon interference between $s1$ and $s2$. Indeed, when $\langle1_{i2}|1_{i1}\rangle = 1$, Eq. (8.35) becomes

$$\langle\Psi|\hat{a}_s^\dagger\hat{a}_s|\Psi\rangle_2 = |\eta|^2(1 + \cos\varphi), \tag{8.36}$$

which shows interference effect in the signal detector counting rate. In the less-than-perfect case of $0 < |\langle1_{i2}|1_{i1}\rangle| < 1$, Eq. (8.35) gives an interference fringe visibility $\mathcal{V} = |\langle1_{i2}|1_{i1}\rangle| < 1$, i.e., partial indistinguishability leads to reduced visibility or coherence. This interference effect, dubbed as "induced coherence without induced emission", was first observed by Zou *et al.* [Zou et al. (1991b)].

The indistinguishability of photons is closely related to the mode of the photons. This is illustrated by the visibility relation $\mathcal{V} = |\langle1_{i2}|1_{i1}\rangle|$

obtained from Eq. (8.35). Although Eq. (8.35) is derived based on the single-mode description of the field, it is straightforward to extend it to multi-mode case when the joint spectral function of the down-converted fields $\Phi(\omega_1, \omega_2) = \phi(\omega_1)\psi(\omega_2)$ is factorized and the state can be written as $|\Psi_2\rangle = |vac\rangle + \eta|\phi\rangle_s|\psi\rangle_i$, same as the cases of independent photons in Section 8.2.1 and the time-bin entangled photons in Section 8.2.2. Then Eq. (8.34) is changed to

$$|\Psi_2\rangle = |vac\rangle + \eta(|\phi_1\rangle_{s1}|\psi_1\rangle_{i1} + e^{i\varphi}|\phi_2\rangle_{s2}|\psi_2\rangle_{i2}) \tag{8.37}$$

and Eq. (8.35) becomes

$$\langle\Psi_2|\hat{E}_s^\dagger\hat{E}_s|\Psi_2\rangle = \Big(\langle\Psi_2|\hat{E}_{s1}^\dagger\hat{E}_{s1}|\Psi_2\rangle + \langle\Psi_2|\hat{E}_{s2}^\dagger\hat{E}_{s2}|\Psi_2\rangle$$
$$+ \langle\Psi_2|\hat{E}_{s1}^\dagger\hat{E}_{s2}|\Psi_2\rangle + \langle\Psi_2|\hat{E}_{s2}^\dagger\hat{E}_{s1}|\Psi_2\rangle\Big)\Big/2$$
$$= |\eta|^2 + |\eta|^2|\langle\psi_2|\psi_1\rangle|\cos\varphi, \tag{8.38}$$

where $\hat{E}_{s1,s2} = (1/\sqrt{2})\int d\omega e^{-i\omega t}\hat{a}_{s1,s2}(\omega)$ and $\hat{E}_s = (\hat{E}_{s1} + \hat{E}_{s2})/\sqrt{2}$. So, the visibility now becomes

$$\mathcal{V} = |\langle\psi_2|\psi_1\rangle| = \left|\int d\omega \psi_2^*(\omega)\psi_1(\omega)\right|, \tag{8.39}$$

which depends on the mode match between $\psi_1(\omega), \psi_2(\omega)$, similar to Eq. (8.20) for the visibility in Hong-Ou-Mandel interference. Thus, the mode functions provide a quantitative description of photon indistinguishability. We will demonstrate more of this in Section 8.4.

There is another way to restore the interference effect through a concept called "quantum eraser" [Scully and Druhl (1982); Scully et al. (1991)], which is a technique that erases the which-path information by projection measurement. Consider a variation of Fig. 8.7(a) shown in Fig. 8.8(a),

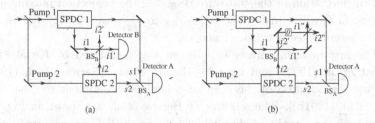

(a) (b)

Fig. 8.8 Quantum erasing of which-path information by projection measurement on detector B to restore interference effect. (a) BS$_B$ to superpose $i1$ and $i2$ seems to prevent us from distinguishing between $i1$ and $i2$. (b) Extra elements are added for the distinction between $i1$ and $i2$.

where we superpose two idler fields with a 50:50 beam splitter (BS_B). The experimental fact is that there is no interference in the superposition of $s1$, $s2$ even with BS_B inserted to superpose the two idler fields, similar to the case of no BS shown in Fig. 8.7(a). It is easy to understand why there is no interference in Fig. 8.7(a), as we discussed earlier. But it is not so obvious to explain the disappearance of interference in Fig. 8.8(a) in terms of complementarity principle since, after all, it seems impossible to obtain which-path information at the output $i1'$ of BS_B. Therefore, it seems that the scheme in Fig. 8.8(a) is similar to the scheme in Fig. 8.7(b) where indistinguishability between $i1$ and $i2$ should lead to interference between $s1$ and $s2$, contradicting the experimental observation. As a matter of fact, there is a way to distinguish between $i1$ and $i2$ without removing BS_B. It is by using the other output $i2'$ of BS_B. As shown in Fig. 8.8(b), we can add an extra BS to recombine the two outputs $i1'$, $i2'$ of BS_B. When the phase difference θ is right, the outputs of the extra BS are exactly the same as the inputs to BS_B: $i1'' = i1$, $i2'' = i2$ due to interference in the Mach-Zehnder interferometer formed with BS_B and the newly placed BS. So, even though $i1$ and $i2$ are mixed after the first BS (BS_B), they are still distinguishable after the second BS. The availability of the second output $i2'$ makes it possible to distinguish $i1$ and $i2$. On the other hand, the technique of quantum eraser is to use detector B at output $i1'$ to make a projection measurement and erase the which-path information for the restoration of the interference between $s1$ and $s2$. To see this more clearly, we write from Eq. (8.34) the two-photon state after the beam splitters as

$$|\Psi_2\rangle_{BS} = \frac{1}{\sqrt{2}} \big[|1\rangle_{s1} + e^{i\varphi}|1\rangle_{s2} \big] |1\rangle_{i1'} + \frac{1}{\sqrt{2}} \big[|1\rangle_{s1} - e^{i\varphi}|1\rangle_{s2} \big] |1\rangle_{i2'}, \quad (8.40)$$

where we used $|1\rangle_{i1} \to (|1\rangle_{i1'} + |1\rangle_{i2'})/\sqrt{2}$ and $|1\rangle_{i2} \to (|1\rangle_{i1'} - |1\rangle_{i2'})/\sqrt{2}$ for the transformation of the states by BS_B. So, the projection measurement at detector B will transform state $|\Psi_2\rangle_{BS}$ to $\frac{1}{\sqrt{2}} \big[|1\rangle_{s1} + e^{i\varphi}|1\rangle_{s2} \big] |1\rangle_{i1'}$ and lead to interference between $s1$ and $s2$.

The first quantum eraser experiment was performed by Kwiat et al. [Kwiat et al. (1992)] and later by Herzog et al. [Herzog et al. (1995)] who used a variation of the induced coherence experiment by Zou et al. [Zou et al. (1991b)]. The schemes by Herzog et al. are shown in Fig. 8.9, where both the signal and idler fields of the first SPDC process are injected back together and are aligned with the signal and idler fields of the second SPDC process in a bow-tie configuration [Herzog et al. (1994)]. When direct reflection is made as in Fig. 8.9(a), both signal and idler fields will

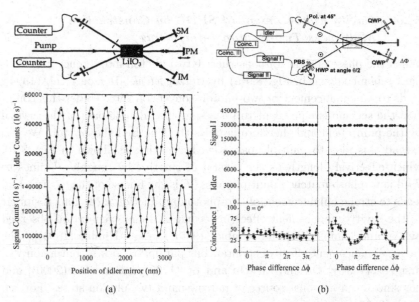

(a)   (b)

Fig. 8.9   (a) Schematic and observation data for enhancement and suppression of two-photon emission in SPDC by two-photon interference. Adapted from [Herzog et al. (1994)]. (b) Schematic and observation data for quantum erasing of which-path information to restore interference effect. Adapted from [Herzog et al. (1995)].

show one-photon interference effect due to indistinguishability in both signal fields and idler fields. The interference pattern in the signal and idler outputs are shown in Fig. 8.9(a). Notice that the interference patterns of the signal and idler fields are completely in phase, indicating the enhancement and suppression of the two-photon process by constructive and destructive interference, respectively. In the quantum eraser scheme shown in Fig. 8.9(b), the polarizations of both the signal and idler field from the first SPDC process are rotated 90 degree before being injected back into the second SPDC process. With the notations in Fig. 8.8, now $s1$ and $s2$ as well as $i1$ and $i2$ are distinguishable due to orthogonal polarizations. In this case, no interference is observed at either signal or idler output field as we project both $s1$ and $s2$ (or $i1$ and $i2$) into 45 degree with a polarization beam splitter, which is equivalent to BS_A (or BS_B) in Fig. 8.8(b). The projection measurement onto i'_1 is equivalent to the detection of a photon at detector B. So, the detection of the superimposed signal fields, when gated upon the detection of a photon at detector B, will show an interference pattern, as observed by Herzog et al. and shown in coincidence count of 45 degree case in Fig. 8.9(b) [Herzog et al. (1995)].

8.2.4 Cavity Enhancement of SPDC by Constructive Multi-pass Two-Photon Interference

When the phase is right for constructive two-photon interference, the feedback scheme shown in Fig. 8.9(a) by Herzog *et al.* [Herzog et al. (1994)] leads to an enhancement of two-photon conversion rate. Under this condition the scheme is equivalent to phase-matched double-pass of the crystal by the pump field and therefore doubles the length of the crystal. We can extend this idea to multiple passes with a cavity, as shown in Fig. 8.10, which effectively lengthens the crystal of length l by roughly $\mathcal{F}$ times to $\mathcal{F}l$ due to phase-matched multiple-pass ($\mathcal{F}$ is the finesse of the cavity), and leads to strong enhancement of two-photon production rate. This is similar to the cavity enhancement effect in second harmonic generation observed by Wu and Kimble [Wu and Kimble (1985); Ou and Kimble (1993)]. The cavity enhancement effect of spontaneous parametric down-conversion was first observed by Ou and Lu [Ou and Lu (1999); Lu and Ou (2000)] and has since been a robust source of narrow-band two-photon states [Fortsch et al. (2013)].

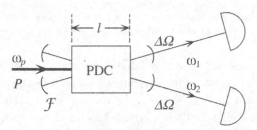

Fig. 8.10   Enhancement of the two-photon emission rate of SPDC with a cavity by lengthening the crystal effective length by $\mathcal{F}$ times.

To understand the enhancement effect quantitatively, we consider the theoretical modeling of parametric process in a cavity which is covered in Section 6.3.4. But here we consider the case of far below threshold where there is only spontaneous process for two-photon generation. From Eq. (6.135) of Section 6.3.4 with $\hat{a}_{out}(\omega_0+\Omega) = \hat{A}_{out}(\Omega)$ and $\zeta^2 \ll (\beta+\gamma)^2/4$ for the far-below threshold case, we find the output operator of a degenerate OPO on resonance ($\Delta_a = 0$, ω_0 is the degenerate frequency of the OPO) is related to the input as follows:

$$\hat{a}_{out}(\omega_0 + \Omega) = G_D(\Omega)\hat{a}_{in}(\omega_0 + \Omega) + g_D(\Omega)\hat{a}_{in}^\dagger(\omega_0 - \Omega)$$

$$+G_L(\Omega)\hat{c}_{in}(\omega_0 + \Omega) + g_L(\Omega)\hat{c}_{in}^\dagger(\omega_0 - \Omega) \quad (8.41)$$

with

$$G_D(\Omega) = \frac{\gamma_1 - \gamma_2 + 2i\Omega}{\gamma_1 + \gamma_2 - 2i\Omega}, \quad g_D(\Omega) = \frac{4\zeta\gamma_1}{(\gamma_1 + \gamma_2 - 2i\Omega)^2}, \quad (8.42)$$

$$G_L(\Omega) = \frac{2\sqrt{\gamma_1\gamma_2}}{\gamma_1 + \gamma_2 - 2i\Omega}, \quad g_L(\Omega) = \frac{4\zeta\sqrt{\gamma_1\gamma_2}}{(\gamma_1 + \gamma_2 - 2i\Omega)^2}. \quad (8.43)$$

Here we set $\beta = \gamma_1, \gamma = \gamma_2$ from Eq. (6.135) as the coupling constants (a.k.a. decay constants) for $\hat{a}_{in}$ and $\hat{c}_{in}$, respectively. $\hat{c}_{in}$ represents the unwanted vacuum mode coupled-in due to losses in the system. ζ is the single-pass parametric amplitude gain and is proportional to the pump amplitude and the nonlinear coefficient. In Eqs. (8.42) and (8.43) we dropped the $|\zeta|^2$-term in the denominator because the OPO is operated far below threshold so that $|\zeta| \ll \gamma_1, \gamma_2$.

Now let us look at the enhancement effect in down-conversion due to resonance. For vacuum inputs, we calculate from Eq. (8.41) the spectrum $S(\omega)$ of the field defined by

$$\langle a_{out}^\dagger(\omega_0 + \Omega)a_{out}(\omega_0 + \Omega')\rangle = \langle A_{out}^\dagger(\Omega)A_{out}(\Omega') \equiv S(\Omega)\delta(\Omega - \Omega'). \quad (8.44)$$

The result is

$$S(\omega) = |g_D(\Omega)|^2 + |g_L(\Omega)|^2 = \frac{16|\zeta|^2\gamma_1(\gamma_1 + \gamma_2)}{[(\gamma_1 + \gamma_2)^2 + 4\Omega^2]^2}. \quad (8.45)$$

The rate of down-conversion can be calculated as

$$R_{cavity} = \langle E_{out}^{(-)}(t)E_{out}^{(+)}(t)\rangle = \frac{1}{2\pi}\int d\Omega S(\Omega), \quad (8.46)$$

where

$$\hat{E}^{(+)}(t) = [\hat{E}^{(-)}(t)]^\dagger = \frac{1}{\sqrt{2\pi}}\int d\omega\hat{a}(\omega)e^{-i\omega t} = \frac{e^{-i\omega_0 t}}{\sqrt{2\pi}}\int d\Omega\hat{A}(\Omega)e^{-i\Omega t}. \quad (8.47)$$

From Eq. (8.45), we have

$$R_{cavity} = \frac{1}{2\pi}\int_{-\infty}^{\infty} d\Omega \frac{16|\zeta|^2\gamma_1(\gamma_1 + \gamma_2)}{[(\gamma_1 + \gamma_2)^2 + 4\Omega^2]^2}$$
$$= |r|^2\mathcal{F}^2/\pi\Delta t\mathcal{F}_0, \quad (8.48)$$

where $r \equiv \zeta\Delta t$ is the single-pass gain parameter with Δt as the round-trip time, and $\mathcal{F} \equiv 2\pi/(\gamma_1 + \gamma_2)\Delta t = 2\pi/\Delta t\Delta\omega_{opo}$ is the finesse of the cavity (which is of the order of the number of bounces of light before it leaves the cavity) and can be measured directly. Here $\Delta\omega_{opo} = \gamma_1 + \gamma_2$ corresponds to the bandwidth of the OPO cavity. $\mathcal{F}_0 \equiv 2\pi/\gamma_1\Delta t$ is the same quantity without the loss ($\gamma_2 = 0$). To find the enhancement factor, we need the signal rate without the cavity. In the single pass case, we simply have $g_D(\omega) = r\eta(\omega)$ and $g_L = 0$. Here $\eta(\omega)$ is the gain spectrum of single-pass spontaneous down-conversion determined by phase-matching condition with normalization $\eta(0) = 1$. In the experiment, we usually have an interference filter (IF) in front of the detector. The frequency bandwidth $\Delta\nu_{IF} = \Delta\omega_{IF}/2\pi$ of the IF is normally smaller than that of down-conversion so that $\eta(\omega) \approx 1$ for ω within $\Delta\omega_{IF}$ and is zero for ω outside $\Delta\omega_{IF}$. Hence, the signal rate without the cavity is

$$R_{single\_pass} = |r|^2 \Delta\nu_{IF} = |r|^2 \Delta\omega_{IF}/2\pi, \qquad (8.49)$$

and the average enhancement factor per mode is

$$B \equiv \frac{R_{resonance}/\Delta\omega_{opo}}{R_{single\_pass}/\Delta\omega_{IF}} = \mathcal{F}^3/\pi\mathcal{F}_0, \qquad (8.50)$$

or roughly the square of the number of bounces of light before it leaves the cavity. This is consistent with the phase-matched multi-pass parametric down-conversion where the conversion rate is proportional to the square of the crystal length and the effective crystal length is increased by the number of passes or $\mathcal{F}$. Here the phase matching is satisfied by the on resonance condition. The square law is a result of two-photon constructive interference. The loss of the system will reduce the effect by a factor of $\mathcal{F}/\mathcal{F}_0$.

8.2.5   Mode-locked Two-Photon States

In the discussion of the cavity enhancement effect in the previous section, we only considered the single degenerate mode of the cavity ($\omega_1 = \omega_2 = \omega_0$). For a type-I degenerate parametric down-conversion, the two down-converted fields have the same frequency and polarization and thus have a wide spectral bandwidth [Boyd (2003)]. So, in addition to the mode with $\omega_1 = \omega_2 = \omega_0$, other longitudinal modes with $\omega_1 = \omega_0 + N\Delta\Omega_{FSR}, \omega_2 = \omega_0 - N\Delta\Omega_{FSR}$ are also on resonance, with $\Delta\Omega_{FSR}$ as the free spectral range of the cavity and N determined by the dispersion of the nonlinear medium. This leads to a two-photon state of many frequency pairs:

$$|\Psi\rangle_{ML} = \sum_{m=-N}^{N} \int d\Omega\, \psi(\Omega)\hat{a}^\dagger(\omega_0 + m\Delta\Omega_{FSR} + \Omega)$$
$$\times \hat{a}^\dagger(\omega_0 - m\Delta\Omega_{FSR} - \Omega)|vac\rangle, \qquad (8.51)$$

where $\psi(\Omega)$ is the spectral function from one cavity mode and is related to $G_D(\Omega)$ in Eq. (8.42).

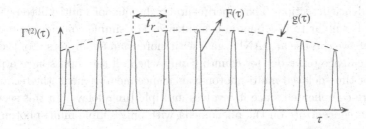

Fig. 8.11   Two-photon time correlation function for a mode-locked two-photon state.

The state in Eq. (8.51) is the so-called mode-locked (ML) two-photon state first discovered by Lu *et al.* [Lu et al. (2003)]. Different modes of photon pairs are in superposition. All the pairs have a common origin (phase) from the pump field, which provides the mechanism for phase (mode) locking. It has a comb-like two-photon spectrum, whose Fourier transformation is the two-photon time correlation function:

$$\Gamma^{(2)}_{ML}(\tau) \equiv \langle \hat{E}^{(-)}(t)\hat{E}^{(-)}(t+\tau)\hat{E}^{(+)}(t+\tau)\hat{E}^{(+)}(t) \rangle$$

$$= \left| g(\tau)F(\tau) \right|^2 \tag{8.52}$$

with

$$g(\tau) \equiv \int d\Omega \psi(\Omega)e^{-i\Omega\tau}, \quad F(\tau) \equiv \frac{\sin[(2N+1)\Delta\Omega\tau/2]}{\sin(\Delta\Omega\tau/2)}.$$

As shown in Fig. 8.11, the two-photon time correlation function has a shape of regular spikes separated by the round-trip time $t_r = 2\pi/\Delta\Omega_{FSR}$ of the cavity. The physical meaning of Eq. (8.52) is very clear: after a photon is detected, we need to wait an integer multiple of the round-trip time t_r for the second photon to come out of the cavity and be detected. This behavior is similar to the pulses coming out of a mode-locked laser. The two-photon time correlation function shown in Fig. 8.11 is a result of frequency-entanglement of the two photons generated. The spike-like two-photon time correlation function of the mode-locked two-photon state was observed by Goto *et al.* [Goto et al. (2003)] and indirectly confirmed with a Hong-Ou-Mandel interferometer where a revival of the Hong-Ou-Mandel dip is observed [Lu et al. (2003); Xie et al. (2015)].

8.3   Multi-Photon Interference Effects

Now we can extend two-photon interference effects to multi-photon interference effects where more than two photons are involved in multi-photon coincidence detection. The experimental technique for multi-photon coincidence is similar to two-photon coincidence. For simple coincidence measurement, we can use an "AND" gate with more than two inputs. So, there is a coincidence pulse out for counting only when all the inputs have a pulse in. For time-delayed multi-photon coincidence measurement, the technique is more complicated since it involves multiple time delays. In this section, we will concentrate on the phenomena with only simple multi-photon coincidence measurement.

8.3.1   Multi-Photon Bunching Effects

We have discussed photon bunching effect before in both the classical wave and quantum optical languages. We showed that these two are actually equivalent. However, there is a deeper physical principle in photon bunching effect buried under the mathematical calculation. The principle is quantum interference. In fact, Fano was the first to associate an interference effect of two-photon amplitudes to the photon bunching effect [Fano (1961)]. Glauber later used a similar argument to explain the photon bunching effect [Glauber (1964)]. In the following, we will show this connection clearly with Hong-Ou-Mandel interferometer and extend it to the case of more than two photons.

8.3.1.1   Photon Bunching Effects and Two-Photon Constructive Interference

Consider a different detection scheme for the Hong-Ou-Mandel interferometer: instead of detecting the $|1, 1\rangle$ state with two detectors placed at two different output ports (Fig. 8.5(a)), we place them at the same output port (Fig. 8.12(a)). This will measure the probability $P_2(2, 0)$ for $|2, 0\rangle$ state instead of the probability of $P_2(1, 1)$ in Hong-Ou-Mandel effect.

From Eq. (8.8), we have

$$P_2^{qu}(2, 0) = 2t^2r^2. \tag{8.53}$$

Here, "qu" denotes the prediction from quantum theory. On the other hand, we may find the classical probability of detecting both photons at the same side of the beam splitter, i.e., case (a) or (b) in Fig. 8.4. The result is simply the product of the single-photon events due to independence:

$$P_2^{cl}(2, 0) = P_2^{cl}(0, 2) = P_1(1, 0)P_1'(1, 0) = t^2r^2, \tag{8.54}$$

where $P_1(1, 0), P_1'(1, 0)$ are the probabilities for one input photon from two sides, respectively. From Eqs. (8.53) and (8.54), we find that the quantum

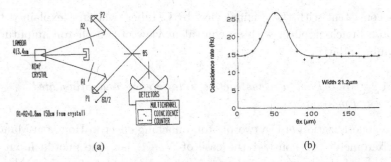

(a) (b)

Fig. 8.12 Photon bunching effect in Hong-Ou-Mandel interferometer: (a) outline of setup; (b) the result of the experiment. Reproduced from [Rarity and Tapster (1989)].

probability is twice the classical probability. This is exactly the same as the photon bunching effect. Its demonstration was first performed by Rarity and Tapster, as shown in Fig. 8.12 [Rarity and Tapster (1989)]. To see that this is the result of two-photon interference, we consider Fig. 8.13, which is the setup to measure $P_2(2,0)$. The two-photon detection measurement for $P_2(2,0)$ is accomplished by a beam-splitter (BS_d) splitting scheme shown in Figs. 8.13(a) and (b). This is similar to HBT experiment (Sections 1.2, 1.6). As seen in Figs. 8.13(a) and (b), there are two possible ways to arrange the detection of the two photons from one output port of the beam splitter (BS). If the incoming two photons are well-separated, they behave like classical particles and we add the probabilities of the two possibilities: $P_2^{cl} = |A|^2 + |A|^2$ where we take A as the amplitude for each of the cases. But if they overlap at the beam splitter, we cannot distinguish the two possibilities, and we add the amplitudes before taking the absolute value square for the overall probability: $P_2^{qu} = |A + A|^2 = 4|A|^2 = 2P_2^{cl}$. Note that the phases for the two cases are the same because the overall paths for the two photons in the two possibilities are the same due to indistinguishability of the two photons. Therefore, it is constructive interference that is responsible for the photon bunching effect in the Hong-Ou-Mandel interferometer. This

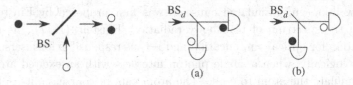

(a) (b)

Fig. 8.13 Two possibilities for two-photon constructive interference in explaining photon bunching effect in Hong-Ou-Mandel interferometer. BS_d is for two-photon detection.

was consistent with the original view by Glauber, who first explained the photon bunching effect with an equivalent view of two-photon amplitudes [Glauber (1964)].

8.3.1.2 *Multi-Photon Constructive Interference and Stimulated Emission*

As a generalization of the two-photon bunching effect in Hong-Ou-Mandel interferometer, we consider the case of $N + 1$, i.e., one photon input to one side of the beam splitter and N photon input at the other side, as shown in Fig. 8.14. From the output state derived in Eq. (6.60), we find the probability for all $N + 1$ photons to exit at one output port as

$$P_{N+1}^{qu}(N+1,0) = (N+1)/2^{N+1}, \qquad (8.55)$$

which is $N + 1$ times of the classical probability $P_{N+1}^{cl}(N + 1, 0) = (1/2) \times (1/2^N) = 1/2^{N+1}$ if the input single photon (white circle) is distinguishable from other input N identical photons (black solid circles). This factor can be understood with the $(N+1)$-photon detection scheme shown in Fig. 8.14, where there are $N + 1$ different possibilities to arrange the input single photon (white circle). $P_{N+1}^{cl}(N + 1, 0) = (N + 1)|A|^2$ corresponds to completely distinguishable situation among the possibilities while $P_{N+1}^{qu}(N+1,0) = |(N+1)A|^2 = (N+1)P_{N+1}^{cl}(N+1,0)$ is due to constructive quantum interference by adding the amplitudes.

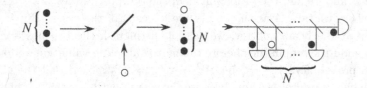

Fig. 8.14   (N+1)-photon bunching effect in Hong-Ou-Mandel interferometer.

The multi-photon bunching effect discussed above can be used to explain the phenomena of stimulated emission of atom by incident photons. As is well-known, stimulated emission was first proposed by Einstein to explain the spectrum of blackbody radiation [Einstein (1917)]. It is the foundation for optical amplification and is thus responsible for lasers. Phenomenologically, when a single photon interacts with an excited atom, it can stimulate the atom to emit. The atom can, of course, emit a photon spontaneously. From Einstein's A- and B-coefficients, the rates of the stimulated emission and spontaneous emission are the same, and are denoted as R. The overall rate is then $2R$. When there are N input photons, each photon may stimulate the atom, and the overall rate is then $(N+1)R$. It seems that the description above has nothing to do with the multi-photon interference effect we discussed in the beginning of this section.

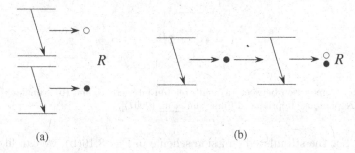

(a)   (b)

Fig. 8.15   Photo-emission from two atoms: (a) spontaneous emission of two independent atoms and (b) stimulated emission of one atom by the emission of another atom.

To make a connection, let us consider two excited atoms, each emitting a photon, as shown in Fig. 8.15. According to the description above, there are two different types in the photo-emission process. In Fig. 8.15(a), the atoms in the excited state independently emit photons due to spontaneous emission, and two-photon detection probability in this case is simply the product of individual emission probability: $P_2^{sp} = P_1^2 = R$. In Fig. 8.15(b), the detected two photons are from stimulated emission, i.e., the photon spontaneously emitted from one atom stimulates the emission of another atom. Since the rate of stimulated emission is the same as that of spontaneous emission, we have $P_2^{st} = R = P_1^2 = P_2^{sp}$. Hence, the overall probability is

$$P_2 = P_2^{st} + P_2^{sp} = 2P_1^2 = 2R = 2P_2^{sp}, \qquad (8.56)$$

which is exactly the ratio of the photon bunching effect.

Since photon bunching effect is due to a two-photon constructive interference effect and the stimulated emission gives the same result as the photon bunching effect, we therefore have reason to speculate that stimulated emission is also a result of the two-photon interference effect. In fact, we look at the two schemes in Fig. 8.16 where the $(N+1)$- photon interference scheme with a beam splitter (Fig. 8.16(a)) is copied from Fig. 8.14. From Eq. (8.55), we find that the $(N+1)$-photon detection probability is $P_{N+1} = (N+1)/2^{N+1}$, which is $N+1$ times the probability $P_{N+1}^{cl} = 1/2^{N+1}$ when the $N+1$ photons were classical particles. The enhancement factor is

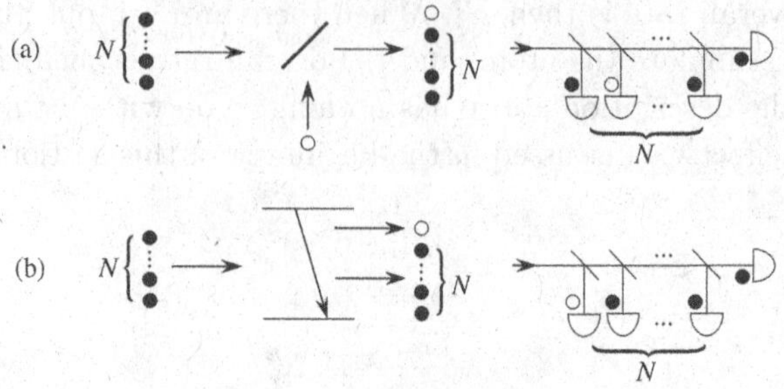

Fig. 8.16   Comparison between (a) multi-photon interference and (b) stimulated emission by N photons. Reproduced from [Sun et al. (2007)].

$N + 1$. For the stimulated emission scheme in Fig. 8.16(b), we can likewise make the same argument as the one for $(N + 1)$-photon constructive interference in Fig. 8.14 since the multi-photon detection scheme cannot make the distinction between the two scenarios in Fig. 8.16. This also leads to an enhancement factor of $N + 1$ as compared to the spontaneous emission, which is equivalent to the case when the atom-emitted photon is completely distinguishable from the N incident photons, i.e., the N incident photons do not interact with the atom at all and are totally independent of its emission.

Although Einstein used energy balance to introduce the stimulated emission but did not present any detail, the quantum theory of light developed later fully explained it. In essence, it is from the Bosonic relation of $\hat{a}^{\dagger}|N\rangle = \sqrt{N+1}|N+1\rangle$, which is consistent with the $N + 1$ enhancement factor given above. But this explanation relies on some complicated operator algebra. Through our discussion above, we find that there is a simpler physical principle underlying the phenomenon of stimulated emission, that is, stimulated emission is a result of multi-photon constructive interference in exactly the same way as the photon bunching effect. Experimentally, this connection between the stimulated emission and multi-photon interference was demonstrated by Sun *et al.* [Sun et al. (2007)].

The ratio of stimulated emission to spontaneous emission discussed above can be used to derive the Schawlow-Townes linewidth of lasers [Schawlow and Townes (1958)]. Suppose that there are N photons in the laser cavity. Then the next photon emitted by the atoms will have a chance of $N/(N + 1)$ to be identical to the N photons and $1/(N + 1)$ chance to be spontaneously emitted photon. The spontaneously emitted photon has a linewidth of the laser cavity $\Delta\nu_c$ while the stimulated photons have the same frequency as the laser. So, upon averaging over all $N + 1$ photons, the laser linewidth is then

$$\Delta\nu_{laser} = \Delta\nu_c \frac{1}{N+1} + 0 \times \frac{N}{N+1} \approx \frac{\Delta\nu_c}{N}. \qquad (8.57)$$

Suppose the laser has an output coupler of transmissivity T. Then the output power of the laser is $P_{out} = TNh\nu/t_r$. Here, $t_r = 2L/c = 1/\Delta\nu_{FSR}$ is the round trip time for the intra-cavity photons with $2L$ as the round trip length of a standing wave cavity and $\Delta\nu_{FSR}$ as the free spectral range of the laser cavity. Taking linewidth of the cavity $\Delta\nu_c = \Delta\nu_{FSR}/\mathcal{F}$ and the finesse $\mathcal{F} = 2\pi/T$, we change Eq. (8.57) to

$$\Delta\nu_{laser} = 2\pi h\nu \frac{(\Delta\nu_c)^2}{P_{out}}, \qquad (8.58)$$

which is the Schawlow-Townes linewidth of a laser. From this simple derivation, we find that the Schawlow-Townes linewidth originates from spontaneous emission and is thus of the quantum nature. Therefore, this linewidth is a fundamental quantum limit for the linewidth of a laser. The linewidth of an actual laser is far above this limit due to classical technical issues such as cavity mechanical vibrations.

8.3.1.3 *More Photon Bunching Effects*

Now let us generalize the case of N-photon + 1-photon to the case of N-photon + M-photon with $M \geq 2$. We start with the $2+2$ case, i.e., two photons enter a 50:50 beam splitter from one side while other two photons enter from the other side. This is similar to the Hong-Ou-Mandel effect for two photons but here, it is for two pairs of photons and we make the observation at only one output port. Using the technique in Section 6.2.2, we find the output state is

$$|\Psi_4\rangle = \sqrt{\frac{3}{8}}(|4,0\rangle + |0,4\rangle) - \frac{1}{2}|2,2\rangle. \qquad (8.59)$$

Hence, the probability to have all four photons exit in one output port is $P_4^{qu}(4,0) = 3/8$. On the other hand, if we treat the photons as classical particles, the probability would simply be $P_4^{cl}(4,0) = (1/2)^4 = 1/16$. So, there is an enhancement factor of 6 for the quantum probability.

We can understand this enhancement factor again by counting the possible ways to arrange the two pairs of photons with four detectors: there are $C_4^2 = 6$ ways, as shown in Fig. 8.17. The phases for all six possibilities are the same due to the same path for the four photons. Then a complete constructive 4-photon interference leads to a factor of 6 enhancement, similar to the argument presented in two-photon bunching effect. This photon

Fig. 8.17   Six possibilities for detecting two pairs of photons in four detectors.

pair bunching effect was first demonstrated by Ou *et al.* with two pairs of photon from parametric down-conversion process [Ou et al. (1999a)]. The experimental schematic and the result are shown in Figs. 8.18(a) and (b), where an approximately 5-fold enhancement was observed, a little smaller than the predicted 6-fold enhancement (see Section 8.4 for the reason).

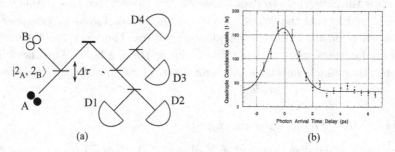

Fig. 8.18   (a) Schematic for demonstrating photon pair bunching effect. (b) The result of photon pair bunching. Adapted from [Ou et al. (1999a)].

For the general case of N-photon + M-photon, it is straightforward to show from the output state in Eq. (6.56) that the enhancement factor is $C_{N+M}^{N} = (N + M)!/N!M!$. A case of 3×3 was demonstrated by Niu *et al.* with the observed enhancement factor of 17, a little short of the theoretical prediction of 20 [Niu et al. (2009)]. The less-than-perfect results in both the two-pair bunching case in Fig. 8.18 and the three-pair bunching case here are due to photon distinguishability effect that we will discuss in Section 8.4. This idea can be applied to a weak coherent state to take out two-photon state for achieving photon anti-bunching.

8.3.2   *Generalized Hong-Ou-Mandel Effects and Destructive Multi-Photon Interference*

The multi-photon bunching effects discussed in the previous section are all the results of constructive interference for enhancement. We will consider multi-photon destructive interference effect in this section. The result is the cancelation or a decrease in multi-photon coincidence count. In fact, Hong-Ou-Mandel effect is the first such effect for two photons. However, we cannot generalize it to the case of more photons with a symmetric beam splitter, as shown in Eq. (8.17) where the $|2, 2\rangle$ state does not disappear as we would want. On the other hand, we may consider its generalization with an asymmetric beam splitter with $T \neq R$.

8.3.2.1   *Three-Photon Hong-Ou-Mandel Interferometer*

The generalization of the Hong-Ou-Mandel interferometer to three-photon case was achieved and demonstrated by Sanaka *et al.* in realizing a non-linear phase gate [Sanaka <u>et al.</u> (2004)], who considered an input state of $|2, 1\rangle$ to a beam splitter with $T \neq R$. With the method discussed in Section 6.2.2, we may easily express the output state as

$$|\Psi_3\rangle = \sqrt{3T^2R}|3, 0\rangle + \sqrt{3TR^2}|0, 3\rangle$$
$$+ \sqrt{T}(T - 2R)|2, 1\rangle + \sqrt{R}(R - 2T)|1, 2\rangle. \quad (8.60)$$

Note that $P_3(2, 1) = T(T - 2R)^2$ and is equal to zero when $T = 2R = 2/3$. Under this condition, Eq. (8.60) becomes

$$|\Psi_3\rangle = \frac{2}{3}|3, 0\rangle + \frac{\sqrt{2}}{3}|0, 3\rangle - \frac{\sqrt{3}}{3}|1, 2\rangle. \quad (8.61)$$

Note that the disappearance of the $|2, 1\rangle$ is due to destructive three-photon interference. This complete cancelation of probability amplitude for the output of $|2, 1\rangle$ is similar to the two-photon Hong-Ou-Mandel effect and can be easily understood with the picture in Fig. 8.19, where there are three possible ways to obtain an output of $|2, 1\rangle$: (a) all three photons are transmitted through the beam splitter with a probability amplitude of $\sqrt{2/3} \times \sqrt{2/3} \times \sqrt{2/3}$; (b) and (c) one of the two photons from one side is transmitted while the other one of the two photons and the single photon from the other side are reflected with each case having a probability amplitude of $-\sqrt{2/3} \times \sqrt{1/3} \times \sqrt{1/3}$. The negative sign is due to an overall π phase shift on the two reflected photons, similar to that in the Hong-Ou-Mandel effect. Thus the overall amplitude is $(\sqrt{2/3})^3 - 2 \times (\sqrt{2/3})(\sqrt{1/3})^2 = 0$, leading to the disappearance of the $|2, 1\rangle$ state. Notice that when $R = 2T = 2/3$, the $|1, 2\rangle$ state will disappear in the same way.

The generalization of the asymmetric beam splitter scheme above to four photons, with two on each side of the BS, is straightforward and will be dealt with in Problem 8.1.

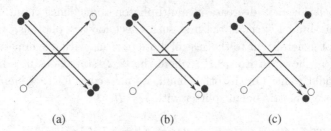

(a)   (b)   (c)

Fig. 8.19   Three possibilities for the output of $|2, 1\rangle$: (a) all photons are transmitted; (b) and (c) one of the two photons from one side is transmitted while the other one and the single photon from the other side are reflected.

8.3.2.2   *Number State Filtering Effect*

The destructive three-photon interference effect in the previous section can be easily generalized to an arbitrary input state of $|N, 1\rangle$. With a beam splitter of T, R, we can easily find the probability amplitude for an output of $|N, 1\rangle$ (Problem 8.4):

$$A_{N+1}(N, 1) = \sqrt{T^{N-1}}(T - NR), \qquad (8.62)$$

which leads to the probability $P_{N+1}(N, 1) = T^{N-1}(T - NR)^2$. $P_{N+1}(N, 1)$ equals to zero when $T = NR = N/(N + 1)$. This effect can be used as a number state filter when conditioned on the single-photon output at port 2. This idea was first proposed and demonstrated by Sanaka *et al.* [Sanaka et al. (2006)]. Consider an arbitrary state $|\psi_{in}\rangle_1 = \sum_n c_n |n\rangle_1$ input at port 1 and a single-photon state at port 2. The single photon at port 2 is often called the ancilla photon. We find from Eq. (8.62) that when conditioned on the detection of a single photon at output port 2, the output state at port 1 is then projected to

$$|\psi_{out}\rangle_1 = \mathcal{N} \sum_n c_n \sqrt{T^{n-1}}(T - nR)|n\rangle_1, \qquad (8.63)$$

where $\mathcal{N}$ is a normalization factor. When we choose T, R so that $T/R = n_0 = $ integer, state $|n_0\rangle$ disappears from the projected output state $|\psi_{out}\rangle_1$. Thus, the number state $|n_0\rangle$ is filtered out at the output.

8.4   Quantum Interference and Photon Indistinguishability

The complementarity principle of quantum mechanics states that distinguishability inevitably degrades the interference effect. For traditional single-photon interference, distinguishability may only occur in paths. But for multi-photon interference, distinguishability among different particles may also lead to degradation of interference, as we have seen in Hong-Ou-Mandel interference dip. In this section, we will use the multi-photon interference effects discussed in the previous sections to quantitatively characterize the degree of temporal distinguishability of photons.

8.4.1   *N-Photon State from N Single-Photon States and Photon Indistinguishability*

We start with the simplest case of two photons and examine how temporal distinguishability of photons arises. Consider the output state of the Hong-Ou-Mandel interferometer for the input of two photons in the single-temporal mode states in Eq. (8.14). The input state of the system is

$$|\Psi_{in}\rangle = |T_1\rangle_1 \otimes |T_2\rangle_2$$
$$= \int d\omega_1 d\omega_2 \phi_1(\omega_1) e^{i\omega_1 T_1} \phi_2(\omega_2) e^{i\omega_2 T_2} \hat{a}_1^\dagger(\omega_1) \hat{a}_2^\dagger(\omega_2)|0\rangle. \quad (8.64)$$

With the method in Section 6.2.2, we find the state of the system after the 50:50 beam splitter as

$$|\Psi_{out}\rangle = \int d\omega_1 d\omega_2 \frac{1}{2} \phi_1(\omega_1) e^{i\omega_1 T_1} \phi_2(\omega_2) e^{i\omega_2 T_2}$$
$$\times \left[\hat{a}_1^\dagger(\omega_1) - \hat{a}_2^\dagger(\omega_1)\right]\left[\hat{a}_1^\dagger(\omega_2) + \hat{a}_2^\dagger(\omega_2)\right]|0\rangle$$
$$= \int d\omega_1 d\omega_2 \frac{1}{2} \phi_1(\omega_1) e^{i\omega_1 T_1} \phi_2(\omega_2) e^{i\omega_2 T_2}$$
$$\times \left[\hat{a}_1^\dagger(\omega_1)\hat{a}_1^\dagger(\omega_2) - \hat{a}_2^\dagger(\omega_1)\hat{a}_2^\dagger(\omega_2)\right.$$
$$\left. + \hat{a}_1^\dagger(\omega_1)\hat{a}_2^\dagger(\omega_2) - \hat{a}_1^\dagger(\omega_2)a_2^\dagger(\omega_1)\right]|0\rangle$$
$$= \frac{1}{2}\left[|\Phi_2\rangle_1 - |\Phi_2\rangle_2 + |\Phi_{1,1}\rangle_{12}\right], \quad (8.65)$$

where

$$|\Phi_2\rangle_j = \int d\omega_1 d\omega_2 \Phi_2(\omega_1, \omega_2)\hat{a}_j^\dagger(\omega_1)\hat{a}_j^\dagger(\omega_2)|0\rangle \quad (j = 1, 2)$$

$$|\Phi_{1,1}\rangle_{12} = \int d\omega_1 d\omega_2 \left[\Phi_2(\omega_1, \omega_2) - \Phi_2(\omega_2, \omega_1)\right]\hat{a}_1^\dagger(\omega_1)\hat{a}_2^\dagger(\omega_2)|0\rangle$$

$$\text{with } \Phi_2(\omega_1, \omega_2) \equiv \phi_1(\omega_1) e^{i\omega_1 T_1} \phi_2(\omega_2) e^{i\omega_2 T_2}. \quad (8.66)$$

Obviously, $|\Phi_{1,1}\rangle_{12}$ gives rise to Hong-Ou-Mandel interference effect but our interest now is on $|\Phi_2\rangle_1$, $|\Phi_2\rangle_2$, which correspond to the cases when both incoming photons exit from the same side of the beam splitter. Apparently they are in a multi-frequency two-photon state. How do they look like in time domain?

To find out, we evaluate the single-photon detection rate:

$$R_1(t) = {}_1\langle \Phi_2|\hat{E}_1^\dagger(t)\hat{E}_1(t)|\Phi_2\rangle_1. \tag{8.67}$$

For this, it is easy to first calculate

$$\hat{E}_1(t)|\Phi_2\rangle_1 = \frac{1}{\sqrt{2\pi}} \int d\omega e^{-j\omega t}\hat{a}_1(\omega) \int d\omega_1 d\omega_2 \Phi_2(\omega_1,\omega_2)\hat{a}_1^\dagger(\omega_1)\hat{a}_1^\dagger(\omega_2)|0\rangle$$

$$= \frac{1}{\sqrt{2\pi}} \int d\omega_1 d\omega_2 \Phi_2(\omega_1,\omega_2)\left[e^{-j\omega_1 t}\hat{a}_1^\dagger(\omega_2) + e^{-j\omega_2 t}\hat{a}_1^\dagger(\omega_1)\right]|0\rangle.$$

Then we have

$$R_1(t) = \langle \Phi_2|\hat{E}_1^\dagger(t)\hat{E}_1(t)|\Phi_2\rangle_1$$

$$= \frac{1}{2\pi} \int d\omega_1' d\omega_2' d\omega_1 d\omega_2 \Phi_2^*(\omega_1',\omega_2')\Phi_2(\omega_1,\omega_2)$$

$$\times \left[e^{j(\omega_1'-\omega_1)t}\delta(\omega_2 - \omega_2') + e^{j(\omega_2'-\omega_2)t}\delta(\omega_1 - \omega_1')\right.$$

$$\left. + e^{j(\omega_1'-\omega_2)t}\delta(\omega_1 - \omega_2') + e^{j(\omega_2'-\omega_1)t}\delta(\omega_2 - \omega_1')\right]. \tag{8.68}$$

We next substitute the explicit form of $\Phi_2(\omega_1,\omega_2)$ given in Eq. (8.66) into the above and arrive at

$$R_1(t) = |g_1(t - T_1)|^2 + |g_2(t - T_2)|^2$$

$$+ c_{12}g_2^*(t - T_2)g_1(t - T_1) + c_{12}^*g_1^*(t - T_1)g_2(t - T_2), \tag{8.69}$$

where we used the normalization $\int d\omega|\phi_j(\omega)|^2 = 1$ and $c_{12} \equiv \int d\omega\phi_1^*(\omega)\phi_2(\omega)e^{j\omega(T_2-T_1)}$, and $g_j(t - T_j)$ with $j = 1, 2$ is the single-photon wavepacket function given in Eq. (4.45) of Section 4.3.2.

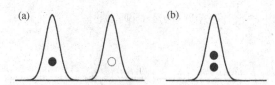

Fig. 8.20   Two different two-photon states: (a) two photons are well separated in time and distinguishable; (b) two photons are in one temporal mode and are indistinguishable.

For the sake of simplicity but without loss of generality, we set $g_1(t) = g_2(t) \equiv g(t)$. There are two extreme cases: (1) $|T_1 - T_2| \gg \Delta T$ and (2) $|T_1 - T_2| \ll \Delta T$ with ΔT as the width of the wavepacket function $g(t)$. The first case corresponds to the situation when the two photons arrive at the beam splitter with well separated time T_1, T_2. This leads to a two-photon state with two photons well separated as shown in Fig. 8.20(a). The two photons in this state are completely distinguishable. With notations in Section 4.3.2, we can write the two-photon state in this case as $|\Phi_2\rangle = \hat{A}^\dagger(T_1)\hat{A}^\dagger(T_2)|0\rangle \equiv |1\rangle_{T_1}|1\rangle_{T_2}$. The second case corresponds to the situation when the two photons arrive at the beam splitter at the same time $T_1 = T_2 \equiv T$ and the two-photon state in this case becomes $|\Phi_2\rangle = [\hat{A}^\dagger(T)]^2|0\rangle \equiv \sqrt{2}|2\rangle_T$. This case is depicted in Fig. 8.20(b), and the two photons in this state are in one single temporal mode and are completely indistinguishable. The coefficient $\sqrt{2}$ in front of $|2\rangle_T$ gives rise to the photon bunching effect (see Section 8.3.1).

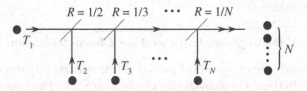

Fig. 8.21   Generation of an N-photon state by superposing N single-photon states with beam splitters.

The argument above can be generalized to an arbitrary number of photons. With N single-photon states, we can combine them with beam splitters to form an N-photon state, as shown in Fig. 8.21. This scheme has its practicality since multi-photon states with $N > 2$ are not easily available in lab[3] but single photons on demand have been widely achieved so far.

With the method in Section 6.3.2, it is straightforward to show that the output state of the port shown in Fig. 8.21 has the form of

$$|\Phi_N\rangle_{out} = \frac{1}{N^{N/2}}\hat{A}^\dagger(T_1)\hat{A}^\dagger(T_2)...\hat{A}^\dagger(T_N)|0\rangle \qquad (8.70)$$

with

$$\hat{A}^\dagger(T) = \int d\omega \phi^*(\omega)e^{-j\omega T}\hat{a}^\dagger(\omega). \qquad (8.71)$$

As we have shown for the two-photon case, the arrival times $T_1, T_2, ..., T_N$ are critical for the indistinguishability of the N-photon state. When $T_1 = T_2 = ... = T_N \equiv T$, we have the N photons completely indistinguishable in one temporal mode:

$$|\Phi_N\rangle_{out} = \sqrt{\frac{N!}{N^N}}\frac{[\hat{A}^\dagger(T)]^N}{\sqrt{N!}}|0\rangle = \sqrt{\frac{N!}{N^N}}|N\rangle_T. \qquad (8.72)$$

---

[3]Multi-photon states can be produced from two-photon states generated in spontaneous parametric processes but with limited quality of indistinguishability. See Section 8.4.2.

The square of the coefficient in front of $|N\rangle_T$ gives the probability for the N-photon state generation:

$$P_N = \frac{N!}{N^N}. \tag{8.73}$$

On the other hand, when $|T_i - T_j| \gg \Delta T$, some of the photons in the N-photon state become distinguishable in time. If a partial group of photons have $T_i = T_j$ among the N photons, we have only partial indistinguishability. According to quantum complementarity principle, this will lead to reduced interference effect in multi-photon interference effect as compared to the case in Eq. (8.72) with complete indistinguishability. We will investigate next how partial indistinguishability affects the multi-photon interference effects.

8.4.2 *Pair Distinguishability and its Characterization*

We have already seen the effect of partial indistinguishability on two-photon interference in Hong-Ou-Mandel effect in Section 8.2.1. The next example is the case of two pairs of photons in the photon pair bunching effect discussed at the end of Section 8.3.1. This is one of the first multi-photon interference phenomena with $N > 2$ that are affected by partial indistinguishability and reveals the challenge encountered in generating higher photon number states ($N > 2$) from spontaneous parametric processes for the applications in quantum information such as quantum state teleportation [Bouwmeester et al. (1997)] and quantum state swapping [Pan et al. (1998)].

After successful studies of two-photon interference phenomena in the 90's of last century, attention started to focus on multi-photon interference phenomena with $N > 2$ due to the discovery of the GHZ states [Greenberger et al. (1989); Yurke and Stoler (1992); Bouwmeester et al. (1999)]. At the time, spontaneous parametric down-conversion processes with $\chi^{(2)}$ materials were popular sources of two-photon states, and with stronger pumping power by pulsed lasers, the higher order processes will produce multiple pairs of photons for higher photon number state generation. However, pair production in spontaneous parametric down-conversion is completely random so that the pairs of photons produced are likely separated in time as shown in Fig. 8.22(a). So, ultra-short pump pulses are required [Zukowski et al. (1995); Rarity (1995)] in order to force the two pairs be produced in one single temporal mode as shown in Fig. 8.22(b). In this way, the four photons become completely indistinguishable. Most likely scenario will be somewhere in between. If the four photons described here are involved in a

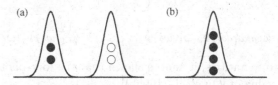

Fig. 8.22 Two different four-photon states: (a) two pairs are separated and distinguishable; (b) all four photons are in one temporal mode and are indistinguishable.

four-photon interference scheme, the interference visibility will be different for the two cases in Fig. 8.22. We consider next the photon pair bunching effect first discussed at the end of Section 8.3.1.

When the two pairs are well-separated in time, as depicted in Fig. 8.22(a), there is still some but less bunching effect. In fact, when the two pairs are distinguishable, the input state becomes

$$|\Phi_{in}^{(4)}\rangle' = |1_1; 1_2\rangle \otimes |1'_1; 1'_2\rangle, \tag{8.74}$$

instead of $|\Phi_{in}^{(4)}\rangle = |2_1, 2_2\rangle$. The output state is then

$$|\Phi_{out}^{(4)}\rangle' = (1/2)(|2_1, 0_2\rangle - |0_1, 2_2\rangle) \otimes (|2'_1, 0'_2\rangle - |0'_1, 2'_2\rangle), \tag{8.75}$$

from which we find the probability $P_4(4, 0) = 1/4$ and the ratio to the classical probability is then

$$P'_4(4, 0)/P_4^{cl}(4, 0) = 4. \tag{8.76}$$

This value is reduced from the maximum value of 6 given in Section 8.3.1, which occurs when the two pairs are indistinguishable from each other, as depicted in Fig. 8.22(b). Thus, distinguishability results in degradation of the interference effect.

For partial distinguishability of the pairs, Ou *et al.* considered a multi-mode four-photon state made from two pairs of correlated photons from spontaneous parametric processes [Ou et al. (1999b)]. The photon state has the form of

$$|\Phi_4\rangle = \frac{\xi^2}{2} \int d\omega_1 d\omega_2 d\omega'_1 d\omega'_2 \Phi_2(\omega_1, \omega_2) \Phi_2(\omega'_1, \omega'_2)$$
$$\times \hat{a}_1^\dagger(\omega_1)\hat{a}_1^\dagger(\omega'_1)\hat{a}_2^\dagger(\omega_2)\hat{a}_2^\dagger(\omega'_2)|vac\rangle, \tag{8.77}$$

which is taken directly from Eq. (6.32) in Section 6.1.5 for a weak parametric process ($|\xi|^2 \ll 1$). Here, $\Phi_2(\omega_1, \omega_2) = \Phi_2(\omega_2, \omega_1)$ is the two-photon wave function that is symmetric with respect to ω_1, ω_2. The distinguishability between pairs is described in terms of the quantity $\mathcal{E}/\mathcal{A}$ with

$$\mathcal{A} \equiv \int d\omega_1 d\omega_2 d\omega'_1 d\omega'_2 |\Phi_2(\omega_1, \omega_2)\Phi_2(\omega'_1, \omega'_2)|^2 \tag{8.78}$$

and

$$\mathcal{E} \equiv \int d\omega_1 d\omega_2 d\omega_1' d\omega_2' \Phi_2(\omega_1, \omega_2) \Phi_2(\omega_1', \omega_2') \Phi_2^*(\omega_1, \omega_2') \Phi_2^*(\omega_1', \omega_2). \quad (8.79)$$

With these two quantities, it can be shown that the four-photon bunching effect takes a value of [Ou et al. (1999b)]

$$P_4(4,0)/P_4^{cl}(4,0) = 4 + \frac{4\mathcal{E}}{\mathcal{A} + \mathcal{E}}, \quad (8.80)$$

which recovers the maximum value of 6 for $\mathcal{E} = \mathcal{A}$ and the value of 4 in Eq. (8.76) for $\mathcal{E} = 0$. The value of 6 corresponds to four-photon bunching effect whereas the value of 4 is only due to two-photon bunching for each pair. Thus, the quantity $\mathcal{E}/\mathcal{A}$ is a good measure of photon indistinguishability between pairs. The experimentally measured value of $P_4(4,0)/P_4^{cl}(4,0) = 5.1 \pm 0.4$ from Fig. 8.18 gives $\mathcal{E}/\mathcal{A} = 0.4$.

With the input state in Eq. (8.77), it is straightforward to find the output state at one output port of the beam splitter as

$$|\Phi_4\rangle_{out} = \frac{\xi^2}{8} \int d\omega_1 d\omega_2 d\omega_1' d\omega_2' \Phi_2(\omega_1, \omega_2) \Phi_2(\omega_1', \omega_2')$$

$$\times \hat{a}_1^\dagger(\omega_1) \hat{a}_1^\dagger(\omega_1') \hat{a}_1^\dagger(\omega_2) \hat{a}_1^\dagger(\omega_2') |vac\rangle. \quad (8.81)$$

This is a temporal four-photon state in the output port 1. We will study the property of photon indistinguishability according to the value of $\mathcal{E}/\mathcal{A}$ in the next section.

8.4.3   Characterization of Photon Indistinguishability by Multi-Photon Bunching Effects

Since photon indistinguishability has such an impact on the multi-photon interference effects, we should be able to use the interference effects to characterize the properties of photon indistinguishability just like optical coherence characterized by the visibility of optical interference. In this way, we should be able to tell the difference between the two four-photon states in Figs. 8.22(a) and (b). This becomes especially important if we want to produce an N-photon state from N single-photon states in the scheme discussed in Section 8.4.1.

Let us first revisit the Hong-Ou-Mandel interference effect discussed in Section 8.2.1. We have shown that it depends on the overlap between the two incoming photons. If the arrival times for the two photons at the beam splitter are quite different, temporal distinguishability between the two photons will diminish the interference effect. Even when the arrival times are the same for the two photons, the difference in their mode functions may also lead to partial distinguishability, as we have demonstrated in the visibility of the Hong-Ou-Mandel interference given in Eq. (8.20) in Section 8.2.1 and of the induced coherence effect given in Eq. (8.39) in Section 8.2.3. For a more general two-photon spectral wave function $\Phi_2(\omega_1; \omega_2)$, which is defined through an arbitrary two-photon state in two one-dimensional fields:

$$|\Phi_{1,1}\rangle_{12}^{(in)} = \int d\omega_1 d\omega_2 \Phi_2(\omega_1, \omega_2) \hat{a}_1^\dagger(\omega_1) \hat{a}_2^\dagger(\omega_2)|vac\rangle, \qquad (8.82)$$

it is straightforward to show that the visibility of the Hong-Ou-Mandel interference effect is related to $\Phi_2(\omega_1; \omega_2)$ as

$$\mathcal{V}_2 = \left| \int d\omega_1 d\omega_2 \Phi_2^*(\omega_1, \omega_2) \Phi_2(\omega_2; \omega_1) \right| \bigg/ \int d\omega_1 d\omega_2 |\Phi_2(\omega_1, \omega_2)|^2, \quad (8.83)$$

where we omit the spatial mode in the one-dimensional approximation for simplicity of argument. The output state $|\Phi_2\rangle_j$ with $j = 1, 2$ for one output port of the beam splitter can be shown to be

$$|\Phi_2\rangle_j = \int d\omega_1 d\omega_2 \Phi_2(\omega_1, \omega_2) \hat{a}_j^\dagger(\omega_1) \hat{a}_j^\dagger(\omega_2)|vac\rangle \quad (j = 1, 2), \qquad (8.84)$$

which is in the same form of Eq. (8.66) but for arbitrary two-photon spectral wave function $\Phi_2(\omega_1, \omega_2)$.

Applying Cauchy-Schwarz inequality to Eq. (8.83), we find that $\mathcal{V}_2 = 1$, or the maximum interference effect occure if and only if [4]

$$\Phi_2(\omega_1, \omega_2) = \Phi_2(\omega_2, \omega_1). \qquad (8.85)$$

This seems to gives the condition for the complete temporal indistinguishability of the two photons in the two-photon state $|\Phi_2\rangle_j (j = 1, 2)$ of Eq. (8.84), for $\mathcal{V}_2 = 1$ gives the maximum two-photon bunching effect or the complete overlap of two photons at the output, corresponding to the situation in Fig. 8.20(b). Indeed, for the factorized $\Phi_2(\omega_1, \omega_2) = \phi(\omega_1)e^{i\omega_1 T_1}\phi(\omega_2)e^{i\omega_2 T_2}$ in Section 8.4.1, condition in Eq. (8.85) is satisfied when $T_1 = T_2$, and the output state is a single-mode two-photon state, as discussed in Section 8.4.1.

However, let us consider a non-factorized $\Phi_2'(\omega_1, \omega_2)$:

$$\Phi_2'(\omega_1, \omega_2) = \phi(\omega_1)\phi(\omega_2)\left(e^{i\omega_1 T_1}e^{i\omega_2 T_2} + e^{i\omega_1 T_2}e^{i\omega_2 T_1}\right) \qquad (8.86)$$

---

[4]This is also obvious from Eq. (8.66). $|\Phi_{1,1}\rangle_{12} = 0$, i.e., maximum Hong-Ou-Mandel effect when condition in Eq. (8.85) is met.

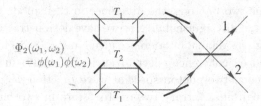

Fig. 8.23 Generation of non-factorized wave function from factorized wave function.

with $|T_1 - T_2| \gg \Delta T$. It satisfies the symmetry condition in Eq. (8.85) and gives rise to $\mathcal{V}_2 = 1$ in Hong-Ou-Mandel interference. The two-photon state of this form can be produced with unbalanced Mach-Zehnder interferometers from the factorized state with $\Phi_2(\omega_1, \omega_2) = \phi(\omega_1)\phi(\omega_2)$, as shown in Fig. 8.23. Notice that the two photons never meet in the beam splitter in this case and yet we have $\mathcal{V}_2 = 1$ and this demonstrates the peculiar quantum nonlocal behavior of multi-photon states [Pittman et al. (1996); Nasr et al. (2003); Lu et al. (2003)].

Now look at the output state at one output port:

$$|\Phi_2\rangle_j = \int d\omega_1 d\omega_2 \phi(\omega_1)\phi(\omega_2) \left(e^{i\omega_1 T_1} e^{i\omega_2 T_2} + e^{i\omega_1 T_2} e^{i\omega_2 T_1}\right)$$
$$\times \hat{a}_j^\dagger(\omega_1)\hat{a}_j^\dagger(\omega_2)|vac\rangle$$
$$= 2|T_1\rangle_j |T_2\rangle_j \quad (j = 1, 2), \tag{8.87}$$

which, with $|T_1 - T_2| \gg \Delta T$, is a two-photon state corresponding to Fig. 8.20(a) with the two photons well-separated in time T_1 and T_2 (ΔT is the width of the wavepacket $\phi(\omega)$).

Therefore, the symmetry condition in Eq. (8.85) is only a necessary condition for two photons in the state of Eq. (8.84) to be indistinguishable in time but not sufficient condition. This is reflected in the fact that for any $\Phi_2(\omega_1, \omega_2)$ that does not satisfy condition in Eq. (8.85), we can always rearrange the order of $\hat{a}_j^\dagger(\omega_1)\hat{a}_j^\dagger(\omega_2)$ in Eq. (8.84) so that it has a new form of

$$|\Phi_2\rangle_j = \int d\omega_1 d\omega_2 \frac{1}{2}\left[\Phi_2(\omega_1, \omega_2) + \Phi_2(\omega_2, \omega_1)\right] \hat{a}_j^\dagger(\omega_1)\hat{a}_j^\dagger(\omega_2)|vac\rangle$$
$$= \int d\omega_1 d\omega_2 \Phi_2^{sym}(\omega_1, \omega_2)\hat{a}_j^\dagger(\omega_1)\hat{a}_j^\dagger(\omega_2)|vac\rangle \quad (j = 1, 2) \tag{8.88}$$

with $\Phi_2^{sym}(\omega_1, \omega_2) \equiv [\Phi_2(\omega_1, \omega_2) + \Phi_2(\omega_2, \omega_1)]/2$, which satisfies the symmetry condition in Eq. (8.85). This is because the number-state representation of a Bosonic system is already symmetrized with respect to particle exchange.

On the other hand, we have $\mathcal{V}_2 = 0$, or no Hong-Ou-Mandel interference effect when

$$\int d\omega_1 d\omega_2 \Phi_2^*(\omega_1, \omega_2) \Phi_2(\omega_2, \omega_1) = 0, \tag{8.89}$$

which gives the criterion for the complete temporal distinguishability of the two photons. This is a sufficient condition but not a necessary one in a similar way as the discussion above for the condition in Eq. (8.85). Equations (8.85) and (8.89) can be generalized to a state with higher number of photons.

For the two pairs of photons from parametric down-conversion, the four-photon state is given by

$$|\Phi_4\rangle = \int d\omega_1 d\omega_2 d\omega_1' d\omega_2' \Phi_2(\omega_1, \omega_2) \Phi_2(\omega_1', \omega_2')$$

$$\times \hat{a}_1^\dagger(\omega_1) \hat{a}_2^\dagger(\omega_2) \hat{a}_1^\dagger(\omega_1') \hat{a}_2^\dagger(\omega_2') |vac\rangle. \tag{8.90}$$

So the four-photon spectral wave function has the form of

$$\Phi_4(\omega_1, \omega_2; \omega_1', \omega_2') = \Phi_2(\omega_1, \omega_2) \Phi_2(\omega_1', \omega_2'). \tag{8.91}$$

We have just shown that the overlap between $\Phi_2(\omega_1, \omega_2)$ and $\Phi_2(\omega_2, \omega_1)$ determines the two-photon indistinguishability. So, let us assume we have total indistinguishability between the two photons within one pair, i.e., $\Phi_2(\omega_1, \omega_2) = \Phi_2(\omega_2, \omega_1)$. From the discussion in Section 8.4.2, we learned that the indistinguishability between two pairs is determined by $\mathcal{E} = \mathcal{A}$. With definition in Eq. (8.91), the condition for indistinguishable pairs, i.e., $\mathcal{E} = \mathcal{A}$, can be rewritten as

$$\Phi_4(\omega_1, \omega_2; \omega_1', \omega_2') = \Phi_4(\omega_1', \omega_2; \omega_1, \omega_2') = \Phi_4(\omega_1, \omega_2'; \omega_1', \omega_2). \tag{8.92}$$

Note that the permutation is between primed and unprimed variables, indicating permutation symmetry between two photons, with one from each pair. Thus, we have pair exchange symmetry. Again, we need to be cautious about the condition in Eq. (8.92) since it is only a necessary condition like the two-photon case.

On the other hand, for the condition for complete distinguishable pairs, i.e., $\mathcal{E} = 0$, we have

$$\int d\omega_1 d\omega_2 d\omega_1' d\omega_2' \Phi_4(\omega_1, \omega_2; \omega_1', \omega_2') \Phi_4^*(\omega_1, \omega_2'; \omega_1', \omega_2) = 0. \tag{8.93}$$

Both Eqs. (8.92) and (8.93) are extensions of Eqs. (8.85) and (8.89) to the four-photon case of two pairs. These can be further generalized to an arbitrary number of photons. But because of the complexity involved, we will not discuss it here. Readers who are interested can find more discussions in [Ou (2008)].

Even though the mathematical description of a temporal multi-photon state such as that in Eq. (8.84) cannot clearly and uniquely determine the temporal distinguishability of the photons, we can resort to multi-photon interference experiments for a solution. A simple scheme is the multi-photon bunching effect discussed in Section 8.3.1. We are specifically interested in the $N + 1$ scheme in Fig. 8.14 where the input N-photon state is a temporal multi-photon state such as that in Eq. (8.84) and the other input is a single-mode single-photon state who serves as the reference. For simplicity of argument, we assume the single-photon state has a form of

$$|\Phi_1\rangle_2 = |T_0\rangle = \int d\omega \phi(\omega) e^{i\omega T_0} \hat{a}_2^\dagger(\omega)|vac\rangle \qquad (8.94)$$

and the N-photon state is the one generated by the scheme in Fig. 8.21 and has a form of

$$|\Phi_N\rangle_1 = |T_1\rangle...|T_N\rangle$$
$$= \int d\omega_1...d\omega_N \phi(\omega_1) e^{i\omega_1 T_1}...\phi(\omega_N) e^{i\omega_N T_N}$$
$$\times \hat{a}_1^\dagger(\omega_1)...\hat{a}_1^\dagger(\omega_N)|vac\rangle. \qquad (8.95)$$

Let us now calculate the $N + 1$-photon coincidence counting probability:

$$P_{N+1} = \int dt_0 dt_1...dt_N \left\langle \hat{E}^\dagger(t_N)...\hat{E}^\dagger(t_0)\hat{E}(t_0)...\hat{E}(t_N) \right\rangle, \qquad (8.96)$$

where we integrate over all time for overall probability since the input states are non-stationary pulses and

$$\hat{E}(t) = [\hat{E}_1(t) + \hat{E}_2(t)]/\sqrt{2} = \frac{1}{\sqrt{2\pi}} \int d\omega \frac{1}{\sqrt{2}} \left[\hat{a}_1(\omega) + \hat{a}_2(\omega)\right] e^{-i\omega t}. \qquad (8.97)$$

It is easier to first calculate:

$$\hat{E}(t_0)\hat{E}(t_1)...\hat{E}(t_N)|\Phi_N\rangle_1|\Phi_1\rangle_2$$
$$= \frac{1}{2^{(N+1)/2}} \sum_{k=0}^{N} \mathbb{P}_{t_0 t_k} \left[\hat{E}_2(t_0)\hat{E}_1(t_1)...\hat{E}_1(t_N)\right]|\Phi_N\rangle_1|\Phi_1\rangle_2$$
$$= \frac{1}{2^{(N+1)/2}} \sum_{k=0}^{N} \mathbb{P}_{t_0 t_k} \left[\sum_{\mathbb{P}} G(t_0; \mathbb{P}\{t_1,...,t_N\})\right]|vac\rangle, \qquad (8.98)$$

where $G(t_0; t_1,...t_N) \equiv g(t_0 - T_0)g(t_1 - T_1)...g(t_N - T_N)$ with

$$g(t) = \frac{1}{\sqrt{2\pi}} \int d\omega \phi(\omega) e^{-i\omega t} \qquad (8.99)$$

having a width of ΔT. $\mathbb{P}_{t_0 t_k}$ is permutation between t_0, t_k and $\mathbb{P}\{t_1, ..., t_N\}$ is any permutation of $t_1, t_2, ..., t_N$. Hence, we have

$$P_{N+1} = \frac{1}{2^{N+1}} \int dt_0 dt_1 ... dt_N \sum_{k,l} \mathbb{P}_{t_0 t_k} \left[\sum_{\mathbb{P}} G^*(t_0; \mathbb{P}\{t_1, ..., t_N\}) \right]$$

$$\times \mathbb{P}_{t_0 t_l} \left[\sum_{\mathbb{P}} G(t_0; \mathbb{P}\{t_1, ..., t_N\}) \right]$$

$$= \sum_{k=l} + \sum_{k \neq l}. \tag{8.100}$$

In Appendix B, we find the first sum as .

$$\sum_{k=l} = (N+1)! \mathcal{N} \tag{8.101}$$

with

$$\mathcal{N} \equiv \frac{1}{2^{N+1}} \int dt_0 dt_1 ... dt_N G^*(t_0; t_1, ..., t_N) \sum_{\mathbb{P}} G(t_0; \mathbb{P}\{t_1, ..., t_N\}). \tag{8.102}$$

For the second sum in Eq. (8.100), let us assume $T_0 = T_i$ with $i = 1, ..., m$ and $|T_0 - T_j| \gg \Delta T$ for $j = m+1, ..., N$, that is, the single photon entering at port 2 completely overlaps with m photons in the N-photon state entering port 1 but is well-separated from other $N - m$ photons. The calculation in Appendix B gives

$$\sum_{k \neq l} = m(N+1)! \mathcal{N}. \tag{8.103}$$

Hence, the total $N + 1$-photon coincidence counting probability is

$$P_{N+1} = (1 + m)(N+1)! \mathcal{N} = (1 + m) P_{N+1}^{\infty}, \tag{8.104}$$

where

$$P_{N+1}^{\infty} \equiv (N+1)! \mathcal{N} \tag{8.105}$$

is the probability when $T_0 = \pm \infty$ or the single photon in port 2 does not overlap with any photon in the N-photon state in port 1, which gives $m = 0$.

With the result in Eq. (8.104), we can proceed as follows to experimentally probe the temporal structure of the N-photon state: we scan the time T_0 of the single photon in port 2 and as it goes through $T_1, ..., T_N$, we will observe a number of peaks in P_{N+1}, as shown in Fig. 8.24. The height of each peak corresponds to the number of photons that overlap with the single photon. Note that when all N photons are in one temporal mode

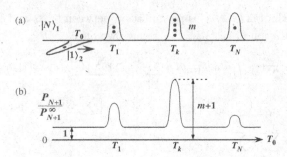

Fig. 8.24   Temporal structure of the N-photon state is scanned by the single photon. (a) Temporal distribution of the photons. (b) Normalized coincidence counting probability as a function of T_0.

with $T_0 = T_n (n = 1, ..., N)$, we have $m = N$ and $P_{N+1}/P_{N+1}^\infty = 1 + N$, exactly the value found in Section 8.3.1 for the $(N + 1)$-photon bunching effect. When $m < N$, photon distinguishability among the N photons will reduce the effect of multi-photon constructive interference. By the method discussed above, the temporal structure of the N-photon state can be revealed experimentally.

On the other hand, Fig. 8.24 only shows an extreme case when photons in the N-photon state are grouped together with no overlap between different groups. The more general case of partial overlap is complicated and was considered in [Ou (2008)]. But to have some idea about how the situation looks like, we treat next the case of $N = 2$. In this case, Eq. (8.98) becomes

$$\hat{E}(t_0)\hat{E}(t_1)\hat{E}(t_2)|\Phi_2\rangle_1|\Phi_1\rangle_2$$
$$= \frac{1}{2^{3/2}}\Big\{ g(t_0 - T_0)[g(t_1 - T_1)g(t_2 - T_2) + g(t_2 - T_1)g(t_1 - T_2)]$$
$$+g(t_1 - T_0)[g(t_0 - T_1)g(t_2 - T_2) + g(t_2 - T_1)g(t_0 - T_2)]$$
$$+g(t_2 - T_0)[g(t_1 - T_1)g(t_0 - T_2) + g(t_0 - T_1)g(t_1 - T_2)] \Big\}|vac\rangle.$$

Then, the three-photon coincidence counting probability is

$$P_3 = \Big| \hat{E}(t_0)\hat{E}(t_1)\hat{E}(t_2)|\Phi_2\rangle_1|\Phi_1\rangle_2 \Big|^2$$
$$= \frac{1}{8}\Big[6 + 6H^2(\Delta T_{12}) + 12H(\Delta T_{01})H(\Delta T_{02})H(\Delta T_{12})$$
$$+6H^2(\Delta T_{01}) + 6H^2(\Delta T_{02})\Big], \qquad (8.106)$$

where

$$H(\Delta T_{ij}) \equiv \int dt g(t - T_i)g(t - T_j) = \int dt g(t)g(t - \Delta T_{ij}) \quad (8.107)$$

with $\Delta T_{ij} \equiv T_i - T_j (i, j = 0, 1, 2)$. Here, we take $g(t)$ as real. Note $H(0) = 1$, $H(\infty) = 0$. When $T_0 = \pm\infty$, we have

$$P_3 = P_3^\infty = 3[1 + H^2(\Delta T_{12})]/4. \qquad (8.108)$$

With this quantity, Eq. (8.106) can be expressed as

$$\frac{P_3}{P_3^\infty} = 1 + \frac{2H(\Delta T_{01})H(\Delta T_{02})H(\Delta T_{12}) + H^2(\Delta T_{01}) + H^2(\Delta T_{02})}{1 + H^2(\Delta T_{12})}$$

$$= \begin{cases} 1 + H^2(\Delta T_{01}) + H^2(\Delta T_{02}) & \text{if } |T_1 - T_2| \gg \Delta T, \\ 1 + 2H^2(T_0 - T_1) & \text{if } T_1 = T_2. \end{cases} \qquad (8.109)$$

When $T_0 = T_1 = T_2$, we have $P_3/P_3^\infty = 3$ and when $T_0 = T_1$ but well separated from T_2, we have $P_3/P_3^\infty = 2$. These values are consistent with the result in Eq. (8.104). In Fig. 8.25, we plot P_3/P_3^∞ as a function of T_0 for three values of ΔT_{12}, which correspond to three different scenarios for the temporal two-photon state $|\Phi_2\rangle$ in Eq. (8.66). Here, we take $g(t)$ as a Gaussian function with a width of ΔT.

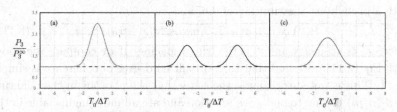

Fig. 8.25 P_3/P_3^∞ as a function of T_0 for (a) $T_1 - T_2 = 0$; (b) $T_1 - T_2 = 7\Delta T$; (c) $T_1 - T_2 = 2\Delta T$.

8.4.4 *Optical Coherence as a Consequence of Photon Indistinguishability*

Classical optical coherence theory, developed in the 1950s [Born and Wolf (1999)], was based on the second-order or single-photon interference effects. In brief, the intensity distribution shows an interference fringe pattern as

$$I(x) \propto I_1 + I_2 + 2\sqrt{I_1 I_2}|\gamma| \cos 2\pi(x - x_0)/L, \qquad (8.110)$$

where I_1, I_2 are the intensities of the two interfering fields, L is the fringe spacing along the x-direction, and

$$\gamma \equiv \langle E_1^* E_2 \rangle / \sqrt{I_1 I_2} \qquad (8.111)$$

is the degree of coherence between the two fields. Here, x_0 in Eq. (8.110) is related to $\arg(\gamma)$.

Quantum coherence theory was later constructed by Glauber primarily along the same line as the classical theory, but with quantum formalism of operators and quantum states. The physics was hidden beneath the complicated mathematical formula. As discussed in previous sections, distinguishability of photons leads to degradation of the visibility of interference. Thus, the two should be related to each other. In the following, we will make an initial attempt to reveal the connection.

In 1996, Javanaainen and Yoo showed that in a single realization, an interference fringe will form in the superposition region of two groups of photons with the same number N, respectively, i.e., with a state of $|N\rangle_1|N\rangle_2$ [Javanaainen and Yoo (1996)]. Later, the study was extended by Ou and Su to the superposition of two groups of photons with different photon numbers n and m, respectively, i.e., with a state of $|n\rangle_1|m\rangle_2$ [Ou and Su (2003)]. A quantum Monte Carlo simulation in the study by Ou and Su shows that for the state of $|n\rangle_1|m\rangle_2$, there is an interference fringe forming with a probability distribution of [Ou and Su (2003)]

$$P(x) \propto n + m + 2\sqrt{nm}\cos 2\pi(x - x_0)/L, \qquad (8.112)$$

where x_0 is arbitrary and L is the fringe spacing. If we compare the above with Eq. (8.111), we find that the normalized degree of coherence is simply $\gamma = 1$. This is not surprising in the sense that the photons in the quantum state $|n, m\rangle$ belong to one wave function and are all indistinguishable in the superposition region. On the other hand, if there is partial indistinguishability among the photons, from the discussion in the previous section, we find that the visibility will drop. Assume that the input state is $|N\rangle_1|M\rangle_2$ but only n photons among the N photons in mode 1 are indistinguishable from m photons among the M photons in mode 2. Therefore, only the $n + m$ photons will give rise to an interference pattern, described by Eq. (8.112). The rest of the photons, i.e., $N - n$ photons from mode 1 and $M - m$ photons from mode 2, are distinguishable and produce no interference fringe. Thus the probability distribution in this case is given by

$$P'(x) \propto [N - n] + [M - m] + [n + m + 2\sqrt{nm}\cos 2\pi(x - x_0)/L]$$
$$= N + M + 2\sqrt{nm}\cos 2\pi(x - x_0)/L$$
$$= (N + M)\left[1 + \frac{2\sqrt{nm}}{N + M}\cos 2\pi(x - x_0)/L\right]. \qquad (8.113)$$

Comparing with Eq. (8.112), we have the degree of coherence

$$\gamma' = \sqrt{nm/NM}. \qquad (8.114)$$

Note that n/N and m/M are the percentages of indistinguishable photons in the two groups, respectively. Thus the degree of coherence is related to the percentage of indistinguishable photons among the photons involved in interference. The same conclusion was made by Mandel in a different argument [Mandel (1991)].

As we demonstrated throughout this Chapter, the indistinguishability of photons is closely related to the modes of the photons. When photons are all in one single mode (special or general), they become indistinguishable and give rise to the maximum effects of interference, which is characterized as $\gamma = 1$ here for interference fringe. However, photon multi-mode excitations of the optical fields should lead to distinguishability and reduced interference effect thus a smaller degree of coherence $\gamma < 1$. On the other hand, if there exist phase correlations among different modes, the optical field can be described by a generalized single mode, leading to indistinguishability and interference. The phase correlation is of course described by the degree of coherence γ. Therefore, photon indistinguishability should be related to the coherence of the field.

8.5 Problems

Problem 8.1 Generalized Hong-Ou-Mandel interferometer for two pairs of photons with an asymmetric beam splitter.

We have seen the generalization of the classic Hong-Ou-Mandel two-photon interferometer to three photons with an asymmetric beam splitter (BS) in Section 8.3.2. We will make a generalization to four photons, two from each side of the BS, i.e., the input state is $|\Phi_{in}\rangle = |2\rangle_1|2\rangle_2$.

(i) Use the technique in Section 6.2.2 to show that the output state is

$$|\Phi_{out}\rangle = TR\sqrt{6}\big(|4,0\rangle + |0,4\rangle\big) + (T^2 + R^2 - 4TR)|2,2\rangle$$
$$+\sqrt{6TR}(T - R)\big(|3,1\rangle - |1,3\rangle\big), \qquad (8.115)$$

where T, R are the transmissivity and reflectivity of the beam splitter, respectively. Notice that the term of $|2,2\rangle$ disappears or $P_4(2,2) = 0$ when $T^2+R^2-4TR = 0$ or $T = (3\pm\sqrt{3})/6, R = (3\mp\sqrt{3})/6$, leading to generalized Hong-Ou-Mandel effect for two pairs of photons.

(ii) Suppose the two pairs of photons are in the state of $|\Phi'_{in}\rangle = |1_1, 1_2\rangle \otimes |1'_1, 1'_2\rangle$ (2×2 case). Use the method in Section 8.4.2 to show

$$P'_4(2,2)/P_4^{cl}(2,2) = 2/3, \qquad (8.116)$$

where $P_4^{cl}(2,2)$ is the probability for classical particles. Here we use $T = (3 \pm \sqrt{3})/6, R = (3 \mp \sqrt{3})/6$. Thus, the visibility is $1/3$ in this case.

Problem 8.2 Three-photon NOON state by projection: Wang-Kobayashi interferometer.

The NOON state introduced in Eq. (4.54) of Section 4.5.1 is of great interest in meteorology because it is capable to produce an interference fringe with $2\pi/N$ period of phase in N-photon coincidence measurement, thus enhancing the sensitivity of phase measurement by a factor of N. On the other hand, pure NOON states of the form in Eq. (4.54) is extremely hard to produce in the lab because states like $|N-k, k\rangle, k = 1, ..., N-1$ are not easy to eliminate. For example, the three-photon state in Eq. (8.61) has the term of $|1, 2\rangle$ besides the NOON state for $N = 3$. However, the contribution of $|1, 2\rangle$-term can be made to zero by using the three-photon Hong-Ou-Mandel effect discussed in Section 8.3.2, as was suggested by Wang and Kobayashi [Wang and Kobayashi (2005)].

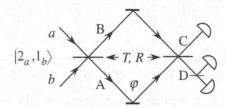

Fig. 8.26 Wang-Kobayashi Interferometer for three-photon de Broglie wavelength.

The schematic of Wang-Kobayashi Interferometer is shown in Fig. 8.26 with $T = 2R = 2/3$. Show that the three-photon coincidence measurement probability has the following dependence on the phase difference $\Delta\varphi$ between the two arms:

$$P_3(2_C, 1_D) \propto 1 + \cos 3\Delta\varphi. \tag{8.117}$$

This dependence on phase $\Delta\varphi$ is equivalent to the situation when the three photons are treated as one entity with an equivalent de Broglie wavelength of $\lambda_0/3$. Here λ_0 is the wavelength of one photon.

Problem 8.3 Four-photon NOON state by projection.

We can generalize Wang-Kobayashi Interferometer to four photons using the four-photon Hong-Ou-Mandel effect discussed in Problem 8.1. Show that the four-photon coincidence measurement probability in the scheme shown in Fig. 8.27 has the following dependence on the phase difference φ between the two arms:

$$P_4(2_C, 2_D) \propto 1 + \cos 4\varphi. \tag{8.118}$$

This leads to four-photon de Broglie wavelength. Here, we take $T = (3 \pm \sqrt{3})/6$, $R = (3 \mp \sqrt{3})/6$.

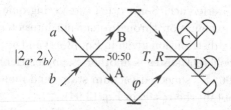

Fig. 8.27   Wang-Kobayashi Interferometer for four-photon de Broglie wavelength.

Problem 8.4 Characterization of N-photon distinguishability by generalized Hong-Ou-Mandel effect of N photons and one photon.

Since multi-photon interference effects depend on the photon indistinguishability, we can in principle employ any one of them for the characterization of photon indistinguishability. Here, we will take a look of the generalized Hong-Ou-Mandel effect for the input of N photons at one side of a beam splitter and a single photon at the other discussed in Section 8.3.2.

(i) For an input state of $|N, 1\rangle$ to a beam splitter of transmissivity T and reflectivity R, use the technique in Section 6.2.2 to show the result in Eq. (8.62). Thus, when $T = NR = N/(N+1)$, we have $P_{N+1}(N, 1) = 0$, i.e., the generalized Hong-Ou-Mandel effect for $N + 1$ photons.

(ii) For the input of the multi-mode N-photon state in Eq. (8.95) and single-photon state in Eq. (8.94), prove, similar to Eq. (8.104),

$$\frac{P_{N+1}(N, 1)}{P_{N+1}^{\infty}(N, 1)} = 1 - \frac{m}{N} \tag{8.119}$$

when $T_0 = T_j (j = 1, 2, ..., m)$ and $|T_0 - T_k| \gg \Delta T (k = m+1, ..., N)$. This is the case when the single photon overlaps with m photons in the N-photon state but is well separated from the other $N - m$ photons. $P_{N+1}^{\infty}(N, 1)$ corresponds to the case when $|T_0 - T_k| \gg \Delta T (k = 1, ..., N)$, i.e., the single photon does not overlap with the N-photon state at all. So, in contrast to the bumps in Fig. 8.24, $P_{N+1}(N, 1)$ shows dips here as T_0 is scanned through $T_k(k = 1, 2, .., N)$.

Problem 8.5 Multi-photon interference effect between a squeezed vacuum and a coherent state.

As we have shown in Section 8.3.1, stimulated emission can be explained by multi-photon interference. Some other interesting quantum phenomena can also have an explanation through quantum interference. The photon number probability distribution given in Eq. (3.94) and shown in Fig. 3.6(a) for the squeezed coherent states is another example. Here, we will use the result in Eq. (3.196) for squeezed vacuum states and multi-photon interference picture to obtain the result in Eq. (3.94).

In order to obtain a coherent component for the squeezed coherent state in Eq. (3.88), we mix the squeezed vacuum state $|-r\rangle$ with a coherent state $|\beta\rangle$ by a beam splitter. We have shown in Section 6.2.4 that when the beam splitter has $T \to 1$ and $R \sim 0$ but with $\beta\sqrt{R} \equiv \alpha = $ constant, the side with squeezed vacuum transmitted will have an output state in the form of a squeezed coherent state. To demonstrate the effect of multi-photon interference, we will assume weak excitation with $|\alpha|^2 \sim \nu \ll 1$ so that we only need to write the squeezed vacuum and the coherent state up to the four-photon terms:

$$|-r\rangle_a = \hat{S}(-r)|0\rangle \approx c_0 \left[|0\rangle_a - \frac{\nu}{\mu\sqrt{2}}|2\rangle_a + \frac{3}{\sqrt{24}}\left(\frac{\nu}{\mu}\right)^2|4\rangle_a + ...\right]$$

$$|\beta\rangle_b = c_0' \left[|0\rangle_b + \beta|1\rangle_b + \frac{\beta^2}{\sqrt{2}}|2\rangle_b + \frac{\beta^3}{\sqrt{6}}|3\rangle_b + \frac{\beta^4}{\sqrt{24}}|4\rangle_b + ...\right] \quad (8.120)$$

where we used Eq. (3.196) for the coefficients in the squeezed vacuum state and $\nu = \sinh r$, $\mu = \cosh r$. So, up to the four-photon terms, the input state to the BS is

$$|\Psi\rangle_{in} = c_0 c_0' \left[|0_{ab}\rangle + \beta|0_a, 1_b\rangle + \frac{\beta^2}{\sqrt{2}}|0_a, 2_b\rangle - \frac{\nu}{\mu\sqrt{2}}|2_a, 0_b\rangle \right.$$
$$+ \frac{\beta^3}{\sqrt{6}}|0_a, 3_b\rangle - \frac{\beta\nu}{\mu\sqrt{2}}|2_a, 1_b\rangle + \frac{3}{\sqrt{24}}\left(\frac{\nu}{\mu}\right)^2|4_a, 0_b\rangle$$
$$\left. + \frac{\beta^4}{\sqrt{24}}|0_a, 4_b\rangle - \frac{\beta^2\nu}{2\mu}|2_a, 2_b\rangle + ...\right]. \quad (8.121)$$

Find the output state $|\Psi\rangle_{out}$ and show that the multi-photon detection probabilities are

$$P(2_a, 0_b) \propto \left(\alpha^2 - \frac{\nu}{\mu}\right)^2$$

$$P(3_a, 0_b) \propto \left(\alpha^3 - 3\frac{\alpha\nu}{\mu}\right)^2$$

$$P(4_a, 0_b) \propto \left(\alpha^4 - 6\frac{\alpha^2\nu}{\mu} + 3\frac{\nu^2}{\mu^2}\right), \tag{8.122}$$

where $\alpha \equiv \beta\sqrt{R}$ with $R \ll 1$ and $T \to 1$. The above is similar to the result in Eq. (3.94) with $\nu \ll 1, \mu \sim 1$. Comparing Eqs. (8.121) and (8.122), we find that $P(2_a, 0_b)$ is from the superposition of two contributions: one is the two photons from the coherent state and the other is the two photons from the squeezed vacuum. $P(3_a, 0_b)$ and $P(4_a, 0_b)$ have similar explanations. Hence, we demonstrate that the result in Eq. (3.94) can be viewed as interference between a squeezed vacuum and a coherent state.

Problem 8.6 Photon anti-bunching effect in a beam splitter with a singlet Bell state.

The photon bunching effect in the Hong-Ou-Mandel interferometer shows the cooperative behavior of the Bosonic nature of photons. It also demonstrates the symmetry between the two photons involved, which is required for Bosons. On the other hand, Pauli's exclusion principle for Fermions will lead to separation of particles at the outputs of the beam splitter or anti-bunching effect [Bocquillon et al. (2013)]. As we have seen in Chapters 1 and 5, photon anti-bunching is allowed in quantum optics. It turns out that we can mimic the Fermionic behavior with photons having some sort of anti-symmetric property.

(i) Consider a two-photon polarization state given with a minus sign in Eq. (4.49) of Section 4.4.2:

$$|\Psi_-\rangle = \left(|1_x\rangle_A|1_y\rangle_B - |1_y\rangle_A|1_x\rangle_B\right)/\sqrt{2}, \tag{8.123}$$

which is a Bell singlet state that changes sign when we switch A and B.

Show that when the state in Eq. (8.123) is input to a 50:50 beam splitter, the output state is

$$|\Psi_-\rangle_{out} = \left(|1_x\rangle_A|1_y\rangle_B - |1_y\rangle_A|1_x\rangle_B\right)/\sqrt{2}, \tag{8.124}$$

where the two photons are separated in the outputs, showing the Fermionic anti-bunching behavior.

(ii) Find the output states for other Bell states in Section 4.4.2.

The properties of the Bell states discussed above were used to distinguish between different Bell states [Braunstein and Mann (1996)] with applications in Bell state projection measurement for quantum state teleportation [Bouwmeester et al. (1997)].

(iii) Consider the reverse process of Hong-Ou-Mandel interferometer (Eq. (8.8)), that is, the input state is

$$|\Psi\rangle_{in} = \left(|2\rangle_A|0\rangle_B - |0\rangle_A|2\rangle_B\right)/\sqrt{2}, \tag{8.125}$$

which is anti-symmetric with respect to A and B. Show that the output state is $|\Psi\rangle_{out} = |1\rangle_A|1\rangle_B$, that is, the two photons are separated from each other like Fermions [Boçquillon et al. (2013)].

Problem 8.7 Purity of single-photon and the visibility of HOM effect.

A single-photon wave packet is described by Eq. (4.43) in Section 4.3.2 as

$$|T\rangle_\phi = \int d\omega \phi(\omega) e^{i\omega T} \hat{a}^\dagger(\omega)|0\rangle. \tag{8.126}$$

But if the peak time T is fluctuating due to uncertainty in emission time, as often happens for the single photon from a quantum dot [Sun and Wong (2009)] or the heralded photon from spontaneous parametric process [Ou (1997b)], the state is described by a mixed state with density operator

$$\hat{\rho}_{1p} = \frac{1}{\Delta T} \int_{\Delta T} dT |T\rangle_\phi\langle T|, \tag{8.127}$$

where ΔT is the uncertainty in T.

Show that if we use the photons described in Eq. (8.127) for the Hong-Ou-Mandel interferometer, the visibility is

$$\mathcal{V}_{HOM} = \mathrm{Tr}\hat{\rho}^2 = \int d\omega d\omega' |\phi(\omega)\phi(\omega')|^2 \mathrm{sinc}^2[(\omega - \omega')\Delta T/2], \tag{8.128}$$

which defines the purity of the single-photon state [Du (2015)]. Obviously, $\mathcal{V}_{HOM} = \mathrm{Tr}\hat{\rho}^2 \approx 1$ if $\Delta T \ll T_c$ with T_c as the width of the single-photon wave packet in Eq. (8.126).

Chapter 9

Experimental Techniques of Quantum Optics II: Detection of Continuous Photo-Currents

We mentioned in Chapter 7 that when the intensity is relatively large, the photoelectric pulses generated by the detector will overlap and form a continuous photoelectric current. This is the situation to be discussed in this chapter. The methods of measurement and analysis will be totally different from those in Chapter 7. For a continuous photocurrent, we cannot use a pulse discriminator to filter out the contribution from the dark current as in Chapter 7. But because of the randomness of the dark current, its contributions are mostly in the low frequency regime in the form of $1/f$ noise. So, we can use the method of spectral analysis to filter it out. Therefore, spectral analysis is the main method in the detection of a strong continuous optical field.

9.1 Photocurrent and Its Relation to Quantum Measurement Theory

When the intensity of the optical field is high, many photoelectrons can be produced from a detector, resulting in the overlap of the amplified photoelectric pulses. Thus a continuous photocurrent is formed. For simplicity, we assume all the photoelectric pulses have an identical response function of $k(t)$, as shown in Fig. 9.1(a). The pulse generated at t_j makes a contribution of $k(t - t_j)$ to the photoelectric current at t. The sum of the total contributions of all photoelectric pulses generated at different times gives the photoelectric current:

$$i(t) = \sum_j k(t - t_j) \tag{9.1}$$

The generation of the photoelectrons is random so t_j is a random variable. Now divide the time axis into small segments of size Δt (Fig. 9.1(b)).

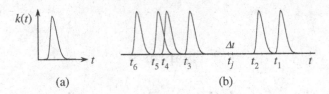

Fig. 9.1   (a) The identical response function $k(t)$ of photoelectric pulses; (b) Photoelectric pulses generated at different times.

From Glauber's photo-detection theory [Glauber (1963b, 1964)], we have the probability of detecting a photoelectron within the small segment as $\Delta P_1(t_j) = p_1(t_j)\Delta t$ with the probability density[1]

$$p_1(t_j) = \eta\langle \hat{E}^{(-)}(t_j)\hat{E}^{(+)}(t_j)\rangle_\Psi = \eta\langle \hat{I}(t_j)\rangle_\Psi. \tag{9.2}$$

Here $0 < \eta < 1$ is the quantum efficiency of the detector. Ψ is the quantum state of the optical field. $\hat{E}^{(+)}(t)$ has the form of

$$\hat{E}^{(+)}(t) = \frac{1}{\sqrt{2\pi}} \int d\omega \hat{a}(\omega)e^{-i\omega t} = \left[\hat{E}^{(-)}(t)\right]^\dagger \tag{9.3}$$

in the one-dimensional quasi-monochromatic approximation (Sections 2.3.4, 2.3.5). $\hat{I}(t) \equiv \hat{E}^{(-)}(t)\hat{E}^{(+)}(t)$ is the intensity operator. Hence, the average of the photo-electric current is

$$\langle i(t)\rangle = \sum_j k(t - t_j)\Delta P_1(t_j) = \sum_j k(t - t_j)p_1(t_j)\Delta t$$

$$= \int dt' k(t - t')p_1(t') = \eta \int dt' k(t - t')\langle \hat{I}(t')\rangle_\Psi. \tag{9.4}$$

We can see from the expression above that, when the quantum efficiency $\eta = 1$, we can define a photoelectric current operator: $\hat{i}(t) \equiv \int dt' k(t - t')\hat{I}(t')$ and the average of the detected photocurrent is simply $\langle i(t)\rangle = \langle \hat{i}(t)\rangle_\Psi$, that is, the expectation value of the photoelectric current operator $\hat{i}(t)$ for the quantum state Ψ. We learn from quantum mechanics that the quantum measurement of any physical observable corresponds to a Hermitian operator. Ou and Kimble proved that photoelectric current operator $\hat{i}(t) = \int dt' k(t - t')\hat{I}(t')$ is exactly the Hermitian operator for the photoelectric measurement process [Ou and Kimble (1995)]. All the statistical properties of photocurrent $i(t)$ can be calculated from the photoelectric

---

[1]The model presented here for the photocurrent can also be applied to Mandel's semiclassical theory of photo-detection for classical waves [Mandel et al. (1964)]. But we need to change the average over the quantum state to the average over the fluctuations of waves.

current operator $\hat{i}(t)$ and the quantum state Ψ of the optical field. The case of quantum efficiency $\eta < 1$ can be modeled by the optical losses (see Section 9.4). In this case, vacuum quantum noise will enter through the lossy channel.

9.2 Spectral Analysis of Photocurrents

Although the photoelectric current is generated by light, it is still an electric current in essence. So, we can study it by using the methods for electric current analysis. Electrical engineering provides us with a unique way for current analysis, that is, the spectral analysis of the current. Because of the existence of the $1/f$ noise in electric currents, the large electronic noise at low frequency forces us to work at high frequency domain where the intrinsic electronic noise is relatively low. The method of spectral analysis can effectively filter out the low frequency electronic noise. This is usually done with an electronic spectrum analyzer, whose output spectrum is related to the fluctuations of the input electric current by

$$S_{sp}(\Omega) = \int d\tau \langle \Delta i(t) \Delta i(t+\tau) \rangle e^{j\Omega\tau}, \qquad (9.5)$$

where $\langle \Delta i(t) \Delta i(t+\tau) \rangle \equiv \langle i(t)i(t+\tau) \rangle - \langle i(t) \rangle \langle i(t+\tau) \rangle$. The correlation function $\langle i(t)i(t+\tau) \rangle$ of the currents can be obtained from Eq. (9.1) as

$$\langle i(t)i(t+\tau) \rangle = \left\langle \sum_i k(t-t_i) \sum_j k(t+\tau-t_j) \right\rangle$$

$$= \left\langle \sum_{i,j} k(t-t_i)k(t+\tau-t_j) \right\rangle$$

$$= \left\langle \sum_j k(t-t_j)k(t+\tau-t_j) \right\rangle$$

$$+ \left\langle \sum_{i \neq j} k(t-t_i)k(t+\tau-t_j) \right\rangle, \qquad (9.6)$$

where, due to the discreteness of the photoelectric pulses, we split the double sum in the second line of the expression above into two terms corresponding to the auto-correlation of each electric pulse and cross-correlation between two electric pulses, respectively. The first term of auto-correlation can be calculated in the same way as we derive Eq. (9.4) and we have

$$\left\langle \sum_j k(t-t_j)k(t+\tau-t_j) \right\rangle = \int dt' k(t-t')k(t+\tau-t')p_1(t'), \qquad (9.7)$$

where $p_1(t')$ is obtained from Eq. (9.2). The second term in Eq. (9.6) concerns cross-correlation between two different times and its calculation requires the joint probability $\Delta P_2(t_i, t_j) = p_2(t_i, t_j)\Delta t_i \Delta t_j$ of detecting two photoelectrons in the small regions $\Delta t_i, \Delta t_j$, respectively. With this, the second term in Eq. (9.6) can be expressed as

$$
\begin{aligned}
\left\langle \sum_{i \neq j} k(t - t_i) \; k(t + \tau - t_j) \right\rangle \\
= \sum_{i \neq j} k(t - t_i)k(t + \tau - t_j)\Delta P_2(t_i, t_j) \\
= \sum_{i \neq j} k(t - t_i)k(t + \tau - t_j)p_2(t_i, t_j)\Delta t_i \Delta t_j \\
= \int dt' dt'' k(t - t')k(t + \tau - t'')p_2(t', t''). \quad (9.8)
\end{aligned}
$$

Combining Eqs. (9.7) and (9.8), we have the correlation function of the photo-electric current as

$$
\begin{aligned}
\langle i(t)i(t + \tau) \rangle = \left\langle \sum_{i,j} k(t - t_i)k(t + \tau - t_j) \right\rangle \\
= \int dt' k(t - t')k(t + \tau - t')p_1(t') \\
+ \int dt' dt'' k(t - t')k(t + \tau - t'')p_2(t', t''). \quad (9.9)
\end{aligned}
$$

Here the probability density $p_2(t', t'')$ can be obtained from Glauber's photo-detection theory [Glauber (1963b, 1964)] as[2]

$$
\begin{aligned}
p_2(t', t'') &= \eta^2 \langle \hat{E}^{(-)}(t')\hat{E}^{(-)}(t'')\hat{E}^{(+)}(t'')\hat{E}^{(+)}(t') \rangle_\Psi \\
&= \eta^2 \langle : \hat{I}(t')\hat{I}(t'') : \rangle_\Psi. \quad (9.10)
\end{aligned}
$$

When $\eta = 1$, Eq. (9.9) with Eqs. (9.2) and (9.10) can also be obtained directly from the current operator $\hat{i}(t) \equiv \int dt' k(t - t')\hat{I}(t')$ introduced earlier

---

[2] See footnote 1.

as

$$
\begin{aligned}
\langle i(t)i(t+\tau)\rangle &= \langle \hat{i}(t)\hat{i}(t+\tau)\rangle_\Psi \\
&= \Big\langle \int dt'k(t-t')\hat{I}(t') \int dt''k(t+\tau-t'')\hat{I}(t'') \Big\rangle_\Psi \\
&= \int dt'dt''k(t-t')k(t+\tau-t'')\langle \hat{I}(t')\hat{I}(t'')\rangle_\Psi \\
&= \int dt'dt''k(t-t')k(t+\tau-t'') \\
&\quad \times \big[\langle : \hat{I}(t')\hat{I}(t'') :\rangle_\Psi + \delta(t'-t'')\langle \hat{I}(t')\rangle_\Psi\big] \\
&= \int dt'k(t-t')k(t+\tau-t')\langle \hat{I}(t')\rangle_\Psi \\
&\quad + \int dt'dt''k(t-t')k(t+\tau-t'')\langle : \hat{I}(t')\hat{I}(t'') :\rangle_\Psi, \quad (9.11)
\end{aligned}
$$

where we used the commutation relation $[\hat{E}^{(+)}(t'), \hat{E}^{(-)}(t'')] = \delta(t'-t'')$. Notice that Eq. (9.11) is exactly the same as Eq. (9.9) with Eqs. (9.2) and (9.10) when $\eta = 1$. For $\eta < 1$, we can also obtain Eq. (9.9) in this way by introducing losses (see Section 9.4).

Using Eqs. (9.4) and (9.11), we have

$$
\begin{aligned}
\langle \Delta i(t)\Delta i(t+\tau)\rangle &= \int dt'k(t-t')k(t+\tau-t')\langle \hat{I}(t')\rangle_\Psi \\
&\quad + \int dt'dt''k(t-t')k(t+\tau-t'')\langle : \Delta\hat{I}(t')\Delta\hat{I}(t'') :\rangle_\Psi,
\end{aligned}
$$
$$(9.12)$$

where $\Delta\hat{I}(t) = \hat{I}(t) - \langle \hat{I}(t)\rangle$. The first term in the expression above is the so-called electronic "shot noise" in electrical engineering and is due to the discreteness of photoelectrons. It is proportional to the average intensity of the optical field. As we mention in footnote 1, the formalism discussed above can also be applied to Mandel's semiclassical theory of photodetection with classical wave fluctuations. So, the shot noise does not come from the quantization of light, as most theoreticians believe (see discussion about Eq. (9.33) in Section 9.4), but rather stems from the discreteness of photoelectrons. The second term, on the other hand, is from the fluctuations of the intensity and is determined by the properties of the optical field. For example, this term is zero for the multi-mode coherent states. So, the measurement on coherent states only results in shot noise, which corresponds to the quantum noise of the coherent states (see Section 3.2.4). For a classical field as defined in Section 5.2.4, the second term in Eq. (9.12) is always bigger than zero and thus produces some extra noise above the shot noise level. Some quantum fields, however, can make this term less than zero so that a sub-shot noise photoelectric current can be produced, whose detected photoelectric current noise level is lower than those of the coherent states. The quantum fields in amplitude squeezed states possess this property (see the end of Section 3.4.3).

For a continuous stationary optical field, the intensity of the field is a constant: $p_1(t) = \eta \langle \hat{E}^{(-)}(t)\hat{E}^{(+)}(t)\rangle_\Psi = \eta I_0$. So, Eq. (9.4) becomes

$$\langle i(t)\rangle = \eta \int dt' k(t-t')I(t') = \eta I_0 \int dt' k(t-t') = \eta' I_0. \qquad (9.13)$$

The first term in Eq. (9.12) for shot noise becomes

$$\langle \Delta i(t)\Delta i(t+\tau)\rangle_{SN} = \int dt' k(t-t')k(t+\tau-t')p_1(t')$$

$$= \eta I_0 \int dt'' k(t'')k(t''+\tau), \qquad (9.14)$$

where $t'' = t - t'$. Hence, the spectrum of the shot noise is

$$S_{SN}(\Omega) = \eta I_0 \int d\tau dt'' k(t'')k(t''+\tau)e^{j\Omega\tau}$$

$$= \eta I_0 |k(\Omega)|^2, \qquad (9.15)$$

where $k(\Omega) \equiv \int d\tau k(\tau)e^{j\omega\tau}$ is the response spectrum of the detection system, which is determined by the spectral response of the detector and the subsequent amplification system. Notice that the shot noise level is linearly proportional to the intensity I_0 of the field. This is the characteristic of the shot noise.

For the second term in Eq. (9.12), we have

$$\langle : \Delta\hat{I}(t')\Delta\hat{I}(t'') : \rangle = I_0^2[g^{(2)}(t'-t'') - 1]. \qquad (9.16)$$

Hence, the second term of Eq. (9.12) becomes

$$\langle \Delta i(t)\Delta i(t+\tau)\rangle_{ex} = \eta^2 I_0^2 \int dt' dt'' k(t-t')k(t+\tau-t'')$$

$$\times [g^{(2)}(t'-t'') - 1], \qquad (9.17)$$

which is determined by the intensity fluctuations of the detected optical field and is sometimes called the "excess noise". For example, we know from Section 4.2.4 that thermal fields have $g^{(2)}(t'-t'') = 1 + |\gamma(t'-t'')|^2$. For the fluorescent light from an atomic gas ensemble, we have $|\gamma(t'-t'')|^2 =$

$e^{-\Gamma|t'-t''|}$ (Γ is the linewidth of the fluorescence). So, the excess noise from the thermal field of the fluorescent light of an atomic gas is

$$\langle \Delta i(t) \Delta i(t+\tau) \rangle_{ex}$$

$$= \eta^2 I_0^2 \int dt' dt'' k(t-t') k(t+\tau-t'') e^{-\Gamma|t'-t''|}. \quad (9.18)$$

Its spectrum is then

$$S_{ex}(\Omega) = (\eta I_0)^2 |k(\Omega)|^2 2\Gamma/(\Gamma^2 + \Omega^2)$$

$$= S_{SN}(\Omega) \eta I_0 2\Gamma/(\Gamma^2 + \Omega^2) > 0. \quad (9.19)$$

It is always larger than zero and is proportional to the square of the intensity of the field. This is the characteristic of the excess noise, which is different from the linear relation for the shot noise. The different dependence on I_0 for the shot noise and excess noise is often used in the experiment to check if the intended shot noise measurement indeed gives the shot noise level or there is some extra noise. Adding Eq. (9.19) to the shot noise, we have the total measured noise level as

$$S_{sp}(\Omega) = S_{SN}(\Omega)[1 + \eta I_0 2\Gamma/(\Gamma^2 + \Omega^2)]. \quad (9.20)$$

This is the spectrum of the photocurrent when we measure a thermal field. Figure 9.2 shows a typical result of the spectral analysis of a photocurrent.

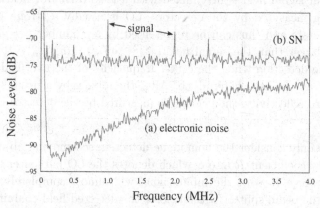

Fig. 9.2 A typical result of the spectral analysis of a photocurrent: (a) the electronic noise level without the illumination of light field on the detector. This is produced by the dark currents of the detector and the noise of the electronic amplifiers. Notice the large $1/f$ noise near zero frequency. (b) The situation after the illumination of light field on the detector: the flat part is the shot noise level (SN). The sharp spikes, especially those at low frequency are from the excess noise due to intensity fluctuations of the light field. They may also be the input signal modulated at a specific frequency and its harmonics. The vertical axis is in log-scale. Courtesy of Jun Jia.

9.3   Homodyne and Heterodyne Detection Techniques

When the optical field to be detected is very weak, the electronic noise of the measurement system such as dark currents and amplifier's pick-up noise will overwhelm the photoelectric signal if we make a direct detection of the optical field. Since the photoelectric currents are continuous, it is impossible to use the pulse shape for discrimination like we did in the photon counting technique. On the other hand, we can apply the technique of homodyne or heterodyne detection to increase the photoelectric signal over the background electronic noise. To see how it works, we first consider the single-mode case.

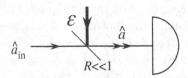

Fig. 9.3   Homodyne and heterodyne detection. An incident weak signal field $\hat{a}_{in}$ is mixed with a strong local oscillator field $\mathcal{E}$ with a beam splitter of reflectivity near zero ($R \ll 1$) so that all the signal field is detected without attenuation.

In homodyne and heterodyne detection schemes shown in Fig. 9.3, a weak input signal field is first mixed with a local oscillator field (LO) before being measured by the detector. LO is usually a strong field in a coherent state and thus can be represented by a c-number $\mathcal{E} = |\mathcal{E}|e^{j\varphi}$. It is much larger than the signal field: $|\mathcal{E}|^2 \gg I_{in} = \langle \hat{a}_{in}^{\dagger}\hat{a}_{in}\rangle$. It is called homodyne detection when the center frequencies of the two fields are the same. Otherwise, it is heterodyne. After the mixing by a beam splitter of nearly zero reflectivity, the field to be measured by the detector is

$$\hat{a} = \hat{a}_{in} + \mathcal{E}, \tag{9.21}$$

where we only consider the homodyne detection scheme and absorbed the reflectivity coefficient R into $\mathcal{E}$ which denotes the LO part after the beam splitter. Since $R \sim 0$, the input signal will almost completely transmit through the beam splitter without loss: the detected field contains all the information of the input field. So, after dropping the small higher order term, we have

$$\hat{a}^{\dagger}\hat{a} = |\mathcal{E}|^2 + \mathcal{E}\hat{a}_{in}^{\dagger} + \mathcal{E}^*\hat{a}_{in} + \hat{a}_{in}^{\dagger}\hat{a}_{in}$$
$$\approx |\mathcal{E}|^2 + \mathcal{E}\hat{a}_{in}^{\dagger} + \mathcal{E}^*\hat{a}_{in}$$
$$= |\mathcal{E}|^2 + |\mathcal{E}|\hat{X}_{in}(\varphi), \tag{9.22}$$

where $\hat{X}_{in}(\varphi) \equiv \hat{a}_{in}^{\dagger}e^{j\varphi} + \hat{a}_{in}e^{-j\varphi}$ is the quadrature-phase amplitude of the input field with its phase determined by the phase φ of the LO and $\hat{X}_{in}(0) = \hat{a}_{in}^{\dagger} + \hat{a}_{in} \equiv \hat{X} \propto \hat{x}$, $\hat{X}_{in}(\pi/2) = (\hat{a}_{in} - \hat{a}_{in}^{\dagger})/j \equiv \hat{Y} \propto \hat{p}$ with $\hat{x}, \hat{p}$ respectively representing the position and momentum operators of the virtual harmonic oscillator for the single-mode input signal field.

So, apart from a constant term, the intensity measured by the detector is proportional to the amplitude-phase quadrature of the input signal field and the proportional constant is just the amplitude of the LO field, which can be controlled in the experiment. Making it large can lift the overall detected signal far above the electronic noise background. In this case, even though the signal from the direct detection of the original weak input field is small and overwhelmed by the electronic noise background, homodyne detection with the aid of a strong LO field can still detect it. Here, because the output from the homodyne detection is the weak input field multiplied by the amplitude $|\mathcal{E}|$ of the LO, the original signal is amplified by the involvement of the LO field. Of course, this amplification effect is only on the optical part but the electronic noise background is not amplified. Obviously, we cannot make the amplitude of the LO infinitely large, for the detector will eventually be saturated by the dominating LO field.

The first constant term in Eq. (9.22) can be filtered out by spectral analysis since it is at zero frequency. In the spectral analysis of the photocurrent, the frequency of our interest in the input signal field is different from that of the LO field. So, to treat this case, we need to consider multi-mode situation. In this case, the LO field is still a single-mode field in the coherent state, set at frequency ω_0. The multi-mode treatment is on the input signal field, which is described as a one-dimensional quasi-monochromatic field $\hat{E}_{in}^{(+)}(t)$. Similar to the single-mode case, we assume $|\mathcal{E}|^2 >> \langle \hat{E}_{in}^{(-)}\hat{E}_{in}^{(+)} \rangle$. The field seen by the detector is then

$$\hat{E}^{(+)}(t) = \hat{E}_{in}^{(+)}(t) + \mathcal{E}e^{-j\omega_0 t}, \qquad (9.23)$$

where $\hat{E}_{in}^{(+)}(t)$ is in the form of Eq. (9.3) but with $\hat{a}(\omega)$ replaced by $\hat{a}_{in}(\omega)$ and the LO field is in a coherent state.[3] Here, because of the normal ordering encountered in the photo-detection theory described in Chapter 5, we replace the field operator of the LO field with a constant. Then, the

---

[3]Here, we assume the signal field has the same spatial mode as the LO field so that we can drop the spatial mode. In Section 9.6, we will consider the situation when the spatial modes are different.

intensity measured by the detector is

$$I_0 = \langle \hat{E}^{(-)} \hat{E}^{(+)} \rangle$$
$$= |\mathcal{E}|^2 + \mathcal{E}^* e^{j\omega_0 t} \langle \hat{E}_{in}^{(-)} \rangle + \mathcal{E} e^{-j\omega_0 t} \langle \hat{E}_{in}^{(+)} \rangle + \langle \hat{E}_{in}^{(-)} \hat{E}_{in}^{(+)} \rangle$$
$$\approx |\mathcal{E}|^2. \tag{9.24}$$

Hence, from Eq. (9.14), the shot noise of the homodyne detection is

$$\langle \Delta i(t) \Delta i(t+\tau) \rangle_{SN} \approx \eta |\mathcal{E}|^2 \int dt'' k(t'') k(t''+\tau) \tag{9.25}$$

with its spectrum as

$$S_{SN}(\omega) = \eta |\mathcal{E}|^2 |k(\omega)|^2. \tag{9.26}$$

This is independent of the input signal field. The part that is related to the input signal field comes from the second term of Eq. (9.12). Using Eq. (9.23) and keeping only terms up to $|\mathcal{E}|^2$, we have

$$p_2(t', t'') \approx \eta^2 |\mathcal{E}|^3 [|\mathcal{E}| + \langle \hat{X}_{in}^\varphi(t') \rangle + \langle \hat{X}_{in}^\varphi(t'') \rangle]$$
$$+ \eta^2 |\mathcal{E}|^2 [\langle : \hat{X}_{in}^\varphi(t') \hat{X}_{in}^\varphi(t'') : \rangle + \langle \hat{I}_{in}(t') \rangle + \langle \hat{I}_{in}(t'') \rangle], \tag{9.27}$$

$$p_1(t') p_1(t'') \approx \eta^2 |\mathcal{E}|^3 [|\mathcal{E}| + \langle \hat{X}_{in}^\varphi(t') \rangle + \langle \hat{X}_{in}^\varphi(t'') \rangle]$$
$$+ \eta^2 |\mathcal{E}|^2 [\langle \hat{X}_{in}^\varphi(t') \rangle \langle \hat{X}_{in}^\varphi(t'') \rangle + \langle \hat{I}_{in}(t') \rangle + \langle \hat{I}_{in}(t'') \rangle]. \tag{9.28}$$

Here,

$$\hat{X}_{in}^\varphi(t) \equiv \hat{E}_{in}^{(-)}(t) e^{j(\varphi - \omega_0 t)} + \hat{E}_{in}^{(+)}(t) e^{-j(\varphi - \omega_0 t)}. \tag{9.29}$$

Notice that in the expression above, the frequency is shifted by ω_0 so that $\hat{X}_{in}^\varphi(t)$ is a slowly-varying part of the field. This can be seen by using Eq. (9.3) to write out $\hat{X}_{in}^\varphi(t)$ explicitly as

$$\hat{X}_{in}^\varphi(t) = \frac{1}{\sqrt{2\pi}} \int d\omega [\hat{a}_{in}^\dagger(\omega) e^{j(\varphi + \omega t - \omega_0 t)} + \hat{a}_{in}(\omega) e^{-j(\varphi + \omega t - \omega_0 t)}]$$
$$\cong \frac{1}{\sqrt{2\pi}} \int d\Omega [\hat{a}_{in}^\dagger(\omega_0 + \Omega) e^{j\varphi + j\Omega t} + \hat{a}_{in}(\omega_0 + \Omega) e^{-j\varphi - j\Omega t}], \tag{9.30}$$

where we shift the frequency to $\Omega \equiv \omega - \omega_0$ in the integral. So, there is no fast oscillating terms like $e^{j\omega_0 t}$ in $\hat{X}_{in}^\varphi(t)$. Combining Eqs. (9.27) and (9.28), we have

$$p_2(t', t'') - p_1(t') p_1(t'') \approx \eta^2 |\mathcal{E}|^2 \langle : \Delta \hat{X}_{in}^\varphi(t') \Delta \hat{X}_{in}^\varphi(t'') : \rangle \tag{9.31}$$

with $\Delta \hat{X}_{in}^\varphi(t) \equiv \hat{X}_{in}^\varphi(t) - \langle \hat{X}_{in}^\varphi(t) \rangle$. Hence, for homodyne detection, the photocurrent correlation function in Eq. (9.12) is changed to

$$\langle \Delta i(t) \Delta i(t+\tau) \rangle \approx \eta |\mathcal{E}|^2 \int dt' k(t-t') k(t-t'+\tau)$$

$$+ \eta^2 |\mathcal{E}|^2 \int dt' dt'' k(t-t') k(t-t''+\tau)$$

$$\times \langle : \Delta \hat{X}_{in}^\varphi(t') \Delta \hat{X}_{in}^\varphi(t'') : \rangle. \tag{9.32}$$

Because of the normal ordering, the second term is zero for vacuum. So, the shot noise from the first term is also regarded as the vacuum quantum noise although it is independent of the input field. Theoreticians usually prefer this association and this viewpoint can be further confirmed by the following arrangement.

9.4 Vacuum Noise and Beam Splitter Model of Losses

When the quantum efficiency $\eta = 1$, using $[E_{in}^{(+)}(t'), E_{in}^{(+)}(t'')] = \delta(t' - t'')$, we have $\langle \Delta \hat{X}_{in}^{\varphi}(t') \Delta \hat{X}_{in}^{\varphi}(t'') \rangle = \langle : \Delta \hat{X}_{in}^{\varphi}(t') \Delta \hat{X}_{in}^{\varphi}(t'') : \rangle + \delta(t' - t'')$. With this, Eq. (9.32) is changed to

$$\langle \Delta i(t) \Delta i(t + \tau) \rangle = |\mathcal{E}|^2 \int dt' dt'' k(t - t') k(t - t'' + \tau) \langle \Delta \hat{X}_{in}^{\varphi}(t') \Delta \hat{X}_{in}^{\varphi}(t'') \rangle$$

$$\equiv \langle \Delta \hat{Z}(t) \Delta \hat{Z}(t + \tau) \rangle, \tag{9.33}$$

where $\hat{Z}(t) \equiv |\mathcal{E}| \int dt' k(t - t') \hat{X}_{in}^{\varphi}(t')$. If the input signal field is in vacuum, we have $\langle \Delta \hat{X}_{in}^{\varphi}(t') \Delta \hat{X}_{in}^{\varphi}(t'') \rangle_{vac} = \delta(t' - t'')$. This is the result of quantum fluctuations of the vacuum fields. Substituting it into Eq. (9.33), we then obtain the first term of shot noise in Eq. (9.32). That is why theoreticians like to regard the shot noise as the contributions from vacuum quantum fluctuations although photo-detection theory based on classical waves can also give rise to the shot noise (see discussion about the origin of the shot noise in Section 9.2). From Eq. (9.33), we find that similar to Eqs. (9.4) and (9.11), homodyne detection realizes a quantum measurement of operator $\hat{Z}(t)$, which is the Hermitian operator corresponding to this quantum measurement process.

When the quantum efficiency $\eta < 1$, we can rewrite Eq. (9.32) as

$$\langle \Delta i(t) \Delta i(t + \tau) \rangle$$

$$= \eta |\mathcal{E}|^2 \int dt' dt'' k(t - t') k(t - t'' + \tau)$$

$$\times \left[(1 - \eta) \langle \Delta \hat{X}_v^{\varphi}(t') \Delta \hat{X}_v^{\varphi}(t'') \rangle_{vac} + \eta \langle \Delta \hat{X}_{in}^{\varphi}(t') \Delta \hat{X}_{in}^{\varphi}(t'') \rangle \right]. \tag{9.34}$$

Here, $\hat{X}_v$ is some independent mode in vacuum. So, there are two contributions in the expression above: $(1 - \eta)$(vacuum noise) + η(input field). The first part is the contribution from vacuum of some independent mode $\hat{X}_v$ whereas the second part is the contribution from the input field. The coefficients in front of them indicate that they are coupled through a beam splitter with transmissivity η as shown in Fig. 9.4: one input side of the beam splitter is the vacuum field while the other is the input field. This

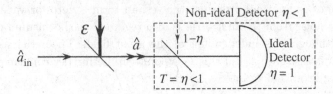

Fig. 9.4   The beam splitter model for non-ideal detector with quantum efficiency $\eta < 1$ due to losses. Vacuum noise is coupled in through the unused port.

beam splitter model shows that the less-than-perfect quantum efficiency $\eta < 1$ is caused by the loss, that is, the mixed field of the input and LO fields does not enter the detector completely. $(1 - \eta)$ part of it is coupled out and thus lost. And vacuum noise is introduced through loss.

On the other hand, the beam splitter model above raises a question about the quantum noise from LO: when we replace it with a non-operator number $\mathcal{E}$, we did not consider the contribution of quantum noise from LO at all in Eqs. (9.33) and (9.34). Is it legitimate to not consider the quantum noise from LO? In fact, since $\langle \Delta^2 \hat{X} \rangle_{\text{coh}} = 1$ and we use a beam splitter with $R \ll 1$ to couple in the LO field, its quantum noise contribution is $R \langle \Delta^2 \hat{X} \rangle_{\text{coh}} = R \to 0$. So, no need for quantum noise of LO.

9.5   Spectral Analysis of Homodyne Detection

Making a Fourier transformation of Eq. (9.32), we obtain the output function of current spectral analysis for the homodyne detection:

$$S_{HD}(\Omega) = \int d\tau \langle \Delta i(t) \Delta i(t + \tau) \rangle e^{j\Omega\tau}. \tag{9.35}$$

For a stationary field, we may define

$$\langle : \Delta \hat{X}_{in}^{\varphi}(t') \Delta \hat{X}_{in}^{\varphi}(t'') : \rangle \equiv \chi^{\varphi}(t' - t''). \tag{9.36}$$

Equation (9.35) is then changed to

$$S_{HD}(\Omega) = \eta |\mathcal{E}|^2 |k(\Omega)|^2 \big[1 + \eta \chi^{\varphi}(\Omega) \big]$$
$$= S_{SN}(\Omega) \big[1 + \eta \chi^{\varphi}(\Omega) \big] \tag{9.37}$$

or

$$\frac{S_{HD}(\Omega)}{S_{SN}(\Omega)} = 1 + \eta \chi^{\varphi}(\Omega), \tag{9.38}$$

where

$$\chi^{\varphi}(\Omega) \equiv \int d\tau \chi^{\varphi}(\tau) e^{j\Omega\tau} \tag{9.39}$$

is determined by the input field and is independent of the LO field. The LO field only determines the size of the shot noise level, making it much higher than the electronic noise. Equation (9.38) indicates that the spectral function of the photocurrent, after being normalized to the shot noise, is only related to $\chi^\varphi(\Omega)$ of the input field, whose ratio to the vacuum noise (the term of 1 in Eq. (9.38)) is the signal-to-noise ratio of the homodyne detection. Therefore, we also regard homodyne detection as a quantum-limited measurement since its noise is purely of quantum nature (vacuum quantum fluctuation). So, it can be used to measure the quantum correlation of optical fields.

To find a specific form for $\chi^\varphi(\Omega)$, we make a change of $\Omega \rightarrow -\Omega$ in the first term in the expression for $\hat{X}^\varphi_{in}(t)$ in Eq. (9.30) and rewrite it as

$$\hat{X}^\varphi_{in}(t) = \frac{1}{\sqrt{2\pi}} \int d\Omega \left[\hat{a}^\dagger(\omega_0 - \Omega)e^{j\varphi}e^{-j\Omega t} + \hat{a}(\omega_0 + \Omega)e^{-j\varphi}e^{-j\Omega t} \right]$$

$$= \frac{1}{\sqrt{2\pi}} \int d\Omega \hat{X}^\varphi(\Omega)e^{-j\Omega t}, \qquad (9.40)$$

where $\hat{X}^\varphi(\Omega) \equiv \hat{a}^\dagger(\omega_0 - \Omega)e^{j\varphi} + \hat{a}(\omega_0 + \Omega)e^{-j\varphi}$ is the multi-mode quadrature-phase amplitude given in Eq. (4.74) of Section 4.6.3. With Eq. (9.40), we can calculate the correlation function

$$\langle \Delta\hat{X}^\varphi_{in}(t')\Delta\hat{X}^\varphi_{in}(t'') \rangle$$

$$= \frac{1}{2\pi} \int d\Omega' d\Omega'' \langle \Delta\hat{X}^\varphi(\Omega')\Delta\hat{X}^\varphi(\Omega'') \rangle e^{-j(\Omega' t' + \Omega'' t'')}$$

$$= \frac{1}{2\pi} \int d\Omega' d\Omega'' \langle \Delta\hat{X}^\varphi(\Omega')\Delta\hat{X}^{\varphi\dagger}(\Omega'') \rangle e^{-j(\Omega' t' - \Omega'' t'')}. \quad (9.41)$$

The second line is due to $[\hat{X}^\varphi(\Omega)]^\dagger = \hat{X}^\varphi(-\Omega)$. Since $\langle \Delta\hat{X}^\varphi_{in}(t')\Delta\hat{X}^\varphi_{in}(t'') \rangle$ only depends on $\tau = t' - t''$ for continuous waves, we must have $\langle \Delta\hat{X}^\varphi(\Omega')\Delta\hat{X}^{\varphi\dagger}(\Omega'') \rangle = S_\varphi(\Omega')\delta(\Omega' - \Omega'')$. Substituting this into Eq. (9.41), we have

$$\langle \Delta\hat{X}^\varphi_{in}(t')\Delta\hat{X}^\varphi_{in}(t'') \rangle = \frac{1}{2\pi} \int d\Omega' S_\varphi(\Omega')e^{-j\Omega'(t'-t'')}. \qquad (9.42)$$

To find the quantity in Eq. (9.36), we use the relation

$$\langle \Delta\hat{X}^\varphi_{in}(t')\Delta\hat{X}^\varphi_{in}(t'') \rangle = \langle : \Delta\hat{X}^\varphi_{in}(t')\Delta\hat{X}^\varphi_{in}(t'') : \rangle + \delta(t' - t''). \quad (9.43)$$

With this and Eqs. (9.36), (9.39) and (9.42), it is straightforward to find

$$\chi^\varphi(\Omega) = S_\varphi(\Omega) - 1. \qquad (9.44)$$

Substituting this into Eq. (9.38), we find the spectral function for homodyne detection

$$\frac{S_{HD}(\Omega)}{S_{SN}(\Omega)} = (1 - \eta) + \eta S_{\varphi}(\Omega). \qquad (9.45)$$

So, we just need to calculate $\langle \Delta \hat{X}^{\varphi}(\Omega') \Delta \hat{X}^{\varphi\dagger}(\Omega'') \rangle$ for the input field in order to find the spectral function in homodyne detection.

As an example, we consider the homodyne detection of the multi-mode squeezed state discussed in Section 4.6.3. From Eq. (4.76), we have

$$\langle \hat{X}(\Omega) \hat{X}^{\dagger}(\Omega') \rangle = S_X(\Omega)\delta(\Omega - \Omega'),$$
$$\langle \hat{Y}(\Omega) \hat{Y}^{\dagger}(\Omega') \rangle = S_Y(\Omega)\delta(\Omega - \Omega'), \qquad (9.46)$$

where $S_X(\Omega) = (|G(\Omega)| - |g(\Omega)|)^2 < 1$ for noise reduction, but $S_Y(\Omega) = (|G(\Omega)| + |g(\Omega)|)^2 > 1$ for noise amplification. For the multi-mode squeezed state, we have $\langle \hat{X}_{in}^{\varphi}(t) \rangle = 0$. Hence, $S_{\varphi}(\Omega) = S_X(\Omega)$ for $\varphi = \theta_0$ and $S_Y(\Omega)$ for $\varphi = \theta_0 + \pi/2$. Then the observed spectral function of homodyne detection for the multi-mode squeezed state is

$$\frac{S_{HD}(\Omega)}{S_{SN}(\Omega)} = (1 - \eta) + \eta(|G(\Omega)| \pm |g(\Omega)|)^2. \qquad (9.47)$$

Here "$-$" is for $\varphi = \theta_0$ and "$+$" is for $\varphi = \theta_0 + \pi/2$. When $\eta = 1$, the spectrum of the detected photocurrent, after normalization to the shot noise, is the spectrum of squeezing $S_{\varphi}(\omega)$. When $\eta < 1$, the first term in Eq. (9.47) is obviously the contribution of vacuum so the vacuum noise is coupled in and reduces the amount of noise reduction effect due to squeezed state. Figure 9.5 shows experimentally observed spectra $S_{HD}(\Omega)$ of photocurrent at $\varphi = \theta_0, \theta_0 + \pi/2$ and the shot noise spectrum $S_{SN}(\Omega)$ for the squeezed state detection. Notice that log-scale is used in Fig. 9.5 so that the ratio of $S_{HD}(\Omega)$ to $S_{SN}(\Omega)$ is simply the difference between the two curves. For $\varphi = \theta_0$, $S_{HD}(\Omega)$ is lower than the shot noise level $S_{SN}(\Omega)$ by about 3 dB at around 2.5 MHz, realizing a quantum noise reduction by a factor of 3 dB = 50%. Notice from Eq. (9.47) that when $\eta = 1$, $S_X(\Omega)S_Y(\Omega) = (|G(\Omega)|^2 - |g(\Omega)|^2)^2 = 1$, which means noise reduction and noise amplification should be the same amount in log-scale. However, Fig. 9.5 shows different amounts for reduction and amplification. This is due to $\eta < 1$. From Fig. 9.5, we find the amount for reduction is -3 dB and that for amplification is 8 dB. From Eq. (9.47), we can deduce the efficiency coefficient $\eta = 0.55$.

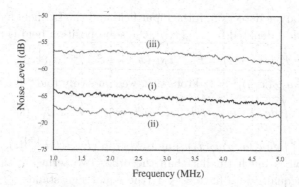

Fig. 9.5   Spectra of photocurrent $S_{HD}(\omega)$ of homodyne detection of a multi-mode squeezed state: (i) spectrum of the shot noise; (ii) $S_{HD}(\omega)$ for $\varphi = \theta_0$; (iii) $S_{HD}(\omega)$ for $\varphi = \theta_0 + \pi/2$. Courtesy of Wei Du.

9.6   Mode Match and Local Oscillator Noise in Homodyne Detection

In all our previous discussions on homodyen detection, we did not consider the spatial modes. This is because we assumed that the input signal field and the LO field have the same spatial modes. However, we know that it is impossible to completely match the spatial modes of two optical fields in the experiment. We will consider this non-ideal case in the following.

Since spatial modes are included, we cannot use the 1-dim approximation. Assume the optical field propagates along z-direction which is perpendicular to the cross-section of the detector and the field has a transverse distribution in $x - y$ plane, that is, $\hat{E}^{(+)} = \hat{E}^{(+)}(x, y, t)$ (we ignore the z coordinate here because we are only interested in the cross-section of the detector). The spatial modes of the field determine the transverse distribution in $x - y$ plane and we can write the field as $\hat{E}^{(+)}(x, y, t) = \hat{E}^{(+)}(t)u(x, y)$ where $u(x, y)$ is the mode function and is normalized: $\int dxdy|u(x, y)|^2 = 1$. In this case, the Glauber photodetection formulas in Eqs. (5.26) and (5.28) are changed to

$$p_1(t) = \eta \int da \langle \hat{E}^{(-)}(x, y, t) \hat{E}^{(+)}(x, y, t) \rangle \tag{9.48}$$

$$p_2(t, t') = \eta^2 \iint dada' \langle : \hat{I}(x, y, t) \hat{I}(x', y', t') : \rangle \tag{9.49}$$

where $da = dxdy$, $da' = dx'dy'$.

In homodyne detection, if the input field and the LO field have different spatial modes, denoted by $u_{1,2}(x,y)$, the superposition field is then

$$\hat{E}^{(+)}(x,y,t) = \hat{E}_{in}^{(+)}(t)u_1(x,y) + \mathcal{E}e^{-j\omega_0 t}u_2(x,y), \qquad (9.50)$$

where $\int da|u_{1,2}(x,y)|^2 = 1$. From this, we can easily find

$$\int da\hat{E}^{(-)}(x,y,t)\hat{E}^{(+)}(x,y,t) \approx |\mathcal{E}|^2 + \beta|\mathcal{E}|\hat{X}_{in}^{\varphi}(t), \qquad (9.51)$$

where $\beta \equiv |\int dau_1^*(x,y)u_2(x,y)|$ and we drop $\langle\hat{E}_{in}^{(-)}(t)\hat{E}_{in}^{(+)}(t)\rangle$ because its contribution is much smaller than the other two terms. Using Cauchy-Schwarz inequality, we have $\beta \leq 1$ (the equal sign stands for $u_1(x,y) = u_2(x,y)$). Substituting the above into Eqs. (9.48) and (9.49), we can prove that η is changed to $\eta|\beta|^2$, that is, the mismatch of the spatial modes leads to the reduction of quantum efficiency. (Readers who are interested in this can prove it as an exercise.)

Although the argument above is for spatial modes, it applies to the mismatch of other modes, such as temporal modes in pulsed fields. In fact, the underlying physical principle of homodyne is the interference between the input field and the LO field and the mode match parameter β is equivalent to the visibility of the interference fringe.

In homodyne detection, there is another issue we need to address, that is, the effect of the intensity fluctuations of the LO field. We can see its effect from the simple single-mode model. Assume the intensity fluctuation of the LO field is from the amplitude of the LO field: $\mathcal{E} \to \mathcal{E} + \Delta\mathcal{E}$. From Eq. (9.22), we have

$$\begin{aligned}\hat{a}^{\dagger}\hat{a} &\approx |\mathcal{E} + \Delta\mathcal{E}|^2 + |\mathcal{E} + \Delta\mathcal{E}|\hat{X}_{in}(\varphi) \\ &\approx |\mathcal{E}|^2 + |\mathcal{E}|[\hat{X}_{in}(\varphi) + 2\Delta\mathcal{E}].\end{aligned} \qquad (9.52)$$

Here, we only take the first term of $\Delta\mathcal{E}$. Since $\mathcal{E}$ is really large, a small relative fluctuation of $\Delta\mathcal{E}/\mathcal{E}$ can make $\Delta\mathcal{E}$ very large, which is added to the signal term of $\hat{X}_{in}(\varphi)$ to become extra noise.

In the experiment, the intensity fluctuations of the LO field will show up in the shot noise spectrum, i.e., the photocurrent spectrum without the input signal. This is equivalent to the case of direct illumination of the detector by only the LO field. From Eq. (9.17), we obtain the excess noise as

$$S_{ex}(\Omega) = (\eta\langle|\mathcal{E}|^2\rangle)^2|k(\Omega)|^2h(\Omega) = S_{SN}(\Omega)\eta\langle|\mathcal{E}|^2\rangle h(\Omega), \qquad (9.53)$$

where $h(\Omega) \equiv \int d\tau[g_{LO}^{(2)}(\tau) - 1]e^{-j\Omega\tau}$ gives the spectrum of the excess noise after being normalized to the shot noise. Notice that the excess noise from the LO is proportional to the square of the LO intensity $\langle|\mathcal{E}|^2\rangle$ whereas the shot noise is only proportional to the intensity of LO. These intensity dependent properties of noise can be used to check if the shot noise level obtained in the experiment, i.e., the vacuum noise level, is truly of the shot noise nature or has some contribution from the intensity fluctuations of the LO field.

The excess noise of the LO field originates from the laser noise. Different lasers have different noise spectra. Solid state lasers like Ti:sapphire and YAG lasers have noise usually in the range within 1 MHz. So, their noise is not a big problem when our working frequency is usually at a few MHz. However, the commonly used semi-conductor lasers have a wide noise spectrum ranging from 10 to 100 MHz. Fortunately, we can use the balanced homodyne detection technique discussed in the following to cancel the excess noise from the LO [Yuen and Chan (1983)].

9.7 Balanced Homodyne Detection

As shown in Fig. 9.6, balanced homodyne measurement is achieved with two detectors. We use a 50:50 beam splitter to combine the input field and the LO field and evenly illuminate on the two detectors. We measure the photocurrent difference from the two detectors. A simple single-mode model gives

$$\hat{a}_1 = (\hat{a}_{in} + \mathcal{E})/\sqrt{2}, \quad \hat{a}_2 = (\mathcal{E} - \hat{a}_{in})/\sqrt{2}. \tag{9.54}$$

So, the difference from the two detector outputs is

$$\hat{a}_1^\dagger \hat{a}_1 - \hat{a}_2^\dagger \hat{a}_2 = |\mathcal{E}|\hat{X}_{in}(\varphi). \tag{9.55}$$

We find from the above that the LO intensity part is cancelled. Hence, its fluctuations will not contribute to the output of the balanced homodyne detection.

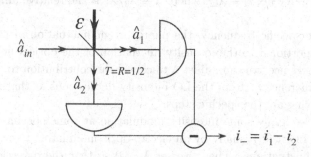

Fig. 9.6 Balanced homodyne detection scheme.

For multi-mode case, we have

$$\hat{E}_1^{(+)}(t) = [\hat{E}_{in}^{(+)}(t) + \mathcal{E}]/\sqrt{2}, \quad \hat{E}_2^{(+)} = [\mathcal{E} - \hat{E}_{in}^{(+)}(t)]/\sqrt{2}. \quad (9.56)$$

We first assume the detectors are identical, i.e., $\eta_1 = \eta_2 \equiv \eta, k_1(t) = k_2(t) \equiv k(t)$ and obtain the same result as the single-mode case:

$$\hat{i}_- = \hat{i}_1 - \hat{i}_2 = \int d\tau k(t-\tau)\eta[\hat{E}_1^{(-)}(\tau)\hat{E}_1^{(+)}(\tau) - \hat{E}_2^{(-)}(\tau)\hat{E}_2^{(+)}(\tau)]$$

$$= \int d\tau k(t-\tau)\eta|\mathcal{E}|\hat{X}_{in}^{(\varphi)}(\tau). \quad (9.57)$$

If the two detectors have different quantum efficiency: $\eta_1 > \eta_2$ but have the same time response[4]: $k_1(t) = k_2(t) \equiv k(t)$, we can balance them by slightly misaligning the detector of higher η_1 to reduce it to match the lower one: $\eta_1' = \eta_2$. We can also achieve balance by selecting a non-50:50 beam splitter with $\eta_2 T = \eta_1 R$:

$$\hat{E}_1^{(+)}(t) = \sqrt{T}\hat{E}_{in}^{(+)}(t) + \sqrt{R}\mathcal{E}, \quad \hat{E}_2^{(+)} = \sqrt{T}\mathcal{E} - \sqrt{R}\hat{E}_{in}^{(+)}(t). \quad (9.58)$$

We also obtain similar result to Eq. (9.57)

$$\hat{i}_- = \hat{i}_1 - \hat{i}_2 = \int d\tau k(t-\tau)[\eta_1\hat{E}_1^{(-)}(\tau)\hat{E}_1^{(+)}(\tau) - \eta_2\hat{E}_2^{(-)}(\tau)\hat{E}_2^{(+)}(\tau)]$$

$$= \int d\tau k(t-\tau)\sqrt{\eta_1\eta_2}|\mathcal{E}|\hat{X}_{in}^{(\varphi)}(\tau). \quad (9.59)$$

But the quantum efficiency is changed to $\eta_{\text{eff}} = \sqrt{\eta_1\eta_2}$, which is better than the misalignment technique.

For the most general case of different time response and quantum efficiency for the two detectors, we cannot achieve balancing in the whole spectral range. But we can balance them at one specific frequency. This requires balancing the amplification gains of the two photocurrents, that is, $i_- = \lambda_1 i_1 - \lambda_2 i_2 \propto i_1 - \lambda i_2$, where $\lambda \equiv \lambda_2/\lambda_1$ is the relative amplification gain.

At one specific frequency, the photocurrent fluctuation of each detector is proportional to the intensity fluctuation of the optical field at that frequency. Hence, we can adjust λ to cancel the contribution to the photocurrent difference i_- from the LO intensity fluctuations at that frequency. The following are the specific steps.

First, we apply some intensity modulation at some specific frequency f_0 on the LO field (usually by an electro-optic modulator) to simulate the intensity fluctuations. Then, we set $\lambda = 0$ so that the detection system

---

[4]This is usually the case when we use detectors made in the same manufacturing run.

is completely unbalanced and we will find a modulation signal at f_0, as shown in Fig. 9.7(a). Now gradually increase λ to see the reduction of the modulation signal at f_0, as shown in Fig. 9.7(b). Fine tune λ to minimize the modulation signal at f_0. The difference between the case of $\lambda = 0$ and the minimized modulation tells us how well the scheme is balanced. A good balancing scheme usually leads to 30 dB reduction of the modulation signal.

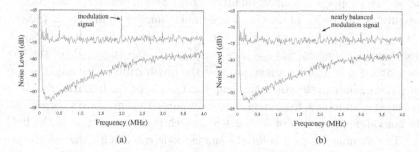

Fig. 9.7   Spectra of homodyne detection for (a) unbalanced case and (b) nearly balanced case. The modulation signal disappears for the completely balanced case.

9.8   Intensity Fluctuations and Self-Homodyne Detection

Measurement of intensity fluctuations is actually more common than homodyne detection in the experiment. A direct detection of an optical field will make a measurement of the intensity of the optical field. But as we will see in this section, when the average intensity of the measured optical field is high, the measurement of intensity fluctuations is equivalent to the homodyne detection of one particular quadrature-phase amplitude with the local oscillator being the coherent component of the optical field itself. Thus, this case is also known as self-homodyne detection.

As a matter of fact, we already encountered this in Section 3.4.3 when we discussed about coherent squeezed state: when we choose the squeezing angle θ so that it is related to the phase angle φ_α of the coherent component by $\theta = \pi + 2\varphi_\alpha$, we obtain the amplitude squeezed state. When we measure intensity (photon number) of the field in this state, we obtain a reduced quantum fluctuation in intensity (see Eq. (9.79) of Problem 9.2). Amplitude squeezed states is a special kind of quadrature-phase amplitude squeezed state. In the measurement of intensity, the coherent component of the field acts as a local oscillator for one particular quadrature-phase amplitude.

To see more clearly the role of coherent component in self-homodyne detection, let us consider a single-mode field with a large coherent component: $\langle \hat{a} \rangle \equiv re^{i\varphi_0}$ with $r \gg 1$. Define a field fluctuation operator $\Delta \hat{a} = \hat{a} - \langle \hat{a} \rangle$ and we have

$$\hat{a}^\dagger \hat{a} = |\langle \hat{a} \rangle|^2 + \langle \hat{a} \rangle \Delta \hat{a}^\dagger + \langle \hat{a} \rangle^* \Delta \hat{a} + \Delta \hat{a}^\dagger \Delta \hat{a}$$

$$\approx r^2 + r\Delta \hat{X}(\varphi_0), \tag{9.60}$$

where we dropped the small quadratic field fluctuation term because it is usually of the order of one, which is much smaller than r^2. $\Delta \hat{X}(\varphi_0) \equiv \Delta \hat{a} e^{-i\varphi_0} + \Delta \hat{a}^\dagger e^{i\varphi_0}$. Comparing the above equation with Eq. (9.22), we see that the coherent part $\langle \hat{a} \rangle$ acts exactly as $\mathcal{E}$, the amplitude of the local oscillator for homodyne detection and the quadrature-phase angle is fixed at φ_0, the phase of the coherent component $\langle \hat{a} \rangle = re^{i\varphi_0}$. In particular, if the coherent component $\langle \hat{a} \rangle$ is real, intensity measurement always corresponds to homodyne detection of $\hat{X} = \hat{a} + \hat{a}^\dagger$, which is the amplitude of the field.

The argument above is for the single-mode case. With the multi-mode case, let us go back to Eqs. (9.9), (9.14) and (9.16) for the full treatment of photo-detection:

$$\langle \Delta i(t) \Delta i(t+\tau) \rangle = \langle \Delta i(t) \Delta i(t+\tau) \rangle_{SN} + \langle \Delta i(t) \Delta i(t+\tau) \rangle_{ex}$$

$$= \int dt' dt'' k(t-t')k(t+\tau-t'')\eta^2 \langle : \Delta \hat{I}(t')\Delta \hat{I}(t'') : \rangle$$

$$+ \int dt' k(t-t')k(t+\tau-t')\eta I_0 \tag{9.61}$$

Now similar to the single-mode case, we write $\Delta \hat{E} \equiv \hat{E} - \langle \hat{E} \rangle$ with the coherent component $\langle \hat{E} \rangle \equiv re^{i\varphi_0 - i\omega_0 t}$ (ω_0 is the center frequency of the measured field) and $r^2 \gg \langle \Delta^2 \hat{E} \rangle$, i.e., coherent component is much larger than the fluctuation of the field. Then $\hat{E} = re^{i\varphi_0 - i\omega_0 t} + \Delta \hat{E}$ and we can make the approximation for the intensity operator:

$$\hat{I}(t) \equiv \hat{E}^\dagger(t)\hat{E}(t) = r^2 + r(\Delta \hat{E}^\dagger e^{i\varphi_0 - i\omega_0 t} + \Delta \hat{E} e^{-i\varphi_0} + i\omega_0 t) + \Delta \hat{E}^\dagger \Delta \hat{E}$$

$$\approx r^2 + r\Delta \hat{X}_{\varphi_0}(t). \tag{9.62}$$

Here $\Delta \hat{X}_{\varphi_0}(t) \equiv \Delta \hat{E}^\dagger e^{i\varphi_0 - i\omega_0 t} + \Delta \hat{E} e^{-i\varphi_0 + i\omega_0 t}$. With $I_0 \equiv \langle \hat{I}(t) \rangle = r(r + \langle \Delta \hat{X}_{\varphi_0}(t) \rangle) \approx r^2$, we have $\Delta \hat{I}(t) \approx r\Delta \hat{X}_{\varphi_0}(t)$ and Eq. (9.61) becomes

$$\langle \Delta i(t) \Delta i(t+\tau) \rangle$$

$$= \eta^2 r^2 \int dt' dt'' k(t-t')k(t+\tau-t'')\langle : \Delta \hat{X}_{\varphi_0}(t')\Delta \hat{X}_{\varphi_0}(t'') : \rangle$$

$$+ \eta r^2 \int dt' k(t-t')k(t+\tau-t'). \tag{9.63}$$

Comparing to Eq. (9.32), we find that the equation above has exactly the same form. So, when the measured field has a large coherent component, the direct intensity fluctuation measurement is equivalent to a homodyne measurement of one particular quadrature-phase amplitude of the measured field. Intensity fluctuation measurement is thus sometimes known as self-homodyne.

We will come back to this in Section 10.1.4 when we discuss about quantum correlation in intensity for twin beams and its relation to Einstein-Podolsky-Rosen entanglement.

9.9 Photo-detection for Ultra-Fast Pulses

9.9.1 *General Consideration*

In quantum optics, generation of nonclassical states of light often relies on nonlinear optical interaction, which requires high power to reach strong interaction. This leads to pulse lasers, which usually have extremely high peak power. Commonly used pulse lasers are mode-locked lasers with a few 100s of femto-second pulse width and a few tens of MHz repetition rate. The quantum fields produced with these lasers also have similar temporal profiles. For example, we treated in Section 6.1.5 the case of pulse-pumped parametric processes, where the modes of the fields are temporal modes. In this section, we will discuss photo-detection of this type of fields.

Compared to the continuous case, the main difference is the detector response bandwidth versus the bandwidth of the optical fields. In the continuous case, optical bandwidth is well within the response bandwidth of the detector and the method of spectral analysis provides the spectrum of the optical field. In the pulsed case, however, the ultra-short optical pulses are too fast for the detector to respond and the photocurrent is then a time average of optical pulses. Furthermore, due to the characteristic of mode-locked lasers, the ultra-short pulses repeat with a repetition rate of a few tens of MHz (80 MHz for a typical Ti:saphire laser) and form a quasi-continuous field. So, the photocurrent has a form of

$$i(t) = \sum_{n=-\infty}^{\infty} k(t - nT_{\text{rep}})I_n, \qquad (9.64)$$

where T_{rep} is the time between pulses, $k(t)$ is the response function of the detector and I_n is the total photon number of the n-th pulse:

$$I_n \equiv \int d\tau \langle \hat{E}^\dagger(\tau)\hat{E}(\tau)_n. \qquad (9.65)$$

I_n may change from pulse to pulse due to intensity fluctuation or modulation. This fluctuation may have classical or quantum origin and the modulation is usually the signal encoded in the pulses. If the detector's response time T_R is much longer than T_{rep}, Eq. (9.64) can be approximated as

$$i(t) \approx \int d\tau k(t - \tau)I(\tau), \qquad (9.66)$$

which is in the same form as the continuous case in Eq. (9.4). Equation (9.64) can be viewed as the discrete version of Eq. (9.4). Then, we can use spectral analysis method to measure the fluctuations of the optical pulses just like the continuous case.

On the other hand, if T_R is comparable to T_{rep}, spectral analysis of the photocurrent will show strong frequency components at the repetition frequency and its harmonics [Slusher et al. (1987)]. These frequency components can be very large to overwhelm the subsequent electronic amplifiers. So, they need to be handled right after the photo-detectors. Two methods can be used. The first one is to directly use a good low pass filter (> 100 dB) to block out these strong frequency components. The other is to use two nearly identical detectors and perform a balanced detection of the current difference of the two detectors. The latter method is commonly employed in intensity difference measurement for the detection of twin beams and in balanced homodyne detection where strong modulation in the LO field is canceled in the difference of the photo-currents. We will address balanced homodyne detection of pulsed fields here.

9.9.2 Homodyne Detection of Pulsed Fields

Although the spectral analysis of the pulsed case is the same as the continuous case after blocking out the repetition frequency and its harmonics, there is one more important issue that does not occur in the continuous case, that is, the temporal mode match. We have seen in Section 9.6 how the spatial mode mismatch can lead to poor detection efficiency. The effect is the same for temporal mode match, which arises only in the pulsed case because of the time integral in Eq. (9.65).

Suppose that the LO field is a transform-limited pulse in the form of

$$\mathcal{E}_{LO}(t) = |\mathcal{E}|e^{i\varphi}\frac{1}{\sqrt{2\pi}} \int A_{LO}(\omega)e^{-i\omega t}d\omega, \qquad (9.67)$$

with $A_{LO}(\omega)$ satisfying the normalization condition $\int |A_{LO}(\omega)|^2 d\omega = 1$. φ is the phase of the LO field and the amplitude of the LO field is strong:

$|\mathcal{E}| \gg 1$. The output current difference operator is similar to the continuous case in Eq. (9.57) but the time integral is over the pulse due to Eq. (9.65). So, the current difference operator in this case takes the form of

$$\hat{i}_-(t) = k(t) \int_{-\infty}^{\infty} \left[\mathcal{E}_{LO}^*(\tau) \hat{E}_{in}(\tau) + h.c. \right] d\tau$$

$$= k(t)|\mathcal{E}| \int d\omega \left[A_{LO}^*(\omega) \hat{a}_{in}(\omega) e^{-i\varphi} + h.c. \right], \quad (9.68)$$

where the input field is a quantum field described by operator

$$\hat{E}_{in} = \frac{1}{\sqrt{2\pi}} \int \hat{a}_{in}(\omega) e^{-i\omega t} d\omega. \quad (9.69)$$

Here, we assume the spatial modes are matched and only concentrate on the temporal part of the fields by using the quasi-monochromatic and one-dimensional approximation for the input field. Equation (9.68) only describes a single pulse contribution and shows that the spectral property of the photocurrent is determined mainly by the detector response function $k(t)$. The field fluctuation is then multiplied on all frequency components in the spectral distribution and is exhibited as pulse-to-pulse variation per explanation of the general expression in Eq. (9.64). Hence, quantum average should also be understood as pulse-to-pulse average.

The pulsed input field is usually described by a set of orthonormal temporal mode functions characterized by $\{\phi_j(\omega)\}$ (e.g., the fields from pulsed -pumped parametric processes in Section 6.1.5) and the LO field amplitude A_{LO} can be decomposed as

$$A_{LO}(\omega) = \sum_j \xi_j \phi_j(\omega), \quad (9.70)$$

where the coefficient ξ_j is given by

$$\xi_j = |\xi_j| e^{i\theta_j} = \int A_{LO}(\omega) \phi_j^*(\omega) d\omega \quad (9.71)$$

with $\sum_j |\xi_j|^2 = 1$. Substituting Eq. (9.70) into Eq. (9.68), the output of homodyne detection can be rewritten as

$$\hat{i}_-(t) = k(t)|\mathcal{E}| \sum_j |\xi_j| \hat{X}_j(\theta_j + \varphi) \quad (9.72)$$

where $\hat{X}_j(\theta)$ is the quadrature-phase amplitude operator for mode j and is defined as

$$\hat{X}_j(\theta) = \hat{B}_j e^{-i\theta} + \hat{B}_j^\dagger e^{i\theta} \quad (9.73)$$

with

$$\hat{B}_j \equiv \int d\omega \phi_j^*(\omega) \hat{a}_{in}(\omega) \qquad (9.74)$$

as the annihilation operator for temporal mode j $[\phi_j(\omega)]$ first defined in Eq. (6.38) in Section 6.1.5.

Notice that the phase of the measured quadrature-phase amplitude $\hat{X}_j(\theta_j + \varphi)$ depends not only on the overall phase φ but also on the phase θ_j of quantity ξ_j, which may be different for different mode j. Furthermore, $|\xi_j|^2$ can be viewed as the mode-matching efficiency for each mode, while θ_j is equivalent to the homodyne detection phase for different temporal mode $\hat{B}_j$ of the input field.

9.9.3   *Temporal Mode Match*

First, let us consider the case when the temporal mode of the LO field matches with one of the temporal modes of the input field, that is, $A_{LO}(\omega) = \phi_{j_0}(\omega)$. Then because of the orthonormal property of $\phi_j(\omega)$, we simply have $\xi_j = \delta_{j,j_0}$ from Eq. (9.71), and Eq. (9.72) becomes

$$\hat{i}_-(t) = k(t)|\mathcal{E}|\hat{X}_{j_0}(\theta_{j_0} + \varphi). \qquad (9.75)$$

The noise spectrum is then

$$S_-(\omega) = S_{SN}(\omega)\langle \Delta^2 \hat{X}_{j_0}(\theta_{j_0} + \varphi)\rangle, \qquad (9.76)$$

where $S_{SN}(\omega) = |k(\omega)\mathcal{E}|^2$ is the shot noise spectrum. So, the homodyne detection measures only the quadrature-phase amplitude of temporal mode j_0 of the input field. Other modes have no contribution. In this case, the LO field acts as a temporal mode filter to pick up only mode j_0 of the field and filter out all others.

Next, for the more general case when the temporal mode of the LO does not match to any particular temporal mode of the input field, instead of responding to one particular temporal mode, the homodyne detection will be the sum of the contributions from all the temporal modes with different homodyne detection phase θ_j and different mode-matching efficiency $|\xi_j|^2$ for each mode.

If the input field is in a single mode, i.e., only one temporal mode, say $j = 1$, is excited and the rest is in vacuum, we then obtain the noise spectrum from Eq. (9.72) as

$$S_-(\omega) = S_{SN}(\omega)\Big[|\xi_1|^2\langle \Delta^2 \hat{X}_{j_0}(\theta_{j_0} + \varphi)\rangle + \sum_{j \neq 1} |\xi_j|^2\Big]$$

$$= S_{SN}(\omega)\Big[|\xi_1|^2\langle \Delta^2 \hat{X}_{j_0}(\theta_{j_0} + \varphi)\rangle + 1 - |\xi_1|^2\Big]. \qquad (9.77)$$

Here, we used $\sum_j |\xi_j|^2 = 1$. So, the consequence of mode mismatch $|\xi_1|^2 < 1$ is equivalent to a loss of $1 - |\xi_1|^2$ or a drop of the quantum efficiency by a factor of $|\xi_1|^2$. This is exactly the same as the case of spatial mode mismatch in Section 9.6.

On the other hand, if there are multiple modes excited in the input field as in the case of pulsed parametric processes, the situation becomes complicated. Suppose we are looking for quantum noise reduction. Assuming there is no correlation between different modes, we obtain from Eq. (9.72)

$$S_-(\omega) = S_{SN}(\omega) \sum_j |\xi_j|^2 \langle \Delta^2 \hat{X}_j(\theta_j + \varphi) \rangle. \tag{9.78}$$

If the phase angles $\{\theta_j\}$ are all same, we can adjust the global phase φ of the LO to achieve minimum values for all modes. In this case, the gains of all the modes are synchronized and quantum noise is squeezed together for all the modes. But if the phase angles $\{\theta_j\}$ are not all same, we have the worst scenario: one mode has optimum squeezing while others may not have or sometimes even have noise increase. In this case, the effect of noise squeezing will be reduced and this reduction effect cannot be accounted for by the simple model of losses, for the unsqueezed modes have noise increasing with the gain or the pump power, eventually leading to no squeezing as a whole. This behavior was observed in some experiments in measuring pulsed squeezing [Guo et al. (2012, 2016b)]. So, temporal mode matching issue is more complicated than spatial mode match issue and needs to be dealt with carefully.

9.10   Problems

Problem 9.1 Measurement of the shot noise and intensity noise of LO.

In the case of ideal balanced homodyne detection, when the input field is in vacuum, prove that the difference between the photocurrents of the two detectors always gives rise to the shot noise of the LO field no matter what quantum state the LO field is in. On the other hand, the sum of the two photocurrents always leads to the intensity noise of the LO field.

Problem 9.2 Self-homodyne of the amplitude-squeezed state.

The amplitude squeezed state $|\alpha, -re^{2i\varphi_\alpha}\rangle \equiv \hat{D}(\alpha)\hat{S}(-re^{2i\varphi_\alpha})|0\rangle$ with $e^{i\varphi_\alpha} = \alpha/|\alpha|$ is depicted in Fig. 3.5(a), showing noise is squeezed in the amplitude of the field. This can be confirmed by direct intensity measurement via self-homodyne discussed in Section 9.8.

(i) Assuming $|\alpha| \gg r$, calculate $\langle \hat{n} \rangle, \langle \hat{n}^2 \rangle$ by using Eq. (3.82).

(ii) Show that direct intensity measurement gives $\langle \Delta^2 \hat{I} \rangle$ with $\hat{I} \equiv \hat{n} = \hat{a}^\dagger \hat{a}$ in the following form

$$\langle \Delta^2 \hat{I} \rangle = \langle \Delta^2 \hat{n} \rangle$$
$$= |\alpha|^2 (1 - 2e^{-r} \sinh r)$$
$$= \langle \hat{I} \rangle (1 - 2e^{-r} \sinh r), \qquad (9.79)$$

which shows intensity fluctuation is smaller than the shot noise level of $\langle \Delta^2 \hat{I} \rangle_{SN} = \langle \hat{I} \rangle$. Here we dropped higher order terms due to $|\alpha|^2 \gg r$.

Note that Eq. (9.79) can be rewritten as

$$\langle \Delta^2 \hat{n} \rangle = \langle \hat{n} \rangle (1 - 2e^{-r} \sinh r) < \langle \hat{n} \rangle, \qquad (9.80)$$

showing sub-Poisson photon statistics (Section 5.4).

Chapter 10

Applications of Homodyne Detection Technique: Quantum Measurement of Continuous Variables

Physics is a science of measurement. In the measurement of physical quantities, many are in the form of continuous variables, which have a continuous
spectrum of measurement outcomes. The amplitude and phase of an optical field are continuous at least in classical sense. Classically, the physical
quantities are well defined and can in principle be measured with arbitrary
precision. This is true even for continuous variables. In quantum mechanics, however, energy is quantized. Intensity, for example, is directly related
to the photon number after quantization of the fields and has discrete values
when we measure it. This discreteness will ultimately lead to measurement
uncertainties of continuous variables. This is somewhat similar to the error
generated in digitizing an analog signal in electrical engineering. Furthermore, measurement of an optical field is usually done with photo-detector
by converting light into electrical signal. As we discussed in the previous
chapter that the origin of shot noise in photocurrents is due to the discreteness of electrons generated by the illumination of light. This has nothing
to do with quantization of the optical field: shot noise exists even in the
semi-classical theory of photo-detection.

   In addition to the experimental difficulties mentioned above, we learn
from the quantum measurement theory, that quantum mechanics has the
Heisenberg uncertainty principle, which seems to give rise to measurement
imprecision in principle. Indeed, in many cases, quantum mechanics is the
culprit for measurement inaccuracy and sets the so-called standard quantum limit. On the other hand, as we will see in this chapter, quantum
mechanics also provides a powerful tool of entanglement to tackle measurement uncertainty problem and reduces the inherent quantum noise to some
extent that is not allowed by the classical theory. This is because quantum
entanglement can lead to higher correlation than what classical physics can

have, as demonstrated in the violation of the classical Bell's inequalities and Cauchy-Schwarz inequalities. Then the magic of quantum interference kicks in. It plays a crucial role in canceling out the correlated quantum noise and gives rise to noise reduction. This even works for the reduction of the shot noise in photocurrents mentioned above.

In this chapter, as an application of the homodyne measurement techniques discussed in the previous chapter, we will find how to reduce quantum noise in the measurement of physical quantities of continuous nature. The main focuses are the quantum noise and its reduction in optical interferometers and in quantum amplification. For an application in quantum measurement, we will find how to use homodyne technique to completely characterize a quantum state of a single-mode optical field by quantum state tomography and then transport it by quantum state teleportation.

10.1   Squeezing and Correlation of Quantum Noise

10.1.1   *Quantum Noise in Intensity*

From the discussion of photo-detection theory in Chapter 5 and direct calculation of the photo-current in Chapter 9, we learned that the photocurrent produced in photo-detection process is directly related to intensity of an optical field. As a matter of fact, it can be shown in [Ou and Kimble (1995)] that the process of photo-detection is a quantum measurement process for the photocurrent operator:

$$\hat{i}(t) = \eta \int d\tau k(t - \tau)\hat{E}^\dagger(\tau)\hat{E}(\tau), \tag{10.1}$$

with η as the quantum efficiency of the detector and $k(\tau)$ as the response function of the detector and

$$\hat{E}(\tau) = \frac{1}{\sqrt{2\pi}} \int d\omega \hat{a}(\omega)e^{-j\omega\tau} \tag{10.2}$$

for a one-dimensional quasi-monochromatic field. So, the photocurrent is directly related to the intensity $\hat{I}(t) = \hat{E}^\dagger(t)\hat{E}(t)$ of the field. From Eqs. (9.14) and (9.17), we find the fluctuation of the photocurrent is

$$\langle \Delta i(t)\Delta i(t + \tau)\rangle = \eta I_0 \int dt' k(t')k(t' + \tau)$$

$$+ \eta^2 I_0^2 \int dt' dt'' k(t - t')k(t + \tau - t'')$$

$$\times [g^{(2)}(t' - t'') - 1], \tag{10.3}$$

which is related to intensity correlation function

$$g^{(2)}(t' - t'') = \langle : \hat{I}(t')\hat{I}(t'') : \rangle / I_0^2, \tag{10.4}$$

with $\hat{I}(t) = \hat{E}^{\dagger}(t)\hat{E}(t)$.

From the discussion in Section 9.2, the first term in Eq. (10.3) is the shot noise contribution which exists for any state of the detected field and only depends on the average intensity. The second term, on the other hand, depends on the state of the field and is related to the intensity fluctuations of the field. It can become negative and leads to a photocurrent noise below the shot noise level. For this to happen, we need $g^{(2)}(t' - t'') < 1$, and anti-bunched light satisfies this.

Intensity noise squeezing (first called amplitude squeezing) was among the first observed effects of quantum noise reduction [Machida et al. (1987); Machida and Yamamoto (1988)]. It is closely related to photon number squeezing or sub-Poisson light [Teich and Saleh (1988)] and is a special case of quadrature-phase amplitude squeezing through the self-homodyne detection discussed in Section 9.8. We presented an example in Problem 9.2.

10.1.2 *Quantum Noise of Quadrature-phase Amplitudes and its Reduction: Squeezed States*

We introduced the squeezed state first for a single-mode field in Section 3.4 and then for a multi-frequency mode field in Section 4.6.3 and a multi-temporal mode field in Section 6.1.5. Experimentally, since homodyne detection measures directly the quadrature-phase amplitude $X(t)$ of an optical field, it is used to measure the quantum noise reduction of squeezed state. Slusher *et al.* first observed a noise level at about 0.3 dB below the shot noise level with a squeezed state generated in a near degenerate four-wave mixing process in atomic sodium [Slusher et al. (1985)]. Soon after, Wu *et al.* obtained a relatively large amount of squeezing from an optical parametric oscillator (OPO) below threshold [Wu et al. (1986)], which has since become a common source of squeezed state of light for a variety of applications. Nowadays, more than 10 dB of squeezing was routinely achieved in the lab [Vahlbruch et al. (2016)] with more or less the same OPO scheme as the original [Wu et al. (1986)].

The schematic of a typical set up for generating the squeezed state of light is shown in Fig. 10.1. The light source is usually a frequency-stabilized laser operating at a single frequency of ω_0. Then it is frequency-doubled to $2\omega_0$ by either intra-laser or external cavity enhancement

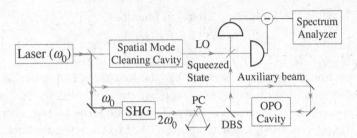

Fig. 10.1   Experimental set-up for generating squeezed states of light. SHG: second harmonic generation; PC: pump mode matching cavity; DBS: dichroic beam splitter; LO: local oscillator.

technique (Section 6.3.2). With the availability of the second harmonic field at $2\omega_0$, we use it to pump the optical parametric oscillator (OPO). The OPO consists of an optical cavity with an efficient nonlinear crystal inside. The cavity is transparent to the pump beam so that the pump passes the nonlinear crystal inside the OPO only once.[1] The auxiliary beam from the original laser serves multiple purposes. The first is to match the spatial mode of the pump to that of the OPO cavity. Since the pump beam is not resonant to the OPO cavity, we have to use the pump mode matching cavity (PC) to accomplish the mode match indirectly. To match PC to the OPO cavity, we use the OPO reversely for generating a beam at $2\omega_0$ from the auxiliary beam. This beam has the same spatial mode as the OPO cavity and can be mode matched to PC. The second purpose is to match the OPO cavity with the spatial mode of the LO field for homodyne detection. Transmitted beam through the OPO cavity by the auxiliary beam will serve this purpose. The third is to use it for locking the OPO cavity to ω_0. This is achieved by shifting the frequency of the locking auxiliary beam to one nearest higher spatial mode of the cavity at an off-set frequency (Eq. (2.61) of Section 2.2.1) and locking the shifted auxiliary beam to the higher spatial mode. Since the higher spatial mode has a large frequency shift (typically 100 MHz) from ω_0, it will not be detected by homodyne measurement.

---

[1]Some designs use double resonance scheme, which has the OPO cavity locked to the pump beam at $2\omega_0$ and then tunes the temperature of the crystal to have the cavity also resonant to ω_0 [Wu et al. (1986); Vahlbruch et al. (2016)]. In this case, mode matching the pump to OPO cavity is straightforward. This design is usually used when the pump power is relatively low but double resonance is also less stable than the single resonance design.

To achieve a good squeezing, we must choose a proper transmissivity T for the output coupler. If the intra-cavity loss (mostly due to crystal absorption and surface reflection) is L, then from Eq. (6.141) in Section 6.3.4, the best squeezing out of the OPO is about $L/(T+L)$ or $10\log[L/(T+L)]$ in dB scale of the vacuum noise or shot noise level. So, we wish to have a large T but this will increase the threshold of the OPO and reduce the amount of squeezing according to Fig. 6.8 in Section 6.3.4. Thus a trade-off must be made depending on the available pump power.

The detection of the squeezed state is done with the balanced homodyne detection scheme and the output photocurrent is analyzed by an electronic spectrum analyzer to obtain a noise level. The local oscillator (LO) for the homodyne detection is from the original laser at ω_0 after the spatial mode cleaning.

A typical result is shown in Fig. 10.2 where Fig. 10.2(a) shows the detected photocurrent noise level in log-scale as the phase of the LO is scanned. The phase insensitive trace Ψ_{01} is the vacuum noise level obtained by simply blocking the light from the OPO. In this diagram, the center frequency of the spectral analyzer is set at a frequency that gives rise to the most squeezing (typically 1-2 MHz) with the frequency span set to zero. In Fig. 10.2(b), we plot the maximum noise Ψ_+ as well as the minimum noise Ψ_- obtained from Fig. 10.2(a) as a function of the quantum noise gain G_q, which is obtained experimentally by averaging the maximum and the minimum of Fig. 10.2(a): $G_q = (\Psi_+ + \Psi_-)/2$. The theory of the OPO is covered in Sections 6.3.4 and 9.5. Theoretically, $G_q = [(G_D + g_D)^2 + (G_L + g_L)^2 + (G_D - g_D)^2 + (G_L - g_L)^2]/2 = G_D^2 + g_D^2 + G_L^2 + g_L^2$, where G_D, g_D, G_L, g_L are from Eq. (6.137) of Section 6.3.4. The solid line in

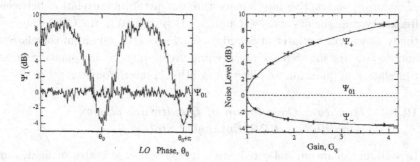

Fig. 10.2   Observed noise level for the photo-current from homodyne detection. (a) Noise level Ψ_I as the phase of the LO is scanned. Ψ_{01} is the shot noise level. (b) Noise as a function of the gain of OPO. Reproduced from [Ou et al. (1992a)].

Fig. 10.2(b) is a theoretical curve of $\Psi_\pm = (G_D \pm g_D)^2 + (G_L \pm g_L)^2$ versus G_q. The good agreement between the theory and experiment supports our understanding of the physics in the process.

In the theory of Sections 4.6.3, 6.3.4, and 9.5, the calculations are straightforward but seem to lack physical intuition about how the noise reduction is achieved. To find this, let us look a little further into Eq. (4.75), whose correlation function leads to the spectrum of squeezing in Eqs. (4.76), (9.45) and (9.46). We rewrite it as

$$\hat{Y}(\Omega) = [\hat{a}(\omega_0 + \Omega) - \hat{a}^\dagger(\omega_0 - \Omega)]/i. \tag{10.5}$$

Here we are only interested in $\hat{Y}(\Omega)$ since it gives noise reduction in $S_Y(\Omega)$.

Let us rewrite the input-output relation obtained in Eq. (6.135) of Section 6.3.4 for degenerate OPO below threshold:

$$\hat{a}_{out}(\omega_0 + \Omega) = G_D(\Omega)\hat{a}_{in}(\omega_0 + \Omega) + g_D(\Omega)\hat{a}_{in}^\dagger(\omega_0 - \Omega)$$
$$\hat{a}_{out}^\dagger(\omega_0 - \Omega) = G_D(\Omega)\hat{a}_{in}^\dagger(\omega_0 - \Omega) + g_D(\Omega)\hat{a}_{in}(\omega_0 + \Omega), \tag{10.6}$$

where we used $G_D^*(-\Omega) = G_D(\Omega), g_D^*(-\Omega) = g_D(\Omega)$. From Eq. (6.136) of Section 6.3.4 for G_D, g_D, we find that when the threshold is approached, G_D, g_D become very large and $G_D(\Omega) \approx g_D(\Omega)$. Then from Eq. (10.6), we find $\hat{a}_{out}(\omega_0 + \Omega)$ and $\hat{a}_{out}^\dagger(\omega_0 - \Omega)$ become almost the same when this happens. Since Eq. (10.6) is an equation for quantum operators, $\hat{a}_{out}(\omega_0 + \Omega)$ and $\hat{a}_{out}^\dagger(\omega_0 - \Omega)$ are highly correlated in their quantum fluctuations.

Now come back to Eq. (10.5) for quantity $\hat{Y}(\Omega)$, which is the difference between two highly correlated quantities $\hat{a}_{out}(\omega_0 + \Omega)$ and $\hat{a}_{out}^\dagger(\omega_0 - \Omega)$. Then quantum fluctuations of $\hat{a}_{out}(\omega_0 + \Omega)$ and $\hat{a}_{out}^\dagger(\omega_0 - \Omega)$ are canceled in $\hat{Y}(\Omega)$. So, the quantum noise reduction in $\hat{Y}(\Omega)$ is a consequence of quantum destructive interference through quantum correlation between frequency components of $\omega_0 + \Omega$ and $\omega_0 - \Omega$ generated in the OPO. Interestingly as we will see later in Sections 10.3.4 and 11.3, quantum correlation and destructive interference are the underlying physics for quantum noise cancelation in quantum amplifiers and SU(1,1) interferometer.

10.1.3   Quantum Correlation of Quadrature-phase Amplitudes: EPR Entangled States

In studying quantum noise reduction in the squeezed states of light, our concern is the quantum fluctuations in the quadrature-phase amplitude of one single-mode field. For measuring this, we employ one set of homodyne detection device, and as we have just showed in the previous section, two correlated frequency components work together for destructive quantum interference between the two correlated amplitudes of the two frequency components. Then next question naturally arises: can we directly observe this type of quantum correlation between the two modes of the field? For this, we need two sets of homodyne detection devices to measure separately the quadrature-phase amplitudes of the two modes and compare the results.

Historically, the first observation of quantum correlation between two continuous variables of quadrature-phase amplitudes was done in the context of demonstration of Einstein-Poldosky-Rosen paradox, leading to the demonstration of non-locality of quantum mechanics. In 1935, Einstein, Podolsky and Rosen (EPR) [Einstein et al. (1935)] proposed a gedanken experiment involving a system of two particles in the following wavefunction:

$$\psi(x_1, x_2) = C\delta(x_1 - x_2 + x_0), \tag{10.7}$$

which can also be written in the momentum space as

$$\phi(p_1, p_2) = C'\delta(p_1 + p_2). \tag{10.8}$$

Here C, C' are normalization constants.[2] So the two particles are spatially separated by x_0 but perfectly correlated in both positions and momenta. From their view of local realism, EPR concluded that quantum mechanics is incomplete since canonically conjugate variables of position and momentum for one of the particles could be assigned some definite values without disturbing it through the perfect correlations from the measurements of the other particle, in apparent conflict with the Heisenberg uncertainty principle. Later, J. S. Bell extended the wavefunctions in Eqs. (10.7) and (10.8) to the Wigner function in $x - p$ phase space [Bell (1987)]:

$$W(x_1, x_2; p_1, p_2) = \iint dy_1 dy_2 e^{-i(p_1 y_1 + p_2 y_2)} \psi\left(x_1 + \frac{y_1}{2}, x_2 + \frac{y_2}{2}\right)$$
$$\times \psi^*\left(x_1 - \frac{y_1}{2}, x_2 - \frac{y_2}{2}\right)$$
$$= C''\delta(x_1 - x_2 + x_0)\delta(p_1 + p_2). \tag{10.9}$$

This demonstrates more directly the perfect correlations between positions and momenta than the two wavefunctions in Eqs. (10.7) and (10.8).

In 1989, Reid showed that the outputs of a nondegenerate optical parametric amplifier (NOPA) possess the same correlation properties of the EPR state except that the two particles are replaced by the virtual harmonic oscillators of two output optical modes of the NOPA and the positions and momenta of the particles by the quadrature-phase amplitudes of the

---

[2]The δ-function in the wavefunction makes it un-normalizable so we use $C, C'(\sim 0)$.

optical modes [Reid (1989)]. Consider the two-mode squeezed state discussed in Section 4.6. The quadrature-phase amplitudes $\hat{X}_A, \hat{Y}_A$ and $\hat{X}_B, \hat{Y}_B$ of the two NOPA output modes A, B are given in Section 4.6.2 and are related to the input modes a, b by

$$\hat{X}_A - \hat{X}_B = (\hat{X}_a - \hat{X}_b)/(G+g), \quad \hat{Y}_A + \hat{Y}_B = (\hat{Y}_a + \hat{Y}_b)/(G+g). \quad (10.10)$$

Here, g is taken to be positive for the simplicity of argument. Problem 10.3 will deal with the case of arbitrary phase for g. So, when the gain G, g become very large, both $\hat{X}_A - \hat{X}_B$ and $\hat{Y}_A + \hat{Y}_B$ go to zero or $\hat{X}_A = \hat{X}_B$ and $\hat{Y}_A = -\hat{Y}_B$. Since these are operator identities, if we make measurement of $\hat{X}_A, \hat{X}_B$ or $\hat{Y}_A, -\hat{Y}_B$, their values will be perfectly correlated or anti-correlated. From Chapter 3, we know that the quadrature-phase amplitude $\hat{X}, \hat{Y}$ are proportional to the position and momentum operators of the harmonic oscillator describing a single-mode of optical field and they are conjugate to each other satisfying commutation relation $[\hat{X}, \hat{Y}] = i$. Therefore, the two output modes of NOPA exhibit the same type of quantum correlation as the two particles described by EPR.

To compare with Eq. (10.9), we can find the Wigner function for the two output modes of NOPA. If the input states to the NOPA are the vacuum states, i.e., the input Wigner function is [Eq. (3.173) in Section 3.7.2]

$$W_{a,b}(x_1, y_1; x_2, y_2) = \left(\frac{1}{2\pi}\right)^2 \exp\left[-\frac{x_1^2 + y_1^2}{2}\right] \exp\left[-\frac{x_2^2 + y_2^2}{2}\right], \quad (10.11)$$

we may derive the output Wigner function from the transformation given in Eq. (10.10) as (or see alternative method in Problem 10.1)

$$\begin{aligned} W_{A,B}(x_1, y_1; x_2, y_2) &= \frac{1}{4\pi^2} \exp\left\{-\frac{1}{4}\left[(x_1 + x_2)^2 + (y_1 - y_2)^2\right]e^{-r}\right. \\ &\quad \left. -\frac{1}{4}\left[(x_1 - x_2)^2 + (y_1 + y_2)^2\right]e^r\right\} \\ &\to C'''\delta(x_1 - x_2)\delta(y_1 + y_2) \quad \text{as } r \to \infty. \quad (10.12) \end{aligned}$$

Here r is related to the amplitude gain by $G = \cosh r$ and C''' is some normalization constant.

So at infinite gain the system can mimic the EPR correlated two-particle system. But we know that infinte gain is impractical, then what happens for the finite gain? Can we still demonstrate the EPR paradox?

At finite gain, the correlation will not be perfect. From Eq. (10.10), we find for a, b in vacuum states

$$\langle \Delta^2(\hat{X}_A - \hat{X}_B)\rangle = 2/(G+g)^2, \quad \langle \Delta^2(\hat{Y}_A + \hat{Y}_B)\rangle = 2/(G+g)^2. \quad (10.13)$$

Now, we can still make the same argument as EPR did, that is, using X_B to infer a value for X_A or $-Y_B$ for Y_A, but inferences are not definite but with errors of $\Delta^2_{inf} X_A \equiv \langle \Delta^2(\hat{X}_A - \hat{X}_B) \rangle = 2/(G+g)^2$ and $\Delta^2_{inf} Y_A \equiv \langle \Delta^2(\hat{Y}_A + \hat{Y}_B) \rangle = 2/(G+g)^2$. As long as $\Delta^2_{inf} X_A \Delta^2_{inf} Y_A = 4/(G+g)^4 < 1$, the inferred values violate the Heisenberg uncertainty relation and the EPR paradox still stands.

The condition $4/(G+g)^4 < 1$ corresponds to a noise squeezing of $1/(G+g)^2 < 1/2$ or 3 dB of noise squeezing. However, Reid suggested a better way of inference of X_A and Y_A by using quantities λX_B or $-\lambda Y_B$, respectively [Reid (1989)]. Then the inference error is given by

$$\Delta^2_{inf} X_A(\lambda) \equiv \langle \Delta^2(\hat{X}_A - \lambda \hat{X}_B) \rangle = (G^2 + g^2)(1 + \lambda^2) - 4\lambda Gg, \quad (10.14)$$

which reaches a minimum value of $1/(G^2 + g^2)$ when $\lambda = 2Gg/(G^2 + g^2)$. Similarly, the inference error $\Delta^2_{inf} Y_A(\lambda) \equiv \langle \Delta^2(\hat{Y}_A + \lambda \hat{Y}_B) \rangle$ has a minimum value of $1/(G^2 + g^2)$ for the same λ value. So, the condition for an EPR paradox is $G^2 + g^2 > 1$ or $g^2 > 0$.

In practice, each field consists of multiple frequency modes. Since spatial modes out of the cavity are well-defined, we can use one-dimensional description with quasi-monochromatic approximation for each field, in a way similar to the multi-frequency mode squeezed state in Section 4.6.3 but for two fields now. The correlation is between quadrature-phase amplitudes defined in frequency domains of the two fields, as in Eq. (4.74). We can proceed with the same argument above.

The first experimental demonstration of EPR paradox in continuous variables was carried out by Ou et al. [Ou et al. (1992b)]. The experimental layout is shown in Fig. 10.3. The EPR correlated fields are generated from a non-degenerate OPO below threshold. This device is theoretically described as a non-degenerate parametric amplifier by Eq. (6.128) in Section 6.3.4. The non-degeneracy of the OPO is achieved via a type-II $\chi^{(2)}$-nonlinear process where the two correlated fields have the same frequency but orthogonal polarizations. So, the two fields can be separated by a polarization beam splitter (P in Fig. 10.3) and directed to two separate homodyne detection schemes for the measurements of $X_A(\theta_1), X_B(\theta_2)$, respectively. θ_1, θ_2 are determined by the phases of LO_1 and LO_2, respectively. The λ parameter from Eq. (10.14) is optimized by controlling the electronic gain g in Fig. 10.3. The experimental results are shown in Fig. 10.4, where measurements of the inferred errors of $\Delta^2_{inf} X$ and $\Delta^2_{inf} Y$ are shown in a log-scale. Both values are below the value of 1, the limit set by the Heisenberg uncertainty relation, thus demonstrating the EPR paradox.

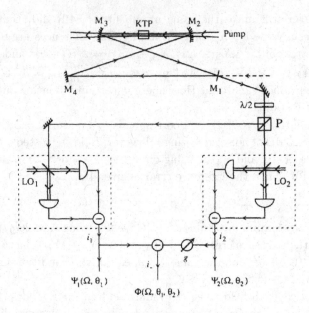

Fig. 10.3 Experimental schematic for demonstrating EPR paradox with a non-degenerate OPO. Reproduced from [Ou et al. (1992b)].

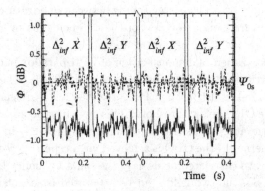

Fig. 10.4 Experimental results that demonstrate EPR paradox for continuous variables. Ψ_{0s} is the vacuum noise for each beam. Reproduced from [Ou et al. (1992b)].

The EPR correlated fields also provide an example of two quantum entangled fields, which cannot be expressed in the form of a separable state:

$$\hat{\rho}_{AB}^{sep} = \sum_j p_j \hat{\rho}_{Aj} \otimes \hat{\rho}_{Bj}. \tag{10.15}$$

Duan *et al.* and Simon each derived independently a necessary and sufficient condition for inseparability of a quantum state [Duan et al. (2000); Simon (2000)], which, when applied to the case here, has the form of

$$\Delta_{inf}^2 X + \Delta_{inf}^2 Y < 2(1 + \lambda^2). \tag{10.16}$$

The results in [Ou et al. (1992b)] give $(\Delta_{inf}^2 X + \Delta_{inf}^2 Y)/2(1 + \lambda^2) = -2.8$ dB ≈ 0.52, clearly satisfying the inseparability condition in Eq. (10.16) and demonstrating the quantum entanglement between the two fields from the non-degenerate OPO.

10.1.4   Quantum Correlation between Intensities: Twin Beams

In Section 4.6.1, we find that there exists a perfect photon number correlation between the two modes in the two-mode squeezed states which, from what we learned in Section 6.3.4, can be generated from a parametric amplifier with no input (spontaneous mode). This perfect photon number correlation persists even at large gain for the parametric amplifier. In this section, we will explore more of this type of correlation.

Historically, nonclassical intensity correlations were first demonstrated by Heidmann *et al.* with an optical parametric oscillator (OPO) above threshold [Heidmann et al. (1987)]. They measured the noise of the intensity difference between the signal and idler fields of the OPO and observed a noise level 30% below the shot noise level of the two intensities, indicating a quantum correlation in the intensity fluctuations of the two fields.

For the twin beams out of a parametric amplifier, the early experiments usually did not have enough gain so that the output intensities are too low in the spontaneous mode to produce a significant photocurrent to overcome the dark currents of the photo-detectors for a direct confirmation of the perfect photon number correlation exhibited in Section 4.6.1. So, in order to boost the intensity, Aytür and Kumar injected a coherent state as input to the parametric amplifier [Aytur and Kumar (1990)] and obtained two bright outputs from the amplifier that have the similar photon number correlation as the twin beams discussed in Section 4.6.1.

Consider an optical parametric amplifier which has the same input-output relation as Eq. (4.58) in Section 4.6 and has a coherent state $|\alpha\rangle$ input at mode a:

$$\hat{A} = \hat{S}_{ab}^\dagger \hat{a} \hat{S}_{ab} = G\hat{a} + g\hat{b}^\dagger, \quad \hat{B} = G\hat{b} + g\hat{a}^\dagger. \tag{10.17}$$

For simplicity of argument, we assume $g > 0$. It is straightforward to show that, with $G^2 - g^2 = 1$, $\hat{A}^\dagger \hat{A} - \hat{B}^\dagger \hat{B} = \hat{a}^\dagger \hat{a} - \hat{b}^\dagger \hat{b}$. Then the photon number difference of the two outputs has a fluctuation of

$$\langle \Delta^2(\hat{N}_A - \hat{N}_B) \rangle = \langle \Delta^2(\hat{N}_a - \hat{N}_b) \rangle_{|\alpha\rangle} = |\alpha|^2. \tag{10.18}$$

Now let us compare this noise level with that of two coherent fields whose photon numbers are at the levels of the two outputs or the shot noise level. This is a fair comparison because two fields in coherent states with noise levels at shot noise level are completely uncorrelated quantum mechanically. In fact, it can be shown (see Problem 10.2) that for two classical fields, we always have $\langle \Delta^2(\hat{I}_A - \hat{I}_B) \rangle_{cl} \geq \langle \Delta^2 \hat{I}_A \rangle_{cs} + \langle \Delta^2 \hat{I}_B \rangle_{cs} = \langle \hat{I}_A \rangle + \langle \hat{I}_B \rangle$. Here, the subscript "$cs$" denotes coherent states.

The two output fields from the parametric amplifier with a coherent state input have photon numbers of $\langle \hat{N}_A \rangle = G^2|\alpha|^2 + g^2 \approx G^2|\alpha|^2$, $\langle \hat{N}_B \rangle = g^2|\alpha|^2 + g^2 \approx g^2|\alpha|^2$, respectively, for $|\alpha|^2 \gg 1$. So, the shot noise level for the number difference of the two fields at the intensity levels of the two outputs is simply

$$\langle \Delta^2(\hat{N}_A - \hat{N}_B) \rangle_{sn} = \langle \Delta^2 \hat{N}_A \rangle_{sn} + \langle \Delta^2 \hat{N}_B \rangle_{sn}$$
$$= \langle \hat{N}_A \rangle + \langle \hat{N}_B \rangle = (G^2 + g^2)|\alpha|^2. \tag{10.19}$$

Note here that the shot noise level is obtained by taking the number statistics as the Poissonian distribution: $\langle \Delta^2 \hat{N}_{A,B} \rangle_{sn} = \langle \hat{N}_{A,B} \rangle$. Hence, compared to the shot noise level, the noise level of the photon number difference of the two output fields has a reduction factor of

$$R \equiv \frac{\langle \Delta^2(\hat{N}_A - \hat{N}_B) \rangle}{\langle \Delta^2(\hat{N}_A - \hat{N}_B) \rangle_{sn}} = \frac{1}{G^2 + g^2}. \tag{10.20}$$

When the gain is infinity, we achieve 100% reduction. This is somewhat similar to the EPR correlation discussed earlier although $\langle \Delta^2(\hat{N}_A - \hat{N}_B) \rangle = |\alpha|^2 \neq 0$. Here we consider relative noise level with reference to the sum of the shot noise levels of the two beams.

At finite gain, the noise reduction is less than 100%. Similar to the calculation of the inference error $\Delta^2_{inf} X_A$, $\Delta^2_{inf} Y_A$ earlier, we should consider $\hat{N}_A - \lambda \hat{N}_B$, whose noise variance, under the approximation of $|\alpha|^2 \gg 1$, is calculated as

$$\langle \Delta^2(\hat{N}_A - \lambda \hat{N}_B) \rangle = [(G^2 + g^2)(G^2 + g^2\lambda^2) - 4\lambda G^2 g^2]|\alpha|^2. \tag{10.21}$$

The corresponding shot noise level is

$$\langle \Delta^2(\hat{N}_A - \lambda \hat{N}_B) \rangle_{sn} = \langle \Delta^2 \hat{N}_A \rangle_{sn} + \lambda^2 \langle \Delta^2 \hat{N}_B \rangle_{sn}$$
$$= \langle \hat{N}_A \rangle + \lambda^2 \langle \hat{N}_B \rangle = (G^2 + \lambda^2 g^2)|\alpha|^2. \tag{10.22}$$

So, the noise reduction factor is

$$\frac{\langle \Delta^2(\hat{N}_A - \lambda \hat{N}_B) \rangle}{\langle \Delta^2(\hat{N}_A - \lambda \hat{N}_B) \rangle_{sn}} = \frac{(G^2 + g^2)(G^2 + g^2\lambda^2) - 4\lambda G^2 g^2}{G^2 + \lambda^2 g^2}. \quad (10.23)$$

When $\lambda = G/g$, the noise reduction factor reaches a minimum value of

$$\frac{\langle \Delta^2(\hat{N}_A - \lambda \hat{N}_B) \rangle}{\langle \Delta^2(\hat{N}_A - \lambda \hat{N}_B) \rangle_{sn}} = (G - g)^2 = \frac{1}{(G + g)^2}, \quad (10.24)$$

which is about 3 dB better than the value in Eq. (10.20) at large gain.

Experimentally, because the scheme by Aytür and Kumar is relatively easy to implement, it has become a standard technique for testing the twin beams property for a number of parametric amplifiers achieved in different nonlinear media including the regular $\chi^{(2)}$ materials [Aytur and Kumar (1990)], four-wave mixing in atomic vapor [McCormick et al. (2007)], and four-wave mixing in optical fiber [Guo et al. (2012)]. The typical experimental arrangement is shown in Fig. 10.5, where a coherent signal is injected in the input port of an optical parametric amplifier (OPA). In addition to the amplified signal, an idler beam is generated conjugate to the signal beam. Both the signal and the idler beams are respectively directed to two identical detectors where the difference of the photo-currents are measured with an electronic spectrum analyzer. With the help of a removable mirror (M) and a beam splitter (BS), an auxiliary coherent beam is used to measure the shot noise level equivalent to the total shot noise level produced by the two twin beams together. To ensure they are equivalent, the DC output levels of the two detectors are used as references and are matched between the twin beams and the auxiliary beams. The results are shown in Fig. 10.6 where trace (i) is the equivalent total shot noise level from the two twin beams and trace (ii) is the noise level for the intensity difference of the twin beams, showing more than 5 dB noise reduction. Trace (iii) is the electronic noise when no light is on.

It is interesting to note that intensity correlations discussed above is a special case of the EPR correlation discussed in the previous section. As we

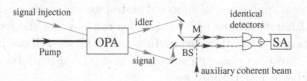

Fig. 10.5 Experimental arrangement for the observation of intensity correlations between two fields of twin beams. M: mirror; BS: beam splitter; SA: spectrum analyzer.

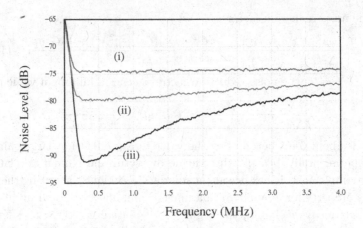

Fig. 10.6   Spectrum of the intensity difference. (i) Total shot noise level for the two beams together; (ii) Noise level for the intensity difference of the twin beams; (iii) Electronic noise. Courtesy of Jun Jia.

discussed in Section 9.8, measurement of intensity fluctuations is basically self-homodyne measurement of the X-quadrature-phase amplitude for the fields with a strong real coherent component.[3] Therefore, the variance of the intensity difference is equivalent to the variance of $\hat{X}_A - \hat{X}_B$. Thus it is not surprising to see noise reduction in the intensity difference. But measurement of Y corresponds to the phase measurement which we did not measure in the study of twin beams. So, we cannot demonstrate the EPR paradox or EPR entanglement through the inseparability criterion of Eq. (10.16) here.

10.2   Quantum Noise and Its Reduction in Linear Interferometers

As a practical application of squeezed states, we will study quantum noise in a linear interferometer and see how we can improve the sensitivity in phase measurement in this device. In Chapter 11, we will discuss in more details the precision phase measurement in a general platform that is not limited to linear interferometers.

---

[3]For a complex coherent component, refer to Problem 10.3 (Eq. (10.110)).

10.2.1   Quantum Noise Analysis for a Mach-Zehnder Interferometer

The quantum noise performance of a linear interferometer was first analyzed by Carl Caves in a seminal paper that studies the sensitivity of using a Michelson interferometer as a tool to detect gravitational waves [Caves (1981)]. In that study, he considered the effect of radiation pressure on suspended mirrors. But for simplicity, we will fix the mirrors at the moment so as to ignore the effect of radiation pressure and only concentrate on the effect of quantum noise of light on the phase measurement sensitivity.

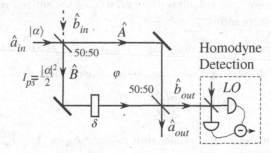

Fig. 10.7   Layout of a Mach-Zehnder interferometer for the measurement of a small phase shift δ with a homodyne detection. I_{ps} is the phase sensing field photon number.

Consider a Mach-Zehnder interferometer depicted in Fig. 10.7. Assume the two beam splitters are 50:50 and lossless. They are described by the input-output relation

$$\hat{A} = (\hat{a}_{in} + \hat{b}_{in})/\sqrt{2}, \quad \hat{B} = (\hat{b}_{in} - \hat{a}_{in})/\sqrt{2};$$
$$\hat{a}_{out} = (\hat{A} - \hat{B}e^{i\varphi})/\sqrt{2}, \quad \hat{b}_{out} = (\hat{B}e^{i\varphi} + \hat{A})/\sqrt{2}, \quad (10.25)$$

where we place the overall phase difference φ of the interferometer on the B arm only. It is straightforward to express the outputs of the interferometer in terms of the inputs:

$$\hat{a}_{out} = t(\varphi)\hat{a}_{in} + r(\varphi)\hat{b}_{in}; \quad \hat{b}_{out} = t(\varphi)\hat{b}_{in} + r(\varphi)\hat{a}_{in}, \quad (10.26)$$

where $t(\varphi) = e^{i\varphi/2}\cos\varphi/2$; $r(\varphi) = -ie^{i\varphi/2}\sin\varphi/2$. So, the interferometer acts as another beam splitter with phase sensitive amplitude transmissivity $t(\varphi)$ and reflectivity $r(\varphi)$.

For a coherent state $|\alpha\rangle$ input at $\hat{a}_{in}$ and vacuum at $\hat{b}_{in}$, the output photon number[4] at output port b of the interferometer is simply

$$I_b = \langle \hat{b}_{out}^\dagger \hat{b}_{out}\rangle = |\alpha|^2(1 - \cos\varphi)/2 = I_{ps}(1 - \cos\varphi), \quad (10.27)$$

---

[4]For single-mode description of the field, the intensity of the field is different from the photon number by a constant. So, we will use intensity symbol for photon number.

where $I_{ps} \equiv \langle \hat{B}^\dagger \hat{B} \rangle = |\alpha|^2/2$ is the photon number of the field subject to the phase shift (phase sensing field). As we will prove later in Chapter 11 that the phase sensing field is what matters in phase measurement accuracy.

From Eq. (10.27), a small change of δ in φ will lead to a change in the output photon number:

$$\delta I_b = I_{ps} \delta \sin \varphi. \tag{10.28}$$

This is the signal due to the phase change δ. The signal size due to the phase change is maximum at $\varphi = \pi/2$: $\delta I_b^M = I_{ps} \delta$. For the noise performance of the interferometer, we calculate the variance of the output at $\varphi = \pi/2$ for the coherent state input:

$$\Delta^2 I_b = \langle \hat{b}_{out}^\dagger \hat{b}_{out} \hat{b}_{out}^\dagger \hat{b}_{out} \rangle_{\pi/2} - \langle \hat{b}_{out}^\dagger \hat{b}_{out} \rangle_{\pi/2}^2 = |\alpha|^2/2 = I_{ps}. \tag{10.29}$$

So, the signal-to-noise ratio of the phase measurement is

$$SNR_\delta = \frac{(\delta I_b)^2}{\Delta^2 I_b} = \frac{(I_{ps}\delta)^2}{I_{ps}} = I_{ps}\delta^2. \tag{10.30}$$

The minimum detectable phase shift is then $\delta_m = 1/\sqrt{I_{ps}}$. This is the so-called standard quantum limit (SQL) of phase measurement, which has a $1/\sqrt{N_{ps}}$-dependence on the phase sensing photon number $N_{ps} = I_{ps}$.

From the derivation above, it is hard to see where the noise in Eq. (10.29) comes from. It seems to come from the photon number fluctuations in the input coherent state. It turns out that Eq. (10.29) stands even for the number state ($|N\rangle$) input with no photon number fluctuation. The noise stems from the interferometer itself: at $\varphi = \pi/2$, the interferometer is basically a 50:50 beam splitter with $|t(\pi/2)| = |r(\pi/2)| = 1/\sqrt{2}$ and this leads to randomness in one of the output even with no randomness at the input. Another way to think about it is that the culprit of the noise is the vacuum input of the unused port as we will soon see in the following.

From Eq. (10.30), we find that the larger I_{ps} is, the better the sensitivity is. But the average detected photon number at the output is I_{ps}, which also becomes large. Sooner or later, it will saturate the detector and it is thus not practical to work at $\varphi = \pi/2$. The way to circumvent this problem is to make the interferometer work at $\varphi = 0$ or at the dark fringe. In this case, Eq. (10.28) does not stand and we need to go to higher order and obtain

$$\delta I_b = I_{ps}\delta^2/2. \tag{10.31}$$

Since the interferometer works at dark fringes now, Eq. (10.29) will lead to zero noise and an infinite SNR. This is not true. At no phase shift, the output is dark, but with a small shift, the output becomes what Eq. (10.31) gives. On the other hand, the detector cannot detect anything less than one photon. So, the minimum detectable phase shift is $\delta_m = 1/\sqrt{I_{ps}/2}$ when $\delta I_b = 1$. This is similar to SQL.

However, for low light level, the dark currents of the detectors will overwhelm the signal (one photon). So, we cannot use direct detection. Fortunately, we know a detection technique that can overcome the dark current problem. This is the homodyne detection method for measuring the quadrature-phase amplitudes that we discussed in the previous chapter (Section 9.3).

With a small phase shift δ for φ and $\alpha = i|\alpha|$, we can easily find

$$\langle \hat{X}^2_{b_{out}} \rangle = \langle \hat{X}^2_{b_{in}}(\delta/2) \rangle \cos^2(\delta/2) + \langle \hat{Y}^2_{a_{in}}(\delta/2) \rangle \sin^2(\delta/2)$$

$$\approx 1 + |\alpha|^2 \delta^2 \quad \text{for } \delta << 1, \, |\alpha|^2 >> 1. \tag{10.32}$$

Here $\hat{X}_b(\delta/2) = \hat{b}e^{i\delta/2} + \hat{b}^\dagger e^{-i\delta/2}$ and $\hat{Y}_a(\delta/2) = (\hat{a}e^{i\delta/2} - \hat{a}^\dagger e^{-i\delta/2})/i$ are the quadrature-phase amplitudes of corresponding fields. Obviously, $|\alpha|^2 \delta^2$ in $\langle \hat{X}^2_{b_{out}} \rangle$ corresponds to the phase signal while the noise of the phase measurement is simply 1, the vacuum noise, in Eq. (10.32). Hence, the signal-to-noise ratio of the linear interferometer for measuring $\hat{X}$ is

$$SNR_X = |\alpha|^2 \delta^2 / 1 = 2I_{ps}\delta^2, \tag{10.33}$$

which leads to the standard quantum limit $\delta_{SQL} = 1/\sqrt{N}$ with $N = 2I_{ps}$.

A close examination of the origin of the noise term in Eq. (10.32) reveals that it is from the vacuum input at $\hat{b}_{in}$, the unused input port. We can confirm this by noting that at $\varphi = 0$, we have $\hat{b}_{out} = \hat{b}_{in}$, which is in vacuum state.

10.2.2 Sub-shot Noise Interferometry with Squeezed States

Knowing the origin of the noise in the interferometers, we can use quantum noise reduction technique to reduce it. Caves was the first to suggest to inject the squeezed vacuum into the unused port of the interferometer [Caves (1981)]. From Eq. (10.32), we can easily see that with a squeezed state at the unused input port b_{in} instead of vacuum, Eq. (10.32) becomes

$$\langle \hat{X}^2_{b_{out}} \rangle_s = e^{-r} + |\alpha|^2 \delta^2 \quad \text{for } |\alpha|^2 >> 1, r. \tag{10.34}$$

Here we choose the phase of the squeezed state so that $\langle \hat{X}^2_{b_{in}}(\delta/2) \rangle$ is at the minimum value of e^{-r} with a squeezing parameter $r > 0$.

The first demonstration of sub-shot noise interferometry was performed soon after the generation of the squeezed states [Xiao et al. (1987); Grangier et al. (1987)]. We will discuss next only the experiment performed by Xiao *et al.* [Xiao et al. (1987)].

Different from the detection schemes we discussed in the previous section, Xiao *et al.* employed a self-homodyne scheme shown in Fig. 10.8. $P_{1,2}$ are two electro-optic modulators (EOM) with one for adjusting overall phase φ and the other for introducing phase signal δ. The input port $\hat{E}_1$ is in a coherent state $|\alpha\rangle$ but the other input port $\hat{E}_s$ can be in vacuum or a squeezed state from OPO. Both outputs are detected and the photocurrents are subtracted before being analyzed by a spectral analyzer. The combination of the interferometer and detection system is similar to the balanced homedyne detection discussed in Section 9.7 except that the 50:50 beam splitter is replaced by the interferometer or a phase dependent beams splitter, as we found in the previous section. From the input-output relation in Eq. (10.26), we find

$$\hat{I}_- \equiv \hat{a}^\dagger_{out}\hat{a}_{out} - \hat{b}^\dagger_{out}\hat{b}_{out}$$
$$= [|t(\varphi)|^2 - |r(\varphi)|^2](\hat{a}^\dagger_{in}\hat{a}_{in} - \hat{b}^\dagger_{in}\hat{b}_{in})$$
$$+ [t(\varphi)r^*(\varphi) - t^*(\varphi)r(\varphi)](\hat{a}_{in}\hat{b}^\dagger_{in} - \hat{a}^\dagger_{in}\hat{b}_{in}). \qquad (10.35)$$

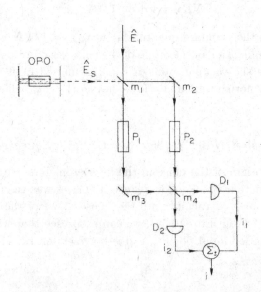

Fig. 10.8   Layout of a Mach-Zehnder interferometer for phase measurement. Reproduced from [Xiao et al. (1987)].

For a small phase change $\delta \ll 1$ and the input port $\hat{a}_{in}$ ($\hat{E}_1$ in Fig. 10.8) in a coherent state ($|\alpha\rangle$), we have the phase signal:

$$\delta I_- = |\alpha|^2 \delta \sin \varphi = 2I_{ps}\delta \sin \varphi. \tag{10.36}$$

So, we have the largest signal when $\varphi = \pi/2$. Notice that the phase signal is twice as large as what Eq. (10.28) gives because we have two detectors here.

For the noise, we find that for $\varphi = \pi/2$, we have $|t(\pi/2)| = |r(\pi/2)| = 1/\sqrt{2}$ and the interferometer is simply a 50:50 beam splitter. Hence, the outcome is the same as a balanced homodyne detection scheme:

$$\langle \Delta^2 \hat{I}_- \rangle = |\alpha|^2 \langle Y_{E_s}^2(\varphi_\alpha) \rangle = 2I_{ps}\langle Y_{E_s}^2(\varphi_\alpha) \rangle \tag{10.37}$$

where $\hat{b}_{in} = \hat{E}_s$ in Fig. 10.8, $\hat{Y}_{E_s}(\varphi_\alpha) = (\hat{b}_{in}e^{-i\varphi_\alpha} - \hat{b}_{in}^\dagger e^{i\varphi_\alpha})/i$ and $e^{i\varphi_\alpha} = \alpha/|\alpha|$. When $\hat{E}_s$ is in vacuum, $\langle Y_{E_s}^2(\varphi_\alpha) \rangle = 1$. So, we obtain $SNR = (\delta I_-)^2/\langle \Delta^2 \hat{I}_- \rangle = 2I_{ps}\delta^2$ and $\delta_m = 1\sqrt{2I_{ps}}$, which is the SQL. Note that in this case, $\langle \Delta^2 \hat{I}_- \rangle_v = |\alpha|^2 = 2I_{ps}$ or the total shot noise from the two detectors. On the other hand, when $\hat{E}_s$ is in a squeezed state and with the right choice of φ_α, we have $\langle \Delta^2 \hat{I}_- \rangle_{sq} = |\alpha|^2 e^{-r} = \langle \Delta^2 \hat{I}_- \rangle_v e^{-r}$ and the noise is reduced below the shot noise level.

Figure 10.9 shows the phase signal level and the noise level with vacuum (Fig. 10.9(a)) or a squeezed state (Fig. 10.9(b)) in the unused port $\hat{E}_s$. Notice that the noise level is about 3 dB below the shot noise level (dashed line at 0 dB) for the squeezed state input and the SNR is improved also about 3 dB, thus achieving sub-shot noise interferometry.

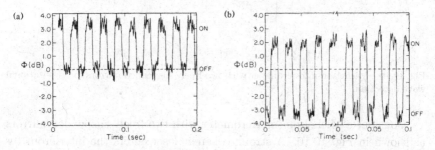

Fig. 10.9 Phase signal level and the noise level for phase measurement with (a) vacuum or (b) a squeezed state in the unused port $\hat{E}_s$. Reproduced from [Xiao et al. (1987)].

10.2.3   *Quantum Noise Analysis for LIGO*

LIGO is an acronym for Light Interferometer Gravitational-wave Obser-
vatory. It employs a long baseline Michelson interferometer to detect a
minute phase change due to space distortion when a gravitational wave
passes by. On September 14, 2015, after over four decades of persistent
endeavor, LIGO detected the first signal generated by the merger of two
black holes [Abbott et al. (2016)]. Since the operation principle of LIGO is
to measure phase shift, it is subject to the same quantum noise discussed
in the previous sections. In addition, there is also the radiation pressure
noise due to randomness of the photon number impinging on the freely suspended mirrors.[5] This noise is also of quantum nature. Carl Caves, in his
seminal paper on quantum noise in interferometers [Caves (1981)], analyzed
the effects of these two sources of noise based on a single-mode model. In
this section, we will make an analysis with a multi-frequency mode model
to calculate the noise spectrum of the interferometer.

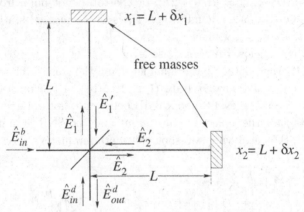

Fig. 10.10   Michelson interferometer with two freely suspended mirrors for gravitational
wave detection.

Consider a Michelson interferometer with two freely suspended mirrors
as shown in Fig. 10.10. A strong coherent laser enters the interferometer
from the bright port $\hat{E}_{in}^b$ together with a vacuum input at the dark port
$\hat{E}_{in}^d$. Assuming that the spatial modes are all matched throughout the
interferometer, we can describe the fields with a one-dimensional quasi-
monochromatic approximation (Section 2.3.5):

---

[5]Free suspension of the mirrors is for seismic isolation consideration.

$$\hat{E}_{in}^d(t) = \frac{1}{\sqrt{2\pi}} e^{-i\omega_0 t} \int d\Omega \hat{a}(\Omega) e^{-i\Omega t},$$

$$\hat{E}_{in}^b(t) = e^{-i\omega_0 t} \left[E_0 + \frac{1}{\sqrt{2\pi}} \int d\Omega \hat{b}(\Omega) e^{-i\Omega t} \right], \tag{10.38}$$

where we treat the strong coherent laser as a c-number E_0 and shift the frequency zero to the center frequency ω_0 of the strong laser. The fields right after the 50:50 beam splitter are

$$\hat{E}_1 = [\hat{E}_{in}^b(t) + \hat{E}_{in}^d(t)]/\sqrt{2}, \quad \hat{E}_2 = [\hat{E}_{in}^d(t) - \hat{E}_{in}^b(t)]/\sqrt{2}. \tag{10.39}$$

After they are bounced back by the mirrors and recombined in the beam splitter, the output fields are

$$\hat{E}_{out}^d = [\hat{E}_1'(t) + \hat{E}_2'(t)]/\sqrt{2}, \quad \hat{E}_{out}^b = [\hat{E}_2'(t) - \hat{E}_1'(t)]/\sqrt{2}, \tag{10.40}$$

where $\hat{E}_j'(t) = \hat{E}_j(t - 2\hat{x}_j(t - L/c)/c)$ with $\hat{x}_j(t - L/c)$ as the position operators of the two freely suspended mirrors. Propagation delays are included in the expression above and L is the length of each arm assuming they are balanced. Substituting Eqs. (10.38) and (10.39) into the above, we obtain

$$\hat{E}_{out}^d = e^{-i\omega_0[t - 2\hat{x}_1(t - L/c)/c]} E_0/2 - e^{-i\omega_0[t - 2\hat{x}_2(t - L/c)/c]} E_0/2 + \Delta\hat{E}(t)$$

$$\approx e^{-i\omega_0(t - 2L/c)}(i\omega_0/c)\Delta\hat{x}(t - L/c)E_0 + \Delta\hat{E}(t), \tag{10.41}$$

where $\Delta\hat{x} \equiv \delta\hat{x}_1 - \delta\hat{x}_2$ with $\delta\hat{x}_j = \hat{x}_j - L(j = 1, 2)$ is the relative displacement for the two mirrors and we made the approximation $e^{-i2\omega_0\delta\hat{x}_j/c} \approx 1 - i2\omega_0\delta\hat{x}_j/c$, and

$$\Delta\hat{E}(t) = \frac{e^{-i\omega_0(t - 2L/c)}}{\sqrt{2\pi}} \int d\Omega \hat{a}(\Omega) e^{-i\Omega(t - 2L/c)}. \tag{10.42}$$

When a gravitational wave passes by, the nature of the wave is such that it stretches the space in one direction and compresses it in the orthogonal direction. So, a disturbance of $h_{GW}(t)/2$ or $-h_{GW}(t)/2$ is added to $\hat{x}_1$ or $\hat{x}_2$, respectively: $\delta\hat{x}_1 = h_{GW}(t)/2 + \hat{x}_1(t) - L, \delta\hat{x}_2 = -h_{GW}(t)/2 + \hat{x}_2(t) - L$. The freely suspended mirrors are illuminated by the light fields and thus their motions are subject to radiation pressure forces:

$$M\frac{d^2\hat{x}_j(t)}{dt^2} = \hat{F}_{ph}^{(j)}(t) \tag{10.43}$$

with

$$\hat{F}_{ph}^{(j)}(t) = \kappa\hat{E}_j^\dagger(t - L/c)\hat{E}_j(t - L/c)$$

$$= \frac{\kappa}{2}\left\{ E_0^* + \frac{1}{\sqrt{2\pi}} \int d\Omega[\hat{b}^\dagger(\Omega) \pm \hat{a}^\dagger(\Omega)]e^{i\Omega(t - L/c)} \right\}$$

$$\times \left\{ E_0 + \frac{1}{\sqrt{2\pi}} \int d\Omega'[\hat{b}(\Omega') \pm \hat{a}(\Omega')]e^{-i\Omega'(t - L/c)} \right\}$$

$$\approx \frac{\kappa|E_0|^2}{2} + \frac{\kappa}{2}\left\{ E_0^* \int d\Omega[\hat{b}(\Omega) \pm \hat{a}(\Omega)]\frac{e^{-i\Omega(t - L/c)}}{\sqrt{2\pi}} + h.c. \right\}. \tag{10.44}$$

Here for one-dimensional approximation, we know from Section 2.3.5 that $|E_0|^2 = R_{in}$ is the input photon flux. The radiation force on mirrors is then $F = 2\hbar k_0 R_{in}$ and $\kappa = 2\hbar\omega_0/c$. Solving Eq. (10.43) in frequency domain with

$$\hat{x}_j(t) = \frac{1}{\sqrt{2\pi}} \int d\Omega \hat{x}_j(\Omega)e^{-i\Omega t}, \quad h_{GW}(t)\frac{1}{\sqrt{2\pi}} \int d\Omega h_{GW}(\Omega)e^{-i\Omega t},$$

$$\hat{F}_{ph}^{(j)}(t) = \frac{1}{\sqrt{2\pi}} \int d\Omega \hat{F}_{ph}^{(j)}(\Omega)e^{-i\Omega t}, \quad (j = 1, 2) \tag{10.45}$$

we have

$$-M\Omega^2 \hat{x}_j(\Omega) = \hat{F}_{ph}^{(j)}(\Omega), \quad (j = 1, 2) \tag{10.46}$$

where

$$\hat{F}_{ph}^{(j)}(\Omega) = \frac{\kappa}{2}\Big\{ E_0[\hat{b}^\dagger(-\Omega) \pm \hat{a}^\dagger(-\Omega)] + E_0^*[\hat{b}(\Omega) \pm \hat{a}(\Omega)]\Big\}e^{i\Omega L/c} \tag{10.47}$$

with $j = 1 \to +$ and $j = 2 \to -$. Hence,

$$(\delta\hat{x}_1 - \delta\hat{x}_2)(\Omega) = -\frac{1}{M\Omega^2}[\hat{F}_{ph}^{(1)}(\Omega) - \hat{F}_{ph}^{(2)}(\Omega)] + h_{GW}$$

$$= h_{GW} - \frac{\kappa}{M\Omega^2}[E_0\hat{a}^\dagger(-\Omega) + E_0^*\hat{a}(\Omega)]e^{i\Omega L/c}. \tag{10.48}$$

Substituting the above into Eq. (10.41) and using Eq. (10.45), we have

$$\hat{E}_{out}^d(t) = \frac{e^{-i\omega_0(t-2L/c)}}{\sqrt{2\pi}} \int d\Omega \ e^{-i\Omega t}\Big\{ \frac{\kappa E_0\omega_0}{iM\Omega^2 c}\big[E_0\hat{a}^\dagger(-\Omega) + E_0^*\hat{a}(\Omega)\big]e^{2i\Omega L/c}$$

$$+\hat{a}(\Omega)e^{2i\Omega L/c} + (i\omega_0/c)E_0 h_{GW}(\Omega)e^{i\Omega L/c}\Big\}. \tag{10.49}$$

Writing $\hat{E}_{out}^d(t)$ in the 1-dim form as in Eq. (10.38), we have

$$\hat{E}_{out}^d(t) = \frac{e^{-i\omega_0(t-2L/c)}}{\sqrt{2\pi}} \int d\Omega \hat{A}(\Omega)e^{-i\Omega(t-2L/c)} \tag{10.50}$$

with

$$\hat{A}(\Omega) \equiv \hat{a}(\Omega) + \frac{\kappa E_0\omega_0}{iM\Omega^2 c}\big[E_0\hat{a}^\dagger(-\Omega) + E_0^*\hat{a}(\Omega)\big]$$

$$+(i\omega_0/c)E_0 h_{GW}(\Omega)e^{-i\Omega L/c}. \tag{10.51}$$

Using the input field $E_0 = |E_0|e^{i\varphi_0}$ as LO for the homodyne detection of the output, we define $\hat{X}_{out}(\Omega) \equiv \hat{A}(\Omega)e^{-i\varphi_0} + \hat{A}^\dagger(-\Omega)e^{i\varphi_0}$ and $\hat{Y}_{out}(\Omega) \equiv [\hat{A}(\Omega)e^{-i\varphi_0} - \hat{A}^\dagger(-\Omega)e^{i\varphi_0}]/i$, which are the operators measured in homodyne detection from Chapter 9. Then we have the input-output relation for the interferometer:

$$\hat{X}_{out}(\Omega) = \hat{X}_{in}(\Omega)$$

$$\hat{Y}_{out}(\Omega) = \hat{Y}_{in}(\Omega) - \mathcal{K}\hat{X}_{in}(\Omega) + \mathcal{M}\delta h_{GW}(\Omega)e^{-i\Omega L/c} \tag{10.52}$$

where we used $h_{GW}^*(-\Omega) = h_{GW}(\Omega)$ and $\delta h_{GW} \equiv h_{GW}/L$ as the strain, and

$$\mathcal{K} \equiv \frac{2\kappa|E_0|^2\omega_0}{M\Omega^2 c}, \quad \mathcal{M} \equiv 2|E_0|\omega_0 L/c. \tag{10.53}$$

$\mathcal{K}$ is the opto-mechanical coupling constant which couples the light fields with the motion of the mirrors through radiation pressure. Taking the power of the input laser as $P_0 = \hbar\omega_0 R_{in} = \hbar\omega_0|E_0|^2$, we can rewrite Eq. (10.53) as

$$\mathcal{K} = \frac{4P_0\omega_0}{M\Omega^2 c^2}, \quad \mathcal{M} = 2L\sqrt{\frac{P_0\omega_0}{\hbar c^2}} \equiv \frac{\sqrt{4\mathcal{K}}}{h_{SQL}} \tag{10.54}$$

with

$$h_{SQL} \equiv \sqrt{\frac{4\hbar}{M\Omega^2 L^2}} = \sqrt{\frac{2\hbar}{M_\mu\Omega^2 L^2}} \tag{10.55}$$

where $M_\mu = M/2$ is the reduced mass for two free masses. This is the standard quantum limit for position measurement of a free mass M_μ [Braginsky et al. (1992)]. With these notations, Eq. (10.52) takes the new form of

$$\hat{X}_{out}(\Omega) = \hat{X}_{in}(\Omega)$$

$$\hat{Y}_{out}(\Omega) = \hat{Y}_{in}(\Omega) - \mathcal{K}\hat{X}_{in}(\Omega) + \sqrt{4\mathcal{K}}\frac{\delta h_{GW}(\Omega)}{h_{SQL}}e^{-i\Omega L/c}. \tag{10.56}$$

From Section 9.5, we find the measured spectrum in homodyne detection is given by

$$S_{HD}(\Omega) = S_{SN}(\Omega)S(\Omega) \tag{10.57}$$

with spectral density $S(\Omega)$ defined through

$$\langle \hat{Y}_{out}(\Omega)\hat{Y}_{out}^\dagger(\Omega')\rangle = S(\Omega)\delta(\Omega - \Omega'). \tag{10.58}$$

For vacuum input, we have $\langle \hat{Y}_{in}(\Omega)\hat{Y}_{in}^\dagger(\Omega')\rangle = \langle \hat{X}_{in}(\Omega)\hat{X}_{in}^\dagger(\Omega')\rangle = \delta(\Omega - \Omega')$. Single-sided spectral density is used for characterizing the gravitational wave strain spectrum so that[6]

$$\langle \delta h_{GW}(\Omega)\delta h_{GW}^*(\Omega')\rangle = \frac{1}{2}S_{GW}(\Omega)\delta(\Omega - \Omega'). \tag{10.59}$$

Combining all of the above, we have the measured spectrum as

$$S(\Omega) = 1 + \mathcal{K}^2 + 2\mathcal{K}S_{GW}(\Omega)/h_{SQL}^2. \tag{10.60}$$

---

[6]Equation (10.58) defines a two-sided spectral density so we take half of that in definition.

Obviously, the last term is the signal while the first two terms are quantum noise contribution. So, the minimum detectable GW signal is

$$S_{GW}(\Omega) = h_{SQL}^2 \frac{1 + \mathcal{K}^2}{2\mathcal{K}} = \frac{h_{SQL}^2}{2} \left(\frac{1}{\mathcal{K}} + \mathcal{K} \right), \tag{10.61}$$

which has a minimum of h_{SQL}^2 when $\mathcal{K} = 1$ or $P_0^{SQL} = M\Omega^2 c^2/4\omega_0$.

To better understand the physical meaning of the quantum noise spectrum in Eq. (10.61), we may rearrange the second expression in Eq. (10.56) as

$$\hat{Y}_{out}(\Omega) = \sqrt{4\mathcal{K}} \frac{\delta h_{GW}(\Omega) + \hat{h}_n}{h_{SQL}} e^{-i\Omega L/c}, \tag{10.62}$$

where we define a quantum noise operator

$$\hat{h}_n \equiv \frac{h_{SQL}}{\sqrt{2}} e^{i\Omega L/c} \frac{\hat{Y}_{in}(\Omega) - \mathcal{K}\hat{X}_{in}(\Omega)}{\sqrt{2\mathcal{K}}}$$

$$\propto \frac{1}{\sqrt{P_0}} \hat{Y}_{in}(\Omega) - \frac{\sqrt{P_0}}{P_0^m} \hat{X}_{in}(\Omega) \tag{10.63}$$

whose single-sided spectral density is the quantum limit for gravitational wave detection. Indeed, we find $S_{h_n}(\Omega) = S_{GW}(\Omega)$. The expression in the second line of Eq. (10.63) helps us identify the sources of the quantum noise. The first term is inversely proportional to $\sqrt{N}$ with N as the photon number. This is just the standard quantum limit for phase measurement due to shot noise in photon detection. The second term is proportional to $\sqrt{N}$ or $\langle \Delta^2 N \rangle$ for coherent states and is thus caused by the radiation pressure noise due to the fluctuations of the photon number.

To have a sense of what LIGO is dealing with, let us take some number from LIGO design: $M_\mu = 30kg/4$ (there are four free masses), $L = 4000m$, $\Omega = 2\pi \times 100 rad/s$ (GW signal frequency $\sim 100Hz$), $\omega_0 = 1.8 \times 10^{15}/s$, we obtain, $h_{SQL} \sim 2 \times 10^{-24}/\sqrt{Hz}$ and $P_0^{SQL} \sim 1.5 \times 10^8 W$. This power is impossible to achieve in the lab. So, the shot noise dominates in the conventional Michelson interferometer. To reduce P_0^{SQL}, LIGO resorts to a design with power enhancement cavities as shown in Fig. 10.11(a).

With the enhancement cavity, it can be shown [Kimble et al. (2001)] that the output field operators take the same form[7] of Eq. (10.56) but with a different frequency response in the opto-mechanical coupling constant $\mathcal{K}$ which is now changed to

$$\mathcal{K} = \frac{2\gamma^4(P_0/\bar{P}_0^{SQL})}{\Omega^2(\gamma^2 + \Omega^2)}. \tag{10.64}$$

---

[7]Because our definition of the quadrature-phase amplitudes in Eq. (10.56) is off by a factor of $\sqrt{2}$ from the cited reference, the last term is off by the same factor.

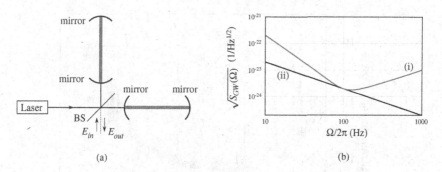

(a)   (b)

Fig. 10.11   (a) Michelson interferometer with power enhancement cavities. (b) Frequency dependence of $\sqrt{S_{GW}}$ (Trace (i)) for $P_0 = P_0^{SQL}$ and $h_{SQL}(\Omega)$ (Trace (ii)).

Here $\gamma \equiv cT/4L$ with T as the transmissivity of the coupling mirrors and $\bar{P}_0^{SQL} = ML^2\gamma^4/4\omega_0$ is the optimum input power to reach h_{SQL} at $\Omega = \gamma$. Take the numbers given before and $T = 0.033$, we have $\bar{P}_0^{SQL} \approx 1.0 \times 10^4$W. This power is more manageable in the lab, especially with a power recycling mirror right before the beam splitter. With $\mathcal{K}$ in Eq. (10.64), the minimum detectable GW spectrum S_{GW} takes the form of

$$S_{GW}(\Omega) = \frac{h_{SQL}(\Omega)}{2}\left[\frac{\Omega^2(\gamma^2 + \Omega^2)}{2\gamma^4} + \frac{2\gamma^4}{\Omega^2(\gamma^2 + \Omega^2)}\right], \qquad (10.65)$$

where $h_{SQL}^2(\Omega) \equiv 8\hbar/M\Omega^2 L^2$. Figure 10.11(b) shows the frequency dependence of $\sqrt{S_{GW}(\Omega)}$, which has a minimum at $\Omega = \gamma$. It is dominated by the shot noise term at high frequency: $S_{GW}(\Omega \gg \gamma) \propto \Omega^2/P_0$, whereas at low frequency, the radiation pressure effect kicks in: $S_{GW}(\Omega \ll \gamma) \propto P_0/\Omega^4$. The current noise performances of two advanced LIGO interferometers at Hanford, WA (H1) and Livingston, LA (L1) are shown in Fig. 10.12 [Martynov et al. (2016)]. Narrow spikes are well-characterized suspension vibrations and calibration lines and 60 Hz electric power grid harmonics. The noise are mostly from photon shot noise at frequencies higher than 150 Hz, and from other sources at low frequency. On September 14, 2015 at 09:50:45 UTC, the two LIGO detectors at Hanford, WA and Livingston, LA observatories both detected in coincidence with the signal GW150914 from the merger of two black holes of masses 36 and 29 solar mass [Abbott et al. (2016)]. The signals from the two detectors are shown in Fig. 10.13.

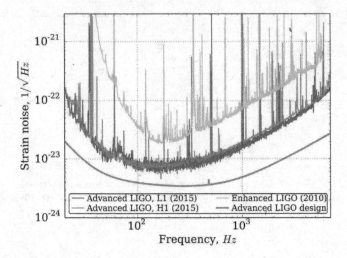

Fig. 10.12   LIGO noise performance data in 2016. Reproduced from [Martynov et al. (2016)].

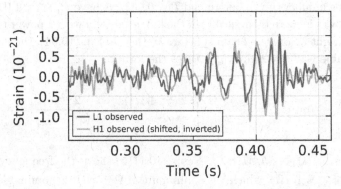

Fig. 10.13   The gravitational signals of GW150914. Adapted from [Abbott et al. (2016)].

10.3   Quantum Noise in Amplifiers

10.3.1   *Quantum Amplifiers in General*

Noise performance of an amplifier is always of great concern in the practical applications of magnifying a small signal. Many techniques were invented to suppress the noise and increase the signal-to-noise ratio. However, even in an ideal amplifier when all the technical noise such as thermal noise is eliminated, there is still quantum noise due to uncertainties generically embedded in quantum mechanics, as we have seen in the previous discussions.

It thus becomes a fundamental problem related to quantum measurement. So, the question is this: what role does quantum noise play in the noise performance of an amplifier?

The study of quantum noise in amplifier started as early as in 1960s [Haus and Mullen (1962); Heffner (1962)]. It was generally believed that quantum noise must be added through the internal degrees of the amplifier in the amplification process to make quantum cloning or simple duplication of microscopic observable impossible [Hong et al. (1985)], for otherwise the microscopic quantum superposition could be amplified to macroscopic scale and Schrödinger cat state could be easily produced [Glauber (1986)]. The added noise is the key in the decoherence process from microscopic quantum world to macroscopic classical world.

Quantum theory of amplification was established around 1980s [Caves (1982); Ley and Loudon (1984)] and experimentally tested [Levenson et al. (1993); Ou et al. (1993)]. A phase insensitive amplifier can be described quantum mechanically by [Caves (1982); Ley and Loudon (1984)]

$$\hat{a}_{out} = G\hat{a}_{in} + \hat{F}, \tag{10.66}$$

where G is the amplitude gain of the amplifier. In Eq. (10.66), the commutation relation for $\hat{a}_{out}$ requires the existence of $\hat{F}$, which is related to the gain medium of the amplifier and thus is also dubbed the term "internal degrees" of the amplifier. These degrees are usually unattended and left in vacuum state, as shown in the conceptual diagram in Fig. 10.14. $\hat{F}$ is responsible for spontaneous emission, which is a source of extra noise added to the signal from the amplifier. Even though $\hat{F}$ may be related to many modes of the internal degrees of the amplifier, we can define[8] a new operator $\hat{a}_0 \equiv \hat{F}^\dagger/g$ with $g \equiv \sqrt{G^2 - 1}$. From Eq. (10.66), we can easily show that $[\hat{a}_0, \hat{a}_0^\dagger] = 1$ so that $\hat{a}_0$ corresponds to a single mode annihilation operator. When $\hat{F}$(or $\hat{a}_0$) is in vacuum, the photon numbers of the input and output are related by

$$N_{out} = G^2 N_{in} + g^2. \tag{10.67}$$

Obviously, g^2 is the spontaneous emission of the amplifier and acts as the noise from the vacuum contribution of the internal degrees.

If signal information is encoded in the quadrature-phase amplitude $\hat{X} = \hat{a} + \hat{a}^\dagger$, Eq. (10.66) becomes

$$\hat{X}_{out} = G\hat{X}_{in} + g\hat{X}_0, \tag{10.68}$$

---

[8]Any phase of g is absorbed in operator $\hat{a}_0$.

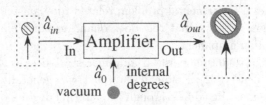

Fig. 10.14　Conceptual diagram for the quantum noise in an amplifier.

where $\hat{X}_j \equiv \hat{a}_j + \hat{a}_j^\dagger (j = out, in, 0)$. Then the noise of the signal field at the output of the amplifier as specified by the variance of $\hat{X}_{out}$ is given by

$$\langle \Delta^2 \hat{X}_{out} \rangle = G^2 \langle \Delta^2 \hat{X}_{in} \rangle + g^2 \langle \Delta^2 \hat{X}_0 \rangle. \tag{10.69}$$

Here we assume the input signal field is independent of the internal mode of the amplifier. Normally, the input is in a coherent state and the internal mode is unattended and is in vacuum. So, $\langle \Delta^2 \hat{X}_{in} \rangle = \langle \Delta^2 \hat{X}_0 \rangle = 1$. This case is best illustrated in Fig. 10.14 by the noise circles in the input and output. There are two contributions in the output: the inside shaded circle represents the directly amplified noise from the input and the outside solid ring is the extra vacuum noise added from the internal degrees of the amplifier. If the input signal is denoted by $\langle \hat{X}_{in} \rangle$, then the output signal is simply $\langle \hat{X}_{out} \rangle = G \langle \hat{X}_{in} \rangle$ and an output signal-to-noise ratio (SNR) is

$$SNR_{out} \equiv \frac{\langle \hat{X}_{out} \rangle^2}{\langle \Delta^2 \hat{X}_{out} \rangle} = \frac{G^2 \langle \hat{X}_{in} \rangle^2}{G^2 \langle \Delta^2 \hat{X}_{in} \rangle + |g|^2 \langle \Delta^2 \hat{X}_0 \rangle}$$

$$= \frac{G^2 \langle \hat{X}_{in} \rangle^2}{G^2 + g^2}. \tag{10.70}$$

We can find the noise figure of the amplification, which is defined as the ratio of the output signal-to-noise ratio (SNR) to that of the input:

$$NF \equiv \frac{SNR_{out}}{SNR_{in}} = \frac{G^2}{G^2 + g^2} = \frac{G^2}{2G^2 - 1}. \tag{10.71}$$

This leads to the famous 3 dB degradation ($1/2 = -3$ dB) in SNR for high gain amplifier ($G \gg 1$). The origin of this degradation is the extra added vacuum noise from the internal modes of the amplifier.

10.3.2　*Quantum Noise Reduction in Amplifiers*

Although the extra added noise cannot be avoided, it can be rearranged. With the availability of the squeezed states for quantum noise reduction,

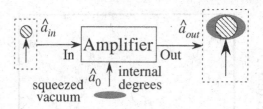

Fig. 10.15   Conceptual diagram for the quantum noise reduction in an amplifier by coupling the internal modes of the amplifier to a squeezed state.

the extra noise in one quadrature-phase amplitude where signal is encoded can be suppressed while most of the extra noise is dumped into the unused conjugate quadrature, as already implied in Eq. (10.69). This idea will work provided the internal modes $\hat{a}_0$ of the amplifier can be accessed from outside, as shown in Fig. 10.15. A parametric amplifier is such an amplifier where the internal mode is simply the idler mode: $\hat{a}_0 = \hat{b}_{idl}$ and Eq. (10.72) becomes

$$\langle \Delta^2 \hat{X}_{out} \rangle = G^2 \langle \Delta^2 \hat{X}_{in} \rangle + g^2 \langle \Delta^2 \hat{X}_{idl}(\theta) \rangle, \qquad (10.72)$$

where $\langle \Delta^2 \hat{X}_{idl}(\theta) \rangle$ is the phase dependent noise from squeezed vacuum and has a form of $\langle \Delta^2 \hat{X}_{idl}(\theta) \rangle = e^{-2r} \cos^2 \theta + e^{2r} \sin^2 \theta$ (Eq. (3.83)) with a squeezing parameter r. When $\theta = 0$, maximum suppression of the extra noise is achieved.

This strategy was first proposed by Milburn, Steyn-Ross, and Walls [Milburn et al. (1987)] and later demonstrated by Ou, Pereira, and Kimble [Ou et al. (1993)], whose results are shown in Fig. 10.16. Here, a homodyne detection scheme is used to measure the quadrature-phase amplitude of the output field of a non-degenerate optical parametric amplifier (NOPA) whose idler mode is injected with a squeezed vacuum state. The NOPA was realized with a type-II optical parametric down-conversion process inside a cavity which is resonant to both polarization components of the down-converted fields for gain enhancement. The theoretical model of such a device is presented in Section 6.3.4. In Fig. 10.16, trace *iii* (Ψ_0) is the vacuum noise level; trace *i* (Φ_G) is the output noise level of the amplifier when the idler field ($\hat{a}_0$) is in vacuum. This, according to Eq. (10.69), is $G^2 + g^2$ and gives the noise gain from vacuum noise of the amplifier. Trace *ii* ($\Phi(\theta)$) is the output noise level when the idler mode is coupled to the squeezed vacuum state. For the right phase θ, $\Phi(\theta)$ is below Φ_G realizing suppression of the extra noise of the amplifier.

Quantum Optics For Experimentalists

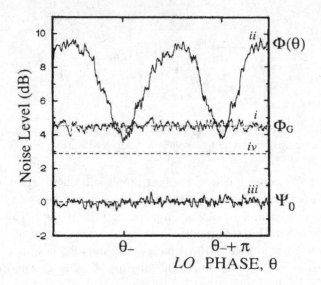

Fig. 10.16   Measured noise levels at the output of a non-degenerate parametric amplifier whose idler mode is coupled to a squeezed state. LO is the local oscillator in the
homodyne detection. Reproduced from [Ou et al. (1993)].

Notice that a dashed line (trace *iv*) is drawn, corresponding to the directly amplified input signal noise or the noise-free case with infinite amount
of squeezing. So, this scheme only suppresses the extra noise and thus at
best preserves the signal-to-noise ratio (SNR) during the amplification process ($NF = 1$), which corresponds to noiseless amplification. Since the
extra noise is added to the other quadrature, the quantum state is altered
during the amplification process and thus is not truly noise-free.

10.3.3   *Quantum Correlation for Quantum Noise Reduction in Amplifiers*

Up to now, in the discussion of quantum noise in an amplifier, we assume
that internal modes ($\hat{F}$) of the amplifier are independent of the signal ($\hat{a}_{in}$)
because the input field is usually arbitrary or the internal modes are not
accessible. On the other hand, quantum mechanics also allows the correlation of quantum fluctuations, which gives rise to quantum entanglement
and can be used to subtract out quantum noise, as we have seen in Sections
10.1.3 and 10.1.4. Therefore, if we make the internal mode of the amplifier
entangled with the input mode, as shown in Fig. 10.17, the correlation in
their quantum noise may lead to the noise cancelation in the output. The

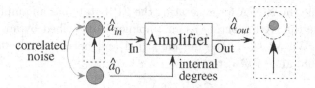

Fig. 10.17 Conceptual diagram for the quantum noise reduction in an amplifier by entangling the internal modes of the amplifier and the input signal field. The output noise level depends on the phase difference between the input and the internal mode.

noise reduction here is applied not only to the excess noise of the internal degrees but also to the input noise. So, the amplified output may have a better SNR than that at the input port.

From Eq. (10.68), we find the output noise as

$$\langle \Delta^2 \hat{X}_{out} \rangle = G^2 \langle \Delta^2 [\hat{X}_{in} + (g/G)\hat{X}_0] \rangle. \tag{10.73}$$

If the internal mode $\hat{a}_0$ and the input field $\hat{a}_{in}$ are correlated and the phase is adjusted correctly, there will be cancelation of the quantum noise between the correlated quantum fluctuations resulting in noise reduction, as we have shown in Section 10.1.3 with the EPR entangled state. In this case, the input cannot be arbitrary but needs to be prepared in the entangled state with the internal mode of the amplifier. Since a non-degenerate optical parametric amplifier (NOPA) can output an EPR entangled state with correlated quantum noise [Ou et al. (1992b)], we use one beam of the EPR entangled state to combine with a coherent state by a beam splitter with $t \ll 1$ to generate a coherent signal $S = t\alpha$ for the amplifier, as shown in Fig. 10.18. The other correlated beam is coupled to the internal mode $\hat{a}_0$ of the amplifier, which in this case is also an NOPA with its idler mode as the internal mode.

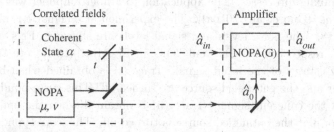

Fig. 10.18 Schematic for an amplification scheme by using an EPR correlated state generated from an NOPA with gain parameters μ, ν.

Assume the NOPA for generating the EPR state has an amplitude gain of μ while the NOPA acting as an amplifier is described by an amplitude gain G. With $t \ll 1$, the output signal and noise are calculated from Eqs. (10.68) and (10.73) to be

$$\langle \hat{X}_{out} \rangle = GS = Gt\alpha$$

$$\langle \Delta^2 \hat{X}_{out} \rangle = (\mu^2 + \nu^2)(G^2 + g^2) - 4\mu\nu Gg, \quad (10.74)$$

where $\nu = \sqrt{\mu^2 - 1}$, $g = \sqrt{G^2 - 1}$ and the relative phase is adjusted so that the noise is minimum. Here, because $t \ll 1$, the contribution to the noise performance from the coherent state is negligible. Then the output signal-to-noise ratio is

$$\begin{aligned} SNR_{out} &= \langle \hat{X}_{out} \rangle^2 / \langle \Delta^2 \hat{X}_{out} \rangle \\ &= \frac{G^2 S^2}{(\mu^2 + \nu^2)(G^2 + g^2) - 4\mu\nu Gg}, \quad (10.75) \end{aligned}$$

which has a maximum value of $G^2 S^2$ when we adjust μ of the EPR state so that $\mu = G$. This is certainly better than the SNR of $G^2 S^2 / (G^2 + g^2)$ when the signal and the idler of the amplifier are uncorrelated and are in coherent state and vacuum, respectively.[9] Notice that when $\mu = G$, the output noise of the amplifier is simply 1, which not only excludes the extra internal noise of the amplifier (the solid part in Figs. 10.14 and 10.15) but is even smaller than the input signal contribution as well (the shaded part in Figs. 10.14 and 10.15 and the first term in Eq. (10.72)). This is because of the destructive interference between the signal and idler input fields, as expressed in the form of the addition of two field amplitudes in Eq. (10.68). This only occurs when the input field is correlated with the idler field.

The proof-of-principle experiment was performed by Kong *et al.* [Kong et al. (2013a)] with two optical parametric amplifiers achieved in atomic 4-wave mixing processes. The application to a fiber amplifier was done by Guo *et al.* [Guo et al. (2016a)]. The experimental schematic is shown in Fig. 10.18. The noise levels and signal levels are shown in Figs. 10.19(a) and (b), respectively. "On" and "Off" in Fig. 10.19(b) denote the "on" and "off" states of the input signal. Trace (i) is obtained when both the amplifier and the entangled source are turned off. This corresponds to the input of the coherent signal. Trace (ii) is obtained when the amplifier is turned on but the entangled source is turned off. This corresponds to the case of uncorrelated input to the amplifier. Trace (iii) is obtained when both the amplifier and the entangled source are turned on and the phase of

---

[9]This SNR can be obtained by setting $\mu = 1, \nu = 0$ in Eq. (10.75).

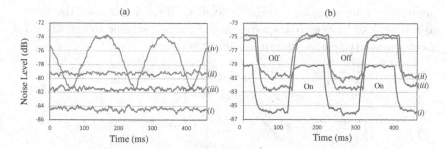

Fig. 10.19 (a) Noise levels for an amplifier with an EPR entangled state coupled between the input signal and the internal mode. (b) Encoded signal levels at the input (Trace (i)) and the output of the amplifier with its internal mode coupled to vacuum (Trace (ii)) or entangled state (Trace (iii)). Adapted from [Kong et al. (2013a)].

the entangled source is such that the noise is minimum. This corresponds to the case of correlated input. Trace (iv) in Fig. 10.19(a) is the same as Trace (iii) but with the phase of the entangled source scanned. The sinusoidal change of the noise level indicates an interference effect between the input signal and the idler fields as discussed before. The data shown in Fig. 10.19 clearly demonstrates the noise cancelation effect due to correlated input to the amplifier. The output noise is reduced about 2 dB between Trace (ii) and Trace (iii) in Fig. 10.19(a) with SNR enhanced about 2 dB as well in Fig. 10.19(b).

Since correlation from entanglement is an extra resource here, it is not fair to find the noise figure of the amplifier by comparing the SNR of the output to that of the input directly. But we can treat the entangled source and the NOPA as one amplification device and see how it performs. This time, the input signal is the coherent state. It is straightforward to find

$$\langle \hat{X}_{out} \rangle = Gt\alpha$$
$$\langle \Delta^2 \hat{X}_{out} \rangle = G^2 t^2 + [\mu G\sqrt{1 - t^2} - \nu g]^2$$
$$+ [\nu G\sqrt{1 - t^2} - \mu g]^2. \qquad (10.76)$$

This device has an effective amplitude gain of $G' = Gt$. The first term in the noise expression is the contribution from the coherent state. The output noise has a minimum value of $2G^2 t^2 - 1$ when $\mu^2 = G^2(1 - t^2)/(G^2 t^2 - 1)$. This gives the output $SNR'_{out} = G^2 t^2 |\alpha|^2/(2G^2 t^2 - 1)$ which is the same as Eq. (10.70) or the case shown in Fig. 10.14 when the amplifier's internal modes are vacuum. So, we cannot use this scheme to amplify an arbitrary signal, which is usually uncorrelated to anything in the amplifier. But we can use the combination of the correlated source and the coherent state as a probe to some samples for signal encoding. If the signal after the sample is too weak to be detected, we can use the current scheme to amplify it for better measurement, as shown in Fig. 10.20.

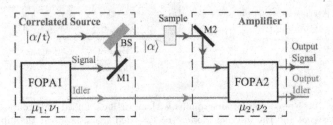

Fig. 10.20 Amplification of a weak signal from a sample probed by correlated sources. Reproduced from [Guo et al. (2016a)].

The effect of quantum noise cancelation in amplification by quantum interference will be applied in Section 11.3.2 to the SU(1,1) interferometer for the enhancement of phase measurement sensitivity.

10.3.4 *Phase-sensitive Amplifiers*

In the general discussion of the amplifier's noise in Section 10.3.1, we considered the output port ($\hat{a}_{out}$) that is directly coupled to the input port ($\hat{a}_{in}$). In fact, there are other output ports related to the internal modes. These other output ports also carry the amplified information of the input signal which we can make use of provided that we can get access to these modes. In the case of a non-degenerate optical parametric amplifier (NOPA), we have access to both the signal and idler (internal) modes in both the input and output sides. Then, Eq. (10.66) for a general amplifier becomes

$$\hat{a}_{out} = G\hat{a}_{in} + g\hat{b}_{in}^{\dagger}$$
$$\hat{b}_{out} = G\hat{b}_{in} + g\hat{a}_{in}^{\dagger}, \tag{10.77}$$

where we replace operator $\hat{F}$ for the internal modes with idler mode operator $\hat{b}$. Since $\hat{b}_{out}$ also contains the information about the input, let us add the two outputs:

$$\hat{X}_{a_{out}} + \hat{X}_{b_{out}} = (G + g)(\hat{X}_{a_{in}} + \hat{X}_{b_{in}}). \tag{10.78}$$

Notice that this is an operator equation so that

$$\langle \hat{X}_{a_{out}} + \hat{X}_{b_{out}} \rangle = (G + g)\langle \hat{X}_{a_{in}} + \hat{X}_{b_{in}} \rangle$$
$$\langle \Delta^2(\hat{X}_{a_{out}} + \hat{X}_{b_{out}}) \rangle = (G + g)^2 \langle \Delta^2(\hat{X}_{a_{in}} + \hat{X}_{b_{in}}) \rangle, \tag{10.79}$$

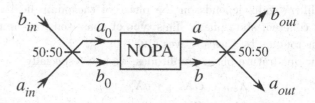

Fig. 10.21   Phase-sensitive amplifier for noiseless amplification.

which means $SNR_{out} = SNR_{in}$ or noiseless amplification. The drawback here is that only half of the input is amplified because b_{in} is in vacuum. But we can solve this by injecting b_{in} with signal, as shown in Fig. 10.21 where the input signal is first equally split into two before sending to both signal and idler input port of an NOPA and the two outputs are combined with another 50:50 beam splitter. Using the notation in Fig. 10.21, we rewrite Eq. (10.77) as

$$\hat{a} = G\hat{a}_0 - ge^{i\varphi}\hat{b}_0^\dagger,$$
$$\hat{b} = G\hat{b}_0 - ge^{i\varphi}\hat{a}_0^\dagger, \tag{10.80}$$

where, for the sake of general argument, we include a phase $-e^{i\varphi}$ in g, which is related to the phase of the pump field to the NOPA. For the beam splitters, we have

$$\hat{a}_0 = (\hat{a}_{in} + \hat{b}_{in})/\sqrt{2}, \quad \hat{b}_0 = (\hat{b}_{in} - \hat{a}_{in})/\sqrt{2},$$
$$\hat{a}_{out} = (\hat{a} - \hat{b})/\sqrt{2}, \quad \hat{b}_{out} = (\hat{b} + \hat{a})/\sqrt{2}. \tag{10.81}$$

Combining the above with Eq. (10.80), we have

$$\hat{a}_{out} = G\hat{a}_{in} + ge^{i\varphi}\hat{a}_{in}^\dagger$$
$$\hat{b}_{out} = G\hat{b}_{in} - ge^{i\varphi}\hat{b}_{in}^\dagger. \tag{10.82}$$

Both $\hat{a}_{out}$ and $\hat{b}_{out}$ are decoupled from each other and are described respectively by a degenerate parametric amplifier of the form of Eq. (6.135) studied in Section 6.3.4 but with their pump fields 180° out of phase. Since they are decoupled, we only need to consider one, say $\hat{a}_{out}$. With the coherent state input, the output photon number is

$$\langle \hat{a}_{out}^\dagger \hat{a}_{out} \rangle = G^2|\alpha|^2 + g^2(|\alpha|^2 + 1) + Gge^{i\varphi}\alpha^{*2} + Gge^{-i\varphi}\alpha^2$$
$$= |G + ge^{i(\varphi - 2\varphi_\alpha)}|^2|\alpha|^2 + g^2. \tag{10.83}$$

The last term is from spontaneous emission. But the equivalent gain $G' \equiv |G + ge^{i(\varphi - 2\varphi_\alpha)}|^2 = G^2 + g^2 + 2Gg\cos(\varphi - \varphi_\alpha)$ is phase dependent. So, this amplifier's gain depends on the phase of the pump field relative to the input coherent state phase. This type of phase-sensitive amplifier was studied thoroughly by Caves [Caves (1982)].

For the quadrature phase amplitudes, we have especially

$$\hat{X}_{out} = G\hat{X}_{a_{in}} + g\hat{X}_{a_{in}}(\varphi),$$
$$\hat{Y}_{out} = G\hat{Y}_{a_{in}} - g\hat{X}_{a_{in}}(\varphi + \pi/2), \qquad (10.84)$$

where $\hat{X}_{a_{in}}(\varphi) = \hat{a}_{in}e^{-i\varphi} + \hat{a}_{in}^{\dagger}e^{i\varphi}$, $\hat{X}_{a_{in}} = \hat{X}_{a_{in}}(0)$, $\hat{Y}_{a_{in}} = \hat{X}_{a_{in}}(\pi/2)$. Take $\varphi = 0$, Eq. (10.84) changes to

$$\hat{X}_{out} = (G + g)\hat{X}_{a_{in}},$$
$$\hat{Y}_{out} = (G - g)\hat{Y}_{a_{in}}. \qquad (10.85)$$

Notice that the above are operator equations so $SNR_{out} = SNR_{in}$ or noiseless amplification for X but attenuation for Y. Therefore, degenerate parametric amplifiers can achieve noiseless amplification for one quadrature-phase amplitude but attenuation for the other orthogonal one. The drawback is that we need to phase lock the input phase relative to the pump phase and the phase noise (classical) of the input can transfer to intensity noise in the output. Experimental demonstration of the noiseless amplification was performed by Choi *et al.* [Choi et al. (1999)].

10.4   Complete Measurement of Quantum States: Quantum State Tomography

In Chapter 9, we discussed homodyne detection and found that the homodyne detection corresponds directly to the quantum measurement of the quadrature-phase amplitude $\hat{X}_{\theta} = \hat{a}e^{-j\theta} + \hat{a}^{\dagger}e^{j\theta}$ where θ is the phase of the local oscillator (LO) of the homodyne detection. Then, we can measure $\hat{X}_{\theta}$ many times and record each outcome to build up the statistics. In terms of the language of statistics, we make an ensemble measurement of $\hat{X}_{\theta}$ on the quantum state of the field, i.e., we measure $\hat{X}_{\theta}$ for a large number of the identical states. An example of the ensemble is a series of pulses and each pulse corresponds to a sample in the ensemble. In this way, we can obtain the probability distribution $P_{\theta}(x)$ for the quantum state of the optical field.

On the other hand, we found from Eq. (3.185) in Section 3.7.2 that the marginal probability distribution of the quadrature-phase amplitude $\hat{X}_{\theta}$ is related to the Wigner function of the quantum state of the optical field:

$$P_{\theta}(x) = \int dy\, W(x\cos\theta - y\sin\theta, y\cos\theta + x\sin\theta), \qquad (10.86)$$

where the Wigner function $W(x, y)$ is defined in Eq. (3.167) for a single-mode field.

The integral in Eq. (10.86) is known as the Radon transformation and can be inverted to obtain the Wigner function [Herman (1980); Leonhardt (1997)]:

$$W(X, Y) = \frac{1}{2\pi^2} \int_0^\pi \int_{-\infty}^\infty P_\theta(x) K(X \cos\theta + Y \sin\theta - x) dx d\theta, \quad (10.87)$$

where the integration kernel is

$$K(x) = \frac{1}{2} \int_{-\infty}^\infty |\xi| e^{j\xi x} d\xi. \quad (10.88)$$

The function above diverges in the regular form so it can only exist as a generalized function inside integrals similar to the Dirac's δ-function. Leonhardt presented a full characterization of this function and its use in the inverse Radon transformation [Leonhardt (1997)].

Therefore, we can measure $P_\theta(x)$ for a number of fixed θ ranging from 0 to π and perform the inverse Radon transformation in Eq. (10.87). This technique is similar to the CT scan in X-ray tomography [Herman (1980)], and is thus dubbed "quantum state tomography".

Alternatively, we can rewrite Eq. (10.87) as

$$W(X, Y) = \frac{1}{2\pi^2} \langle K(X \cos\theta + Y \sin\theta - x) \rangle_{(x,\theta)}. \quad (10.89)$$

Strictly from the mathematical expression in Eq. (10.89), the Wigner function $W(X, Y)$ is just an ensemble average of the kernel $K(X \cos\theta + Y \sin\theta - x)$ over a pair of variables (x, θ), which can be measured experimentally by homodyne detection: the LO phase gives θ and the photocurrent output of the HD measurement gives x. Thus, we can make N measurements of (x, θ) by monitoring x while changing θ in a controlled way such as phase scanning of the LO field. From these data, we may reconstruct the Wigner function as

$$W(X, Y) = \lim_{N \to \infty} \frac{1}{2\pi^2 N} \sum_{m=1}^N K(X \cos\theta_m + Y \sin\theta_m - x_m). \quad (10.90)$$

In practice, there are a number of more efficient and accurate approaches to measure experimentally the Wigner function [Leonhardt (1997)]. The technique of quantum tomography was thoroughly reviewed by Lvovsky and Raymer [Lvovsky and Raymer (2009)]. The first experimental measurement of the Wigner function of a quantum state was performed by Smithey *et al.*

on the squeezed state generated from a pulsed parametric down-conversion [Smithey et al. (1993)]. The Wigner function of a single-photon state was measured by Lvovsky and Mlnyek, which showed for the first time the nonclassical negativeness of the Wigner function [Lvovsky et al. (2001)].

10.5 Complete Quantum State Teleportation

Quantum state teleportation is a transfer of the quantum state of a system to another different (likely remote) system without physically transporting the original system. Sounds like some magic tricks? Yes, it is the magic play of quantum entanglement. It is possible because of the existence of nonlocal correlations due to quantum entanglement [Bell (1987)]. The idea was first proposed by Bennett et al. for teleporting an unknown spin state of a particle to another [Bennett et al. (1993)] and demonstrated experimentally by Bouwmeester et al. using photons in a polarization entangled states [Bouwmeester et al. (1997)]. However, the teleported quantum states of this scheme are in a discrete Hilbert space with finite dimension. Teleportation of quantum states with continuous variables was first proposed by Vaidman [Vaidman (1994)] and analyzed for a more practical system by Braustein and Kimble [Braunstein and Kimble (1998)]. The first experimental teleportation of quantum states with continuous variables was performed by Furusawa et al. [Furusawa et al. (1998)].

Both quantum state teleportation schemes by Bennett et al. and by Vaidman require the sharing of quantum entanglement between the sender and receiver. The discrete scheme is aided by the Bell polarization entangled states (Section 4.4.2) while the scheme by Vaidmen is via the EPR entangled states (Section 4.6.2). The implementation and confirmation of a state teleportation in the discrete scheme by Bennett et al. can be performed with the photon counting techniques discussed in Chapter 7 [Braunstein and Mann (1996); Bouwmeester et al. (1997)]. The scheme of continuous variables by Vaidman can be realized [Furusawa et al. (1998)] with the homodyne detection technique discussed in Chapter 9, which we will consider here.

To begin with, Alice and Bob share two fields (fields 2 and 3) of EPR entangled sources from a non-degenerate OPA (NOPA) of amplitude gain $G = \cosh r$ and Alice wants to teleport an unknown state in field 1 described by its Wigner function W_{in} to Bob. As shown in Fig. 10.22, Alice combines one EPR entangled field (field 2) that she owns with the unknown state by a 50:50 beam splitter. The other EPR field (field 3) that is owned by Bob

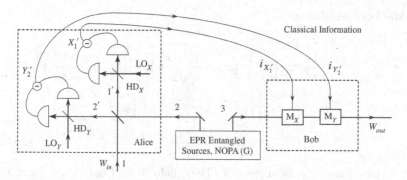

Fig. 10.22 Schematic of the quantum state teleportation of an unknown state with continuous variables by EPR entangled sources from a non-degenerate OPA (NOPA) of amplitude gain $G = \cosh r$.

is the output field to which the unknown state will be teleported. The EPR entangled fields provide the quantum channel that transmits the quantum entanglement. Homodyne detections (HD_X and HD_Y) are made by Alice to measure $\hat{X}_1'$ of the mixed field $1'$ and $\hat{Y}_2'$ of the mixed field $2'$ after the beam splitter. The results of the measurement are the photocurrents $i_{X_1'}$, $i_{Y_2'}$, which are transmitted to Bob through classical communication channel. Bob uses this classical information to modify field 3 in order to recover the input unknown state W_{in}.

From Eq. (10.12), we have the Wigner function for the EPR sources as

$$
\begin{aligned}
W_{EPR}&(x_2, y_2; x_3, y_3) \\
&= \frac{1}{4\pi^2} \exp \Big\{ -\frac{1}{4} \big[(x_3 + x_2)^2 + (y_3 - y_2)^2\big] e^r \\
&\qquad -\frac{1}{4} \big[(x_3 - x_2)^2 + (y_3 + y_2)^2\big] e^{-r} \Big\},
\end{aligned} \tag{10.91}
$$

where we changed the variables $x_1, y_1; x_2, y_2$ to $x_2, y_2; x_3, y_3$ to suit the notation in Fig. 10.22 and r to $-r$ for convenience later.

From Eq. (6.72) in Section 6.2.3 for the transformation of the Wigner function for a 50:50 beam splitter, we have the Wigner function after the

50:50 beam splitter as

$$W_{BS}(x_1', y_1'; x_2', y_2'; x_3, y_3)$$

$$= \frac{1}{4\pi^2} \exp\left\{ -\frac{1}{4}\left[\left(x_3 + \frac{x_1' + x_2'}{\sqrt{2}}\right)^2 + \left(y_3 - \frac{y_1' + y_2'}{\sqrt{2}}\right)^2\right]e^r \right.$$

$$\left. -\frac{1}{4}\left[\left(x_3 - \frac{x_1' + x_2'}{\sqrt{2}}\right)^2 + \left(y_3 + \frac{y_1' + y_2'}{\sqrt{2}}\right)^2\right]e^{-r} \right\}$$

$$\times W_{in}\left(\frac{x_1' - x_2'}{\sqrt{2}}, \frac{y_1' - y_2'}{\sqrt{2}}\right). \tag{10.92}$$

If we discard the field going to Bob (field 3) and only care about the fields for homodyne detection, we trace out field 3 by integrating over x_3, y_3 in Eq. (10.92) and the result is

$$W_{BS}(x_1', y_1'; x_2', y_2')$$

$$= \frac{1}{2\pi \cosh r} \exp\left[-\frac{(x_1' + x_2')^2 + (y_1' + y_2')^2}{4 \cosh r} \right]$$

$$\times W_{in}\left(\frac{x_1' - x_2'}{\sqrt{2}}, \frac{y_1' - y_2'}{\sqrt{2}}\right). \tag{10.93}$$

When Alice makes the homodyne measurement of $\hat{X}_1'$ and $\hat{Y}_2'$, she obtains photocurrents of $i_{X_1'}$ and $i_{Y_2'}$ which have a joint marginal probability of

$$P(i_{X_1'}, i_{Y_2'}) = \int dy_1' dx_2' W_{BS}(i_{X_1'}, y_1'; x_2', i_{Y_2'})$$

$$= \frac{1}{2\pi \cosh r} \int dy_1' dx_2' \exp\left[-\frac{(i_{X_1'} + x_2')^2 + (y_1' + i_{Y_2'})^2}{4 \cosh r} \right]$$

$$\times W_{in}\left(\frac{i_{X_1'} - x_2'}{\sqrt{2}}, \frac{y_1' - i_{Y_2'}}{\sqrt{2}}\right)$$

$$= \frac{1}{2\pi \cosh r} \int dx dy \exp\left[-\frac{(i_{X_1'}\sqrt{2} - x)^2 + (i_{Y_2'}\sqrt{2} - y)^2}{2 \cosh r} \right]$$

$$\times W_{in}(x, y), \tag{10.94}$$

where we made changes of $x = (i_{X_1'} - x_2')/\sqrt{2}$ and $y = (y_1' - i_{Y_2'})/\sqrt{2}$. This is a convolution between W_{in} and a Gaussian of width $\cosh r$. When $r \to \infty$, the Gaussian is much broader than W_{in} and can be pulled out of the integral and Eq. (10.94) becomes

$$P(i_{X_1'}, i_{Y_2'}) \approx \frac{1}{2\pi \cosh r} \exp\left(-\frac{i_{X_1'}^2 + i_{Y_2'}^2}{\cosh r} \right) \int dx dy W_{in}(x, y)$$

$$= \frac{1}{2\pi \cosh r} \exp\left(-\frac{i_{X_1'}^2 + i_{Y_2'}^2}{\cosh r} \right), \tag{10.95}$$

which is dominated by field 2 and thus contains little information about the input state. This is because the field from the EPR entangled sources is in a thermal state of average photon number $\sinh^2 r$ if the other field is discarded [Yurke and Potasek (1987)] and the superposed fields are dominated by the large thermal state when $r \to \infty$.

However, when measurements are made on X_1', Y_2' with results $i_{X_1'}$, $i_{Y_2'}$, the system is projected to a state of

$$\hat{\rho}_{proj} = \text{Tr}_{1'2'} \left(|i_{X_1'}, i_{Y_2'}\rangle\langle i_{X_1'}, i_{Y_2'}| \hat{\rho}_{sys} \right). \tag{10.96}$$

The Wigner function of the projected state is then

$$W_{proj}(x_3, y_3) = \int dx_2' dy_1' W_{BS}(x_1', y_1'; x_2', y_2'; x_3, y_3)|_{x_1' = i_{X_1'}, y_2' = i_{Y_2'}}$$

$$= \frac{1}{4\pi^2} \exp\left(-\frac{x_3^2 + y_3^2}{2\cosh r} \right) \int dx dy W_{in}(x, y)$$

$$\times \exp\left\{ -\frac{\cosh r}{2} \left[(x - i_{X_1'}\sqrt{2} - x_3 \tanh r)^2 \right.\right.$$

$$\left.\left. + (y + i_{Y_2'}\sqrt{2} - y_3 \tanh r)^2 \right] \right\}$$

$$= \frac{1}{4\pi^2} \exp\left(-\frac{x_3^2 + y_3^2}{2\cosh r} \right) \int dx' dy' \exp\left[-\frac{\cosh r}{2} (x'^2 + y'^2) \right]$$

$$\times W_{in}(x' + i_{X_1'}\sqrt{2} + x_3 \tanh r, y' - i_{Y_2'}\sqrt{2} + y_3 \tanh r)$$

$$= \frac{1}{2\pi\cosh r} \exp\left(-\frac{x_3^2 + y_3^2}{2\cosh r} \right) \int dx' dy' \delta_r(x', y')$$

$$\times W_{in}(x' + i_{X_1'}\sqrt{2} + x_3 \tanh r, y' - i_{Y_2'}\sqrt{2} + y_3 \tanh r)$$

$$\tag{10.97}$$

where $\delta_r(x', y') \equiv (\cosh r/2\pi) \exp\left[-\frac{\cosh r}{2} (x'^2 + y'^2) \right] \to \delta(x')\delta(y')$ as $r \to \infty$. Then, we have for large r

$$W_{proj}(x_3, y_3) \approx \frac{1}{2\pi\cosh r} \exp\left(-\frac{x_3^2 + y_3^2}{2\cosh r} \right) W_{in}(x_3 + i_{X_1'}\sqrt{2}, y_3 - i_{Y_2'}\sqrt{2})$$

$$= P(x_3, y_3) W_{in}(x_3 + i_{X_1'}\sqrt{2}, y_3 - i_{Y_2'}\sqrt{2}), \tag{10.98}$$

which is just the shifted input Wigner function weighted on the detection probability in Eq. (10.95).

If Bob doesn't do anything on field 3, the output state is an integration over $i_{X_1'}, i_{Y_2'}$ and is simply the thermal state described by $P(x_3, y_3)$, which is just one side of the EPR entangled state with the other side discarded. However, if Bob, upon receiving the information of $i_{X_1'}, i_{Y_2'}$ from Alice after her homodyne detection measurement, makes displacements of $x_3 \to x_3 +$

$i_{X_1'}\sqrt{2}$ and $y_3 \to y_3 - i_{Y_2'}\sqrt{2}$ on the transmitted field 3, he will obtain $W_{proj}(x_3, y_3) \propto W_{in}(x_3, y_3)$ in field 3, thus recover the input state and completely teleport the state from field 1 to field 3.

The displacement in Wigner phase space can be implemented by a coherent state through a low-coupling beam splitter, in a similar way to that discussed in Section 6.2.4 where a coherent squeezed state is produced by combining a squeezed vacuum state with a coherent state with a beam splitter. The coupled-in coherent state is modulated in both amplitude and phase by an amplitude modulation for the displacement of $x_3 \to x_3 + i_{X_1'}\sqrt{2}$ and a phase modulation for the displacement of $y_3 \to y_3 - i_{Y_2'}\sqrt{2}$. This strategy was used by Furusawa *et al.* in the first continuous-variable quantum state teleportation experiment [Furusawa et al. (1998)]. Since then, teleportations of a number of non-trivial quantum states have been achieved in the lab, including squeezed vacuum states [Yonezawa et al. (2007)], a single-photon state [Lee et al. (2011)], and a "Schrödinger kitten" state [Takeda et al. (2013)].

10.6   Problems

Problem 10.1 Wigner function of the state from a non-degenerate parametric amplifier with vacuum input.

From Section 6.1.2, we found that the output state from a nondegenerate parametric amplifier with vacuum input is simply a two-mode squeezed state, which can also be produced from two single-mode squeezed vacuum states by a 50:50 beam splitter with 180° phase shift and $r/2$ squeezing parameter (Section 6.2.4). So, we can start by writing down the Wigner functions for the two single-mode squeezed states as (Eq. (3.176) in Section 3.7.2)

$$W_1(x,y) = \frac{1}{2\pi} \exp\left[-\frac{1}{2}\left(x^2 e^r + y^2 e^{-r}\right) \right],$$
$$W_2(x,y) = \frac{1}{2\pi} \exp\left[-\frac{1}{2}\left(x^2 e^{-r} + y^2 e^r\right) \right]. \tag{10.99}$$

Here, we changed the notations x_1, x_2 to x, y. Using the input-output relation in Eq. (6.72) of Section 6.2.3 for a 50:50 beam splitter here, show that the output Wigner function is

$$W_{out}(x_1, y_1; x_2, y_2) = \frac{1}{4\pi^2} \exp \left\{ - \frac{1}{4} \left[(x_1 + x_2)^2 + (y_1 - y_2)^2 \right] e^{-r} \right.$$
$$\left. - \frac{1}{4} \left[(x_1 - x_2)^2 + (y_1 + y_2)^2 \right] e^r \right\}$$
$$\rightarrow C\delta(x_1 - x_2)\delta(y_1 + y_2) \quad \text{as } r \rightarrow \infty . \quad (10.100)$$

Problem 10.2 Limit for the fluctuations of intensity difference of two classical fields.

Consider the intensities of two single-mode fields: $\hat{I}_A = \eta \hat{A}^\dagger \hat{A}, \hat{I}_B = \eta \hat{B}^\dagger \hat{B}$ with η being some constant.

(i) Prove the following result of normal ordering operation:

$$(\hat{I}_A - \hat{I}_B)^2 = \ : (\hat{I}_A - \hat{I}_B)^2 : + \eta(\hat{I}_A + \hat{I}_B), \quad (10.101)$$

which leads to

$$\langle \Delta^2(\hat{I}_A - \hat{I}_B) \rangle = \langle : \Delta^2(\hat{I}_A - \hat{I}_B) : \rangle + \eta(\langle \hat{I}_A \rangle + \langle \hat{I}_B \rangle). \quad (10.102)$$

(ii) Use the Glauber-Sudarshan P-representation to show

$$\langle : \Delta^2(\hat{I}_A - \hat{I}_B) : \rangle_{cl} \geq 0 \quad (10.103)$$

for any classical states of fields A, B with equal sign for coherent states.

This gives the classical limit for the fluctuations of intensity difference of two fields:

$$\langle \Delta^2(\hat{I}_A - \hat{I}_B) \rangle_{cl} \geq \langle \Delta^2(\hat{I}_A - \hat{I}_B) \rangle_{cs} = \eta(\langle \hat{I}_A \rangle + \langle \hat{I}_B \rangle). \quad (10.104)$$

Here "cs" stands for the coherent state. The right-hand side is also the shot noise level of intensity fluctuations.

Problem 10.3 Quantum entanglement and correlation in a non-degenerate parametric amplifier with an arbitrary phase of the amplitude gain g.

Consider a non-degenerate parametric amplifier whose input and output relations are given by

$$\hat{A} = G\hat{a} + g\hat{b}^\dagger; \hat{B} = G\hat{b} + g\hat{a}^\dagger \quad (10.105)$$

with $g = |g|e^{i\varphi_g}$. Here, the phase φ_g is set to be arbitrary and is usually determined by the phase of the pump fields in parametric processes.

(i) Construct the quadrature-phase amplitudes: $\hat{X}_A(\theta_A) \equiv \hat{A}e^{-i\theta_A} + \hat{A}^\dagger e^{i\theta_A}$
and $\hat{X}_B(\theta_B) \equiv \hat{B}e^{-i\theta_B} + \hat{B}^\dagger e^{i\theta_B}$. If inputs $\hat{a}, \hat{b}$ are in vacuum, show that

$$\langle \hat{X}_A(\theta_A)\hat{X}_B(\theta_B)\rangle = 2G|g|\cos(\theta_A + \theta_B - \varphi_g), \qquad (10.106)$$

and

$$\langle [\hat{X}_A(\theta_A) - \hat{X}_B(\theta_B)]^2 \rangle$$
$$= 2(G^2 + |g|^2) - 4G|g|\cos(\theta_A + \theta_B - \varphi_g), \qquad (10.107)$$

which has a minimum of $2(G-|g|)^2$ when $\theta_A + \theta_B = 2n\pi + \varphi_g$ (n = integer).

(ii) We now define

$$\hat{X}_A \equiv \hat{X}_A(\theta_A - \varphi_g/2), \quad \hat{X}_B \equiv \hat{X}_B(\theta_B - \varphi_g/2);$$
$$\hat{Y}_A \equiv \hat{X}_A(\theta_A - \varphi_g/2 + \pi/2),$$
$$\hat{Y}_B \equiv \hat{X}_B(\theta_B - \varphi_g/2 + \pi/2) = -\hat{X}_B(\theta_B - \varphi_g/2 - \pi/2). \quad (10.108)$$

Calculate $\langle \Delta^2(\hat{X}_A - \hat{X}_B)\rangle$ and $\langle \Delta^2(\hat{Y}_A + \hat{Y}_B)\rangle$ and show $\langle \Delta^2(\hat{X}_A - \hat{X}_B)\rangle\langle \Delta^2(\hat{Y}_A + \hat{Y}_B)\rangle = 4(G - |g|)^4 < 1$ if $(G - |g|)^2 < 1$, exhibiting the EPR paradox.

(iii) For an injection of coherent state $|\alpha\rangle$ with $|\alpha|^2 \gg |g|^2 \sim 1$, calculate $\langle \hat{A}\rangle, \langle \hat{B}\rangle$ and show the self-homodyne relation (Section 9.8)

$$\hat{A}^\dagger\hat{A} \approx |\langle\hat{A}\rangle|^2 + |\langle\hat{A}\rangle|[\hat{X}_A(\varphi_\alpha) - \langle\hat{X}_A(\varphi_\alpha)\rangle],$$
$$\hat{B}^\dagger\hat{B} \approx |\langle\hat{B}\rangle|^2 + |\langle\hat{B}\rangle|[\hat{X}_B(\varphi_g - \varphi_\alpha) - \langle\hat{X}_B(\varphi_g - \varphi_\alpha)\rangle]. \quad (10.109)$$

(iv) From Eqs. (10.109) and (10.107), show the intensity correlation:

$$\langle \Delta^2(\hat{I}_A - \lambda\hat{I}_B)\rangle = \frac{\langle\hat{I}_A\rangle + \lambda^2\langle\hat{I}_B\rangle}{(G + |g|)^2}, \qquad (10.110)$$

where $\hat{I}_A = \hat{A}^\dagger\hat{A}$, $\hat{I}_B = \hat{B}^\dagger\hat{B}$, and $\lambda = G/|g|$. This is exactly the same as Eq. (10.24). So, as we discussed at the end of Section 10.1.4, intensity correlation in twin beams corresponds to one special EPR-type quadrature-phase amplitude correlation (it is $\hat{X}_A - \hat{X}_B$ for real injection and positive amplitude gain of $g > 0$).

Problem 10.4 Phase-sensitive amplifier for quantum information tapping.

The phase-sensitive amplifier discussed in Section 10.3.4 can be used to split quantum information without the degradation of signal-to-noise ratio. This type of device was dubbed "quantum information tap" [Shapiro

(1980)]. In contrast, quantum information splitting by a beam splitter will have vacuum noised added from the unused vacuum port and have output SNR degraded by 3 dB for shot-noise limited input [Shapiro (1980)].

Consider a variation of the phase-sensitive amplifier shown in Fig. 10.21 where the output beam splitter is taken out. We study the two outputs $\hat{a}$, $\hat{b}$ of the NOPA as a function of the inputs $\hat{a}_{in}$, $\hat{b}_{in}$.

(i) Find the relationship between $\hat{a}$, $\hat{b}$ and $\hat{a}_{in}$, $\hat{b}_{in}$.

(ii) For the coherent state $|\alpha\rangle$ input at $\hat{a}_{in}$ and vacuum at $\hat{b}_{in}$, find the photon numbers at the two outputs $\hat{a}$, $\hat{b}$. Show that they are phase-sensitive.

(iii) Show the noise at the outputs is phase-insensitive.

(iv) At the phase with optimum outputs, show that the ratios of the output SNRs to the input SNR are given by [Levenson et al. (1993)]

$$\frac{SNR_a}{SNR_{in}} = \frac{SNR_b}{SNR_{in}} = \frac{1}{2}\left(1 + \frac{2Gg}{G^2 + g^2}\right). \tag{10.111}$$

Thus, we have

$$\frac{SNR_a}{SNR_{in}} + \frac{SNR_b}{SNR_{in}} = 1 + \frac{2Gg}{G^2 + g^2}, \tag{10.112}$$

which is larger than the value 1 of the classical limit for quantum information splitting [Shapiro (1980)] and approaches the maximum value of 2 when $g \to \infty$.

Chapter 11

Quantum Noise in Phase Measurement

We have already discussed phase measurement sensitivity and the standard quantum limit to it in Chapter 10 with a linear interferometer. But phase measurement is not limited to linear interferometers. There are many schemes for phase measurement. On the one hand, phase of an optical field is introduced classically to describe the wave oscillation states of the field in a continuous way. On the other hand, we find in quantum mechanics that phase is not associated with any physical observable that can be represented by a Hermitian operator. So, we can only treat phase as a parameter in a similar way as time in quantum mechanics. Furthermore, the concept of phase is applied to any wave to describe its state of oscillation, which can be sensitive to a variety of physical quantities. For example, matter waves for atoms or electrons can form a matter wave interferometer whose phase difference is sensitive to gravity. Therefore, measurement of phase has a broad significance and is intimately related to precision measurement of many physical quantities.

Since homodyne measurement is one type of phase measurement in which the input field and the LO field interfere to produce a phase sensitive photocurrent, phase measurement is an application of this quantum optical technique. Indeed, we studied phase measurement in Mach-Zehnder and Michelson interferometers in Chapter 10. But because of its broad impact, we shall devote this whole chapter to the discussion of general phase measurement. We will present a general argument leading to the ultimate quantum limit, i.e., the Heisenberg limit, for phase measurement and make some general discussions. A new type of interferometers based on nonlinear interaction will be introduced. Finally, the technique for phase measurement can also be applied to amplitude measurement of optical fields, which will lead to the topic of joint measurement of both phase and amplitude.

11.1   Phase Measurement in General

It is well-known that quantum nature of electromagnetic fields leads to limitations on how precise a physical quantity of an optical field can be measured. It is generally believed that given an arbitrary state of light, Heisenberg uncertainty relation sets the lower bound on the sensitivity of the measurement. On the other hand, if we are allowed to prepare the system in some specific states, according to quantum theory of measurement, a physical quantity can be measured to arbitrary precision, provided that the states are the eigen-states of the operator representing the physical quantity in quantum mechanics. For the phase of an optical field, however, the answer is not so straightforward, mainly because of the fact that there does not exist a Hermitian operator for phase in an infinite-dimensional state space for a quantized optical mode [Dirac (1927); Heitler (1954); Susskind and Glogower (1964); Caruthers and Nieto (1968)]. A more modern approach [Pegg and Barnett (1988, 1989); Barnett and Pegg (1990)] in identifying a quantum mechanical operator for the phase in a finite-dimensional state space leads to the following limiting state as the eigen-state of a phase operator (which is also defined by a limiting process):

$$|\theta\rangle = \lim_{s \to \infty} (s+1)^{-1/2} \sum_{m=0}^{s} e^{im\theta} |m\rangle, \qquad (11.1)$$

which resembles the eigen-state of the position operator (defined through a limiting process). One problem with the phase state in Eq. (11.1), however, is that the average photon number of this phase state is infinite in the limiting process. Thus it is not a physical state, reflecting the difficulty encountered in the search for a physical phase operator. It becomes a common consensus that with unlimited resource of energy, it is possible to measure a phase shift to arbitrary precision but for a real system of finite energy, the precision of phase measurement is limited no matter how the phase is defined.

In addition to the theoretical approach in solving the difficulty in defining a quantum mechanical phase, another approach from the experimental point of view was adapted in defining an operational phase as related to a measurement process [Noh et al. (1991, 1992, 1993)]. However, it encounters some problems at low average photon numbers. It turns out that the problem lies in the discrete output in the photon-counting technique used for the measurement [Ou and Su (2003)], which is an intrinsic property of quantized fields (we will show later that this also leads to the Heisenberg limit in precision phase measurement). Thus, this provides another perspective on the quantum mechanical phase problem: phase is a classical concept based on the continuous wave picture for the description of the state of wave oscillation. This continuous wave picture of phase is incompatible with the discrete energy of quantized fields in quantum description unless the total energy tends to infinity so that the discreteness is transformed to continuum, as we demonstrated in Section 2.3.3 and quantum is transformed to classical by correspondence principle.

On the other hand, we do see a phase-like quantity in the quantum superposition state description of the system:

$$|\Psi\rangle = c_1 e^{i\varphi_1}|\psi_1\rangle + c_2 e^{i\varphi_2}|\psi_2\rangle = e^{i\varphi_1}\left(c_1|\psi_1\rangle + c_2 e^{i\Delta\varphi}|\psi_2\rangle\right), \qquad (11.2)$$

where $c_1, c_2 \geq 0$, and quantity $\Delta\varphi$ has all the properties associated with the classical phase we are familiar with in optics because its change can lead to a shift in the interference fringe in the measurement for the superposition state in Eq. (11.2). But this phase depends on other physical quantities such as the propagation distance for optical fields and magnetic field for atomic states. Since interferometer output is very sensitive to the change of this phase quantity, its determination is the basis for the precision measurement of a variety of physical quantities.

Usually, the phase quantity discussed above comes into play as a part of the mode function such as e^{ikz} for optical fields or time evolution function $e^{i\omega t}$ with ω depending on the magnetic field for the atomic states. In these cases, the phase quantity is a parameter that depends on other physical quantities. Therefore, everything comes down to the problem of estimating the phase parameter $\Delta\varphi$ through its action on the state of the system, as shown in Fig. 11.1.

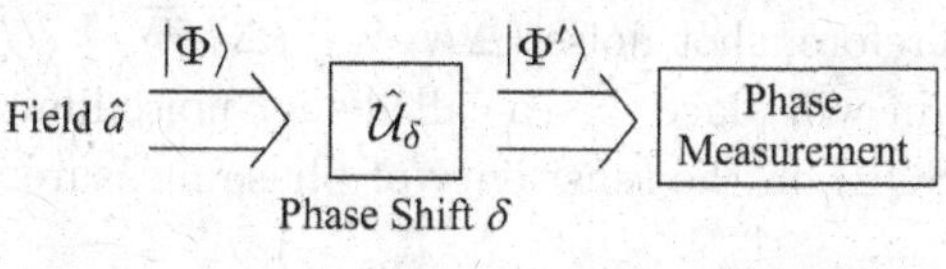

Fig. 11.1 Phase shift δ on an optical field $\hat{a}$ to change its state from $|\Phi\rangle$ to $|\Phi'\rangle$.

Particularly in the quantum theory of light, if a phase shift δ is induced by a linear optical element on a single mode optical field, it can be described by a state evolution unitary operator:

$$\hat{\mathcal{U}}_\delta = \exp(i\hat{n}\delta), \qquad (11.3)$$

where $\hat{n} = \hat{a}^\dagger \hat{a}$ is the number operator with $\hat{a}$ as the annihilation operator for the optical mode (see Section 3.2.3 for the evolution of a single-mode field). If the optical field is in state $|\Phi\rangle$, the state after the phase shift is then $|\Phi'\rangle = \hat{\mathcal{U}}_\delta |\Phi\rangle$ (Fig. 11.1). We can then make a quantum measurement on $|\Phi'\rangle$ to estimate the phase shift δ. In fact, the measurement for the estimation of δ plays an essential role in precision measurement and has been widely used in practical applications as well as in fundamental studies.

The most common way to measure δ is the interferometric method by comparing $|\Phi'\rangle$ with some reference field. Homodyne measurement is a kind of interference between the input field and a strong local oscillator (LO) with the LO acting as the reference. In Section 11.3, we will introduce some general schemes of phase measurement which are different from traditional interferometric method such as a Mach-Zehnder interferometer.

11.1.1 Ultimate Quantum Limit in Precision Phase Measurement

Since we only have finite amount of energy in a realistic physical world, we will limit our discussion under the finite energy constraint throughout this chapter. But can we still measure phase to an arbitrary precision with finite energy? In some sense, the failure to find an eigen-state of phase with finite photon number implies a negative answer to the question. The traditional and straightforward argument for the limit on the precision in phase measurement comes from the Heisenberg uncertainty principle for phase and photon number [Dirac (1927); Heitler (1954)]:

$$\Delta\phi\Delta N \geq 1, \tag{11.4}$$

where $\Delta\phi$ and ΔN are the fluctuations for phase and photon number, respectively. Therefore, shot noise ($\Delta N = \sqrt{\langle \Delta^2 N \rangle} \sim \sqrt{\langle N \rangle}$) due to particle nature of light will place the so-called shot-noise limit or the standard quantum limit (SQL) on the sensitivity of phase measurement:

$$\Delta\phi_{SQL} \gtrsim \frac{1}{\sqrt{\langle N \rangle}}. \tag{11.5}$$

On the other hand, quantum mechanics does not set any restriction on the fluctuation ΔN of photon number. Intuitively, one would argue that because of energy constraint, ΔN should be bound by the mean number of photons, that is, $\langle \Delta^2 N \rangle \sim O(\langle N \rangle^2)$. Thus given total mean number of photons, the limit on precision phase measurement should be the so-called

Heisenberg limit:

$$\Delta\phi_{HL} \gtrsim \frac{1}{\langle N \rangle}. \tag{11.6}$$

Note that Heisenberg limit should be understood as an asymptotic limit at large mean photon number, that is, the phase uncertainty approaches the order of $\langle N \rangle^{-1}$ for large $\langle N \rangle$. We treat it in this way throughout the chapter.

However, Shapiro *et al.* proposed the following state

$$|\Phi\rangle_{ssw} = A \sum_{m=0}^{M} \frac{1}{m+1}|m\rangle \quad (M \gg 1, \ A \simeq \sqrt{6/\pi^2}) \tag{11.7}$$

as an optimum state in precision phase measurement [Shapiro et al. (1989)]. Surprisingly, this state has a photon number fluctuation $\langle \Delta^2 N \rangle$ in the order of $\exp(\langle N \rangle/A^2)$ for large M. Hence, Eq. (11.4) would lead to a limit of $\Delta\phi \gtrsim \exp(-\langle N \rangle/A^2)$, which is much better than the Heisenberg limit of Eq. (11.6) for large $\langle N \rangle$.

As a matter of fact, the validity of Eq. (11.4) is not general [Susskind and Glogower (1964)]. For example, for the state of vacuum, the left-hand side of Eq. (11.4) is obviously zero thus violating the inequality. Therefore, the arguments based on the Heisenberg uncertainty relation in Eq. (11.4) cannot hold in general and the question remains: What is the limit in precision phase measurement given the available total mean number of photons? In the following, we will show that the ultimate quantum limit in the precision of phase measurement is precisely the Heisenberg limit given in Eq. (11.6).

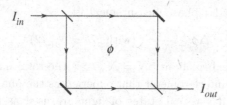

Fig. 11.2 Phase measurement by a Mach-Zehnder interferometer.

A simple argument

Classically, phase is just the argument of the complex field amplitude used to describe an optical field. Many factors may change the value of phase. The traditional method of measuring phase shift is interferometry. This method relies on optical interference effect for the comparison of phases in two paths. If we fix the phase delay of one path, any detected change in the output intensity of the interferometer will indicate a phase shift experienced in the other path leading to a measurement of the phase shift. To be more specific, as shown in Fig. 11.2, a coherent optical field is split by a beam splitter into two fields which are later recombined to form interference fringes. If the interferometer is properly balanced, the output intensity has the form of

$$I_{out} = I_{in}(1 - \cos\phi)/2, \tag{11.8}$$

where I_{in} is the intensity of the input field and ϕ is the relative phase shift between the two interfering paths. If we have a well-defined amplitude in the input field, any change ΔI_{out} in the output intensity must come from the change $\Delta\phi$ in the relative phase. The sensitivity is highest when we set $\phi = \pi/2$:

$$\Delta I_{out} = I_{in}\Delta\phi/2. \tag{11.9}$$

Classically, there is no limit on how small the change ΔI_{out} in intensity can be. Therefore, in principle, there is no limit on how small a phase shift $\Delta\phi$ can be measured. In quantum theory, however, the particle nature of light does not allow infinite division of energy thus setting a lower limit on ΔI_{out}. We can rewrite Eq. (11.9) in terms of photon number as

$$\Delta N_{out} = N_{in}\Delta\phi/2, \tag{11.10}$$

where N_{in} is the total input photon number and ΔN_{out} is the change in output photon number. The minimum ΔN_{out} that is allowed by quantum theory is simply one, corresponding to the change of one quanta. Therefore, the quantum limit for phase measurement is

$$\Delta\phi \geq \frac{1}{N}, \quad (\text{with } N = N_{in}/2) \tag{11.11}$$

which is the Heisenberg limit. $N = N_{in}/2$ is the total number of photons in the arm of the interferometer that experiences the phase shift.

Furthermore, if classical states of light are used as the input to the interferometer, photon statistics at the output is at best the Poisson distribution, i.e., $\Delta N_{out} \geq \sqrt{N_{out}}$. But for optimum sensitivity at $\phi = \pi/2$, $N_{out} = N_{in}/2 = N$. Hence, from Eq. (11.10), we arrive at the standard quantum limit in a conventional interferometer [Caves (1981)]:

$$\Delta\phi_{SQL} \geq 1/\sqrt{N}. \tag{11.12}$$

Thus nonclassical state of light must be employed in order to surpass the standard quantum limit to reach the Heisenberg limit.

The semiclassical argument above for the Heisenberg limit is independent of the input field and runs equally well if we describe the optical field quantum mechanically. However, it is limited to the specific scheme of interferometry for phase comparison and to the detection scheme of intensity measurement. Nevertheless, the semiclassical argument above clearly demonstrates that it is the quantization of optical fields that is responsible for the Heisenberg limit in precision phase measurement. Next, we will present another more general argument for the Heisenberg limit that is independent of phase measurement scheme.

A general argument by the complementarity principle of quantum mechanics

The complementarity principle of quantum mechanics [Bohr (1983)] concerns the particle and wave duality of light. Although light exhibits both wave-like and particle-like behavior, it is impossible to observe both of them simultaneously. When we apply the complementarity principle to the phenomena of interference, we find that it is impossible to obtain the complete which-path information for the two possible interfering paths of a photon and to observe in the meantime the interference effect in a single experiment. In other words, the interference effect will disappear if we know exactly from which one in the two possible interfering paths the photon is coming to the detector, whereas the appearance of interference is always a manifestation of intrinsic indistinguishability of the path of the photon. In a more quantitative language, the fringe visibility of interference and the which-way information are related through an inequality which sets an upper bound on the amount of which-way information and the visibility of the interference effect [Mandel (1991); Englert (1996)]. The visibility will be zero if we know exactly which path the photon goes through whereas 100% visibility in the interference pattern leads to no knowledge of the which-path information at all. If we have some partial information about which path the photon goes through, the visibility of interference will lie between 0 and 1. Furthermore, if, without disturbing the interference system, there exists a possibility, even in principle, for the distinction of two interfering paths, all interference is wiped out. Notice that it is not necessary to actually carry out an experiment for the distinction in order for the interference to disappear. The mere possibility that it can be performed is sufficient to suppress the interference effect [Zajonc et al. (1991)]. This supplement of the complementarity principle is the key in our next argument for the fundamental quantum limit in precision phase measurement.

Consider a single-photon interferometer shown in Fig. 11.3. In one of the interfering paths (Field $\hat{A}$), we add a device which couples to another field labeled as $\hat{a}$ and allows the measurement of the photon number in $\hat{A}$ without destroying the photons. This is a kind of quantum nondemolition measurement (QND) of photon number [Braginsky et al. (1992)]. In this way, it is possible to obtain which-path information for the single photon without destroying it (no disturbance to the photon number of the interference system). It is known [Milburn and Walls (1983); Imoto et al. (1985); Kartner and Haus (1993)] that the optical Kerr effect can be used to implement a QND measurement of the photon number. In this case, the field to be measured ($\hat{A}$ in Fig. 11.3) imposes a phase shift on another beam called the probe beam ($\hat{a}$ in Fig. 11.3). Measurement of the phase shift on the probe beam provides the information about the photon number in $\hat{A}$ and will influence the interference pattern. Thus, this interference system provides us a platform for the discussion of the precision of phase measurement in connection with the complementarity principle. Sanders and Milburn were the first to make such a connection [Sanders and Milburn (1989)]. We will use it to derive the limit for precision phase measurement.

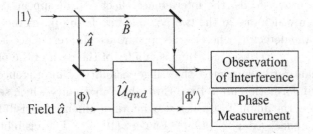

Fig. 11.3　Single-photon Mach-Zehnder interferometer with a QND measurement device for photon number in one of the arms to obtain which-path information.

In the QND measurement of photon number by optical Kerr interaction, two fields $\hat{A}$ and $\hat{a}$ (one is called the signal while the other the probe) are coupled through a Kerr medium and the state evolution is determined by the unitary operator [Kartner and Haus (1993)]

$$\hat{\mathcal{U}}_{qnd} = e^{i\kappa \hat{a}^{\dagger}\hat{a}\hat{A}^{\dagger}\hat{A}}, \tag{11.13}$$

where κ is a parameter that depends on the strength of the interaction and is adjustable. To see further the physical meaning of κ, let the input state to the QND device be a single-photon state for the signal field $\hat{A}$ and a general state $|\Phi\rangle$ similar to Fig. 11.1 for the probe field $\hat{a}$. Then the output state for the two fields is

$$\hat{\mathcal{U}}_{qnd}|1\rangle_A|\Phi\rangle_a = |1\rangle_A e^{i\kappa\hat{a}^\dagger\hat{a}}|\Phi\rangle_a = |1\rangle_A|\Phi'\rangle_a. \tag{11.14}$$

Thus, according to Eq. (11.3), the probe field $\hat{a}$ is subject to a phase shift κ imposed by the input of a single photon in the signal field $\hat{A}$. Moreover, if the input state to field $\hat{A}$ is an N-photon state $|N\rangle_A$, it is straightforward to see that the phase shift on field $\hat{a}$ will be $N\kappa$. So, κ is simply the phase shift on field $\hat{a}$ induced by a single photon in field $\hat{A}$.

Next, we perform some measurement on the probe field $\hat{a}$ to estimate the phase shift (Fig. 11.1). If we can detect the phase shift in field $\hat{a}$ with precision better than κ by whatever means, we will be able to tell whether a photon is in path $\hat{A}$ or not. Hence, if we use this device in one arm of the single-photon interferometer, we will know the which-path information, and according to the complementarity principle, the interference effect will disappear. On the other hand, if we can observe a 100% visibility in the single-photon interferometer in Fig. 11.3, it implies that it is impossible to detect the phase shift κ in field $\hat{a}$ no matter what kind of method or strategy we use for the extraction of the phase shift. Therefore, the visibility of the interferometer is directly related to our ability to resolve the phase shift κ due to a single photon.

Let us now examine the visibility of the single-photon interferometer with the QND device in path $\hat{A}$. Assume that a single-photon state is fed into one of the input ports of the interferometer, and to be more general, we assign a mixed state described by the density operator $\hat{\rho}_a$ to the probe field $\hat{a}$. Thus the input state for the total system is described by the density operator

$$\hat{\rho}_{tot} = |1\rangle\langle 1| \otimes \hat{\rho}_a. \tag{11.15}$$

After the first beam splitter, the state for the system becomes

$$\hat{\rho}'_{tot} = |\psi\rangle\langle\psi| \otimes \hat{\rho}_a, \tag{11.16}$$

with

$$|\psi\rangle = \frac{1}{\sqrt{2}}\Big(|1\rangle_A|0\rangle_B + |0\rangle_A|1\rangle_B\Big). \tag{11.17}$$

After passing the QND device, the state of the system has the form of

$$\hat{\rho}''_{tot} = \hat{\mathcal{U}}_{qnd}\hat{\rho}'_{tot}\hat{\mathcal{U}}^\dagger_{qnd} = e^{i\kappa\hat{a}^\dagger\hat{a}\hat{A}^\dagger\hat{A}}\hat{\rho}'_{tot}e^{-i\kappa\hat{a}^\dagger\hat{a}\hat{A}^\dagger\hat{A}}$$

$$= \frac{1}{2}\Big(|1_A, 0_B\rangle\langle 1_A, 0_B|e^{i\kappa\hat{a}^\dagger\hat{a}}\hat{\rho}_a e^{-i\kappa\hat{a}^\dagger\hat{a}} + |0_A, 1_B\rangle\langle 0_A, 1_B|$$

$$+ |0_A, 1_B\rangle\langle 1_A, 0_B|\hat{\rho}_a e^{-i\kappa\hat{a}^\dagger\hat{a}} + |1_A, 0_B\rangle\langle 0_A, 1_B|e^{i\kappa\hat{a}^\dagger\hat{a}}\hat{\rho}_a\Big). \tag{11.18}$$

From this state, we can calculate the probability of detecting a photon at one of the output ports of the interferometer: $P\pm = \langle \hat{A}_\pm^\dagger \hat{A}_\pm \rangle$ with $\hat{A}_\pm = (\hat{A} \pm e^{i\phi}\hat{B})/\sqrt{2}$. It has the form of

$$P_\pm = \frac{1}{2}\left[1 \pm v\cos(\phi - \epsilon)\right], \qquad (11.19)$$

with the visibility

$$v = \left|\mathrm{Tr}(e^{i\kappa\hat{a}^\dagger\hat{a}}\hat{\rho}_a)\right|, \qquad (11.20)$$

and ϵ as the phase of $\mathrm{Tr}(e^{i\kappa\hat{a}^\dagger\hat{a}}\hat{\rho}_a)$ [Sanders and Milburn (1989)]. For a single mode field, $\hat{\rho}_a$ has a general form of

$$\hat{\rho}_a = \sum_{m,n} \rho_{mn}|m\rangle\langle n| \qquad (11.21)$$

in the number state base. Therefore, the visibility of the interference pattern is

$$v = \left|\sum_m P_m e^{im\kappa}\right| \qquad (11.22)$$

with $P_m = \rho_{mm}$. For the purpose of comparison with the unit visibility, let us evaluate the quantity $1 - v$ as follows:

$$1 - v = 1 - \left|\sum_m P_m e^{im\kappa}\right| \leq \left|1 - \sum_m P_m e^{im\kappa}\right|$$

$$= 2\left|\sum_m P_m e^{im\kappa/2}\sin\frac{m\kappa}{2}\right| \leq 2\sum_m P_m \left|\sin\frac{m\kappa}{2}\right|, \qquad (11.23)$$

where we used $\sum_m P_m = 1$. By using the inequality $\sin x < x$ in Eq. (11.23), we end up with the following inequality

$$1 - v < \langle N\rangle\kappa \quad \text{or} \quad \langle N\rangle > \frac{1-v}{\kappa} \qquad (11.24)$$

with $\langle N\rangle \equiv \sum_m mP_m$ as the average photon number in field $\hat{a}$. This inequality sets a lower limit on the total mean number of photon required in field $\hat{a}$ in order to resolve the phase shift of κ in the phase measurement of field $\hat{a}$. The argument runs as follows: When it is possible, by whatever means, to resolve the phase shift κ in field $\hat{a}$, we can tell whether the photon entering the interferometer passes through path $\hat{A}$ or $\hat{B}$ in the single-photon interferometer. Since we know the which-path information, according to complementarity principle, the interference effect in the interferometer will disappear or equivalently, $v \sim 0$. Thus from Eq. (11.24), we find that the total mean photon number in field $\hat{a}$ must satisfy $\langle N \rangle \gtrsim 1/\kappa$, which provides a lower bound on the photon number required in field $\hat{a}$ in order to resolve a phase shift κ. Moreover, Eq. (11.24) can also be written as

$$\kappa > \frac{1-v}{\langle N \rangle}, \qquad (11.25)$$

which sets a lower limit on the minimum detectable phase shift, given the total mean number of photons available in field $\hat{a}$. If a phase shift κ can be resolved by whatever means, as the previous argument shows, this will result in the disappearance of the interference pattern, or $v \sim 0$. From Eq. (11.25), we have $\kappa \gtrsim 1/\langle N \rangle$. So the minimum detectable phase shift in field $\hat{a}$ is of the order of $1/\langle N \rangle$ or the Heisenberg limit.

The argument above is for the case when there is only a single-mode in field $\hat{a}$. With a multi-mode field probing the phase shift, the problem is equivalent to multiple phase measurement schemes [Shapiro et al. (1989); Braunstein et al. (1992); Lane et al. (1993); Braunstein (1992)]. The multiple measurement schemes divide the available energy into multiple fields with each sensing the same phase shift. Optimization of the measurement strategy can be performed based on quantum information theory for the estimation of the phase shift. For this measurement scheme, we can make a similar argument as the single-mode case but a modification of the unitary operator for QND measurement is needed to include the coupling to multiple fields. The modified unitary operator has the following form:

$$\hat{U}_{qnd} = \exp\left[i\kappa \hat{A}^\dagger \hat{A} \sum_j \hat{a}_j^\dagger \hat{a}_j\right], \qquad (11.26)$$

where a multi-mode field with modes characterized by the annihilation operators $\{\hat{a}_j\}$ is coupled to a single-mode of $\hat{A}$ that is one arm of a single-photon interferometer. Note that the unitary operator in Eq. (11.26) involves a hypothetical interaction that is used here purely for the sake of argument. It can be easily checked that a single photon in field $\hat{A}$ will induce a phase shift κ in all the modes $\{\hat{a}_j\}$. Joint measurements on all the modes can be performed to estimate the phase shift. By following the same line of argument as the single mode case above, we can easily show that the precision in the joint phase measurement cannot be better than $\langle N_{tot} \rangle^{-1}$ with $\langle N_{tot} \rangle = \sum_j \langle \hat{a}_j^\dagger \hat{a}_j \rangle$, the total mean photon number in all the modes. Therefore, we have generalized the proof for the fundamental limit to the multi-mode case. Furthermore, we did not specify here how the energy is distributed among different modes. Thus the argument applies to the case of uneven distribution of energy as well as to the case of equal partition of energy in all fields involved [Shapiro et al. (1989); Braunstein et al. (1992); Lane et al. (1993); Braunstein (1992)].

Notice that although the loss of interference ($v \sim 0$) relies on the ability to resolve the phase shift κ, it does not require actually performing the measurement of the phase shift. Actually, the mere passing of the probe field $\hat{a}$ is sufficient to wipe out the single-photon interference effect. To see how it works, note that while the single photon in field $\hat{A}$ can cause a phase shift of κ in field $\hat{a}$, the effect of field $\hat{a}$ on field $\hat{A}$ is the same: it will also induce a phase shift depending on the photon numbers in field $\hat{a}$. If there is a large fluctuation in the photon number of field $\hat{a}$, it will impose on the phase of field $\hat{A}$ and if the phase fluctuation is large enough due to either the size of κ or the photon number fluctuation in $\hat{a}$, the interference fringe in the single-photon interferometer will be averaged out, leading to $v \sim 0$. This effect can be seen by the following examples where we find the explicit form of visibility for some known states. It is straightforward to calculate the visibility for various known states by using Eq. (11.22):

(i) For the coherent state $|\alpha\rangle$, $v = e^{-|\alpha|^2(1-\cos\kappa)} \approx e^{-\langle N\rangle \kappa^2/2}$ for $\kappa << 1$ [Sanders and Milburn (1989)]. $v \sim 0$ when $\langle N\rangle >> 1/\kappa^2$, which is consistent with the shot-noise limit of $1/\sqrt{\langle N\rangle}$ in phase measurement sensitivity for coherent state interferometry. In this case, the photon number fluctuation is $\langle \Delta^2 N\rangle = \langle N\rangle$ which leads to a phase fluctuation of $\Delta\phi = \kappa\sqrt{\langle N\rangle}$ for the single-photon interferometer. When $\Delta\phi = \kappa\sqrt{\langle N\rangle} \sim \pi$, the interference fringe will diminish.

(ii) For the thermal state described by the density matrix $\hat{\rho}_{th} = \sum_n P_n |n\rangle\langle n|$ with $P_n = \langle N\rangle^n/(\langle N\rangle + 1)^{n+1}$,

$$v = \frac{1}{[1 + 4\langle N\rangle(\langle N\rangle + 1)\sin^2 \kappa/2]^{1/2}}$$

$$\simeq \frac{1}{[1 + \langle N\rangle^2 \kappa^2]^{1/2}} \quad \text{for } \langle N\rangle >> 1 \text{ and } \kappa << 1. \quad (11.27)$$

Notice that v is significantly different from unity only when $\kappa \gtrsim 1/\langle N\rangle$, which is consistent with the Heisenberg limit. Likewise the photon number fluctuation of a thermal state is $\langle \Delta^2 N\rangle = \langle N\rangle^2$ and the induced phase fluctuation is $\Delta\phi = \kappa\langle N\rangle$. This is consistent with Eq. (11.27).

(iii) For the phase state in Eq. (11.1),

$$v = \lim_{s\to\infty} \frac{1}{s+1} \left| \frac{\sin(s+1)\kappa/2}{\sin \kappa/2} \right| = 0 \quad \text{for any } \kappa \neq 0, \quad (11.28)$$

which reflects the fact that it is possible to make a precise measurement of phase in this state no matter how small the phase shift is. With a finite s, on the other hand, we have $v = |\text{sinc}[(s + 1)\kappa/2]/\text{sinc}(\kappa/2)|$ and v is different from 1 only when $\kappa \gtrsim 2/s = 1/\langle N \rangle_\theta$.

(iv) For a number state $|M\rangle$, $v = 1$, and it is impossible to resolve a phase shift no matter how large $\langle N \rangle = M$ is. This reflects the fact that no phase fluctuation is induced by the photon number state and that the number state is not good state to measure the phase shift κ due to its random phase nature.

(v) From the examples in (i) and (ii), we find that photon number fluctuations play some role for a vanishing visibility in the single-photon interferometer. But there is no direct connection. For example for the phase state of Eq. (11.7), which has a large photon number fluctuation, $v \simeq 1 - 6\kappa/\pi$ when $\langle N \rangle >> 1$ and $\kappa << 1$. Therefore, for $\kappa << 1$, $v \simeq 1$, which means that the state in Eq. (11.7) is not suitable for the probe field $\hat{a}$ to sense a small phase shift. Indeed, there are a number of studies [Schleich et al. (1991); Braunstein et al. (1992); Lane et al. (1993)] showing that the state in Eq. (11.7) has a phase distribution that is not appropriate for measuring the phase shift in high precision.

It is puzzling to notice from example (ii) above that for the thermal state, the disappearance of interference pattern ($v \sim 0$) is not necessarily related to the existence of a scheme of measurement on the state to resolve the phase shift κ, for we have from example (ii) $v \sim 0$ when $\langle N \rangle \kappa >> 1$ but the phase-shifted thermal state $\rho' = \hat{U}\hat{\rho}\hat{U}^\dagger = \rho$ does not contain any information about the phase shift. This fact seems to contradict the complementarity principle, which states that interference should always occur whenever there does not exist in principle a method to find the which-path information. To solve this puzzle, we point to the fact that mixed states are obtained after tracing out some other correlated states as a result of our lack of interest or ability to know these states (e.g. reservoir fields for thermal state). Once we enlarge the state space to bring in these correlated states to make a pure state, the whole system will carry the information of the phase shift. The question is then: Does there always exist a phase measurement scheme that can resolve the phase shift κ whenever this modified state is utilized in field a for sensing the phase shift and causes $v = 0$ in the single-photon interferometer? We will address this question in Section 11.1.3 when we discuss the general scheme of phase measurement.

11.1.2   A Necessary Condition for the Heisenberg Limit

It is known that squeezed state interferometry can achieve the Heisenberg limit [Bondurant and Shapiro (1984)]. Some other schemes [Yurke et al. (1986); Holland and Burnett (1993); Jacobson et al. (1995)] were discovered that have the same sensitivity. But it is not common for a phase measurement scheme to achieve the Heisenberg limit. For example, coherent state interferometry only reaches $1/\sqrt{\langle N \rangle}$ sensitivity, or the standard quantum limit (SQL). As we proved in Eq. (11.12), if classical sources are used in a conventional interferometer shown in Fig. 11.2, the sensitivity is always limited by SQL or $1/\sqrt{\langle N \rangle}$. To achieve the Heisenberg limit, nonclassical sources must be used. So what are the general requirements for the optical fields that can achieve the Heisenberg limit when they are employed in a phase measurement scheme?

Let us now consider those states which have relatively small photon number fluctuations so that

$$\langle \Delta^2 N \rangle \ll \langle N \rangle^2 \quad \text{for large } \langle N \rangle. \tag{11.29}$$

We will use these states in the probe field $\hat{a}$ in the single-photon interferometer with a QND measurement device having a coupling constant

$$\kappa \sim \frac{1}{\langle N \rangle}, \tag{11.30}$$

which is also the phase shift in field $\hat{a}$ induced by a single photon in field $\hat{A}$. Assume further that photon distribution P_m for these states is smooth so that $P_m \sim 0$ for those m with $|m - \langle N \rangle| > \sqrt{\langle \Delta^2 N \rangle}$. Then the contribution to the sum in the visibility formula in Eq. (11.22) only comes from those terms with $|m - \langle N \rangle| \lesssim \sqrt{\langle \Delta^2 N \rangle}$ and we can approximate Eq. (11.22) as

$$v \approx \left| \sum_{|m-\langle N \rangle| \lesssim \sqrt{\langle \Delta^2 N \rangle}} P_m e^{im\kappa} \right|. \tag{11.31}$$

But for $|m - \langle N \rangle| \lesssim \sqrt{\langle \Delta^2 N \rangle}$, because $\kappa \sqrt{\langle \Delta^2 N \rangle} \ll 1$ as a result of Eqs. (11.29) and (11.30), we can approximate $e^{im\kappa}$ with $e^{i\langle N \rangle \kappa}$ and Eq. (11.31) becomes

$$v \approx \left| \sum_{|m-\langle N \rangle| \lesssim \sqrt{\langle \Delta^2 N \rangle}} P_m e^{i\langle N \rangle \kappa} \right| \approx \left| e^{i\langle N \rangle \kappa} \sum_m P_m \right| = 1. \tag{11.32}$$

Therefore, with those states satisfying Eq. (11.29) in field $\hat{a}$ and a phase shift of size $\kappa \sim 1/\langle N \rangle$, we can observe interference pattern with 100%

visibility, indicating that it is impossible to resolve the phase shift of size $\kappa \sim 1/\langle N \rangle$ no matter what we do on field $\hat{a}$. Thus in order to obtain the sensitivity at Heisenberg limit in phase measurement, we must utilize states satisfying

$$\langle \Delta^2 N \rangle \gtrsim \langle N \rangle^2 \tag{11.33}$$

for sensing the phase shift. Notice that the condition in Eq. (11.33) is only a necessary condition, because we have a counter-example in the state in Eq. (11.7), which has a large photon number fluctuation but gives $v \sim 1$ as we showed in example (v) above. It can be easily checked that the phase measurement schemes that have been discovered so far to achieve Heisenberg limit all utilize states satisfying the condition in Eq. (11.33).

11.1.3 *General Consideration in the Search for Schemes Reaching the Fundamental Limit*

From the necessary condition derived in the previous section, we find that in the search for phase measurement schemes that have the sensitivity reaching the Heisenberg limit, we must first look for those states that satisfy the necessary conditions in Eq. (11.33). Then we need to construct a scheme in which we employ these states in field $\hat{a}$ for sensing the phase shift. So far, there have been a number of schemes that reach the ultimate limit in phase measurement [Bondurant and Shapiro (1984); Yurke et al. (1986); Holland and Burnett (1993); Jacobson et al. (1995)]. Among them, some are the conventional interferometers with different detection methods [Bondurant and Shapiro (1984); Holland and Burnett (1993)] while others utilize unconventional interferometers which do not use beam splitters as their wave divider [Yurke et al. (1986); Jacobson et al. (1995)]. In the following, we will make a general discussion and thus will not limit it to a particular type of interferometer.

In order to detect the phase shift, we will take measurement on field $\hat{a}$. However, direct photo-detection does not reveal any information about the phase of the field. Therefore, we need to first transform the state of the field $\hat{a}$ into some other state for which photo-detection is sensitive to the phase (e.g. homodyne). Let the state of field $\hat{a}$ be $|\Phi'\rangle$ with or $|\Phi\rangle$ without the phase shift. Consider a unitary operator $\hat{U}$ which corresponds to some phase measurement scheme. It operates on the state $|\Phi\rangle$ or $|\Phi'\rangle$ and results in the state:

$$|\Psi\rangle = \hat{U}|\Phi\rangle, \quad \text{or} \quad |\Psi'\rangle = \hat{U}|\Phi'\rangle, \tag{11.34}$$

which will be phase sensitive, that is, detection on $|\Psi'\rangle$ will result in significantly different outcome from $|\Psi\rangle$. Our goal now is to detect the difference between $|\Psi\rangle$ and $|\Psi'\rangle$. This can be easily achieved if we select those states for $|\Psi\rangle$ such that detection on it will yield null result whereas detection on $|\Psi'\rangle$ gives nonzero result. One such state that can serve as $|\Psi\rangle$ is simply the vacuum state. Thus any detection of a photon in the state $|\Psi'\rangle$ will be an indication of a phase shift.

Furthermore, phase is a relative quantity. We often need a reference in order to find the change in phase. Therefore, we will bring in another field called b as the reference (e.g. the field in the other arm of the conventional optical interferometers with beam splitters or the local oscillator in homodyne detection). After we find the state (or mixed state) $|\Phi\rangle$ satisfying the condition in Eq. (11.33), it is useful to enlarge the state space to include the field b so that fields a, b are correlated. Therefore, the state for the total system has the general form:

$$|\Phi\rangle_{tot} = \sum_{m,n} c_{mn} |m\rangle_a |n\rangle_b \qquad (11.35)$$

in the number state base. The unitary operator $\hat{U}_{ab}$ acts on the enlarged state space of two modes. Normally, the output will also consist of two modes. Detection can then be performed on the two output modes and comparison is made between the two modes for the extraction of the phase shift.

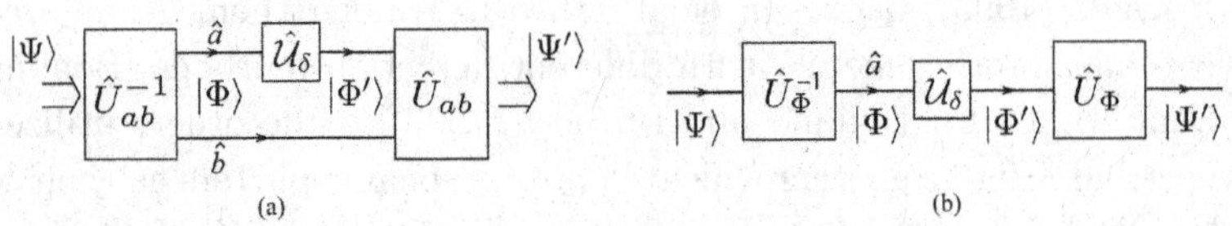

(a)   (b)

Fig. 11.4   General scheme of phase measurement with some unitary operation $\hat{U}_{ab}, \hat{U}_{ab}^{-1}$ as the generalized beam splitters. (a) The scheme with a reference field b. (b) The scheme with a vacuum input state $|\Psi\rangle = |vac\rangle$ and no reference field.

If the state $|\Psi\rangle$ is easily available such as the vacuum state, we can generate the special state $|\Phi\rangle$ as the phase sensing state by the inverse process of $\hat{U}_{ab}$. Then, we form a general type of interferometers as shown in Fig. 11.4(a). Notice that $\hat{U}_{ab}$ is a general type of unitary operator that satisfies our requirement for producing a unique state $|\Psi\rangle$ from $|\Phi\rangle$. Thus we have generalized our discussion to a broad class of unconventional interferometers. Specifically for a conventional interferometer with beam splitters, $\hat{U}_{ab} = \exp[\theta(\hat{a}^\dagger \hat{b} - \hat{a}\hat{b}^\dagger)]$ given in Appendix A.

Let us now go back to the question raised at the end of Section 11.1.1 about whether there always exist a phase measurement scheme that can resolve the phase shift κ whenever $v = 0$ in the single-photon interferometer. We will look for a scheme for phase measurement. Consider first the case with a pure state in the general form of

$$|\Phi\rangle = \sum_m c_m |m\rangle \qquad (11.36)$$

in the number state base for field $\hat{a}$. For any state $|\Phi\rangle$ with nonzero norm in a Hilbert space, it is possible to find a unitary transformation $\hat{U}_\Phi$ so that

$$\hat{U}_\Phi|\Phi\rangle = |0\rangle \equiv |\Psi\rangle \quad \text{or} \quad |\Phi\rangle = \hat{U}_\Phi^{-1}|0\rangle, \qquad (11.37)$$

where $|0\rangle$ is the vacuum state and contains no photon. So it can serve as the state $|\Psi\rangle$ with special feature for distinction. Thus the interferometer has the form shown in Fig. 11.4(b). Notice that only single-mode field is used in this interferometer which is different from a conventional interferometer (see more discussion on unconventional interferometer in Section 11.3.).

Obviously, we have from Eqs. (11.36) and (11.37)

$$c_m = \langle m|\hat{U}_\Phi^{-1}|0\rangle = \langle m|\hat{U}_\Phi^\dagger|0\rangle = \langle 0|\hat{U}_\Phi|m\rangle^*. \qquad (11.38)$$

With a phase shift on the state $|\Phi\rangle$, we find the output state becomes

$$|\Psi'\rangle = \hat{U}_\Phi|\Phi'\rangle = \hat{U}_\Phi e^{i\hat{n}\delta}\hat{U}_\Phi^{-1}|0\rangle, \qquad (11.39)$$

where Eq. (11.37) is used. With no phase shift, the output field is simply in $|0\rangle$ and has no photon, but with a nonzero phase shift, the output state is no longer the vacuum state and will contain photons. Thus detection of any photon in the output field is an indication of nonzero phase shift. A better measure for this will be the probability $\bar{P}$ of detecting any photon in the output. Obviously, we have $\bar{P} = 1 - P_0$ with P_0 being the probability of no photon. With the output state in Eq. (11.39), we find

$$P_0 = |\langle 0|\Psi'\rangle|^2 = |\langle 0|\hat{U}_\Phi e^{i\hat{n}\delta}\hat{U}_\Phi^{-1}|0\rangle|^2 = \left|\sum_m |c_m|^2 e^{im\delta}\right|^2, \qquad (11.40)$$

where we used the closure relation $\sum_m |m\rangle\langle m| = 1$ and Eq. (11.38) in the last equality. Therefore, we have for the probability of detecting any photon:

$$\bar{P} = 1 - P_0 = 1 - \left|\sum_m P_m e^{im\delta}\right|^2 = 1 - v^2, \qquad (11.41)$$

where we used Eq. (11.22) with $\kappa = \delta$ for the visibility v of the single-photon interferometer. Therefore, if we use this scheme for detecting the phase shift on the probe field $\hat{a}$ induced by a photon in field $\hat{A}$ of the single-photon interferometer in Fig. 11.3 in Section 11.1.1, we find that whenever $v = 0$, $\bar{P} = 1$ indicating that we are able to detect the phase shift of δ. Thus we have shown that whenever the interference disappears ($v = 0$), we will have at least in principle a method to know whether the single photon passes the path $\hat{A}$ or not with 100% probability. A good example for the pure state is the phase state in Eq. (11.1) with finite s. As a matter of fact, such a scheme achieves the Heisenberg limit. From Eq. (11.41), we see that the quantity v as expressed in Eq. (11.22) is a good measure in the search for optimum phase measurement schemes.

Next, let us consider a more general case with a mixed state of

$$\hat{\rho}_A = \sum_{mn} \rho_{mn} |m\rangle\langle n|. \tag{11.42}$$

As we have discussed at the end of Section 11.1.1, let us assume that, when we enlarge the state space, we are able to obtain a pure state of the form:

$$|\Phi\rangle_{ab} = \sum_{m,\lambda} c_m(\lambda)|m\rangle_a|\lambda\rangle_b, \tag{11.43}$$

which, after tracing over the fields b, will reproduce the mixed state in Eq. (11.42). The states $\{|\lambda\rangle_b\}$ characterize the other states in fields b that are correlated with field $\hat{a}$. It is always possible to make the states $\{|\lambda\rangle_b\}$ a set of orthonormal states with $\langle\lambda'|\lambda''\rangle = \delta_{\lambda'\lambda''}$. Therefore, after tracing over the fields b and comparing with Eq. (11.42), we have

$$\rho_{mn} = \sum_{\lambda} c_m(\lambda)c_n^*(\lambda). \tag{11.44}$$

Consider now the vacuum state $|0\rangle_a|0\rangle_b$ for all the relevant modes in fields a, b. As before, it is possible to find a unitary operator $\hat{U}_{ab}$ so that $\hat{U}_{ab}|\Phi\rangle_{ab} = |0\rangle_a|0\rangle_b$. We can then run through the same argument as the case of a pure state for field $\hat{a}$. The only thing different here is that the criterion for finding the phase shift is the detection of any photon in any mode of the fields a, b. Therefore, we have proved that if we can write the state of the system in the form of a pure state after enlarging the state space, it is always possible to find a measurement scheme to resolve the phase shift due to a single photon whenever the visibility of the single-photon interferometer is zero.

11.2 Phase Measurement Schemes Reaching Heisenberg Limit

Among the states that are available in the lab, only the thermal states and the single- and two-mode squeezed states have the variance of photon number in the order of the square of the mean, thus satisfying the condition in Eq. (11.33). More specifically, the variance is $\langle \Delta^2 N \rangle = 2\langle N \rangle (1 + \langle N \rangle)$ for the squeezed states of total mean photon number of $\langle N \rangle$ and $\langle \Delta^2 N \rangle = \langle N \rangle (1 + \langle N \rangle)$ for the thermal states or the two-mode squeezed states before tracing out one mode (Section 4.6.1). In this section, we will discuss a number of schemes that utilize the squeezed states and two-mode squeezed states and other nonclassical states to achieve Heisenberg limit in precision phase measurement.

11.2.1 *Schemes Reaching Heisenberg Limit with Conventional Interferometers*

Let us first consider squeezed states. There are a number of ways to utilize squeezed states to form interferometer. As a matter of fact, the first interferometer that beats the shot noise limit for the sensitivity in precision phase measurement employs a squeezed vacuum state in the unused input port of a conventional interferometer [Xiao et al. (1987); Grangier et al. (1987)]. We have discussed this scheme in detail in Section 10.2.2. In this section, we will consider a number of phase measurement schemes based on the conventional interferometers such as Mach-Zehnder (MZ) interferometers.

However, using only squeezed vacuum in the unused port cannot reach the sensitivity set by the Heisenberg limit. Bondurant and Shapiro proposed to use a squeezed coherent state instead of the coherent state together with a squeezed vacuum in the unused port and proved that such a scheme can reach the Heisenberg limit [Bondurant and Shapiro (1984)]. Their argument runs as follows.

Consider a Mach-Zehnder interferometer in Fig. 11.5, where a squeezed vacuum state of $|-r\rangle$ and a squeezed coherent state $|r, -i\alpha\rangle$ (Section 3.4) are injected into the two input ports. Referring to Fig. 11.5, for identical 50:50 beam splitters, we have

$$\hat{A} = (\hat{a}_{in} + \hat{b}_{in})/\sqrt{2}, \quad \hat{B} = (\hat{b}_{in} - \hat{a}_{in})/\sqrt{2},$$
$$\hat{A}' = \hat{A}e^{i\varphi}, \quad \hat{B}' = \hat{B}e^{i\theta},$$
$$\hat{a}_{out} = (\hat{A}' - \hat{B}')/\sqrt{2}, \quad \hat{b}_{out} = (\hat{B}' + \hat{A}')/\sqrt{2}. \qquad (11.45)$$

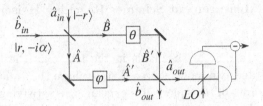

Fig. 11.5   Mach-Zehnder interferometer with squeezed vacuum and squeezed coherent states input.

For a dark fringe output at $\hat{a}_{out}$, we set $\theta = \pi$, $\varphi = \delta \ll 1$ and

$$\hat{a}_{out} = ie^{i\delta/2}(\hat{a}_{in} \sin \delta/2 - i\hat{b}_{in} \cos \delta/2)$$
$$\approx (i\delta/2)\hat{a}_{in} + (1 + i\delta/2)\hat{b}_{in}. \tag{11.46}$$

We measure $\hat{X}_a = \hat{a}_{out} + \hat{a}_{out}^\dagger$ by homodyne detection and have

$$\hat{X}_a = -(\delta/2)(\hat{Y}_{a_{in}} + \hat{Y}_{b_{in}}) + \hat{X}_{b_{in}}, \tag{11.47}$$

where $\hat{X}_{b_{in}} = \hat{b}_{in} + \hat{b}_{in}^\dagger$, $\hat{Y}_{b_{in}} = (\hat{b}_{in} - \hat{b}_{in}^\dagger)/i$. For the input states of $\hat{a}_{in}$ in $|-r\rangle$ and $\hat{b}_{in}$ in $|r, -i\alpha\rangle$, we have

$$\langle \hat{X}_a \rangle = \delta(\mu + \nu)\alpha,$$
$$\langle \Delta^2 \hat{X}_a \rangle = (\mu - \nu)^2 \tag{11.48}$$

with $\mu = \cosh r$, $\nu = \sinh r$. So, the signal-to-noise ratio is

$$SNR \equiv \frac{\langle \hat{X}_a \rangle^2}{\langle \Delta^2 \hat{X}_a \rangle} = \frac{\delta^2(\mu + \nu)^2\alpha^2}{(\mu - \nu)^2}. \tag{11.49}$$

We now define the phase sensing photon number as $N_{ps} \equiv \langle \hat{A}^\dagger \hat{A} \rangle = \nu^2 + \alpha^2(\mu + \nu)^2/2$. If we keep N_{ps} constant and assume $\nu \gg 1$ so that $\mu - \nu = 1/(\mu + \nu) \approx 1/2\nu$, Eq. (11.49) becomes

$$SNR = 8\delta^2(N_{ps} - \nu^2)\nu^2 \leq 2\delta^2 N_{ps}^2. \tag{11.50}$$

The maximum SNR of the right-hand side is reached when $\nu^2 = N_{ps}/2$. When SNR ~ 1, we obtain the minimum measurable $\delta_m \sim 1/N_{ps}$, or the Heisenberg limit.

Another scheme using a Mach-Zehnder interferometer for reaching the Heisenberg limit employs the photon number states at the inputs. Photon number states can be considered as amplitude squeezed states. It has no photon number fluctuations. In 1993, Holland and Burnett proposed to inject the state $|N, N\rangle$ into a Mach-Zehnder interferometer [Holland and Burnett (1993)]. In Section 6.2.2, we have found the output state when

$|N, N\rangle$ is input to a 50:50 beam splitter, which is now the state to probe a phase shift δ. From Eq. (6.58), we have the photon number distribution for the output state of the first BS:

$$P(k) = \frac{(2k)!(2N - 2k)!}{2^{2N}[k!(N - k)!]^2}. \tag{11.51}$$

This photon number distribution leads to a photon number fluctuation $\langle \Delta^2 N \rangle \sim N^2/2$, as shown in Fig. 11.6 where we plot $\langle \Delta^2 N \rangle$ as a function of N^2 in log-log scale and it approaches a linear dependence of $\langle \Delta^2 N \rangle = N^2/2$ (straight line). In fact, for large N, the distribution in Eq. (11.51) approaches

$$P(x) = \frac{1}{\pi\sqrt{x(1 - x)}} \tag{11.52}$$

with $x = k/N$, from which we find $\langle \Delta^2 x \rangle = 1/2$ or $\langle \Delta^2 N \rangle = N^2/2 \sim N^2$. This satisfies the necessary condition in Eq. (11.33) for reaching the Heisenberg limit when this state is used to probe a phase shift.

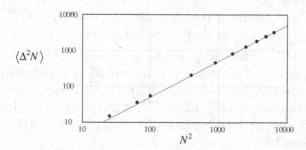

Fig. 11.6 Photon number fluctuation $\langle \Delta^2 N \rangle$ as a function of N^2 in log-log scale. The straight line is $\langle \Delta^2 N \rangle = N^2/2$.

Now we need to find a phase measurement scheme and check if the phase measurement precision reaches Heisenberg limit. For this, we set the working point of the Mach-Zehnder interferometer such that if the phase shift δ is zero, the output state is the same as the input state and the photon number difference between the two outputs is zero. This is achieved when we set $\theta = 0, \varphi = \delta \ll 1$ in the MZ interferometer in Fig. 11.5. From Eq. (11.45), the input-output relation then has the form of

$$\hat{a}_{out} = \hat{a}_{in} \cos \delta/2 + i\hat{b}_{in} \sin \delta/2,$$
$$\hat{b}_{out} = \hat{b}_{in} \cos \delta/2 + i\hat{a}_{in} \sin \delta/2. \tag{11.53}$$

When there is a phase shift, the output photon difference will not be zero. This can be confirmed by calculating the variance of the fluctuation in photon number difference $\langle \Delta^2 N_-^{(out)}(\delta) \rangle$ with $\hat{N}_-^{(out)} \equiv \hat{a}_{out}^\dagger \hat{a}_{out} - \hat{b}_{out}^\dagger \hat{b}_{out}$. For this, we first calculate

$$\langle \hat{N}_-^{(out)}(\delta) \rangle = \langle \hat{a}_{in}^\dagger \hat{a}_{in} - \hat{b}_{in}^\dagger \hat{b}_{in} \rangle \cos \delta$$
$$+ i \langle \hat{a}_{in}^\dagger \hat{b}_{in} - \hat{b}_{in}^\dagger \hat{a}_{in} \rangle \sin \delta. \qquad (11.54)$$

For the input state of $|N, N\rangle$, we can easily find $\langle \hat{N}_-^{(out)} \rangle(\delta) = 0$, which means that the photon number distribution is symmetric for the two outputs. Next we calculate in the same way $\langle \hat{N}_-^{(out)2}(\delta) \rangle$:

$$\langle \hat{N}_-^{(out)2}(\delta) \rangle = -\langle (\hat{a}_{in}^\dagger \hat{b}_{in} - \hat{b}_{in}^\dagger \hat{a}_{in})^2 \rangle \sin^2 \delta$$
$$= \langle \hat{a}_{in}^\dagger \hat{b}_{in} \hat{b}_{in}^\dagger \hat{a}_{in} + \hat{b}_{in}^\dagger \hat{a}_{in} \hat{a}_{in}^\dagger \hat{b}_{in} \rangle \sin^2 \delta$$
$$= 2N(N+1) \sin^2 \delta, \qquad (11.55)$$

where we used the twin photon state property: $\hat{N}_-^{(in)} |N, N\rangle = 0$.

In this case, since $\langle N_-^{(out)} \rangle = 0$ for any value of δ, we can't use it as signal for the detection of the phase shift δ. However, $\langle \hat{N}_-^{(out)2}(\delta) \rangle \neq 0$ for nonzero δ, so it can be used as the signal. But the noise is not simply $\sqrt{\langle \Delta^2 \hat{N}_-^{(out)}(\delta) \rangle}$. In fact, when $\delta = 0$, we have $\langle (\hat{N}_-)^m \rangle = 0$ for any nonzero integer m so that any detection of nonzero N_- is an indication of $\delta \neq 0$. But N_- is quantized and only takes $0, \pm 1, \pm 2, \ldots$. Thus the noise of N_- is simply 1. Therefore, $SNR = \langle \Delta^2 \hat{N}_-^{(out)} \rangle \simeq 2N^2 \delta^2$ for $\langle N \rangle \gg 1$ and $\delta \ll 1$, and the minimum detectable phase shift is $\delta_{min} \sim 1/N = 1/\langle N \rangle$, which is the Heisenberg limit.

The above situation can also be understood in terms of the distribution of photon number difference. For $\delta = 0$, the output has a state same as the input $|N, N\rangle$. When $\delta \neq 0$, we will have other states like $|N - m, N + m\rangle$ with $m \neq 0$. Whenever the variance $\langle \Delta^2 \hat{N}_-^{(out)}(\delta) \rangle = \langle \hat{N}_-^{(out)2}(\delta) \rangle \gtrsim 1$, there will be some significant chance to have $N_-^{(out)} = 2m \neq 0$, indicating a non-zero phase shift δ. This requires $\delta \gtrsim 1/N$. Thus, the minimum detectable δ is at the Heisenberg limit.

To further demonstrate the validity of the argument above, let us consider the probability of detecting an imbalance in the photon number at the two outputs, that is, the probability to have $|N - m, N + m\rangle (m \neq 0)$ in the output. For this, we need to find the output state of the interferometer. Since the interferometer is equivalent to a beam splitter, we can find its evolution operator in Eq. (A.6) from Appendix A:

$$\hat{U}(\delta) = \exp[\delta(\hat{b}\hat{a}^\dagger - \hat{a}\hat{b}^\dagger)/2], \qquad (11.56)$$

where we take $t = \cos\delta/2, r = \sin\delta/2$ for the MZ interferometer with a phase difference of δ. The probability to have the output state $|N-m, N+m\rangle$ for the interferometer is then

$$P_{2m}(\delta) = |\langle N-m, N+m|\hat{U}(\delta)|N,N\rangle|^2$$
$$= |\langle N-m, N+m|e^{\delta(\hat{b}\hat{a}^\dagger - \hat{a}\hat{b}^\dagger)/2}|N,N\rangle|^2. \quad (11.57)$$

When N is large and $m \ll N$, Holland and Burnett showed that $P_{2m}(\delta) \approx J_m^2(N\delta)$ with J_m as the mth-order Bessel function [Holland and Burnett (1993)]. We are particularly interested in $P_0(\delta)$ because the probability $\bar{P}$ of detecting any non-zero m is simply $\bar{P}(\delta) = 1 - P_0(\delta) = 1 - J_0^2(N\delta)$. When $N\delta \ll 1$, $J_0 \sim 1$ and $\bar{P}(\delta) \sim 0$, indicating we are unable to detect any non-zero m. But J_0 starts to drop for appreciable $N\delta$. Especially, the first root of $J_0(x) = 0$ is $x_1 = 2.405$, that is, when $N\delta \approx 2.4$, we have $\bar{P}(\delta) = 1$. This means that we will be able to detect the phase shift of $\delta \approx 2.4/N$ and this is the Heisenberg limit for phase measurement.

However, measurement of the photon number distribution in the output is not easily performed in the experiment. To circumvent this shortcoming, Campos et al. proposed to employ parity measurement at only one of the outputs of the interferometer [Campos et al. (2003)]. The parity operator is defined as $\hat{O} = (-1)^{\hat{a}^\dagger\hat{a}} = \exp(i\pi\hat{a}^\dagger\hat{a})$. It was shown that $\langle\hat{O}\rangle_N = P_N[\cos(2\delta)]$ with $P_N(x)$ as the Legendre polynomial and this measurement scheme also leads Heisenberg limit in phase measurement sensitivity [Campos et al. (2003)].

In practice, state $|N,N\rangle$ is hard to generate in the lab. But the twin-beam state $|\eta_{ab}\rangle$ discussed in Section 4.6.1 possesses the same twin photon property $\hat{N}_-|\eta_{ab}\rangle = 0$ that is required here. Let us now evaluate it for this scheme.

Suppose the two input ports $\hat{a}_{in}, \hat{b}_{in}$ are injected with a twin-beam state discussed in Section 4.6.1 and given by

$$|\eta_{ab}\rangle = \hat{S}_{ab}(\eta)|vac\rangle \quad (11.58)$$

with

$$\hat{S}_{ab}(\eta) = e^{\eta\hat{a}_{in}^\dagger\hat{b}_{in}^\dagger - h.c.}. \quad (11.59)$$

Here, we changed the notation $a, b \to a_{in}, b_{in}$. From Section 4.6.1, we know that $(\hat{a}_{in}^\dagger\hat{a}_{in} - \hat{b}_{in}^\dagger\hat{b}_{in})^m|\eta_{ab}\rangle = 0$ for any non-zero integer m.

Similar to the case of $|N,N\rangle$ state input, we can find from the input-output relation in Eq. (11.53) that the average photon number difference at the output is $\langle N_-^{(out)}\rangle = 0$ but the variance is $\langle\Delta^2 N_-^{(out)}(\delta)\rangle =$

$4\mu^2\nu^2\sin^2\delta = 4((\langle N\rangle + 1)\langle N\rangle\sin^2\delta$ where $\mu^2 = \cosh^2|\eta|, \nu^2 = \sinh^2|\eta|$, and $\langle N\rangle = |\nu|^2$ is the total mean number of photons in one arm of the interferometer. When $\langle N\rangle \gg 1$ and $\delta \ll 1$, $\langle \Delta^2 N_-^{(out)}(\delta)\rangle \simeq 4\langle N\rangle^2\delta^2$.

Using an argument similar to the case of $|N, N\rangle$ injection, we find that for the twin-beam state $|\eta_{ab}\rangle$, when $\langle \Delta^2 N_-^{(out)}(\delta)\rangle \simeq 4\langle N\rangle^2\delta^2 \sim 1$, there is a significant chance to obtain $N_- \neq 0$. Then, the minimum detectable phase shift is $\delta_{min} \sim 1/2\langle N\rangle$, which is the Heisenberg limit. This can be confirmed by considering the probability $\bar{P}$ of detecting $N_- \neq 0$. Obviously, $\bar{P} = 1 - P_0$ with P_0 as the probability of detecting $N_- = 0$. We can find P_0 by first calculating the characteristic function for N_-:

$$C(r) \equiv \langle e^{ir\Delta N}\rangle$$
$$\approx 1/(1 + 4\langle N\rangle^2\delta^2\sin^2 r)^{\frac{1}{2}} \quad (\langle N\rangle \gg 1, \delta \ll 1) \qquad (11.60)$$

and then making a finite Fourier transformation:

$$P_m = \frac{2}{\pi}\int_0^{\frac{\pi}{2}} dr \frac{\cos 2mr}{(1 + 4\langle N\rangle^2\delta^2\sin^2 r)^{\frac{1}{2}}}. \qquad (11.61)$$

Thus $P_0 = 2K(-4\langle N\rangle^2\delta^2)/\pi$ with $K(x)$ as the complete elliptic function of the first kind. Figure 11.7 shows $\bar{P}$ as a function of $\langle N\rangle\delta$. It is obvious that $\bar{P} \simeq 0$ when $\langle N\rangle\delta \ll 1$ and starts to rise when $\langle N\rangle\delta \sim 1$, which is an indication of nonzero phase shift δ. So we find in this figure that the minimum detectable phase shift is of the order of $1/\langle N\rangle$ and the interferometer is operated at the Heisenberg limit.

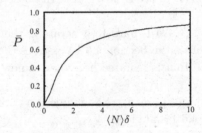

Fig. 11.7   Probability $\bar{P}$ of detecting any non-zero N_- as a function of $\langle N\rangle\delta$ for a twin-beam interferometer. Reproduced from [Ou (1997a)].

11.2.2   Schemes Reaching Heisenberg Limit with Unconventional Interferometers

Let us go back to the general scheme of phase measurement discussed in Section 11.1.3. We have demonstrated in the previous section that a number of phase measurement schemes using beam splitters as the unitary transformation in Fig. 11.4 can reach the Heisenberg limit. In this section, we will replace the beam splitters with some more general types of unitary transformation.

The first scheme is to use the squeezing operator

$$\hat{S}(\xi) = \exp\left[\frac{\xi}{2}(\hat{a}^2 - h.c.)\right] \quad (\xi = \text{real}) \tag{11.62}$$

to form a single-mode interferometer. As is known, when acting on vacuum state, the squeezing operator produces a squeezed vacuum state that has the variance of photon number fluctuations as $\langle\Delta^2 N\rangle \sim \langle N\rangle^2$ (Problem 3.5), thus satisfying the condition in Eq. (11.33) for reaching the Heisenberg limit. So, we can choose $|\Phi\rangle = \hat{S}(\xi)|vac\rangle$ for probing the phase shift δ. Then we have $\hat{U}_\Phi = \hat{S}^{-1}(\xi)$ and $|\Psi\rangle = |vac\rangle$ in the general scheme in Fig. 11.4(b). The single-mode squeezed vacuum state interferometer takes the form shown in Fig. 11.8.

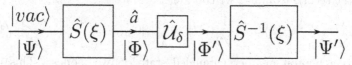

Fig. 11.8　Single-mode squeezed state interferometer.

Without the phase shift, the output state of the system is simply $|\Psi\rangle_{out} = \hat{S}^{-1}\hat{S}|vac\rangle = |vac\rangle$. With a phase shift δ on the state $|\Phi\rangle = \hat{S}|vac\rangle$, the output state becomes

$$|\Psi'\rangle_{out} = \hat{S}^{-1}e^{i\hat{a}^\dagger \hat{a}\delta}\hat{S}|vac\rangle, \tag{11.63}$$

and the average photon number at the output of the interferometer is then

$$\langle\Psi'|\hat{a}^\dagger\hat{a}|\Psi'\rangle_{out} = 4\langle N\rangle(1 + \langle N\rangle)\sin^2\delta, \tag{11.64}$$

where $\langle N\rangle = \sinh^2\xi$ is the average photon number in the state $|\Phi\rangle$. Since the output of the interferometer is vacuum when there is no phase shift, whenever we detect a photon at the output, we may conclude that there is a phase shift. So, as before, we take the noise in the output simply as one photon. Thus we have the signal-to-noise ratio as

$$SNR = 4\langle N\rangle(1 + \langle N\rangle)\sin^2\delta$$
$$\approx 4\langle N\rangle^2\delta^2 \quad (\langle N\rangle >> 1, \delta << 1). \tag{11.65}$$

Therefore, the SNR is significant only when $\delta \sim 1/\langle N\rangle$. Since the detection of a single photon will indicate a nonzero phase shift, a better quantity to characterize the sensitivity of the interferometer is the probability $\bar{P}$ of finding any photon in the output. Obviously $\bar{P} = 1 - P_0$ with P_0 being the probability of finding no photon in the output. P_0 can be calculated as

$$
\begin{aligned}
P_0 &= \langle \Psi' | : e^{-\hat{a}^\dagger \hat{a}} : |\Psi' \rangle_{out} \\
&= \frac{1}{\sqrt{1 + 4\langle N \rangle (1 + \langle N \rangle) \sin^2 \delta}} \\
&\approx \frac{1}{\sqrt{1 + 4\langle N \rangle^2 \delta^2}}, \quad \text{for } \langle N \rangle \gg 1, \ \delta \ll 1.
\end{aligned} \tag{11.66}
$$

This expression can also be derived from Eq. (11.41) through the visibility v for squeezed states [Sanders and Milburn (1989)]. Thus $\bar{P}$ is significantly different from zero when $\delta \gtrsim 1/\langle N \rangle$ and the interferometer is operated at the Heisenberg limit. This scheme was also discussed by Yurke et al. who used signal-to-noise ratio for the sensitivity [Yurke et al. (1986)].

As the second example, we consider the state

$$
|\Phi\rangle_M = \frac{1}{\sqrt{2}} (|M\rangle_a |0\rangle_b + |0\rangle_a |M\rangle_b), \tag{11.67}
$$

which is a photon number entangled state of two modes. This is the so-called NOON state (Section 4.5.1) [Kok et al. (2002)]. Obviously, $\langle \Delta^2 N \rangle_a = M^2/4 = \langle N \rangle_a^2$, thus it also satisfies the necessary condition in Eq. (11.33). However, it is not so easy to find an evolution process to produce a distinctive state $|\Psi\rangle$ as discussed in Section 11.1.3. But the M-photon entangled state above looks similar to the one-photon entangled state $|1\rangle_{ab} = (|1\rangle_a |0\rangle_b + |0\rangle_a |1\rangle_b)/\sqrt{2}$ discussed in Section 4.3.1, which can be produced by a 50:50 beam splitter from a single-photon state. From the form of the evolution operator of a beam splitter $\hat{U}_{BS} = \exp[\theta(\hat{b}\hat{a}^\dagger - \hat{a}\hat{b}^\dagger)]$ (see Appendix A), we find that it is a one-photon creation and annihilation process. For M-photon creation and annihilation as one entity, we consider the evolution operator

$$
\hat{U}_M = \exp\left(\frac{\pi}{4M!} [(\hat{a}\hat{b}^\dagger)^M - h.c.] \right), \tag{11.68}
$$

which comes from the Hamiltonian given by

$$
\hat{H}_I = i\hbar\xi[(\hat{a}\hat{b}^\dagger)^M - h.c.]. \tag{11.69}
$$

It is easy to check that

$$
\begin{aligned}
\hat{U}_M |0\rangle_a |M\rangle_b &= \tfrac{1}{\sqrt{2}} (|0\rangle_a |M\rangle_b - |M\rangle_a |0\rangle_b), \\
\hat{U}_M |M\rangle_a |0\rangle_b &= \tfrac{1}{\sqrt{2}} (|0\rangle_a |M\rangle_b + |M\rangle_a |0\rangle_b),
\end{aligned} \tag{11.70}
$$

so that
$$\hat{U}_M|\Phi\rangle_M = |0\rangle_a|M\rangle_b \equiv |\Psi\rangle_M. \tag{11.71}$$
With a phase shift of δ in field a, the state $|\Phi'\rangle$ becomes
$$|\Phi'\rangle = e^{i\delta\hat{a}^\dagger\hat{a}}|\Phi\rangle = \frac{1}{\sqrt{2}}(|0\rangle_a|M\rangle_b + e^{iM\delta}|M\rangle_a|0\rangle_b), \tag{11.72}$$
and the output state then has the form of
$$|\Psi'\rangle_M = \hat{U}_M|\Phi'\rangle$$
$$= e^{iM\delta/2}\Big(\cos\frac{M\delta}{2}|0\rangle_a|M\rangle_b + i\sin\frac{M\delta}{2}|M\rangle_a|0\rangle_b\Big). \tag{11.73}$$
Thus, if we measure photon number at output port a, any detection of photon will indicate a phase shift. The probability of detecting any photon in output port a is
$$\bar{P} = \sin^2 M\delta/2 = \frac{1}{2}(1 - \cos M\delta) \tag{11.74}$$
which is significantly different from zero only when $\delta \sim 1/(M/2) = 1/\langle N\rangle_a$. Therefore, such a scheme reaches the Heisenberg limit.

The criterion here for the detection of a phase shift is different from the general discussion in Section 11.1.3 because of the form of the unitary operator in Eq. (11.68). It does not annihilate all the photons to produce a vacuum state as required for the general scheme but rather preserves the total photon number.

Because of the unusual form of the unitary operator (which depends on the total input photon number M), this scheme is not likely to be practical as compared to the scheme discussed earlier with single-mode squeezed states. It is used here as another example of unconventional interferometers which can achieve the Heisenberg limit. But a unitary operator in a slightly different form from Eq. (11.68) can be simulated in a trapped ion system [Leibfried et al. (2002)] and is given as
$$\hat{U}_{1-M} = e^{\xi(\hat{a}^{\dagger M}\hat{b} - h.c.)}. \tag{11.75}$$
This is a one-to-M photon beam splitter. We have encountered this type of interaction in Section 6.1.4 where the evolution operator in Eq. (6.19) for the photon number doubler is of the same form as Eq. (11.75) but with $M = 2$. So, the unitary operator in Eq. (11.75) describes a one-to-M photon converter. Using this type of interaction but with only partial conversion ($\xi\sqrt{M!} = \pi/4$), we can transform a photon in b-field into a superposition state of the form
$$|\Phi\rangle_{1-M} = \frac{1}{\sqrt{2}}(|M\rangle_a|0\rangle_b + |0\rangle_a|1\rangle_b). \tag{11.76}$$

If we use field a to sense the phase shift, the phase shifted state is

$$|\Phi'\rangle_{1-M} = \frac{1}{\sqrt{2}}(e^{iM\delta}|M\rangle_a|0\rangle_b + |0\rangle_a|1\rangle_b). \qquad (11.77)$$

The output state is then

$$|\Psi'\rangle_{1-M} = \hat{U}|\Phi'\rangle$$

$$= e^{iM\delta/2}\left(\cos\frac{M\delta}{2}|0\rangle_a|1\rangle_b + i\sin\frac{M\delta}{2}|M\rangle_a|0\rangle_b\right), \quad (11.78)$$

which is similar to Eq. (11.73). The probability of detecting photons in field a is then

$$P_a(\delta) = \sin^2\frac{M\delta}{2} = \frac{1}{2}(1 - \cos M\delta). \qquad (11.79)$$

Notice that P_a shows $2\pi/M$ period and is M times more sensitive to δ than a traditional MZ interferometer. Then we can use the same argument before to claim that the interferometer based on the unitary transformation in Eq. (11.75) can reach the Heisenberg limit. Experiment was performed by Leibfried et al. to simulate the unitary operator in Eq. (11.75) with $M = 2,3$ and confirm the phase dependence in Eq. (11.79) [Leibfried et al. (2002)].

Next, let us consider the thermal state because its photon number fluctuation satisfies the necessary condition in Eq. (11.33) for reaching the Heisenberg limit. But it is impossible to implement an interferometer based on the thermal state alone due to its lack of phase coherence. However, as we discussed before, we can always enlarge the state space to include field b to form a pure state $|\Phi\rangle_{ab}$ so that we may find coherence between fields a and b. We have shown in Section 4.6 that a two-mode squeezed state from the nondegenerate parametric down-conversion process has two correlated fields with each having the thermal statistics, or in other words, $\langle\Delta^2 N\rangle_{a,b} = \langle N\rangle(\langle N\rangle + 1)$. Such a state can be generated from vacuum by applying the operator $\hat{S}_{ab}(\eta)$ (see Section 4.6)

$$\hat{S}_{ab}(\eta) = e^{\eta\hat{a}^\dagger\hat{b}^\dagger - h.c.}. \qquad (11.80)$$

Now let us reverse the process by assigning the two-mode squeezed state fields as the fields a, b in Fig 11.4(a) and set $|\Phi\rangle_{ab} = \hat{S}_{ab}(\eta)|vac\rangle_{ab}$ and $|\Psi\rangle = |vac\rangle_{ab}$. Then the unitary operator required in Eq. (11.34) is $\hat{U}_{ab} = \hat{S}_{ab}^{-1}(\eta)$. The interferometer for the whole system then has the same form as the two-mode interferometer shown in Fig. 11.4(a). In this case, $|\Psi\rangle = \hat{S}_{ab}^{-1}(\eta)|\Phi\rangle_{ab} = |vac\rangle_{ab}$ when the phase shift is zero. But when we have a nonzero phase shift δ, the output state becomes

$$|\Psi'\rangle = \hat{S}_{ab}^{-1}(\eta)e^{i\delta\hat{a}^\dagger\hat{a}}\hat{S}_{ab}(\eta)|vac\rangle_{ab}. \qquad (11.81)$$

Obviously, the state $|\Psi'\rangle$ has a nonzero average photon number. By measuring the photon number in one of the outputs of the interferometer, we can detect the phase shift δ.

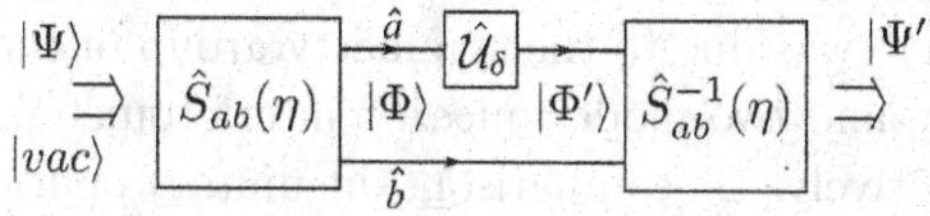

Fig. 11.9   SU(1,1) interferometer with vacuum input.

To find out the average photon number in the output of the interferometer, we find it easier to consider the evolution of the operators than that of the states. Referring to Fig. 11.9, fields a, b are related to the input vacuum fields a_0, b_0 as

$$\hat{a} = \mu \hat{a}_0 + \nu \hat{b}_0^\dagger, \quad \hat{b} = \mu \hat{b}_0 + \nu \hat{a}_0^\dagger, \tag{11.82}$$

where $\mu = \cosh|\eta|, \nu = \sinh|\eta|$. Assume field a experiences a phase shift of δ. Then the output fields a_{out}, b_{out} of the interferometer are related to the fields a, b as

$$\hat{a}_{out} = \mu \hat{a} e^{i\delta} - \nu \hat{b}^\dagger, \quad \hat{b}_{out} = \mu \hat{b} - \nu \hat{a}^\dagger e^{-i\delta}. \tag{11.83}$$

Hence,

$$\hat{a}_{out} = \left(1 + 2ie^{i\delta/2}\mu^2 \sin\frac{\delta}{2}\right) \hat{a}_0 + 2ie^{i\delta/2}\mu\nu \sin\frac{\delta}{2} \hat{b}_0^\dagger$$

$$\equiv G \hat{a}_0 + g \hat{b}_0^\dagger, \tag{11.84}$$

with $G \equiv 1 + 2ie^{i\delta/2}\mu^2 \sin\delta/2$ and $g \equiv 2ie^{i\delta/2}\mu\nu \sin\delta/2$. The photon number operator at output field a_{out} is then

$$\hat{n}_{a_{out}} = |G|^2 \hat{a}_0^\dagger \hat{a}_0 + |g|^2 \hat{b}_0 \hat{b}_0^\dagger + Gg^* \hat{a}_0 \hat{b}_0 + G^*g \hat{a}_0^\dagger \hat{b}_0^\dagger. \tag{11.85}$$

It is easily found that $\langle \hat{n}_{a_{out}} \rangle = |g|^2 \simeq 4\langle N \rangle^2 \sin^2 \delta/2$ for $(\langle N \rangle \gg 1)$ with $\langle N \rangle \equiv |\nu|^2$ being the average photon number in field a. Similar to the squeezed state interferometers discussed earlier, the noise $\Delta n_{a_{out}}$ is one photon so that $SNR = 4\langle N \rangle^2 \sin^2 \delta/2$. Therefore, $SNR \sim 1$ only when $\delta \sim 1/\langle N \rangle$. Furthermore, similar to the squeezed state interferometers, we find the probability of detecting any photon in the output field a_{out} as

$$\bar{P} = 1 - P_0 = 1 - \langle : e^{-\hat{n}_{a_{out}}} : \rangle$$

$$= \frac{4\langle N \rangle(\langle N \rangle + 1) \sin^2 \delta/2}{1 + 4\langle N \rangle(\langle N \rangle + 1) \sin^2 \delta/2}$$

$$\simeq \frac{\langle N \rangle^2 \delta^2}{1 + \langle N \rangle^2 \delta^2}, \quad \text{for } \langle N \rangle \gg 1, \ \delta \ll 1. \tag{11.86}$$

Therefore, the probability of detecting a phase shift of size δ is significantly different from zero only if $\delta \gtrsim 1/\langle N \rangle$. Thus the minimum detectable phase shift is $1/\langle N \rangle$, or the Heisenberg limit. Such a scheme was first discussed by Yurke *et al.* along the line of SU(1,1) interferometer [Yurke et al. (1986)]. The word "SU(1,1)" was due to the symmetry group of the unitary operators used in single- and two-mode squeezed interferometers in Eqs. (11.62) and (11.80), respectively. By comparison, the unitary operator for a lossless beam splitter in Eq. (A.6) falls in the SU(2) symmetry group.

Notice that the criterion here for detecting a phase shift is simply the detection of any photon in field a_{out} alone without considering field b_{out}. This is quite different from the criterion discussed earlier in Section 11.1.3 for the general case of mixed state. The general criterion is the detection of any photon in any relevant modes which will include both modes a_{out} and b_{out}. The reason for the difference is that the output fields a_{out}, b_{out} are actually in a twin photon state as Eq. (11.84) indicates and the photon numbers of modes a_{out}, b_{out} are perfectly correlated in such a way that the two modes have exactly the same photon number at all time. Thus detection of photon in mode a_{out} alone is equivalent to the detection of photon in any of the two fields.

11.3 Non-conventional Interferometers

We have already discussed in Section 11.2.2 a number of phase measurement schemes using some devices other than the traditional beam splitters in interferometers. They can reach a phase measurement sensitivity at Heisenberg limit. In this section, we will make a general discussion about phase measurement with non-conventional interferometers and concentrate on a specific type, i.e., SU(1,1) interferometers with coherent state boost, which is more practical to work with in the lab.

11.3.1 *General Consideration*

As we found before, in order to increase the sensitivity of phase measurement in a traditional interferometer such as the MZ interferometer, squeezed states are employed to reduce the quantum noise in the measurement. Some other nonclassical states with quantum correlation such as the photon number correlated twin-beam state may also be used. So, the attention is focused on the quantum states that are used in the interferometer but with the hardware structure of the interferometer unchanged. This is straightforward thinking after Caves pointed out, in a seminal paper [Caves (1981)], that quantum noise is the limiting factor in a traditional interferometer. So, most research effort is on finding the correct quantum states with various quantum correlations by reducing the noise for the increase of the phase measurement sensitivity. On the other hand, the measurement sensitivity, which is usually characterized by the signal-to-noise ratio (SNR), depends not only on the noise but also on the signal, that is,

$$\text{SNR} = \frac{\text{Signal}}{\text{Noise}}. \tag{11.87}$$

Signal increase will also improve the the measurement sensitivity even with the same noise. As a matter of fact, we already encountered an example of signal increase in Eqs. (11.74) and (11.79) in Section 11.2.2 where the change in the measured signal of δN due to a phase change δ is $\delta N = N_0 \delta P = M N_0 \delta$ (N_0 is total number of photons). This is M times as large as the case of a traditional interferometer. This increase in phase signal is because the interference fringe depends on $\sin(M\delta)$ which has a slope at $\delta = 0$, M times as large as $\sin \delta$ for a traditional interferometer. The achievement of this is by the employment of a non-traditional beam splitter described by Eqs. (11.68) and (11.75).

The increase in the signal thus relies on the hardware or structure change in an interferometer, which is on the unitary transformation $\hat{U}_{ab}, \hat{U}_\Phi$ in Fig. 11.4. On the other hand, the quantum state change can be viewed as software programming. Therefore, both strategy should be considered in order to find the optimum phase measurement scheme. In the following, we will revisit the SU(1,1) interferometer which shows the advantages in both aspects.

11.3.2 Coherent State Boosted SU(1,1) Interferometers

Although both single- and two-mode squeezed states used in SU(1,1) interferometers can be produced in the lab, there is a serious problem that makes their practical application difficult. For input in vacuum, we can find the photon number of the squeezed states as $\langle N \rangle = |\nu|^2 = \sinh^2 r$ with r as the squeezing parameter. However, it is usually hard to achieve very large squeezing parameter in the lab so that the fields used to sense the phase have a quite low photon number to be competitive with the traditional interferometry involving lasers which usually have a large number of photons. To increase the photon number for phase measurement, we can use a coherent state from a laser to inject into the SU(1,1) interferometer. This idea was first proposed by Plick *et al.* and dubbed as "coherent-state boosted SU(1,1) interferometer" [Plick et al. (2010)]. Furthermore, the detection scheme in the discussion of previous sections is mostly the photon number counting, whose sensitivity is limited by the dark counts of photo-detectors. As we have shown in Chapter 9, homodyne detection can overcome the dark count problem and achieve quantum-limited measurement. We will analyze the performance of the coherent-state boosted SU(1,1) interferometers and compare it with the traditional MZ interferometers in this section.

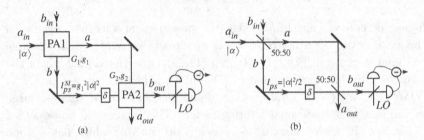

(a)   (b)

Fig. 11.10   (a) An SU(1,1) interferometer with parametric amplifiers (PA1, PA2) as the equivalent beam splitters. (b) A traditional Mach-Zehnder interferometer.

So, referring to the general scheme in Fig. 11.4(a), we have $|\Psi\rangle = |\alpha\rangle_a |vac\rangle_b$ as the input state and $\hat{U}_{ab}$ is the parametric amplifier from which a two-mode squeezed state can be produced (Section 4.6). With these, we redraw the configuration in Fig. 11.10(a) together with a traditional MZ interferometer in Fig. 11.10(b) for comparison. Thus, this scheme is similar to a traditional interferometer except that the wave splitting and combination elements are not regular beam splitters but parametric amplifiers. For general discussion, we assume the two parametric amplifiers (PA1,PA2) have different gains denoted by G_1, g_1 and G_2, g_2, respectively. A coherent state is input at both interferometers in field a_{in} while field b_{in} is in vacuum.

For the MZ interferometer in Fig. 11.10(b), with input-output relation for a lossless beam splitter given by

$$\hat{a} = (\hat{a}_{in} + \hat{b}_{in})/\sqrt{2}, \quad \hat{b} = (\hat{b}_{in} - \hat{a}_{in})/\sqrt{2};$$
$$\hat{a}_{out} = (\hat{a} - \hat{b}e^{i\delta})/\sqrt{2}, \quad \hat{b}_{out} = (\hat{b}e^{i\delta} + \hat{a})/\sqrt{2}, \quad (11.88)$$

it is straightforward to express the outputs of the interferometer in terms of the inputs:

$$\hat{a}_{out} = t(\delta)\hat{a}_{in} + r(\delta)\hat{b}_{in}; \quad \hat{b}_{out} = t(\delta)\hat{b}_{in} + r(\delta)\hat{a}_{in}, \quad (11.89)$$

where $t(\delta) = e^{i\delta/2}\cos\delta/2$; $r(\delta) = -ie^{i\delta/2}\sin\delta/2$. Here δ is the phase shift and we assume the beam splitters are identical with 50:50 transmissivity and reflectivity.

For a coherent state $|\alpha\rangle$ input at $\hat{a}_{in}$ and vacuum at $\hat{b}_{in}$, $\hat{b}_{out}$ is at the dark fringe at $\delta = 0$ and the output of this port is simply

$$\langle \hat{b}_{out}^{\dagger}\hat{b}_{out}\rangle = |\alpha|^2(1-\cos\delta)/2 = I_{ps}(1-\cos\delta), \tag{11.90}$$

where $I_{ps} \equiv \langle \hat{b}^{\dagger}\hat{b}\rangle = |\alpha|^2/2$ is the photon number of the field subject to the phase shift (phase sensing field). Note that the phase sensing field is what matters in phase measurement accuracy as shown in Section 11.1.1. We usually make homodyne measurement at the dark port, i.e., $\hat{b}_{out}$ of the interferometer with $\delta = 0$. With a small phase shift $\delta \ll 1$ and $\alpha = i|\alpha|$, we can easily find

$$\langle \hat{X}_{b_{out}}^2\rangle = \langle \hat{X}_{b_{in}}^2(\delta/2)\rangle\cos^2(\delta/2) + \langle \hat{Y}_{a_{in}}^2(\delta/2)\rangle\sin^2(\delta/2)$$
$$\approx 1 + |\alpha|^2\delta^2 \quad \text{for} \quad \delta \ll 1, \ |\alpha|^2 \gg 1. \tag{11.91}$$

Here $\hat{X}_{b_{out}} = \hat{b}_{out} + \hat{b}_{out}^{\dagger}$, $\hat{X}_b(\delta/2) = \hat{b}e^{i\delta/2} + \hat{b}^{\dagger}e^{-i\delta/2}$, and $\hat{Y}_a(\delta/2) = (\hat{a}e^{i\delta/2} - \hat{a}^{\dagger}e^{-i\delta/2})/i$ are the quadrature-phase amplitudes of corresponding fields. Obviously, $|\alpha|^2\delta^2$ in $\langle \hat{X}_{b_{out}}^2\rangle$ corresponds to the phase signal while the noise of the phase measurement is simply 1 from vacuum quantum noise. Hence, the signal-to-noise ratio of the Mach Zehnder interferometer is

$$SNR_{MZ} = |\alpha|^2\delta^2/1 = 2I_{ps}\delta^2, \tag{11.92}$$

which leads to the standard quantum limit $\delta_{SQL} = 1/\sqrt{N}$ with $N = 2I_{ps}$.

For the SU(1,1) interferometer in Fig. 11.10(a), a parametric amplifier (PA1) now acts as a beam splitter to split the input signal beam (a_{in}) into the amplified signal beam (a) and the accompanying beam (b). Another parametric amplifier (PA2) acts as a beam combiner to complete the interferometer. Even though there is no injection at mode b_{in} for the amplifier, vacuum still contributes with quantum noise. The full input-output relation for the amplifiers is given by (Section 4.6)

$$\hat{a} = G_1\hat{a}_{in} + g_1\hat{b}_{in}^{\dagger}, \quad \hat{b} = G_1\hat{b}_{in} + g_1\hat{a}_{in}^{\dagger};$$
$$\hat{a}_{out} = G_2\hat{a} - g_2\hat{b}^{\dagger}e^{-i\delta}, \quad \hat{b}_{out} = G_2\hat{b}e^{i\delta} - g_2\hat{a}^{\dagger}, \tag{11.93}$$

where we assume the amplifiers have amplitude gains G_1, G_2 with $|G_i|^2 - |g_i|^2 = 1 (i = 1, 2)$. We choose the sign of g_2 so that PA2 acts as a deamplifier. We introduce a phase shift of δ on field $\hat{b}$. Therefore, the output-input relation of the interferometer is

$$\hat{a}_{out} = [G_T(\delta)\hat{a}_{in} + g_T(\delta)\hat{b}_{in}^{\dagger}]e^{-i\delta},$$
$$\hat{b}_{out} = G_T(\delta)\hat{b}_{in} + g_T(\delta)\hat{a}_{in}^{\dagger}, \tag{11.94}$$

with $G_T(\delta) = G_1 G_2 e^{i\delta} - g_1 g_2$, $g_T(\delta) = G_2 g_1 e^{i\delta} - G_1 g_2$. The interferometer works best at dark fringe when $\delta = 0$ with $G_T(0) = G_1 G_2 - g_1 g_2$ and $g_T(0) = G_1 g_2 - G_2 g_1$, which has a unit overall gain $G_T(0) = 1$ for $G_1 = G_2$ [Yurke et al. (1986)].

With a coherent state input at a_{in} and no input at b_{in} as shown in Fig. 11.10(a), similar to Eq. (11.90), we can easily find the dark output port (b_{out}) intensity as

$$\langle \hat{b}_{out}^\dagger \hat{b}_{out} \rangle = |G_T(\delta)|^2 |\alpha|^2 + |g_T(\delta)|^2 (|\alpha|^2 + 1)$$
$$\approx 2G^2 g^2 |\alpha|^2 (1 - \cos\delta) \quad \text{for } |\alpha|^2 \gg 1 \text{ and } G_1 = G_2 \equiv G$$
$$= 2G^2 I_{ps}^{SI} (1 - \cos\delta), \qquad (11.95)$$

where $I_{ps}^{SI} = \langle \hat{b}^\dagger \hat{b} \rangle = g_1^2(|\alpha|^2 + 1) \approx g_1^2 |\alpha|^2 (|\alpha|^2 \gg 1)$ is the photon number of the phase sensing field $\hat{b}$ of the SU(1,1) interferometer (SI). Comparing the above with Eq. (11.90) for the MZ interferometer, we find that the fringe size is increased by a factor of $2G^2$ for the same I_{ps} and I_{ps}^{SI}.

For homodyne detection around the dark fringe with $\delta \ll 1$ and $\alpha = i|\alpha|$, we have from Eq. (11.94)

$$\langle \hat{X}_{b_{out}}^2 \rangle \equiv |G_T(\delta)|^2 \langle \hat{X}_{b_{in}}^2(\varphi_G) \rangle + |g_T(\delta)|^2 \langle \hat{X}_{a_{in}}^2(-\varphi_g) \rangle$$
$$\approx 1 + G^2 g^2 (4|\alpha|^2 + 2)\delta^2 \quad \text{for } \delta \ll 1, G_1 = G_2 \equiv G, \quad (11.96)$$

where $e^{i\varphi_G} \equiv G_T(\delta)/|G_T(\delta)|$, $e^{i\varphi_g} \equiv g_T(\delta)/|g_T(\delta)|$. Hence, the signal-to-noise ratio for the SU(1,1) interferometer is

$$SNR_{SI} = G^2 g^2 (4|\alpha|^2 + 2)\delta^2/1$$
$$\approx 4G^2 I_{ps}^{SI} \delta^2 \quad \text{for } |\alpha|^2 \gg 1. \qquad (11.97)$$

Comparing this with Eq. (11.92), we find that under the condition of the same number of photons for the phase sensing field, i.e., $I_{ps} = I_{ps}^{SI}$, the SU(1,1) interferometer has a better signal-to-noise ratio than the MZ interferometer with an enhancement factor of

$$SNR_{SI}/SNR_{MZ} \approx 2G^2 \quad \text{for } |\alpha|^2 \gg 1. \qquad (11.98)$$

The sensitivity of phase measurement for the MZ interferometer with coherent state and vacuum inputs is at the so-called standard quantum limit. Then the SU(1,1) interferometer improves upon the standard quantum limit by $2G^2$ fold. The physical picture of this enhancement in sensitivity is straightforward if we compare the output fringe intensities in Eqs. (11.90) and (11.95) for the two interferometers: the fringe size is increased by a factor of $2G^2$. But in the meantime, at $\delta = 0$ (dark output) with $G_1 = G_2$, we have $\hat{b}_{out} = \hat{b}_{in}$, which is in vacuum, so the noise is simply vacuum noise, just as in the MZ interferometer. Thus the improvement in sensitivity is achieved not by reducing the vacuum quantum noise at the unused input port, which is usually done with a MZ interferometer but rather by enhancing the signal level via amplification in the SU(1,1) interferometer.

Since the vacuum noise at the unused input port ($\hat{b}_{in}$) is the noise source of the SU(1,1) interferometer discussed above, we can reduce it with a squeezed state input at this port, i.e., $\langle \hat{X}_{b_{in}}^2 \rangle = e^{-r}$. Then Eq. (11.96) is modified to

$$\langle \hat{X}_{b_{out}}^2 \rangle = e^{-r} + 4G^2 g^2 |\alpha|^2 \delta^2. \tag{11.99}$$

Then the sensitivity of phase measurement can be further increased from the standard quantum limit by a factor of

$$SNR_{SI}^{sq}/SNR_{MZ} = 2G^2 e^r. \tag{11.100}$$

This enhancement factor combines the signal increase by amplification and the noise reduction by squeezed states at the input.

Another interesting case is when there is no injection of the coherent state at all. Setting $|\alpha|^2 = 0$ in Eq. (11.97), we have the signal-to-noise ratio without coherent injection

$$SNR_{SI}^{nc} = 2G^2 g^2 \delta^2 = 2I_{ps}^{SI}(I_{ps}^{SI} + 1)\delta^2, \tag{11.101}$$

where $I_{ps}^{SI} = g^2$. This leads to the Heisenberg limit [Yurke et al. (1986)]:

$$\delta_m = 1 \Big/ \sqrt{2I_{ps}^{SI}(I_{ps}^{SI} + 1)} \sim 1/N, \tag{11.102}$$

with $N \equiv I_{ps}^{SI} \gg 1$ as the number of photons probing the phase shift. Squeezed vacuum injection at the idler port can further increase the SNR but the photon number also increases because squeezed vacuum contains photons.

For the SU(1,1) interferometer, it is not surprising that signal is increased since the field containing the phase signal is amplified by PA2. However, as we have shown in Section 10.3.1, amplification of the signal is also accompanied by the amplification of noise and extra noise is usually added leading to a worse SNR. So, amplification alone cannot explain the enhancement of the SNR in the SU(1,1) interferometer. In fact, the role of the first parametric amplifier (PA1) is crucial here. It produces a pair of entangled fields (a, b) that have their noise correlated. We have demonstrated in Section 10.3.3 that SNR can be improved with correlated fields input to the amplifier. Therefore, it is noise cancelation due to quantum interference that leads to signal amplification without noise amplification and thus improvement in SNR.

To understand better the noise cancelation due to quantum interference, we calculate the output noise at dark port at an arbitrary phase δ. It is straightforward to obtain for coherent state and vacuum state input

$$\langle \Delta^2 \hat{X}_{b_{out}} \rangle = |G_T(\delta)|^2 + |g_T(\delta)|^2$$
$$= (G_1^2 + g_1^2)(G_2^2 + g_2^2) - 4G_1 G_2 g_1 g_2 \cos \delta. \quad (11.103)$$

The first term corresponds to the case when there is no correlation between the inputs to PA2 and the second one is the interference term from the quantum correlation between the inputs to PA2.

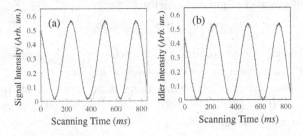

Fig. 11.11   Interference fringes observed at two outputs of an SU(1,1) interferometer. Reproduced from [Jing et al. (2011)].

Experimental realization of an SU(1,1) interferometer was first achieved by Jing *et al.* with atomic four-wave mixing processes as the two parametric amplifiers [Jing et al. (2011)]. Figure 11.11 shows the interference fringes at the two output ports. Notice that the two interference fringes are in phase in contrast to the situation of 180° out of phase for a MZ interferometer. This is typical of SU(1,1) interferometers [Chen et al. (2015)].

The schematic with noise behavior of an SU(1,1) interferometer is depicted in Fig. 11.12(b) as compared to an uncorrelated case in Fig. 11.12(a). The circles at each stage represent the noise sizes with labels corresponding to the noise levels labeled in Fig. 11.12(c). The quantum noise performance of an SU(1,1) interferometer was first measured by Hudelist *et al.* whose results are shown in Fig. 11.12(c) [Hudelist et al. (2014)]. Trace (iv) represents the noise level as the phase of the interferometer is scanned in time. It shows the interference pattern predicted in Eq. (11.103). The SNR improvement of 4 dB by the SU(1,1) interferometer as compared to a MZ interferometer was also demonstrated by Hudelist *et al.*

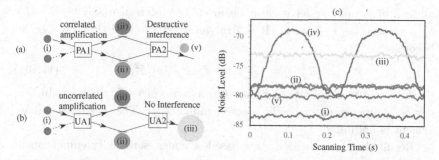

Fig. 11.12 (a) Schematic for the case of quantum-correlated amplifiers forming the SU(1,1) interferometer. (b) Schematic for the case of uncorrelated amplifiers with amplified noise. The circles represent sizes of noise levels with labels corresponding to part (c). (c) Quantum noise measured by homodyne detection for the *b*-output port of the SU(1,1) interferometer in Fig. 11.10(a). Trace (i): shot noise; Trace (ii): noise levels for each parametric amplifier alone; Trace (iii): noise level for two uncorrelated amplifiers in series; Traces (iv): noise level for correlated parametric amplifiers in SU(1,1) interferometer with phase scanned in time; Trace (v): same as (iv) but with phase locked at dark fringe. Adapted from [Hudelist et al. (2014)].

From Eq. (11.103), we find that $\langle \Delta^2 \hat{X}_{b_{out}} \rangle = 1$ for $\delta = 0$ when $G_1 = G_2$, i.e., it is at the shot noise level. However, the lowest noise level shown in Trace (v) in Fig. 11.12(c) is about 3 dB above the shot noise level (Trace (i)). This is due to the existence of losses inside the interferometer. The losses inside the SU(1,1) interferometer will destroy the quantum correlation with the introduction of uncorrelated vacuum noise. These uncorrelated vacuum noise will not be canceled but amplified above the shot noise level. We will make a quantitative analysis of the effect of losses next.

11.3.3 *Loss Analysis for the SU(1,1) Interferometer*

It is well-known that losses are the limiting factor that hinders the application of squeezed states in precision measurement. We will examine next the effect of losses on the sensitivity of the SU(1,1) interferometer. There are two types of losses: inside and outside the interferometer. We start with loss outside the interferometer first.

Effect of loss outside the interferometer

Loss outside the interferometer may come from propagation loss, less-than-perfect homodyne mode match, and most likely the finite quantum efficiency of the detectors. We can place all these losses into an overall loss of L and model it by a beam splitter of transmissivity $(1 - L)$: $\hat{b}'_{out} = \hat{b}_{out}\sqrt{1 - L} + \hat{b}_0\sqrt{L}$. It is straightforward to find that Eq. (11.96) becomes

$$\langle \hat{X}^2_{b'_{out}} \rangle = 1 + 4(1 - L)G^2 g^2 |\alpha|^2 \delta^2. \tag{11.104}$$

So the signal-to-noise ratio is reduced by a factor of $1 - L$. Such a reduction can be compensated by the increase in gains (G, g). The sensitivity enhancement is unlimited.

Recall that for squeezed state-based schemes, sensitivity enhancement is limited by the loss even for large amount of squeezing. In this sense, the scheme with the SU(1,1) interferometer is less prone to detection loss than squeezed state-based interferometers. The underlying physics for this insensitivity to loss is that the SU(1,1) interferometer is operated at vacuum noise level $(G_1 = G_2)$ or above $(G_1 \neq G_2)$. So the introduction of vacuum noise through losses will not change its noise performance too much. On the other hand, if we inject squeezed vacuum into the unused input port $(\hat{b}_{in})$ to further increase SNR as in Eq. (11.100), loss will limit the effect of squeezing just like squeezed state-based schemes.

Effect of loss inside the interferometer

For losses inside the interferometer, the situation is not as good. We consider two situations: (i) losses in the propagation between the two parametric amplifiers; (ii) losses inside the parametric amplifiers. For the first case, we again model the losses by beam splitters: $\hat{a}' = \sqrt{1 - L_1}\hat{a} + \sqrt{L_1}\hat{a}_0, \hat{b}' = \sqrt{1 - L_2}\hat{b} + \sqrt{L_2}\hat{b}_0$. We also assume the amplifiers have different gains of G_1, G_2, respectively. Then the idler output port is

$$\hat{b}_{out} = G'(\delta)\hat{b}_{in} + g'(\delta)\hat{a}^\dagger_{in}$$
$$+ g_2\sqrt{L_1}\hat{a}^\dagger_0 + G_2\sqrt{L_2}\hat{b}_0, \tag{11.105}$$

with

$$G'(\delta) = G_1 G_2 e^{i\delta}\sqrt{1 - L_2} - g_1 g_2\sqrt{1 - L_1}$$
$$g'(\delta) = g_1 G_2 e^{i\delta}\sqrt{1 - L_2} - g_2 G_1\sqrt{1 - L_1}. \tag{11.106}$$

For strong coherent state input with $|\alpha|^2 \gg 1$ and

$$g_2 G_1\sqrt{1 - L_1} = g_1 G_2\sqrt{1 - L_2} \equiv gG\sqrt{1 - L} \tag{11.107}$$

for 100% visibility, we find the output intensity as

$$\langle \hat{b}^\dagger_{out}\hat{b}_{out} \rangle \approx 2G^2 g^2 (1 - L)|\alpha|^2 (1 - \cos\delta)$$
$$= 2G_2^2 I^{SI}_{ps}(1 - L_2)(1 - \cos\delta). \tag{11.108}$$

So with loss, the fringe size is only reduced by $1-L_2$. However, the quantum noise is not so. With $\delta \ll 1$, we find from Eq. (11.105)

$$\langle \hat{X}_{b_{out}}^2 \rangle = \Big[|g_1 g_2 \sqrt{1 - L_1} - G_1 G_2 e^{i\delta} \sqrt{1 - L_2}|^2$$
$$+ |g_2 G_1 \sqrt{1 - L_1} - g_1 G_2 e^{i\delta} \sqrt{1 - L_2}|^2 \Big] 4|\alpha|^2$$
$$+ G_2^2 L_2 + g_2^2 L_1. \tag{11.109}$$

With condition in Eq. (11.107) and $\delta \ll 1$, Eq. (11.109) becomes

$$\langle \hat{X}_{b_{out}}^2 \rangle = 1 + 2g_2^2 L_1 + (4|\alpha|^2 + 2)g_1^2 G_2^2 (1 - L_2)\delta^2. \tag{11.110}$$

Hence, the signal-to-noise ratio is

$$SNR'_{NL} = (4|\alpha|^2 + 2)g_1^2 G_2^2 (1 - L_2)\delta^2/(1 + 2g_2^2 L_1)$$
$$\approx 4I_{ps}^{SI} G_2^2 (1 - L_2)\delta^2/(1 + 2g_2^2 L_1). \tag{11.111}$$

Compared to Eq. (11.97) for the case without losses, the SNR is reduced by a factor of $(1 - L_2)/(1 + 2g_2^2 L_1)$ and compared to the MZ interferometer, the enhancement in SNR is $2G_2^2 (1 - L_2)/(1 + 2g_2^2 L_1) \approx 2(1 - L_2)/L_1$ for large g_2^2. Thus, like the MZ interferometer with squeezed state, the enhancement is limited by the loss L_1.

Furthermore, for the case of no coherent state injection, the SNR becomes

$$SNR_{NL}^{nc'} = 2g_1^2 G_2^2 (1 - L_2)\delta^2/(1 + 2g_2^2 L_1) \tag{11.112}$$

and if we set $g_1^2 = g_2^2 \equiv I_{ps}^{SI}$ and $L_1 = L_2 \equiv L$,

$$SNR_{NL}^{nc'} = 2I_{ps}^{SI}(I_{ps}^{SI} + 1)(1 - L)\delta^2/(1 + 2I_{ps}^{SI} L). \tag{11.113}$$

So the minimum measurable phase is at the Heisenberg limit for small photon number with $I_{ps}^{SI} \ll 1/L$ but for large photon number of $I_{ps}^{SI} \gg 1/L$, Eq. (11.113) becomes

$$SNR_{NL}^{nc'} = (I_{ps}^{SI} + 1)(1 - L)\delta^2/L, \tag{11.114}$$

which only improves upon the standard quantum limit by $(1 - L)/L$, similar to the case of strong coherent injection. This shows that loss is the limiting factor for reaching Heisenberg limit in precision phase measurement.

The second type of losses of the interferometer is the losses inside the parametric amplifiers. This type of losses cannot be modeled as beam splitters but was considered as coupling to the outside vacuum of a non-degenerate OPO in Eq. (6.153) in Problem 6.6. In Problem 11.1, we treat

the SU(1,1) interferometer with this type of parametric amplifier and find
the SNR for phase measurement is given in Eq. (11.138) as

$$SNR''_{SI} = 4I^{SI}_{ps} \bar{G}^2 \delta^2 / (1 + 4\bar{g}'^2)$$
$$\approx I^{SI}_{ps} \delta^2 (\beta/\gamma) \quad \text{for} \ \bar{G} \gg 1$$
$$= 2I^{SI}_{ps} \delta^2 (T/2L), \tag{11.115}$$

where we used Eq. (6.154) for $\bar{G}$, $\bar{g}$, $\bar{G}'$, $\bar{g}'$. This SNR is enhanced from
SNL_{MZ} in Eq. (11.92) by a factor of $T/2L$. Notice that $L/T \approx S$ is the
maximum squeezing from one of the two NOPAs. So the enhancement is limited by the overall vacuum noise leaked into the two parametric amplifiers
through intra-cavity losses.

11.4   Joint Measurement of Conjugate Observables

Heisenberg uncertainty relation in quantum mechanics sets the limit on the
measurement precision of two conjugate observables. With squeezed states,
one can measure one observable more precisely than the standard quantum
limit at the expense of worse precision in the conjugate observable. In some
applications, however, we need to obtain the information embedded in two
conjugate observables. For example, the real and imaginary parts of the
linear susceptibility of an optical medium correspond respectively to the
phase and amplitude modulation of an optical field passing through the
medium. In this section, we will discuss the problem of joint measurement
of two conjugate variables.

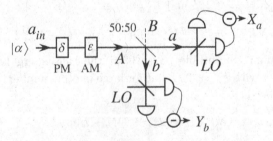

Fig. 11.13   Joint measurement of two orthogonal observables by using a beam splitter.

11.4.1   *Classical Measurement Schemes*

For the joint measurement of two quadrature-phase amplitudes, one simple
method is to split the modulated beam into two parts by a beam splitter of

transmissivity T and measure each quadrature-phase amplitude respectively, as shown in Fig. 11.13. Suppose the beam for signal encoding is in a coherent state $|\alpha\rangle$ with $\alpha = |\alpha|e^{j\varphi_0}$. A phase modulation of $\delta \ll 1$ and an amplitude modulation $\epsilon \ll 1$ are applied to the input beam simultaneously, as expressed in terms of the phase of $e^{j\delta} \approx 1 + j\delta$ and the amplitude transmission of $e^{-\epsilon} \approx 1 - \epsilon$, the modulated field[1] is then $\hat{A} = \hat{a}_{in}e^{j\delta}e^{-\epsilon} \approx \hat{a}_{in}(1 + j\delta - \epsilon)$.

The outputs of the 50:50 beam splitter are given by

$$\hat{a} = (\hat{A} + \hat{B})/\sqrt{2}, \quad \hat{b} = (\hat{B} - \hat{A})/\sqrt{2}, \tag{11.116}$$

where $\hat{B}$ is in vacuum. Now we measure $\hat{X}_a = \hat{a}e^{-j\varphi_0} + \hat{a}^\dagger e^{j\varphi_0}$ at one output and $\hat{Y}_b = (\hat{b}e^{-j\varphi_0} - \hat{b}^\dagger e^{j\varphi_0})/j$ at the other by homodyne measurement. Since the input is in a coherent state, the noise is simply the vacuum noise for both $\hat{X}_a$ and $\hat{Y}_b$:

$$\langle \Delta^2 \hat{X}_a \rangle = 1 = \langle \Delta^2 \hat{Y}_b \rangle. \tag{11.117}$$

The measured signals can be calculated as

$$\langle \hat{Y}_b \rangle = -[|\alpha|e^{j\varphi_0}(1 + i\delta - \epsilon)e^{-j\varphi_0}/\sqrt{2} - c.c.]/j = -\sqrt{2}|\alpha|\delta$$

$$\langle \hat{X}_a \rangle = |\alpha|e^{j\varphi_0}(1 + i\delta - \epsilon)e^{-j\varphi_0}/\sqrt{2} + c.c. = \sqrt{2}|\alpha|(1 - \epsilon). \tag{11.118}$$

Hence, the signals due to modulations are $S_{X_a} = 2|\alpha|^2\epsilon^2$, $S_{Y_b} = 2|\alpha|^2\delta^2$ and the signal-to-noise ratios are

$$SNR_X = 2I_{ps}\epsilon^2, \quad SNR_Y = 2I_{ps}\delta^2, \tag{11.119}$$

where $I_{ps} \equiv |\alpha|^2$ is the photon number of the probe sensing beam.

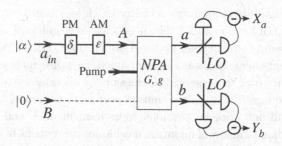

Fig. 11.14 Joint measurement of two orthogonal observables by using a non-degenerate parametric amplifier for splitting the incoming field for simultaneous measurement.

However, the actually measured SNRs can be smaller than those given above due to detection losses. One way to circumvent this is to use a

---

[1]Because $\epsilon \ll 1$, we dropped the vacuum term that resulted from transmission loss of $e^{-\epsilon}$.

parametric amplifier as the beam splitter before the homodyne detection, as shown in Fig. 11.14. A parametric amplifier is described by the following input-output relation:

$$\hat{a} = G\hat{A} + g\hat{B}^\dagger, \quad \hat{b} = G\hat{B} + g\hat{A}^\dagger, \tag{11.120}$$

where $\hat{A}$ is the modulated signal beam in the coherent state and $\hat{B}$ is in vacuum. It is straightforward to show that the output signals are

$$\langle X_a \rangle^2 = 4G^2 I_{ps}\epsilon^2, \langle Y_b \rangle^2 = 4g^2 I_{ps}\delta^2 \tag{11.121}$$

and the output noise are

$$\langle \Delta^2 X_a \rangle = G^2 + g^2 = \langle \Delta^2 Y_b \rangle. \tag{11.122}$$

The output signal to noise ratios for the amplification scheme are then

$$SNR_X^{amp} = \frac{4G^2 I_{ps}\epsilon^2}{G^2 + g^2}, \quad SNR_Y^{amp} = \frac{4g^2 I_{ps}\delta^2}{G^2 + g^2}. \tag{11.123}$$

The superscript "amp" is meant for the amplification scheme. At large gain of $g^2 \gg 1$, the results in Eq. (11.123) are the same as that in Eq. (11.119).

The above schemes are classical schemes for joint measurement in the sense that their performance is at shot noise limit and joint measurement leads to 3 dB = 1/2 reduction in SNR (for $G \gg 1$) due to vacuum noise introduced (from field B) in equal information splitting.

11.4.2   *Joint Measurement with EPR Correlated States*

With the availability of quantum resources such as squeezed states, we have demonstrated in Section 10.2.1 that it is possible to enhance the sensitivity of phase measurement beyond classical limit. However, the noise decrease in the X-component is always accompanied by noise increase in the Y-component in order to preserve the Heisenberg uncertainty relation. Thus injecting a squeezed state in the unused port in the schemes in Figs. 11.13 and 11.14 will not work for the joint measurement of X and Y. It seems that the fundamental law of quantum mechanics prevents us from improving beyond classical limit the sensitivity of joint measurement of the modulations encoded in two orthogonal conjugate observables.

On the other hand, quantum mechanics allows quantum correlations that are stronger than classical correlations through the quantum magic of entanglement, as beautifully presented in the Einsein-Poldosky-Rosen (EPR) paradox. In Section 10.1.3, we demonstrated that we can infer through EPR correlations two conjugate observables with precision better

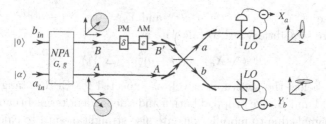

Fig. 11.15 Joint measurement of two orthogonal observables with entangled fields generated from a non-degenerate parametric amplifier.

than what is allowed by Heisenberg uncertainty relation. Next, we exploit this quantum correlation for the improvement of the sensitivity of joint measurement of the modulations encoded in two orthogonal conjugate observables.

One way to make use of this is to encode information on one of the EPR correlated beams and combine the two EPR correlated beams with a 50:50 beam splitter. We have shown in Section 6.2.4 that when two squeezed states (Eq. (6.86)) are combined in a 50:50 beam splitter, a two-mode squeezed state or an EPR entangled state can be produced (Eq. (6.90)). If we reverse the process with two EPR entangled beams combined in the beam splitter, as shown in the beam splitter part of Fig. 11.15, the two output beams will be in squeezed states with one squeezed in X while the other in Y (see also Problem 10.1). Thus, if we can encode signal in both X and Y on the input beam and measure X on one output and Y on the other, the encoded signal is only split half for two sides but the noise can be squeezed on both X and Y, leading to joint measurement of X and Y with improved SNR.

However, the parametric amplifier in the spontaneous mode has small photon number outputs for signal encoding. So, we resort to a coherent state input to the PA for photon number boosting, similar to the case in SU(1,1) interferometer discussed in Section 11.3.2. This scheme is shown in Fig. 11.15. The parametric amplifier is again described by

$$\hat{A} = G\hat{a}_{in} - g\hat{b}_{in}^{\dagger}, \quad \hat{B} = G\hat{b}_{in} - g\hat{a}_{in}^{\dagger}, \tag{11.124}$$

where $\hat{a}_{in}$ is in a coherent state $|\alpha\rangle$ and $\hat{b}_{in}$ is in vacuum. We changed the sign of g so that the corresponding quadrature-phase amplitudes are squeezed in noise. We use the output field $\hat{B}$ to probe the modulations: $\hat{B}' = \hat{B}(1 + j\delta - \epsilon)$. The outputs of the BS are related to the inputs by

$$\hat{a} = (\hat{A} + \hat{B}')/\sqrt{2}, \quad \hat{b} = (\hat{B}' - \hat{A})/\sqrt{2}. \tag{11.125}$$

The noise of $\hat{X}_a = \hat{a}e^{-j\varphi_0} + \hat{a}^\dagger e^{j\varphi_0}$ and $\hat{Y}_b = (\hat{b}e^{-j\varphi_0} - \hat{b}^\dagger e^{j\varphi_0})/j$ ($\varphi_0 \equiv \alpha/|\alpha|$) are straightforward to calculate as

$$\langle \Delta^2 \hat{X}_a \rangle = (G-g)^2 = \langle \Delta^2 \hat{Y}_b \rangle, \tag{11.126}$$

which is below the shot noise level, as expected for EPR entangled state between $\hat{A}$ and $\hat{B}$. We dropped both δ and ϵ dependent terms because $\delta, \epsilon \ll 1$. The signals due to modulations are also straightforward to calculate:

$$\langle X_a \rangle^2 = 2I_{ps}\epsilon^2, \langle Y_b \rangle^2 = 2I_{ps}\delta^2, \tag{11.127}$$

where $I_{ps} \equiv g^2|\alpha|^2$. The SNRs are then

$$SNR_X^{EPR} = \frac{2I_{ps}\epsilon^2}{(G-g)^2}, \quad SNR_Y^{EPR} = \frac{2I_{ps}\delta^2}{(G-g)^2}, \tag{11.128}$$

This result improves upon the classical SNRs in Eq. (11.119) by a factor of $1/(G-g)^2 = (G+g)^2$.

11.4.3   *Joint Measurement with SU(1,1) Interferometer*

The scheme in the previous section involves reduction of the shot noise in photo-detection and is thus sensitive to losses at detection. To avoid this problem, we can use the amplification scheme for the replacement of the BS and combine the schemes in Fig. 11.15 and Fig. 11.14. The new scheme is shown in Fig. 11.16, which is just the SU(1,1) interferometer that we discussed in Section 11.3.2. We will next analyze the performance of the SU(1,1) interferometer on the joint measurement of two quadrature-phase amplitudes.

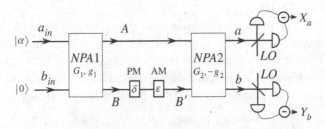

Fig. 11.16   Joint measurement of two orthogonal observables with an SU(1,1) interferometer.

The first parametric amplifier (NPA1) amplifies the input signal field $\hat{a}_{in}$ at coherent state $|\alpha\rangle$ to field $\hat{A}$ and produces a correlated idler field $\hat{B}$. The idler input field $\hat{b}_{in}$ is in vacuum. The idler field $\hat{B}$ then passes through a phase modulator (PM) and an amplitude modulator (AM) for signal encoding at orthogonal observables. The second parametric amplifier (NPA2) acts as a beam combiner to complete the interferometer, and the output fields of signal mode and idler mode are denoted by $\hat{a}$ and $\hat{b}$. Take the amplitude gains of the two parametric amplifiers as G_1, g_1 and G_2, g_2, respectively. Then the operators are related by

$$\hat{A} = G_1\hat{a}_{in} + g_1\hat{b}_{in}^{\dagger}, \quad \hat{B} = G_1\hat{b}_{in} + g_1\hat{a}_{in}^{\dagger}$$
$$\hat{a} = G_2\hat{A} - g_2\hat{B}'^{\dagger}, \quad \hat{b} = G_2\hat{B}' - g_2\hat{A}^{\dagger}. \tag{11.129}$$

$\hat{B}'$ is the modulated field and $\hat{B}' = \hat{B}(1+j\delta-\epsilon)$. The signs of the amplitude gains are chosen so that the interferometer works at a dark fringe. Under this condition, correlated noise generated by NPA1 is canceled and noise variance is reduced because of the phase sensitive de-amplification of NPA2. To evaluate the noise of the outputs, we can ignore the modulation terms by setting $\delta = 0 = \epsilon$ since $\delta, \epsilon \ll 1$. Then, the outputs are related to the inputs by

$$\hat{a} = \tilde{G}\hat{a}_{in} + \tilde{g}\hat{b}_{in}^{\dagger}, \quad \hat{b} = \tilde{G}\hat{b}_{in} + \tilde{g}\hat{a}_{in}^{\dagger}, \tag{11.130}$$

where $\tilde{G} = G_2G_1 - g_2g_1$, $\tilde{g} = G_2g_1 - g_2G_1$ are the phase sensitive gain at the dark fringe.

For arbitrary quadrature angles θ_1, θ_2, the noise levels of the outputs of the interferometer can be calculated as

$$\left\langle \Delta^2\hat{X}_a(\theta_1) \right\rangle = (G_2G_1 - g_1g_2)^2 + (G_1g_2 - G_2g_1)^2$$
$$\left\langle \Delta^2\hat{X}_b(\theta_2) \right\rangle = (G_2G_1 - g_1g_2)^2 + (G_1g_2 - G_2g_1)^2. \tag{11.131}$$

Note that the noise variances of different quadrature-phase amplitudes of the output fields are determined by the gain g_1 and g_2 of the parametric amplifiers only and are independent of the angles θ_1, θ_2, which is different from the squeezed state interferometry [Xiao et al. (1987)]. When $g_1 = g_2$, the output noise variances are minimized to $\left\langle \Delta^2\hat{X}_{a,b}(\theta) \right\rangle = 1$, which means that the output noise levels of the two-amplifier combination can be reduced to shot noise level on both output sides of the interferometer at the same time. This is quite different from the single amplifier case where output noise level is amplified from the shot noise level. This is because of a destructive quantum interference effect for noise cancelation due to quantum correlation [Ou (1993); Kong et al. (2013a)]. Thus, the SU(1,1) interferometer can be applied to the joint measurement at two different outputs of two arbitrary quadrature-phase amplitudes which are not necessarily orthogonal to each other.

For the output signals at two output ports, we find

$$\langle X_a \rangle^2 = 4g_2^2 I_{ps} \delta^2, \quad \langle Y_b \rangle^2 = 4G_2^2 I_{ps} \epsilon^2, \tag{11.132}$$

where $I_{ps} = \left\langle \hat{B}^+ \hat{B} \right\rangle = g_1^2 |\alpha|^2 (|\alpha|^2 \gg 1)$ is the photon number of the probe sensing field $\hat{B}$.

Combining Eqs. (11.131) and (11.132), we obtain the SNRs of outputs at $\hat{a}$ and $\hat{b}$ as

$$SNR_{SI}(X_a) = \frac{4g_2^2 I_{ps} \delta^2}{(G_2 G_1 - g_1 g_2)^2 + (G_1 g_2 - G_2 g_1)^2}$$

$$SNR_{SI}(Y_b) = \frac{4G_2^2 I_{ps} \epsilon^2}{(G_2 G_1 - g_1 g_2)^2 + (G_1 g_2 - G_2 g_1)^2} \tag{11.133}$$

where subscript SI represents the measurement scheme of SU(1,1) interferometer. With a fixed amplitude gain g_1, SNR_{SI}'s of quadrature-phase amplitudes have the maximum value when $g_2 \to \infty$. In this situation, $SNR_{SI}(X_a) = 2(G_1 + g_1)^2 I_{ps} \delta^2$, $SNR_{SI}(Y_b) = 2(G_1 + g_1)^2 I_{ps} \epsilon^2$. Notice that the optimum condition here is different from the minimum output noise condition ($g_1 = g_2$) discussed in Eq. (11.131). It is because the modulated signal is amplified more than the noise with the increase of gain g_2.

Comparing to the results in Eq. (11.119) for the classical joint measurement schemes in Figs. 11.13 and 11.14, we have an SNR improvement factor of

$$\frac{SNR_{SI}(X(\theta))}{SNR_c(X(\theta))} = (G_1 + g_1)^2 \qquad (\theta = 0, \pi/2). \tag{11.134}$$

This improvement factor is the same as the one for the scheme using the EPR states. However, the scheme with EPR states can only be applied to joint measurement of orthogonal variables of X_a and Y_b because of the property of squeezed states. The other quadrature-phase amplitudes will not have minimum noise. The situation is different for the scheme with the SU(1,1) interferometer, which can be applied to the joint measurement of two arbitrary quadrature-phase amplitudes. This is so because the noise outputs of the interferometer given in Eq. (11.131) do not depend on the phase angle θ, i.e., the noise is canceled for all quadrature-phase amplitudes due to a destructive quantum interference in the interferometer.

The experimental demonstration of joint measurement of two orthogonal observables with precision beating the classical limit was performed by Li *et al.* [Li et al. (2002)] with a scheme proposed by Zhang and Peng (see Problem 11.3) [Zhang and Peng (2000)].

11.5 Problems

Problem 11.1 Effect of losses inside the parametric amplifiers on the performance of SU(1,1) interferometer.

Losses inside the parametric amplifiers were considered in Eq. (6.153) in Problem 6.6. Use that result and assume the two parametric amplifier in the SU(1,1) interferometer in Fig. 11.10(a) are identical.

(i) With a phase shift δ on b-field, prove that the output fields of the interferometer are

$$\hat{b}_{out} = \bar{G}_T(\delta)\hat{b}_{in} + \bar{g}_T(\delta)\hat{a}_{in}^\dagger + \bar{G}_T'(\delta)\hat{b}_{01}$$
$$+ \bar{g}_T'(\delta)\hat{a}_{01}^\dagger + \bar{G}'\hat{b}_{02} + \bar{g}'\hat{a}_{02}^\dagger, \qquad (11.135)$$

where $\hat{a}_{01}, \hat{b}_{01}, \hat{a}_{02}, \hat{b}_{02}$ are the vacuum modes coupled in through the losses inside the two NOPAs and

$$\bar{G}_T(\delta) = \bar{g}^2 - \bar{G}^2 e^{i\delta}, \quad \bar{g}_T(\delta) = \bar{g}\bar{G}(1 - e^{i\delta}),$$
$$\bar{G}_T'(\delta) = \bar{g}\bar{g}' - \bar{G}\bar{G}'e^{i\delta}, \quad \bar{g}_T'(\delta) = \bar{G}'\bar{g} - \bar{G}\bar{g}'e^{i\delta}. \qquad (11.136)$$

$\bar{G}, \bar{g}, \bar{G}', \bar{g}'$ are given in Eq. (6.154).

(ii) For $\delta \ll 1$ and a strong coherent state at $\hat{a}_{in}$, prove that

$$\langle \hat{X}_{b_{out}}^2 \rangle = (\bar{g}^2 - \bar{G}^2)^2 + 4I_{ps}^{SI}\bar{G}^2\delta^2 + (\bar{g}\bar{g}' - \bar{G}\bar{G}')^2$$
$$+ (\bar{g}\bar{G}' - \bar{G}\bar{g}')^2 + \bar{G}'^2 + \bar{g}'^2$$
$$= 4I_{ps}^{SI}\bar{G}^2\delta^2 + 1 + 2(\bar{g}\bar{G}' - \bar{G}\bar{g}')^2 + 2\bar{g}'^2 \qquad (11.137)$$

where $I_{ps}^{SI} = \langle \hat{b}^\dagger \hat{b} \rangle \equiv \bar{g}^2|\alpha_{in}|^2$. Hence, the SNR is

$$SNR_{SI}'' = 4I_{ps}^{SI}\bar{G}^2\delta^2/(1 + 4\bar{g}'^2), \qquad (11.138)$$

where we used Eq. (6.154).

Problem 11.2 An unconventional interferometer with a parametric amplifier and a beam splitter [Kong et al. (2013b)].

The joint measurement scheme depicted in Fig. 11.15 is actually an unconventional interferometer in which the beam splitting is done with a parametric amplifier but the beam recombination is achieved with a regular beam splitter. This interferometer scheme works as long as the two outputs A, B from the parametric amplifier (NPA) are degenerate in frequency. For the generality, assume the beam splitter has a transmissivity of T. Next, we will assume a strong coherent state injection at a_{in} with $|\alpha| \gg 1$.

(i) Replace the modulators in Fig. 11.15 with a phase shifter of phase φ. Show that the intensities at the two outputs of the BS exhibit interference fringes as φ changes. Find the visibility of the interference as a function of g, α, and T and show that the visibility becomes 100% on one side when $T = g^2/(G^2 + g^2)$ or on the other side when $T = G^2/(G^2 + g^2)$.

(ii) Let us concentrate on the homodyne detection on output a only. Show that the SNR has a maximum of

$$SNR_{Max} = 4I_{ps}\delta^2(G^2 + g^2) \tag{11.139}$$

when $T = 4G^2g^2/(8G^2g^2 + 1) \to 1/2$ for $G \gg 1$. Given that the phase measurement SNR with a classical interferometer is $SNR_{cl} = 2I_{ps}\delta^2$, the improvement over classical scheme is then

$$\frac{SNR_{Max}}{SNR_{cl}} = 2(G^2 + g^2). \tag{11.140}$$

This is a small improvement over Eq. (11.128) because we optimized T.

Problem 11.3 A scheme of direct detection for joint measurement with an EPR source and a beam splitter [Zhang and Peng (2000)].

The joint measurement scheme depicted in Fig. 11.15 involves homodyne detections which require extra beams for local oscillators (LO) and phase lock to observe $\hat{X}_a$ and $\hat{Y}_b$. This makes the experimental implementation complicated. In Section 9.8, we discussed the scheme of self-homodyne detection by direct detection without local oscillators. We will use it here to simplify the detection scheme.

In Section 9.8, we showed that direct detection of a real bright field ($\langle \hat{a} \rangle = $ real) measures $\hat{X}$ and in the discussion part at the end of Section 10.1.4, we find that the twin beams generated by the injection of a real coherent state ($|\alpha\rangle$ with $\alpha = $ real) in a parametric amplifier have intensity correlation that is equivalent to noise reduction of $\hat{X}_A - \hat{X}_B$. This occurs for a parametric amplifier with all positive gain parameters: $g > 0$. Now consider the case of negative gain, that is,

$$\hat{A} = G\hat{a}_{in} - g\hat{b}_{in}^\dagger, \quad \hat{B} = G\hat{b}_{in} - g\hat{a}_{in}^\dagger, \tag{11.141}$$

which leads to

$$\hat{X}_A + \hat{X}_B = (G - g)(\hat{X}_{a_{in}} + \hat{X}_{b_{in}}),$$
$$\hat{Y}_A - \hat{Y}_B = (G - g)(\hat{Y}_{a_{in}} - \hat{Y}_{b_{in}}). \tag{11.142}$$

If $\hat{a}_{in}$, $\hat{b}_{in}$ are in vacuum or coherent state, the relations above give rise to noise reduction in $\hat{X}_A + \hat{X}_B$ and $\hat{Y}_A - \hat{Y}_B$ in contrast to the noise reduction in $\hat{X}_A - \hat{X}_B$ and $\hat{Y}_A + \hat{Y}_B$ (Eq. (10.10)) for the case of positive gain ($g > 0$) discussed in Section 10.1.3. At the end of Section 10.1.4, we discussed the scheme of injection at only one input of the parametric amplifier and found it always leads to noise reduction in $\hat{X}_A - \hat{X}_B$. Things are different for injection at two inputs.

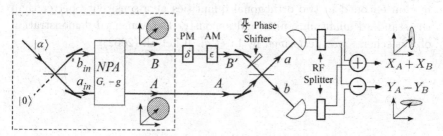

Fig. 11.17   Joint measurement of two orthogonal observables by direct detection.

(i) Consider now the injection scheme shown on the left part of Fig. 11.17 (dashed box). Show that for real α, the outputs have $\langle \hat{A} \rangle = (G-g)\alpha/\sqrt{2} = \langle \hat{B} \rangle$ = real. So, direct detection of fields A, B will yield $\hat{X}_A$, $\hat{X}_B$.

(ii) Consider the detection scheme on the right part of Fig. 11.17 (outside dashed box). Let us take out the PM and AM modulators first. Show that for large $\langle \hat{A} \rangle$, $\langle \hat{B} \rangle$ or $\bar{\alpha} \equiv (G - g)\alpha/\sqrt{2} \gg 1$, we have

$$\hat{a}^\dagger \hat{a} + \hat{b}^\dagger \hat{b} = \hat{A}^\dagger \hat{A} + \hat{B}^\dagger \hat{B} \approx \bar{\alpha}(\hat{X}_A + \hat{X}_B) + 2\bar{\alpha}^2,$$

$$\hat{a}^\dagger \hat{a} - \hat{b}^\dagger \hat{b} \approx \bar{\alpha}(\hat{Y}_A - \hat{Y}_B). \tag{11.143}$$

Thus, the sum and difference currents make measurements of $\hat{X}_A + \hat{X}_B$, $\hat{Y}_A - \hat{Y}_B$, respectively and we achieve self-homodyne detection of these quantities together. Show that these currents have fluctuations below the shot noise levels by a factor of $1/(G + g)^2$.

Contrary to the scheme of bright twin-beams generation in Section 10.1.4, where only $\hat{X}_A - \hat{X}_B$ is measured via self-homodyne, the scheme presented here achieves measurement of both $\hat{X}_A + \hat{X}_B$ and $\hat{Y}_A - \hat{Y}_B$ via self-homodyne and they have the EPR-type correlations, which demonstrates the EPR paradox with self-homodyne detection.

(iii) Now let us move the PM and AM modulators in place as shown. Calculate the signal sizes for the sum and difference currents and show that the sum current measures amplitude modulation ϵ (AM) and the difference current gives the phase modulation δ (PM).

(iv) Show the signal-to-noise ratios at the two output currents are

$$SNR_+ = \frac{2I_{ps}\epsilon^2}{(G-g)^2}; \quad SNR_- = \frac{2I_{ps}\delta^2}{(G-g)^2} \qquad (11.144)$$

with $I_{ps} \equiv \bar{\alpha}^2$. Therefore, we can achieve joint measurement of the information encoded in two orthogonal quantities at a sensitivity better than the standard quantum limit simultaneously. Experimental demonstration of this scheme was performed by Li *et al.* [Li et al. (2002)].

Appendix A

Derivation of the Explicit Expression for $\hat{U}$ of a Lossless Beam Splitter

This derivation of $\hat{U}$ bears some resemblance to angular momentum operators in rotation in quantum mechanics.

For a lossless beam splitter, since we have $t^2 + r^2 = 1$, we can assign t as $\cos\theta$ and r as $\sin\theta$. With this, we can rewrite the operator relations in Eq. (6.46) as:

$$\begin{cases} \hat{b}_1 = \hat{U}^\dagger \hat{a}_1 \hat{U} = \cos\theta \hat{a}_1 + \sin\theta \hat{a}_2, \\ \hat{b}_2 = \hat{U}^\dagger \hat{a}_2 \hat{U} = \cos\theta \hat{a}_2 - \sin\theta \hat{a}_1, \end{cases} \tag{A.1}$$

which is similar to the transformation of the two-dimensional rotation of angle θ. As in any transformation, we consider an infinitesimal transformation of $\delta\theta << 1$ and make a linear approximation of

$$\hat{U}(\delta\theta) \approx 1 + i\delta\theta\hat{I}, \tag{A.2}$$

where $\hat{I}$ is a function of $\hat{a}_1, \hat{a}_2$ to be determined. Because $\hat{U}\hat{U}^\dagger = \hat{U}^\dagger\hat{U} = 1$, we have $\hat{I} = \hat{I}^\dagger$ or $\hat{I}$ is a Hermitian operator. For the transformation of finite θ, we have:

$$\hat{U}(\theta) = \hat{U}(\theta/N)^N$$

$$= \lim_{N\to\infty} (1 + i\hat{I}\theta/N)^N$$

$$= \exp(i\theta\hat{I}). \tag{A.3}$$

Substituting Eq. (A.2) into Eq. (A.1), we obtain:

$$[\hat{a}_1, \hat{I}] = -i\hat{a}_2, \quad [\hat{a}_2, \hat{I}] = i\hat{a}_1. \tag{A.4}$$

From the commutators: $[\hat{a}_1, \hat{a}_1^\dagger] = [\hat{a}_2, \hat{a}_2^\dagger] = 1$, we have:

$$\hat{I} = -i\hat{a}_2\hat{a}_1^\dagger + i\hat{a}_1\hat{a}_2^\dagger + f(\hat{a}_1, \hat{a}_2). \tag{A.5}$$

But $\hat{I}$ is an Hermitian operator, hence $f(\hat{a}_1, \hat{a}_2) = 0$ and the final expression for $\hat{U}$ is

$$\hat{U} = \exp[\theta(\hat{a}_2\hat{a}_1^\dagger - \hat{a}_1\hat{a}_2^\dagger)], \tag{A.6}$$

with $t = \cos\theta, r = \sin\theta$. The above expression is the same as that in [Campos et al. (1989)], which is based on angular momentum theory.

Appendix B

Evaluation of the Two Sums in Eq. (8.100)

It is straightforward to find the first sum in Eq. (8.100) as

$$
\sum_{k=l} = \frac{1}{2^{N+1}} \int dt_0 dt_1 ... dt_N \sum_k \mathbb{P}_{t_0 t_k} \left[\Big| \sum_{\mathbb{P}} G(t_0; \mathbb{P}\{t_1, ..., t_N\}) \Big|^2 \right]
$$

$$
= \frac{N+1}{2^{N+1}} \int dt_0 dt_1 ... dt_N \sum_{\mathbb{P}'} G^*(t_0; \mathbb{P}'\{t_1, ..., t_N\}) \sum_{\mathbb{P}} G(t_0; \mathbb{P}\{t_1, ..., t_N\})
$$

$$
= \frac{N+1}{2^{N+1}} \sum_{\mathbb{P}'} \int \mathbb{P}'\{dt_0 dt_1 ... dt_N\} G^*(t_0; t_1, ..., t_N)
$$

$$
\times \sum_{\mathbb{P}} G(t_0; \mathbb{P}\{\mathbb{P}'\{t_1, ..., t_N\}\})
$$

$$
= \frac{N+1}{2^{N+1}} N! \int dt_0 dt_1 ... dt_N G^*(t_0; t_1, ..., t_N) \sum_{\mathbb{P}} G(t_0; \mathbb{P}\{t_1, ..., t_N\})
$$

$$
= (N+1)! \mathcal{N} \tag{B.1}
$$

with

$$
\mathcal{N} \equiv \frac{1}{2^{N+1}} \int dt_0 dt_1 ... dt_N G^*(t_0; t_1, ..., t_N) \sum_{\mathbb{P}} G(t_0; \mathbb{P}\{t_1, ..., t_N\}).
$$

Here, we made a change of variables: $\mathbb{P}'\{t_1, ..., t_N\} \to \{t_1, ..., t_N\}$ in evaluating the integral.

To evaluate the second sum in Eq. (8.100), let us assume $T_0 = T_n$ with $n = 1, ..., m$ and $|T_0 - T_n| \gg \Delta T$ for $n = m + 1, ..., N$, that is, the single photon entering at port 2 completely overlaps with m photons in the N-photon state entering port 1 but is well-separated with other $N - m$

photons. We calculate one arbitrary term as follows:

$$P_{kl} \equiv \frac{1}{2^{N+1}} \int dt_0 dt_1 ... dt_N \mathbb{P}_{t_0 t_k} \Big[\sum_{\mathbb{P}'} G^*(t_0; \mathbb{P}'\{t_1, ..., t_N\}) \Big]$$
$$\times \mathbb{P}_{t_0 t_l} \Big[\sum_{\mathbb{P}} G(t_0; \mathbb{P}, \{t_1, ..., t_N\}) \Big]$$
$$= \frac{1}{2^{N+1}} \int dt_0 dt_1 ... dt_N \sum_{\mathbb{P}'} G^*(t_k; \mathbb{P}'\{t_1, ...t_0..., t_N\})$$
$$\times \sum_{\mathbb{P}} G(t_l; \mathbb{P}\{t_1, ...t_0..., t_N\}). \qquad (B.2)$$

Since $k \neq l$, we can make a change of variable: $t_0 \leftrightarrow t_l$ without the change of the integral. This is because t_0, t_l are both inside the permutation $\mathbb{P}'$ of $G^*(t_k; \mathbb{P}'\{t_1, ...t_0..., t_N\})$ and permutation inside permutation does not change the result. Hence, we have

$$P_{kl} = \frac{1}{2^{N+1}} \int dt_0 dt_1 ... dt_N \sum_{\mathbb{P}'} G^*(t_k; \mathbb{P}'\{t_1, ...t_0..., t_N\})$$
$$\times \sum_{\mathbb{P}} G(t_0; \mathbb{P}\{t_1, ...t_l..., t_N\}). \qquad (B.3)$$

To evaluate the integral, we note that $G(t_0; t_1, ...t_N) \equiv g(t_0 - T_0)g(t_1 - T_1)...g(t_N - T_N)$ with $g(t)$ having a width of ΔT. Now since t_0 is not in the permutation of $G(t_0; \mathbb{P}\{t_1, ...t_l..., t_N\})$, this function always has a term of $g(t_0 - T_0)$, which will determine the result of the integral depending on the location of t_0 in $G^*(t_k; \mathbb{P}'\{t_1, ...t_0..., t_N\})$ in the first sum in Eq. (B.3). If t_0 is outside the first m terms, e.g., $\{t_1, ..., t_m, ..., t_0, ..., t_N\}$, the integral will be zero because $\int dt_0 g^*(t_0 - T_j)g(t_0 - T_0) = 0$ for $|T_0 - T_j| \gg \Delta T (j = m + 1, ..., N)$. But if t_0 is inside the first m terms, e.g., $\{t_1, ..., t_0, ..., t_m, ..., t_N\}$, we have $G^*(t_k; \mathbb{P}'_0\{t_1, ...t_0..., t_N\}) = G^*(t_0; \mathbb{P}'_k\{t_1, ...t_k..., t_N\})$ because $T_0 = T_j$ for $j = 1, ..., m$. Here $\mathbb{P}'_0$, $\mathbb{P}'_k$ are the permutations excluding t_0 and t_k, respectively. There are totally m such non-zero terms and Eq. (B.3) becomes

$$P_{kl} = \frac{m}{2^{N+1}} \int dt_0 dt_1 ... dt_N \sum_{\mathbb{P}'_0} G^*(t_k; \mathbb{P}'_0\{t_1, ... t_0 ..., t_N\})$$

$$\times \sum_{\mathbb{P}} G(t_0; \mathbb{P}\{t_1, ... t_l ..., t_N\})$$

$$= \frac{m}{2^{N+1}} \int dt_0 dt_1 ... dt_N \sum_{\mathbb{P}'_k} G^*(t_0; \mathbb{P}'_k\{t_1, ... t_k ..., t_N\})$$

$$\times \sum_{\mathbb{P}} G(t_0; \mathbb{P}\{t_1, ... t_l ..., t_N\})$$

$$= \frac{m}{2^{N+1}} \sum_{\mathbb{P}'_k} \int dt_0 dt_1 ... dt_N G^*(t_0; t_1, ... t_k ..., t_N)$$

$$\times \sum_{\mathbb{P}} G(t_0; \mathbb{P}\{t_1, ... t_l ..., t_N\})$$

$$= m(N-1)! \mathcal{N}. \tag{B.4}$$

Again, we made a change of variables $\mathbb{P}'_k\{t_1, ... t_k, ..., t_N\} \rightarrow \{t_1, ..., t_N\}$ in evaluating the integral in the last line of Eq. (B.4) and $\mathbb{P}'_k$ has $(N-1)!$ terms. So, we have the final sum

$$\sum_{k \neq l} = m(N+1)N(N-1)! \mathcal{N} = m(N+1)! \mathcal{N}. \tag{B.5}$$

Here, the sum has a total of $N(N+1)$ terms.

Bibliography

Abbott, B. P. *et al.* (2016). Observation of gravitational waves from a binary black hole merger, Phys. Rev. Lett. **116**, p. 061102.

Agarwal, G. S. (2013). Quantum Optics (Cambridge University Press, New York, NY).

Andrews, D. L. and Babiker, M. (2013). The Angular Momentum of Light (Cambridge University Press, New York, NY).

Ashkin, A., Boyd, G. D., and Dziedzic, J. M. (1966). Resonant optical second harmonic generation and mixing, IEEE J. Quantum Electron. **2**, p. 109.

Aytur, O. and Kumar, P. (1990). Pulsed twin beams of light, Phys. Rev. Lett. **65**, p. 1551.

Bachor, H. A. and Ralph, T. C. (2004). A Guide to Experiments in Quantum Optics, 2nd edn. (Wiley-Vch, New York).

Barnett, S. M. and Pegg, D. T. (1990). Quantum theory of optical phase correlations, Phys. Rev. A **42**, p. 6713.

Bell, J. S. (1964). On the Einstein-Podolsky-Rosen paradox, Physics **1**, p. 195.

Bell, J. S. (1987). Speakable and unspeakable in quantum mechanics (Cambridge University Press, Cambridge).

Bennett, C. H., Brassard, G., Crepeau, C., Jozsa, R., Peres, A., and Wootters, W. K. (1993). Teleporting an unknown quantum state via dual classical and Einstein-Podolsky-Rosen channels, Phys. Rev. Lett. **70**, p. 1895.

Bennink, R. S., Bentley, S. J., and Boyd, R. W. (2002). Two-photon coincidence imaging with a classical source, Phys. Rev. Lett. **89**, p. 113601.

Beugnon, J., Jones, M. P. A., Dingjan, J., Darquie, B., Messin, G., Browaeys, A., and Grangier, P. (2006). Quantum interference between two single photons emitted by independently trapped atoms, Nature **440**, p. 779.

Bocquillon, E., Freulon, V., Berroir, J.-M., Degiovanni, P., Placais, B., Cavanna, A., Jin, Y., and Feve, G. (2013). Coherence and indistinguishability of single electrons emitted by independent sources, Science **339**, p. 1054.

Bohr, N. (1983). in J. A. Wheeler and W. H. Zurek (eds.), Quantum Theory and Measurement (Princeton University Press, Princeton).

Bondurant, R. S. and Shapiro, J. H. (1984). Squeezed states in phase-sensing interferometers, Phys. Rev. D **30**, p. 2548.

Born, M. and Wolf, E. (1999). Principle of Optics, 7th edn. (Pergamon, Oxford).

Boto, A. N., Kok, P., Abrams, D. S., Braunstein, S. L., Williams, C. P., and Dowling, J. P. (2000). Quantum interferometric optical lithography: Exploiting entanglement to beat the diffraction limit, Phys. Rev. Lett. **85**, p. 2733.

Bouwmeester, D., Pan, J. W., Daniell, M., Weinfurter, H., and Zeilinger, A. (1999). Observation of three-photon Greenberger-Horne-Zeilinger entanglement, Phys. Rev. Lett. **82**, p. 1345.

Bouwmeester, D., Pan, J. W., Mattel, K., Eibl, M., Weinfurter, H., and Zeilinger, A. (1997). Experimental quantum teleportation, Nature **390**, p. 575.

Boyd, R. W. (2003). Nonlinear Optics, 2nd edn. (Academic Press, San Diego, CA).

Braginsky, V. B., Khalili, F. Y., and Thorne, K. S. (1992). Quantum Measurement (Cambridge University Press, Cambridge, England).

Braunstein, S. L. (1992). Quantum limits on precision measurements of phase, Phys. Rev. Lett. **69**, p. 3598.

Braunstein, S. L. and Kimble, H. J. (1998). Teleportation of continuous quantum variables, Phys. Rev. Lett. **80**, p. 869.

Braunstein, S. L., Lane, A. S., and Caves, C. M. (1992). Maximum-likelihood analysis of multiple quantum phase measurements, Phys. Rev. Lett. **69**, p. 2153.

Braunstein, S. L. and Mann, A. (1996). Measurement of the Bell operator and quantum teleportation, Phys. Rev. A **51**, p. R1727.

Brendel, J., Gisin, N., Tittel, W., and Zbinden, H. (1999). Pulsed energy-time entangled twin-photon source for quantum communication, Phys. Rev. Lett. **82**, p. 2594.

Brune, M., Hagley, E., Dreyer, J., Maître, X., Maali, A., Wunderlich, C., Raimond, J. M., and Haroche, S. (1996). Observing the progressive decoherence of the "meter" in a quantum measurement, Phys. Rev. Lett. **77**, p. 4887.

Buck, J. A. (1995). Fundamentals of Optical Fibers (Wiley, New York).

Burnham, D. C. and Weinberg, D. L. (1970). Phys. Rev. Lett. **25**, p. 84.

Campos, R. A., Gerry, C. C., and Benmoussa, A. (2003). Optical interferometry at the Heisenberg limit with twin Fock states and parity measurements, Phys. Rev. A **68**, p. 023810.

Campos, R. A., Saleh, B. E. A., and Teich, M. C. (1989). Quantum-mechanical lossless beam splitter: Su(2) symmetry and photon statistics, Phys. Rev. A **40**, p. 1371.

Caruthers, P. and Nieto, M. M. (1968). Rev. Mod. Phys. **40**, p. 411.

Casimir, H. B. G. (1948). On the attraction between two perfectly conducting plates, Proc. Kon. Nederland. Akad. Wetensch. **B51**, pp. 793–795.

Caves, C. M. (1981). Quantum-mechanical noise in an interferometer, Phys. Rev. D **23**, p. 1693.

Caves, C. M. (1982). Phys. Rev. D **26**, p. 1817.

Caves, C. M. and Schumaker, B. L. (1985). New formalism for two-photon quantum optics. I. Quadrature phases and squeezed states, Phys. Rev. A **31**, p. 3068.

Chen, B., Qiu, C., Chen, S., Guo, J., Chen, L. Q., Ou, Z. Y., and Zhang, W. (2015). Atom-light hybrid interferometer, Phys. Rev. Lett. **115**, p. 043602.

Choi, S.-K., Vasilyev, M., and Kumar, P. (1999). Noiseless optical amplification of images, Phys. Rev. Lett. **83**, p. 1938.

Clauser, J. F. (1974). Experimental distinction between the quantum and classical field-theoretic predictions for the photoelectric effect, Phys. Rev. D **9**, p. 853.

Collett, M. J. and Walls, D. F. (1985). Squeezing spectra for nonlinear optical systems, Phys. Rev. A **32**, p. 2887.

Dagenais, M. and Mandel, L. (1978). Investigation of two-time correlations in photon emissions from a single atom, Phys. Rev. A **18**, p. 2217.

Dayan, B., Pe'er, A., Friesem, A. A., and Silberberg, Y. (2005). Nonlinear interactions with an ultrahigh flux of broadband entangled photons, Phys. Rev. Lett. **94**, p. 043602.

de Riedmatten, H., Marcikic, I., Tittel, W., Zbinden, H., and Gisin, N. (2003). Quantum interference with photon pairs created in spatially separated sources, Phys. Rev. A **67**, p. 022301.

de Riedmatten, H., Scarani, V., Marcikic, I., Acin, A., Tittel, W., Zbinden, H., and Gisin, N. (2004). J. Mod. Opt. **51**, p. 1637.

Deléglise, S., Dotsenko, I., Sayrin, C., Bernu, J., Brune, M., Raimond, J. M., and Haroche, S. (2008). Nature **455**, p. 510.

Diedrich, F. and Walther, H. (1987). Nonclassical radiation of a single stored ion, Phys. Rev. Lett. **58**, p. 203.

Ding, Y. and Ou, Z. Y. (2010). Frequency downconversion for a quantum network, Opt. Lett. **35**, p. 2591.

Dirac, P. A. M. (1927). Proc. R. Soc. London Ser. A **114**, p. 243.

Dirac, P. A. M. (1930). The Principles of Quantum Mechanics, 1st edn. (Clarendon, Oxford).

Du, S. (2015). Quantum-state purity of heralded single photons produced from frequency-anticorrelated biphotons, Phys. Rev. A **92**, p. 043836.

Duan, L.-M., Giedke, G., Cirac, J. I., and Zoller, P. (2000). Inseparability criterion for continuous variable systems, Phys. Rev. Lett. **84**, p. 2722.

Dur, W. (2001). Multipartite entanglement that is robust against disposal of particles, Phys. Rev. A **63**, p. 020303(R).

Einstein, A. (1905). On a heuristic point of view about the creation and conversion of light, Ann. der Phys. **18**, pp. 132–148.

Einstein, A. (1909). Phys. Z. **10**, pp. 185–193.

Einstein, A. (1917). Zur quantentheorie der strahlung, Phys. Z. **18**, p. 121.

Einstein, A., Podolsky, B., and Rosen, N. (1935). Can quantum-mechanical description of physical reality be considered complete? Phys. Rev. **47**, p. 777.

Ekert, A. K. (1991). Quantum cryptography based on Bell's theorem, Phys. Rev. Lett. **67**, p. 661.

Englert, B. G. (1996). Fringe visibility and which-way information: An inequality, Phys. Rev. Lett. **77**, p. 2154.

Fan, J., Dogariu, A., and Wang, L. J. (2005). Opt. Lett. **30**, p. 1530.

Fano, U. (1961). Quantum theory of interference effects in the mixing of light
    from phase independent sources, Am. J, Phys. 29, p. 539.

Fearn, H. and Loudon, R. (1987). Quantum theory of the lossless beam splitter,
    Opt. Comm. 64, p. 485.

Fearn, H. and Loudon, R. (1989). Theory of two-photon interference, J. Opt. Soc.
    Am. B 6, p. 917.

Fortsch, M., Forst, J. U., Wittmann, C., Strekalov, D., Aiello, A., Chekhova,
    M. V., Silberhorn, C., Leuchs, G., and Marquardt, C. (2013). A versatile
    source of single photons for quantum information processing, Nat. Comm.
    4, p. 1818.

Foster, G. T., Mielke, S. L., and Orozco, L. A. (2000). Intensity correlations in
    cavity QED, Phys. Rev. A 61, p. 053821.

Franson, J. D. (1989). Bell inequality for position and time, Phys. Rev. Lett. 62,
    p. 2205.

Friberg, S. R., Hong, C. K., and Mandel, L. (1985a). Phys. Rev. Lett. 54, p. 2011.

Friberg, S. R., Hong, C. K., and Mandel, L. (1985b). Intensity dependence of the
    normalized intensity correlation function in parametric down-conversion,
    Opt. Comm. 54, p. 311.

Furusawa, A., Sorensen, J. L., Braunstein, S. L., Fuchs, C. A., Kimble, H. J., and
    Polzik, E. S. (1998). Unconditional quantum teleportation, Science 282,
    p. 706.

Gardiner, C. W. and Collett, M. J. (1985). Input and output in damped quan-
    tum systems: Quantum stochastic differential equations and the master
    equation, Phys. Rev. A 31, p. 3761.

Gatti, A., Brambilla, E., Bache, M., and Lugiato, L. A. (2004). Correlated ima-
    ging, quantum and classical, Phys. Rev. A 70, p. 013802.

Gentle, J. E. (1998). Numerical Linear Algebra for Applications in Statistics
    (Springer-Verlag).

Ghosh, R. and Mandel, L. (1987). Observation of nonclassical effects in the in-
    terference of two photons, Phys. Rev. Lett. 59, p. 1903.

Glauber, R. J. (1963a). Quantum theory of optical coherence, Phys. Rev. 130,
    p. 2529.

Glauber, R. J. (1963b). Coherent and incoherent states of the radiation field,
    Phys. Rev. 131, p. 2766.

Glauber, R. J. (1963c). Photon correlations, Phys. Rev. Lett. 1, p. 84.

Glauber, R. J. (1964). Quantum coherence, in C. deWitt, A. Blandin, and
    C. Cohen-Tannoudji (eds.), Quantum Optics and Electronics (Les Houches
    Lectures) (Gordon and Breach, New York), p. 63.

Glauber, R. J. (1986). in E. R. Pike and S. Sarker (eds.), Frontiers in Quantum
    Optics (Institute of Physics, Bristol).

Goodman, J. W. (2015). Statistical Optics, 2nd edn. (Wiley, New York, NY).

Goto, H., Yanagihara, Y., Haibo Wang, T. H., and Kobayashi, T. (2003). Obser-
    vation of an oscillatory correlation function of multimode two-photon pairs,
    Phys. Rev. A 68, p. 015803.

Grangier, P., Slusher, R. E., Yurke, B., and LaPorta, A. (1987). Squeezed-light-
    enhanced polarization interferometer, Phys. Rev. Lett. 59, p. 2153.

Greenberger, D. M., Horne, M. A., and Zeilinger, A. (1989). in M. Katafos (ed.), Bell's Theorem, Quantum Theory, and Conceptions of the Universe (Kluwer Academic, Dordrecht, The Netherlands).

Guo, X., Li, X., Liu, N., and Ou, Z. Y. (2016a). Quantum information tapping using a fiber optical parametric amplifier with noise figure improved by correlated inputs, Sci. Rep. **6**, p. 30214.

Guo, X., Li, X., Liu, N., Yang, L., and Ou, Z. Y. (2012). An all-fiber source of pulsed twin beams for quantum communication, Appl. Phys. Lett. **101**, p. 261111.

Guo, X., Liu, N., Liu, Y., Li, X., and Ou, Z. Y. (2016b). Generation of continuous variable quantum entanglement using a fiber optical parametric amplifier, Opt. Lett. **41**, p. 653.

Hanbury Brown, R. and Twiss, R. Q. (1956a). Correlation between photons in two coherent beams of light, Nature **177**, pp. 27–29.

Hanbury Brown, R. and Twiss, R. Q. (1956b). A test of a new type of stellar interferometer on sirius, Nature **178**, pp. 1046–1048.

Hanbury Brown, R. and Twiss, R. Q. (1958). Interferometry of intensity fluctuations in light II. An experimental test of the theory for partially coherent light, Proc. Royal Soc. London A **243**, pp. 291–319.

Haroche, S. and Kleppner, D. (1989). Cavity quantum electrodynamics, Phys. Today **31(1)**, p. 24.

Haus, H. A. and Mullen, J. A. (1962). Phys. Rev. **128**, p. A2407.

Heffner, H. (1962). Proc. IRE **50**, p. 1604.

Heidmann, A., Horowicz, R. J., Reynaud, S., Giacobino, E., Fabre, C., and Camy, G. (1987). Observation of quantum noise reduction on twin laser beams, Phys. Rev. Lett. **59**, p. 2555.

Heitler, W. (1954). The Quantum Theory of Radiation, 3rd edn. (Oxford University Press, London).

Herman, G. T. (1980). Image Reconstruction from Projections: The Fundamentals of Computerized Tomography (Academic, New York, NY).

Herzog, T. J., Kwiat, P. G., Weinfurter, H., and Zeilinger, A. (1995). Complementarity and the quantum eraser, Phys. Rev. Lett. **75**, p. 3034.

Herzog, T. J., Rarity, J. G., Weinfurter, H., and Zeilinger, A. (1994). Frustrated two-photon creation via interference, Phys. Rev. Lett. **72**, p. 629.

Holland, M. J. and Burnett, K. (1993). Interferometric detection of optical phase shifts at the Heisenberg limit, Phys. Rev. Lett. **71**, p. 1355.

Hong, C. K., Friberg, S. R., and Mandel, L. (1985). J. Opt. Soc. Am. B **2**, p. 494.

Hong, C. K. and Mandel, L. (1986). Phys. Rev. Lett. **56**, p. 58.

Hong, C. K., Ou, Z. Y., and Mandel, L. (1987). Measurement of subpicosecond time intervals between two photons by interference, Phys. Rev. Lett. **59**, p. 2044.

Huang, J. and Kumar, P. (1992). Observation of quantum frequency conversion, Phys. Rev. Lett. **68**, p. 2153.

Hudelist, F., Kong, J., Liu, C., Jing, J., Ou, Z., and Zhang, W. (2014). Quantum metrology with parametric amplifier-based photon correlation interferometers, Nat. Comm. **5**, p. 3049.

Imoto, N., Haus, H. A., and Yamamoto, Y. (1985). Quantum nondemolition measurement of the photon number via the optical kerr effect, Phys. Rev. A **32**, p. 2287.

Jacobson, J., Bjork, G., Chuang, I., and Yamamoto, Y. (1995). Photonic de Broglie waves, Phys. Rev. Lett. **74**, p. 48355.

Javanaainen, J. and Yoo, S. M. (1996). Quantum phase of a Bose-Einstein condensate with an arbitrary number of atoms, Phys. Rev. Lett. **76**, p. 161.

Jing, J., Liu, C., Zhou, Z., Ou, Z. Y., and Zhang, W. (2011). Realization of a nonlinear interferometer with parametric amplifiers, Appl. Phys. Lett. **99**, p. 011110.

Kartner, F. X. and Haus, H. A. (1993). Quantum-nondemolition measurements and the "collapse of the wave function", Phys. Rev. A **47**, p. 4585.

Kelley, P. L. and Kleiner, W. H. (1964). Theory of electromagnetic field measurement and photoelectron counting, Phys. Rev. **136**, p. A316.

Kimble, H. J., Dagenais, M., and Mandel, L. (1977). Photon antibunching in resonance fluorescence, Phys. Rev. Lett. **39**, pp. 691–695.

Kimble, H. J., Levin, Y., Matsko, A. B., Thorne, K. S., and Vyatchanin, S. P. (2001). Conversion of conventional gravitational-wave interferometers into quantum nondemolition interferometers by modifying their input and/or output optics, Phys. Rev. D **65**, p. 022002.

Knill, E., Laflamme, R., and Milburn, G. J. (2001). A scheme for efficient quantum computation with linear optics, Nature **409**, p. 46.

Kok, P., Lee, H., and Dowling, J. P. (2002). Creation of large-photon-number path entanglement conditioned on photodetection, Phys. Rev. A **65**, p. 052104.

Kong, J., Hudelist, F., Ou, Z., and Zhang, W. (2013a). Cancellation of internal quantum noise of an amplifier by quantum correlation, Phys. Rev. Lett. **111**, p. 033608.

Kong, J., Ou, Z., and Zhang, W. (2013b). Phase-measurement sensitivity beyond the standard quantum limit in an interferometer consisting of a parametric amplifier and a beam splitter, Phys. Rev. A **87**, p. 023825.

Kozlovsky, W. J., Nabors, C. D., and Byer, R. L. (1988). Efficient second harmonic generation of a diode-laser-pumped CW ND : YAG laser using monolithic $MGO : LiNbO_3$ external resonant cavities, IEEE J. Quantum Electron. **24**, p. 913.

Kwiat, P., Vareka, W., Hong, C., Nathel, H., and Chiao, R. (1990). Correlated two-photon interference in a dual-beam Michelson interferometer, Phys. Rev. A **41**, p. 2910.

Kwiat, P. G., Steinberg, A. M., and Chiao, R. Y. (1992). Observation of a "quantum eraser": A revival of coherence in a two-photon interference experiment, Phys. Rev. A **45**, p. 7729.

Lamb, W. E. (1995). Anti-photon, Appl. Phys. B **60**, pp. 77–84.

Lamb, W. E. and Retherford, R. C. (1947). Fine structure of the hydrogen atom by a microwave method, Phys. Rev. **72**, pp. 241–243.

Lane, A. S., Braunstein, S. L., and Caves, C. M. (1993). Maximum-likelihood statistics of multiple quantum phase measurements, Phys. Rev. A **47**, p. 1667.

Lee, N., Benichi, H., Takeno, Y., Takeda, S., Webb, J., Huntington, E., and Furusawa, A. (2011). Teleportation of non-classical wave packets of light, Science **332**, p. 330.

Legero, T., Wilk, T., Hennrich, M., Rempe, G., and Kuhn, A. (2004). Quantum beat of two single photons, Phys. Rev. Lett. **93**, p. 070503.

Leibfried, D., DeMarco, B., Meyer, V., Rowe, M., Ben-Kish, A., Britton, J., Itano, W. M., Jelenković, B., Langer, C., Rosenband, T., and Wineland, D. J. (2002). Trapped-ion quantum simulator: Experimental application to nonlinear interferometers, Phys. Rev. Lett. **89**, p. 247901.

Leonhardt, U. (1997). Measuring the Quantum State of Light (Cambridge University Press, Cambridge).

Levenson, J. A., Abram, I., Rivera, T., Fayolle, P., Garreau, J. C., and Grangier, P. (1993). Quantum optical cloning amplifier, Phys. Rev. Lett. **70**, p. 267.

Ley, M. and Loudon, R. (1984). Opt. Acta **33**, p. 371.

Li, X., Chen, J., Voss, P., Sharping, J., and Kumar, P. (2004). Opt. Express **12**, p. 3737.

Li, X., Pan, Q., Jing, J., Zhang, J., Xie, C., and Peng, K. (2002). Quantum dense coding exploiting a bright Einstein-Podolsky-Rosen beam, Phys. Rev. Lett. **88**, p. 047904.

Li, X., Yang, L., Ma, X., Cui, L., Ou, Z. Y., and Yu, D. (2009). All-fiber source of frequency-entangled photon pairs, Phys. Rev. A **79**, p. 033817.

Liu, B. H., Sun, F. W., Gong, Y. X., Huang, Y. F., Guo, G. C., and Ou, Z. Y. (2007). Four-photon interference with asymmetric beam splitters, Opt. Lett. **32**, p. 1320.

Liu, N., Liu, Y., Guo, X., Yang, L., Li, X., and Ou, Z. Y. (2016). Approaching single temporal mode operation in twin beams generated by pulse pumped high gain spontaneous four wave mixing, Opt. Express **24**, p. 001096.

Loudon, R. (2000). The Quantum Theory of Light, 3rd edn. (Oxford University Press, Oxford).

Lu, Y. J., Campbell, R. L., and Ou, Z. Y. (2003). Mode-locked two-photon states, Phys. Rev. Lett. **91**, p. 163602.

Lu, Y. J. and Ou, Z. Y. (2002). Observation of nonclassical photon statistics due to quantum interference, Phys. Rev. Lett. **88**, p. 023601.

Lu, Y. J. and Ou, Z. Y. (2000). Optical parametric oscillator far below threshold: Experiment versus theory, Phys. Rev. A **62**, p. 033804.

Lukens, J. M., Dezfooliyan, A., Langrock, C., Fejer, M. M., Leaird, D. E., and Weiner, A. M. (2013). Phys. Rev. Lett. **111**, p. 193603.

Lvovsky, A. I., Hansen, H., Aichele, T., Benson, O., Mlynek, J., and Schiller, S. (2001). Quantum state reconstruction of the single-photon Fock state, Phys. Rev. Lett. **87**, p. 050402.

Lvovsky, A. I. and Raymer, M. G. (2009). Continuous-variable optical quantum-state tomography, Rev. Mod. Phys. **81**, p. 299.

Machida, S. and Yamamoto, Y. (1988). Ultrabroadband amplitude squeezing in a semiconductor laser, Phys. Rev. Lett. **60**, p. 792.

Machida, S., Yamamoto, Y., and Itaya, Y. (1987). Observation of amplitude squeezing in a constant-current-driven semiconductor laser, Phys. Rev. Lett. **58**, p. 1000.

Mandel, L. (1983). Photon interference and correlation effects produced by independent quantum sources, Phys. Rev. A **28**, p. 929.

Mandel, L. (1991). Coherence and indistinguishability, Opt. Lett. **16**, p. 1882.

Mandel, L. (1999). Quantum effects in one-photon and two-photon interference, Rev. Mod. Phys. **71**, p. S274.

Mandel, L., Sudarshan, E. C. G., and Wolf, E. (1964). On a heuristic point of view about the creation and conversion of light, Proc. Phys. Soc. **84**, p. 435.

Mandel, L. and Wolf, E. (1997). Optical Coherence and Quantum Optics (Cambridge University Press, New York, NY).

Martynov, D. V. *et al.* (2016). Sensitivity of the advanced ligo detectors at the beginning of gravitational wave astronomy, Phys. Rev. D **93**, p. 112004.

Maunz, P., Moehring, D. L., Olmschenk, S., Younge, K. C., Matsukevich, D. N., and Monroe, C. (2007). Quantum interference of photon pairs from two remote trapped atomic ions, Nat. Phys. **3**, p. 538.

McCormick, C. F., Boyer, V., Arimondo, E., and Lett, P. D. (2007). Strong relative intensity squeezing by four-wave mixing in rubidium vapor, Opt. Lett. **32**, p. 178.

Mehta, C. L. and Sudarshan, E. C. G. (1965). Relation between quantum and semiclassical description of optical coherence, Phys. Rev. **138**, p. B274.

Michelson, A. A. (1890). Phil. Mag. **30**, p. 1.

Michelson, A. A. (1920). Astrophys. J. **51**, p. 257.

Michelson, A. A. and Pease, F. G. (1921). Astrophys. J. **53**, p. 249.

Milburn, G. J., Steyn-Ross, M. L., and Walls, D. F. (1987). Phys. Rev. A **35**, p. 4443.

Milburn, G. J. and Walls, D. F. (1983). Quantum nondemolition measurements via quantum counting, Phys. Rev. A **28**, p. 2646.

Monroe, C., Meekhof, D. M., King, B. E., and Wineland, D. J. (1996). Science **272**, p. 1131.

Myatt, C. J., King, B. E., Turchette, Q. A., Sackett, C. A., Kielpinski, D., Itano, W. H., Monroe, C., and Wineland, D. J. (2000). Nature **403**, p. 269.

Nasr, M. B., Saleh, B. E. A., Sergienko, A. V., and Teich, M. C. (2003). Demonstration of dispersion-canceled quantum-optical coherence tomography, Phys. Rev. Lett. **91**, p. 083601.

Niu, X.-L., Gong, Y.-X., Liu, B.-H., Huang, Y.-F., Guo, G.-C., and Ou, Z. Y. (2009). Observation of a generalized bunching effect of six photons, Opt. Lett. **34**, p. 1297.

Noh, J., Fougerés, A., and Mandel, L. (1991). Measurement of the quantum phase by photon counting, Phys. Rev. Lett. **67**, p. 1426.

Noh, J., Fougerés, A., and Mandel, L. (1992). Operational approach to the phase of a quantum field, Phys. Rev. A **45**, p. 424.

Noh, J., Fougerés, A., and Mandel, L. (1993). Measurements of the probability distribution of the operationally defined quantum phase difference, Phys. Rev. Lett. **71**, p. 2579.

Ou, Z. Y. (1988). Quantum theory of fourth-order interference, Phys. Rev. A **37**, p. 1607.

Ou, Z. Y. (1993). Quantum amplification with correlated quantum fields, Phys. Rev. A **48**, p. R1761.

Ou, Z. Y. (1996). Quantum multi-particle interference due to a single-photon state, Quantum and Semicl. Opt. **8**, p. 315.

Ou, Z. Y. (1997a). Fundamental quantum limit in precision phase measurement, Phys. Rev. A **55**, p. 2598.

Ou, Z. Y. (1997b). Parametric down-conversion with coherent pulse pumping and quantum interference between independent fields, Quantum and Semi-class. Opt. **9**, p. 599.

Ou, Z. Y. (2008). Characterizing temporal distinguishability of an n-photon state by a generalized photon bunching effect with multiphoton interference, Phys. Rev. A **77**, p. 043829.

Ou, Z. Y., Hong, C. K., and Mandel, L. (1987). Relation between input and output states for a beam splitter, Opt. Comm. **63**, p. 118.

Ou, Z. Y. and Kimble, H. J. (1993). Enhanced conversion efficiency for harmonic generation with double resonance, Opt. Lett. **18**, p. 1053.

Ou, Z. Y. and Kimble, H. J. (1995). Probability distribution of photoelectric currents in photodetection processes and its connection to the measurement of a quantum state, Phys. Rev. A **52**, p. 3126.

Ou, Z. Y. and Lu, Y. J. (1999). Cavity enhanced spontaneous parametric down-conversion for the prolongation of correlation time between conjugate photons, Phys. Rev. Lett. **83**, p. 2556.

Ou, Z. Y. and Mandel, L. (1988). Observation of spatial quantum beating with separated photodetectors, Phys. Rev. Lett. **61**, p. 54.

Ou, Z. Y. and Mandel, L. (1989). Derivation of reciprocity relations for a beam splitter from energy balance, Am. J. Phys. **57**, p. 66.

Ou, Z. Y., Pereira, S. F., and Kimble, H. J. (1992a). Realization of the Einstein-Podolsky-Rosen paradox for continuous variables in nondegenerate parametric amplification, Appl. Phys. B **55**, p. 265.

Ou, Z. Y., Pereira, S. F., and Kimble, H. J. (1993). Quantum noise reduction in optical amplification, Phys. Rev. Lett. **70**, p. 3239.

Ou, Z. Y., Pereira, S. F., Kimble, H. J., and Peng, K. C. (1992b). Realization of the Einstein-Podolsky-Rosen paradox for continuous variables, Phys. Rev. Lett. **68**, p. 3663.

Ou, Z. Y., Rhee, J. K., and Wang, L. J. (1999a). Observation of four-photon interference with a beam splitter by pulsed parametric down-conversion, Phys. Rev. Lett. **83**, p. 959.

Ou, Z. Y., Rhee, J. K., and Wang, L. J. (1999b). Photon bunching and multiphoton interference in parametric down-conversion, Phys. Rev. A **60**, p. 593.

Ou, Z. Y. and Su, Q. (2003). Uncertainty in determining the phase for an optical field due to the particle nature of light, Laser Phys. **13**, p. 1175.

Ou, Z. Y., Zou, X. Y., Wang, L. J., and Mandel, L. (1990a). Observation of nonlocal interference in separated photon channels, Phys. Rev. Lett. **65**, p. 321.

Ou, Z. Y., Zou, X. Y., Wang, L. J., and Mandel, L. (1990b). Experiment on nonclassical fourth-order interference, Phys. Rev. A **42**, p. 2957.

Ou, Z. Y. J. (2007). Multi-Photon Quantum Interference (Springer, New York, NY).

Pan, J. W., Bouwmeester, D., Weinfurter, H., and Zeilinger, A. (1998). Experimental entanglement swapping: Entangling photons that never interacted, Phys. Rev. Lett. **80**, p. 3891.

Pegg, D. T. and Barnett, S. M. (1988). Europhys. Lett. **6**, p. 483.

Pegg, D. T. and Barnett, S. M. (1989). Phys. Rev. A **39**, p. 1665.

Pfleegor, R. L. and Mandel, L. (1967a). Phys. Lett. **24A**, p. 766.

Pfleegor, R. L. and Mandel, L. (1967b). Phys. Rev. **159**, p. 1084.

Pittman, T. B., Shih, Y. H., Strekalov, D. V., and Sergienko, A. V. (1995). Optical imaging by means of two-photon quantum entanglement, Phys. Rev. A **52**, p. R3429.

Pittman, T. B., Strekalov, D. V., Migdall, A., Rubin, M. H., Sergienko, A. V., and Shih, Y. H. (1996). Can two-photon interference be considered the interference of two photons, Phys. Rev. Lett. **77**, p. 1917.

Planck, M. (1900). On an improvement of wien's equation for the spectrum, Verhandlungen der Deutschen Physikalischen Gesellschaft **2**, pp. 202–204.

Plick, W. N., Dowling, J. P., and Agarwal, G. S. (2010). Coherent-light-boosted, sub-shot noise, quantum interferometry, New J. Phys. **12**, p. 083014.

Polzik, E. S. and Kimble, H. J. (1991). Frequency doubling with $KNbO_3$ in an external cavity, Opt. Lett. **16**, p. 731.

Prasad, S., Scully, M. O., and Martienssen, W. (1987). A quantum description of the beam splitter, Opt. Comm. **62**, p. 139.

Rarity, J. G. (1995). in D. M. Greenberger and A. Zeilinger (eds.), Fundamental Problems in Quantum Theory (Ann. NY Acad. Sci. vol. 755), p. 624.

Rarity, J. G. and Tapster, P. R. (1989). Fourth-order interference in parametric downconversion, J. Opt. Soc. Am. B **6**, p. 1221.

Reid, M. D. (1989). Demonstration of the Einstein-Podolsky-Rosen paradox using nondegenerate parametric amplification, Phys. Rev. A **40**, p. 913.

Rice, P. R. and Carmichael, H. J. (1988). Single-atom cavity-enhanced absorption. i. photon statistics in the bad-cavity limit, IEEE J. Quantum Electron. **24**, p. 1351.

Richter, G. (1977). Abh. Acad. Wiss. DDR **7N**, p. 245.

Sanaka, K., Jennewein, T., Pan, J.-W., Resch, K., and Zeilinger, A. (2004). Experimental nonlinear sign shift for linear optics quantum computation, Phys. Rev. Lett. **92**, p. 017902.

Sanaka, K., Resch, K. J., and Zeilinger, A. (2006). Filtering out photonic Fock states, Phys. Rev. Lett. **96**, p. 083601.

Sanders, B. C. and Milburn, G. J. (1989). Complementarity in a quantum non-demolition measurement, Phys. Rev. A **39**, p. 694.

Santori, C., Fattal, D., Vuckovic, J., Solomon, G. S., and Yamamoto, Y. (2002). Indistinguishable photons from a single-photon device, Nature **419**, p. 594.

Schawlow, A. L. and Townes, C. H. (1958). Infrared and optical masers, Phys. Rev. **112**, p. 1940.

Schleich, W., Dowling, J. P., and Horowicz, R. J. (1991). Exponential decrease in phase uncertainty, Phys. Rev. A **44**, p. 3365.

Schleich, W. and Wheeler, J. A. (1987). Oscillations in photon distribution of squeezed states and interference in phase space, Nature **326**, p. 574.

Schleich, W. P. (2001). Quantum Optics in Phase Space (Wiley-Vch, Berlin).

Schrodinger, E. (1926). Naturwissenschaften **14**, p. 664.

Schrodinger, E. (1935). Naturwissenschaften **23**, pp. 807, 823, 844.

Scully, M. O. and Druhl, K. (1982). Quantum eraser: A proposed photon correlation experiment concerning observation and "delayed choice" in quantum mechanics, Phys. Rev. A **25**, p. 2208.

Scully, M. O., Englert, B.-G., and Walther, H. (1991). Nature **351**, p. 111.

Scully, M. O. and Zubairy, M. S. (1995). Quantum Optics (Cambridge University Press, New York, NY).

Shafiei, F., Srinivasan, P., and Ou, Z. Y. (2004). Generation of three-photon entangled state by quantum interference between a coherent state and parametric down-conversion, Phys. Rev. A **70**, p. 043803.

Shapiro, J. H. (1980). Optical waveguide tap with infinitesimal insertion loss, Opt. Lett. **5**, p. 351.

Shapiro, J. H., Shepard, S. R., and Wong, N. C. (1989). Measurements of the probability distribution of the operationally defined quantum phase difference, Phys. Rev. Lett. **62**, p. 2377.

Short, R. and Mandel, L. (1983). Observation of sub-Poissonian photon statistics, Phys. Rev. Lett. **51**, p. 384.

Siegman, A. E. (1986). Lasers (University Science Books, Mill Valley, CA).

Simon, R. (2000). Peres-Horodecki separability criterion for continuous variable systems, Phys. Rev. Lett. **84**, p. 2726.

Slusher, R. E., Grangier, P., LaPorta, A., Yurke, B., and Potasek, M. J. (1987). Pulsed squeezed light, Phys. Rev. Lett. **59**, p. 2566.

Slusher, R. E., Hollberg, L. W., Yurke, B., Mertz, J. C., and Valley, J. F. (1985). Observation of squeezed states generated by four-wave mixing in an optical cavity, Phys. Rev. Lett. **55**, p. 2409.

Smiles-Mascarenhas, K. (1991). Am. J. Phys. **59**, p. 1150.

Smithey, D. T., Beck, M., Raymer, M. G., and Faridani, A. (1993). Measurement of the Wigner distribution and the density matrix of a light mode using optical homodyne tomography: Application to squeezed states and the vacuum, Phys. Rev. Lett. **70**, p. 1244.

Stoler, D. (1970). Equivalence classes of minimum uncertainty packets, Phys. Rev. D **1**, p. 3217.

Sudarshan, E. C. G. (1963). Equivalence of semiclassical and quantum mechanical descriptions of statistical light beams, Phys. Rev. Lett. **10**, p. 277.

Sun, F. W., Liu, B. H., Huang, Y. F., Ou, Z. Y., and Guo, G. C. (2006). Observation of the four-photon de Broglie wavelength by state-projection measurement Phys. Rev. A **74**, 033812.

Sun, F. W., Liu, B. H., Gong, Y. X., Huang, Y. F., Ou, Z. Y., and Guo, G. C. (2007). Stimulated emission as a result of multiphoton interference, Phys. Rev. Lett. **99**, p. 043601.

Sun, F. W. and Wong, C. W. (2009). Indistinguishability of independent single photons, Phys. Rev. A **79**, p. 013824.

Susskind, L. and Glogower, J. (1964). Physics **1**, p. 49.

Takeda, S., Mizuta, T., Fuwa, M., van Loock, P., and Furusawa, A. (2013). Deterministic quantum teleportation of photonic quantum bits by a hybrid technique, Nature **500**, p. 315.

Takesue, H. (2010). Single-photon frequency down-conversion experiment, Phys. Rev. A **82**, p. 013833.

Taylor, G. I. (1909). Proc. Camb. Phil. Soc. **15**, p. 114.

Teich, M. C. and Saleh, B. E. A. (1988). Photon bunching and antibunching, in E. Wolf (ed.), Progress in Optics, Vol. XXVI (Elsevier Science Publishers B.V.), pp. 1–104.

Vahlbruch, H., Mehmet, M., Danzmann, K., and Schnabel, R. (2016). Detection of 15 dB squeezed states of light and their application for the absolute calibration of photoelectric quantum efficiency, Phys. Rev. Lett. **117**, p. 110801.

Vaidman, L. (1994). Teleportation of quantum states, Phys. Rev. A **49**, p. 1473.

van Cittert, P. H. (1934). Physica **1**, p. 201.

Vandevender, A. P. and Kwiat, P. G. (2004). High efficiency single photon detection via frequency up-conversion, J. Mod. Opt. **51**, p. 1433.

Walls, D. F. and Milburn, G. J. (1985). Phys. Rev. A **31**, p. 2403.

Walls, D. F. and Milburn, G. J. (2008). Quantum Optics, 2nd edn. (Springer-Verlag, Berlin).

Wang, H. and Kobayashi, T. (2005). Phase measurement at the Heisenberg limit with three photons, Phys. Rev. A **71**, p. 021802(R).

Wang, L. J. and Rhee, J. K. (1999). in Quantum Electronics and Laser Science Conference, OSA Technical Digest (Optical Society of America, Washington DC), p. 143.

Wigner, E. P. (1932). On the quantum correction for thermodynamic equilibrium, Phys. Rev. **40**, p. 749.

Wu, L. A. and Kimble, H. J. (1985). Interference effects in second-harmonic generation within an optical cavity, J. Opt. Soc. Am. B **2**, p. 697.

Wu, L.-A., Kimble, H. J., Hall, J., and Wu, H. (1986). Generation of squeezed states by parametric down conversion, Phys. Rev. Lett. **57**, p. 2520.

Wu, L. A., Xiao, M., and Kimble, H. J. (1987). Squeezed states of light from an optical parametric oscillator, J. Opt. Soc. Am. B **4**, p. 1465.

Xiao, M., Wu, L.-A., and Kimble, H. J. (1987). Precision measurement beyond the shot-noise limit, Phys. Rev. Lett. **59**, p. 278.

Xie, Z., Zhong, T., Shrestha, S., Xu, X., Liang, J., Gong, Y., Bienfang, J. C., Restelli, A., Shapiro, J. H., Wong, F. N. C., and Wong, C. W. (2015). Harnessing high-dimensional hyperentanglement through a biphoton frequency comb, Nat. Photon. **9**, p. 536.

Yonezawa, H., Braunstein, S. L., and Furusawa, A. (2007). Experimental demonstration of quantum teleportation of broadband squeezing, Phys. Rev. Lett. **99**, p. 110503.

Yuen, H. P. (1976). Two-photon coherent states of the radiation field, Phys. Rev. A **13**, p. 2226.

Yuen, H. P. and Chan, V. W. S. (1983). Noise in homodyne and heterodyne detection, Opt. Lett. **8**, p. 177.

Yurke, B., McCall, S. L., and Klauder, J. R. (1986). SU(2) and SU(1, 1) interferometers, Phys. Rev. A **33**, p. 4033.

Yurke, B. and Potasek, M. (1987). Obtainment of thermal noise from a pure quantum state, Phys. Rev. A **36**, p. 3464.

Yurke, B. and Stoler, D. (1992). Einstein-Podolsky-Rosen effects from independent particle sources, Phys. Rev. Lett. **68**, p. 1251.

Zajonc, A. G., Wang, L. J., Zou, X. Y., and Mandel, L. (1991). Quantum eraser, Nature **353**, p. 507.

Zeilinger, A. (1981). General properties of lossless beam splitters in interferometry, Am. J. Phys. **49**, p. 882.

Zeilinger, A. (1999). Experiment and the foundations of quantum physics, Rev. Mod. Phys. **71**, p. S288.

Zernike, F. (1938). The concept of degree of coherence and its application to optical problems, Physica **5**, p. 785.

Zhang, J. and Peng, K. (2000). Quantum teleportation and dense coding by means of bright amplitude-squeezed light and direct measurement of a Bell state, Phys. Rev. A **62**, p. 064302.

Zou, X. Y., Wang, L. J., and Mandel, L. (1991a). Violation of classical probability in parametric down-conversion, Opt. Comm. **84**, p. 351.

Zou, X. Y., Wang, L. J., and Mandel, L. (1991b). Induced coherence and indistinguishability in optical interference, Phys. Rev. Lett. **67**, p. 318.

Zukowski, M., Zeilinger, A., and Weinfurter, H. (1995). in D. M. Greenberger and A. Zeilinger (eds.), Fundamental Problems in Quantum Theory (Ann. NY Acad. Sci. vol. 755), p. 91.

Index

Printed in the United States
By Bookmasters